T0185696

Knot Theory
Second Edition

Knot Theory
Second Edition

By

Vassily Manturov

CRC Press
Taylor & Francis Group
Boca Raton London New York

CRC Press is an imprint of the
Taylor & Francis Group, an **informa** business

A CHAPMAN & HALL BOOK

CRC Press
Taylor & Francis Group
6000 Broken Sound Parkway NW, Suite 300
Boca Raton, FL 33487-2742

First issued in paperback 2020

© 2018 by Taylor & Francis Group, LLC
CRC Press is an imprint of Taylor & Francis Group, an Informa business

No claim to original U.S. Government works

ISBN-13: 978-1-138-56124-3 (hbk)
ISBN-13: 978-0-367-65729-1 (pbk)

**Visit the Taylor & Francis Web site at
http://www.taylorandfrancis.com**

**and the CRC Press Web site at
http://www.crcpress.com**

To my mother, Elena Ivanovna Manturova

Gelegentlich ergreifen wir die Feder
Und schreiben Zeichen auf ein weißes Blatt,
Die sagen dies und das, es kennt sie jeder,
Es ist ein Spiel, das seine Regeln hat.

Doch wenn ein Wilder oder Mondmann käme
Und solches Blatt, solch furchig Runenfeld
Neugierig forschend vor die Augen nähme,
Ihm starrte draus ein fremdes Bild der Welt,
Ein fremder Zauberbildersaal entgegen.
Er sähe A und B als Mensch und Tier,
Als Augen, Zungen, Glieder sich bewegen,
Bedächtig dort, gehetzt von Trieben hier,
Er läse wie im Schnee den Krähentritt,
Er liefe, ruhte, litte, flöge mit
Und sähe aller Schöpfung Möglichkeiten
Durch die erstarrten schwarzen Zeichen spuken,
Durch die gestabten Ornamente gleiten,
Säh Liebe glühen, sähe Schmerzen zucken.
Er würde staunen, lachen, weinen, zittern,
Da hinter dieser Schrift gestabten Gittern
Die ganze Welt in ihrem blinden Drang
Verkleinert ihm erschiene, in die Zeichen
Verzwergt, verzaubert, die in steifem Gang
Gefangen gehn und so einander gleichen,
Daß Lebensdrang und Tod, Wollust und Leiden
Zu Brüdern werden, kaum zu unterscheiden...

Und endlich würde dieser Wilde schreien
Vor unerträglicher Angst, und Feuer schüren
Und unter Stirnaufschlag und Litaneien
Das weiße Runenblatt den Flammen weihen.
Dann würde er vielleicht einschlummernd spüren,
Wie diese Un-Welt, dieser Zaubertand,
Dies Unerträgliche zurück ins Niegewesen
Gesogen würde und ins Nirgendland,
Und würde seufzen, lächeln und genesen.

        Hermann Hesse, "Buchstaben" (Das Glasperlenspiel).

Preface

Knot theory now plays a large role in modern mathematics, and the most significant results in this theory have been obtained in the last two decades. For scientific research in this field, Jones, Witten, Drinfeld, and Kontsevich received the highest mathematical award, the Fields medals. Even after these outstanding achievements, new results were obtained and even new theories arose as ramifications of knot theory. Here we mention Khovanov's categorification of the Jones polynomial, virtual knot theory proposed by Kauffman and the theory of Legendrian knots.

The aim of the present monograph is to describe the main concepts of modern knot theory together with full proofs that would be both accessible for beginners and useful for professionals. Thus, in the first chapter of the second part of the book (concerning braids) we start from the very beginning and in the same chapter construct the Jones two-variable polynomial and the faithful representation of the braid groups. A large part of the present title is devoted to rapidly developing areas of modern knot theory, such as virtual knot theory and Legendrian knot theory.

In the present book, we give both the "old" theory of knots, such as the fundamental group, Alexander's polynomials, the results of Dehn, Seifert, Burau, and Artin, and the newest investigations in this field due to Conway, Matveev, Jones, Kauffman, Vassiliev, Kontsevich, Bar–Natan and Birman. We also include the most significant results from braid theory, such as the full proof of Markov's theorem, Alexander's and Vogel's algorithms, Dehornoy algorithm for braid recognition, etc. We also describe various representations of braid groups, e.g., the famous Burau representation and the newest (1999–2000) faithful Krammer–Bigelow representation. Furthermore, we give a description of braid groups in different spaces and simple newest recognition algorithms for these groups. We also describe the construction of the Jones two–variable polynomial.

In addition, we pay attention to the theory of coding of knots by d–diagrams, described in the author's papers [Man1, Man4, Man5]. Also, we give an introduction to virtual knot theory, proposed recently by Louis H. Kauffman [Kau5]. A great part of the book is devoted to the author's results in the theory of virtual knots.

Proofs of theorems involve some constructions from other theories, which have their own interest; i.e., quandle, product integral, Hecke algebras, connection theory and the Knizhnik–Zamolodchikov equation, Hopf algebras and quantum groups, Yang–Baxter equations, LD–systems, etc.

The contents of the book are not covered by existing monographs on knot theory; the present book has been taken a lot out of the author's Russian lecture notes book [Man'2] on the subject. The latter describes the lecture course that has been being delivered by the author since 1999 for undergraduate students, graduate students, and professors of the Moscow State University.

The present monograph contains many new subjects (classical and modern) which are not represented in [Man'2].

While describing the skein polynomials we have added the Przytycki–Traczyk approach and Conway algebra. We have also added the complete knot invariant, the distributive groupoid, also known as a quandle, and its generalisation. We have rewritten the virtual knot and link theory chapter. We have added some recent author's achievements on knots, braids, and virtual braids. We also describe the Khovanov categorification of the Jones polynomial, the Jones two–variable polynomial via Hecke algebras, the Krammer–Bigelow representation, etc.

The book is divided into thematic parts. The first part describes the state of "pre-Vassiliev" knot theory. It contains the simplest invariants and tricks with knots and braids, the fundamental group, the knot quandle, known skein polynomials, Kauffman's two–variable polynomial, some pretty properties of the Jones polynomial together with the famous Kauffman–Murasugi theorem and a knot table.

The second part discusses braid theory, including Alexander's and Vogel's algorithms, Dehornoy's algorithm, Markov's theorem, Yang–Baxter equations, Burau representation and the faithful Krammer–Bigelow representation. In addition, braids in different spaces are described, and simple word recognition algorithms for these groups are presented. We would like to point out that the first chapter of the second part (Chapter 9) is central to this part. This gives a representation of the braid theory in total: from various definitions of the braid group to the milestones in modern knot and braid theory, such as the Jones polynomial constructed via Hecke algebras and the faithfulness of the Krammer–Bigelow representation.

The third part is devoted to the Vassiliev knot invariants. We give all definitions, prove that Vassiliev invariants are stronger than all polynomial invariants, study structures of the chord diagram and Feynman diagram algebras, and finally present the full proof of the invariance for Kontsevich's integral. Here we also present a sketchy introduction to Bar-Natan's theory on Lie algebra representations and knots. We also give estimates of the dimension growth for the chord diagram algebra.

In the fourth part we describe a new way for encoding knots by d–diagrams proposed by the author. This way allows us to encode topological objects (such as knot, links, and chord diagrams) by words in a finite alphabet. Some applications of d–diagrams (the author's proof of the Kauffman–Murasugi theorem, chord diagram realisability recognition, etc.) are also described.

The fifth part contains virtual knot theory together with "virtualisations" of knot and link invariants. Here we describe Kauffman's results (basic definitions, foundation of the theory, Jones and Kauffman polynomials, quandles, finite–type invariants), the work of Goussarov, Polyak, and Viro (finite–type invariants), and Vershinin (virtual braids and their representation). We also include our own results concerning new invariants of virtual knots: those coming from the "virtual quandle", matrix formulae and invariant polynomials in one and several variables, generalisation of the Jones polynomials via curves in 2–surfaces, invariants of "long virtual link" and virtual braids.

The final part gives a sketchy introduction to two theories: knots in 3-manifolds (e.g., knots in $\mathbb{R}P^3$ with Drobotukhina's generalisation of the Jones polynomial), the introduction to Kirby's calculus and Witten's theory, and Legendrian knots and links after Fuchs and Tabachnikov. We recommend the newest book on 3-manifolds by Matveev [Mat5].

At the end of the book, a list of unsolved problems in knot and link theory and the knot table is given.

The description of the mathematical material is sufficiently closed; the monograph is quite accessible for undergraduate students of younger courses, thus it can be used as a course book on knots. The book can also be useful for professionals because it contains the newest and the most significant scientific developments in knot theory. Some technical details of proofs which are not used in the sequel are either omitted or printed in small type.

Besides the special course at the Moscow State University, I have also held a seminar "Knots and Representation Theory" since 2000, where many aspects of modern knot theory were discussed. Until 2002 this seminar was held together with Professor Valery Vladimirovich Trofimov (1952–2003). I am deeply indebted to him for his collaboration over many years and for fruitful advice.

It is a great pleasure to express my gratitude to all those who helped me at different stages of the present book. I am grateful to my father, Professor Oleg V. Manturov [ManO1, ManO2, ManO3, ManO4, ManO5, ManO6], for attention to my mathematical work during my entire life. I wish to thank Professor Victor A. Vassiliev for constant attention to my scientific papers on knots, fruitful ideas and comments.

I am glad to express my gratitude to Profs. Louis H. Kauffman, Roger A.Fenn, Heiner Zieschang, Alexey V. Chernavsky, Sergei V. Matveeev, Nikita Yu. Netsvetaev, Joan Birman, Patrick Dehornoy, Kent E. Orr, Michiel Hazewinkel, Vladimir V. Vershinin, Drs. Vladimir P. Lexin, Sergei K. Lando, Sergei V. Duzhin, Alexei V. Shchepetilov for many useful comments concerning my book and papers.

I am also grateful to the participants of the seminar "Knots and Representation Theory", especially to Evgeny V. Teplyakov.

It is a pleasure for me to thank my friend and colleague Dr. Rutwig Campoamor–Stursberg for constant consultations and correspondence.

The book was written using knots.tex fonts containing special symbols from knot theory, such as ⏾⏽ and ⏦ created by Professor M.M. Vinogradov and A.B. Sossinsky. I am grateful to them for these fonts.

I am also very grateful to A.Yu. Abramychev for helpful advices about typesetting this book.

Vassily Olegovich Manturov,
May 2003.

Preface to the second edition

The modern knot theory has undergone yet more combinatorial developments over the last few decades. Here one has to mention the "hard" techniques coming from instanton, Seiberg–Witten theory [Ati1, KrMr3, Hil, Akb, Sav, GoSt, Moo01, Sco, Wit1, Wit2, Wit3], and Heegaard–Floer homology due to Ozsvath–Szabo [OS1, OS2, OS3, OS4]. These developments led to various "unknot detectors": knot Floer homology [OS5], the A-polynomial [DG], and finally, the Khovanov homology [KrMr4], which, being defined combinatorially, turned out to have various mysterious relations with the Heegaard–Floer homology (many of them are not yet understood, but the spectral sequence from Khovanov to Heegaard–Floer allows one to detect the unknot).

On the other hand, a wealth of new ideas arose from virtual knot theory [Kau6]; this theory arose as a combinatorial generalisation of the classical knot theory and works as a source of new ideas for the latter.

Various results on virtual knot theory belonging to the author are collected in [MI]; among them there is Khovanov homology theory for virtual knots with arbitrary coefficients, algorithmic recognizability of virtual links and many other results.

The present book is the second edition of "Knot Theory" published in 2004. We tried not to change the style of exposition drastically and not to start writing a new book. Nevertheless, we have added several topics devoted to

- Heegard Floer homology

- Rasmussen invariant

- Khovanov homology for virtual knots

- A-polynomial

- Solution to braid conjugacy problem

Some material (like knot tables that are available online [BNk]) was removed, and some other chapters and appendices were rearranged. Nevertheless, some classical aspects of knot theory are still missing (e.g., Milnor's link homotopy theory [HL, Lev, Lin1, Lin2, Mil2], here we refer the reader to Milnor's original book [Mil1]).

The list of unsolved problems is somewhat random. On one hand, we did not touch a lot of the list written 14 years ago (many problems still remain unsolved).

On the other hand, we added one separate problem about picture-valued invariants for classical knots (such invariants give lots of powerful results for virtual knots and we hope they will lead to picture-valued invariants of classical knots).

It gave us a great pleasure to look through knot theory again. We corrected some existing misprints and small mistakes; in particular, I am grateful to Carlo Petronio who mentioned a gap in the construction of Conway algebra invariants.

I am extremely grateful to the editor of this book, my friend and colleague Igor Mikhailovich Nikonov, who has undertaken an enormous job as a scientist, proofreader, typesetter.

Long discussions of different topics with him led me to a better understanding of various aspects of modern knot theory.

Vassily Olegovich Manturov,
October, 2017

Contents

Part I

Knots, links, and invariant polynomials

Chapter 1

Introduction

As a mathematical theory, knot theory appeared at the end of the 18th century. It should be emphasised that for more than two hundred years knot theory was studied by A.T. Vandermonde, C.-F. Gauss (who found the famous electromagnetic link coefficient formula [Gau]), F. Klein, and M. Dehn [Dehn2]. Systematic study of knot theory begins at the end of the 19th century, when mathematicians and physicists started to tabulate knots. A very interesting (but incorrect!) idea belonged to W. Thompson (later known as Lord Kelvin). He thought that knots should correspond to chemical elements [Kel]. However, the most significant results in knot theory took place in the second part of the 20th century. These achievements are closely connected with the names of J.H. Conway, V.F.R. Jones, V.A. Vassiliev, M.L. Kontsevich, V.G. Turaev, M.N. Goussarov, J.H. Birman, L.H. Kauffman, D. Bar–Natan, M. Khovanov, P.S. Ozsváth, Z. Szabó, J. Rasmussen and many others.

Knot theory originates from a beautiful and quite simple (or so it can seem) topological problem [ChFa, Per, Reid, Rol]. In order to solve this problem, one should involve a quite complicated and sophisticated mathematical approach, coming from topology, discriminant theory, Lie theory, product integral theory, quantum algebras, and so on. Knot theory is rapidly developing; it stimulates constructions of new branches of mathematics. One should mention that in recent years Jones, Witten, Drinfeld (1990) and Kontsevich (1998) were awarded their Fields medals for work in knot theory.

And even after these works, new directions of knot theory appeared. Here we would like to mention the beautiful construction of Khovanov [Kho1] who proposed a categorification of the Jones polynomial — a new knot invariant based on brand new ideas. We shall also touch on the theory of Legendrian knots, lying at the junction of knot theory and contact geometry and topology. The most significant contributions to this (very young) science were made in the last few years (Fuchs, Tabachnikov, Chekanov, Eliashberg).

Another beautiful ramification is virtual knot theory proposed by Kauffman in 1996. The main results in this area are still to be obtained.

Besides this, knot theory is instrumental in constructing other theories; a very important example is the Kirby calculus; i.e., the theory of encoding 3–manifolds by means of links with a special structure — framed links.

1.1   Basic definitions

Here, *knot* means a smooth embedding of the circle S^1 in $\mathbb{R}^3$ (or in the sphere S^3)[1] as well as the image of this embedding.

While deforming the ambient space $\mathbb{R}^3$, our knot (image) will be deformed as well and hence be embedded. Two knots are called *isotopic*, if one of them can be transformed to the other by a diffeomorphism of the ambient space onto itself; here we require that this homeomorphism should be homotopic to the identical one in the class of diffeomorphisms.

The main question of knot theory is the following: which two knots are (*isotopic*) and which are not? This problem is called *the knot recognition problem*. Having an isotopy equivalence relation, one can speak about *knot isotopy classes*. When seen in this context, we shall say "knot" when referring to the knot isotopy class. One can also talk about *knot invariants*; i.e., functions on knot isotopy classes or functions on knots invariant under isotopy.

A partial case of the knot recognition problem is the *trivial knot recognition problem*. Here, *trivial knot* means the simplest knot that can be represented as the boundary of a 2-disc embedded in $\mathbb{R}^3$.

Both questions are very difficult. Though they are solved their solution requires many techniques (see [Hem]) and cannot in fact be implemented for practical purposes. The main stages of the complete solutions of these (and some other) problems can be read in [Mat1].

In the present book we shall give partial answers to these questions. As usual, in order to prove that two knots are isotopic, one should present a step-by-step isotopy transforming one knot to the other. Later we shall present the list of Reidemeister moves, which are indeed step–by–step isotopy moves. To show that two knots are not isotopic, one usually finds an invariant having different values on these two knots.

Usually, knots are encoded as follows. Fix a knot; i.e., a map $f : S^1 \to \mathbb{R}^3$. Consider a plane $h \subset \mathbb{R}^3$ and the projection of the knot on it. Without loss of generality, one can assume that $h = Oxy$. In the general position case, this projection is a quadrivalent graph embedded in the plane. Usually, we shall call a part (the image of an interval) of a knot *a branch* of it. Each vertex V of this graph (also called a *crossing*) is endowed with the following structure. Let a, b be two branches of a knot, whose projections intersect in the point V. Since a and b do not intersect in $\mathbb{R}^3$, the two preimages of V have different z-coordinates. So, we can say which branch (a or b) comes over, or forms an *overcrossing*; the other one forms an *undercrossing* (see Fig. 1.1). The quadrivalent graph obtained is called a *knot diagram*.

The quadrivalent projection graph without an over– and undercrossing structure is called the *shadow* of the knot. The *complexity* of a knot is the

---

[1]These two theories are equivalent because of codimension reasons. In the sequel, we shall deal with knots in $\mathbb{R}^3$, unless otherwise specified.

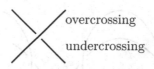

FIGURE 1.1: Local structure of a crossing

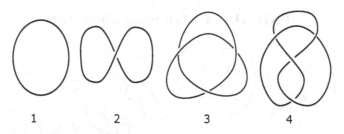

1   2   3   4

FIGURE 1.2: The simplest knots

minimal number of crossings for knots of given isotopy type. There are also other parameters that can measure knot complexity (bridge number, knot genus etc.).

The following exercise is left for the reader.

Exercise 1.1. *Show that any two knots having the same combinatorial structure of planar diagrams (i.e. isomorphic embeddings with the same crossing structure) are isotopic.*

Let us now give some examples.

Example 1.1. *The knot having a diagram without crossings (see Fig. 1.2.1) is called the* unknot *or the* trivial knot. *Figure* **1.2.2.** *represents another planar diagram of the unknot. The knot shown in Fig.* **1.2.3.** *is called the* trefoil, *and that in Fig.* **1.2.4.** *is called the* figure eight knot. *Both knots are not trivial; they are not isotopic to each other.*

Definition 1.1. A knot diagram L is called *ascending* (starting from a point A on it different from any vertex) if while walking along L from A (in some direction) each crossing is first passed under and then over.

Exercise 1.2. *Show that each ascending diagram represents an unknot.*

We shall use this fact in Chapter 5.

For each knot, one can construct its *mirror image*; i.e. the knot obtained from the initial one by reflecting it in some plane. Typical diagrams of a knot and its mirror image can be obtained by switching all crossing types (overcrossing replaces undercrossing and vice versa). A knot is called *amphicheiral* if it is isotopic to its mirror image.

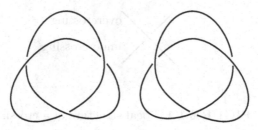

FIGURE 1.3: a.Left trefoil b.Right trefoil

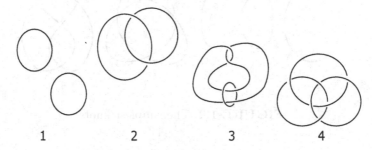

1   2   3   4

FIGURE 1.4: The simplest links

For example, the trefoil knot is not amphicheiral. We shall prove this fact later. This allows us to speak about two trefoil knots, the *right* one and the *left one*; see Fig. 1.3.

Exercise 1.3. *Show that figure eight is an amphicheiral knot.*

One can also speak about *oriented* knots; i.e. smooth mappings (images) of an oriented circle in $\mathbb{R}^3$. By an *isotopy* of oriented knots is meant an isotopy of knots preserving orientation.

Considering several circles instead of one circle, one comes to the notion of a link. A *link* is a smooth embedding (image) of several disjoint circles in $\mathbb{R}^3$. Each knot, representing the image of one of these circles is called a *link component*. One can naturally define *link isotopy* (by using an orientation-preserving diffeomorphism of the ambient space), *link planar diagrams* and *link invariants*. By an *oriented link* is meant a smooth mapping (image) of the disjoint union of several oriented circles.

There is another approach to link isotopy, when each link component is allowed to have self–intersections during the isotopy, but intersections of different components are forbidden. This theory is described in a beautiful paper [Mil1] of John Willard Milnor, who introduced his famous μ–invariants for classification of links up to this "isotopy".

The *trivial n–component link* or *n–unlink* is a link represented by a diagram consisting of n circles without crossings.

Example 1.2. *Figure* **1.4.1.** *represents the trivial two–component link. Fig-*

ures **1.4.2.**, **1.4.3.**, *and* **1.4.4.** *show us links, representing the* Hopf link, *the* Whitehead link *and the* Borromean rings, *respectively. The latter link is named in honour of the famous Italian family Borromeo, whose coat of arms was decorated by these rings. All these three links are not trivial (this will be shown later, when we are able to calculate values of some invariants). Borromean links demonstrate an interesting effect: while deleting any of three link components, one obtains the trivial link, whence the total link is not trivial.*

Exercise 1.4. *Show that the Whitehead link has component symmetry: there is an isotopy to itself permuting the link components.*

Let us now talk about knot (and link) invariants. The first well–known link invariant (after the Gauss electromagnetic linking coefficient) is the *fundamental group of the complement to the knot (link)*. This invariant is purely topological; it distinguishes different knots quite well (in particular, it recognises the unknot as well as the trivial link with arbitrary many components). However, this "solution" of the knot recognition problem is not complete because we only reduce this problem to the group recognition problem, which is, generally, indecidable.

In 1923, the famous American mathematician James Alexander [Ale1, Ale3] derived a polynomial invariant of knots and links from the fundamental group. This invariant is, certainly, weaker than the fundamental group itself, but the invariant polynomial is much easier to recognise: one can easily compare two polynomials (unlike groups given by their presentation).

In 1932, the German topologist K. Reidemeister published his book *Knotentheorie* (English translation: [Reid]), in which he presented a list of local moves (known as Reidemeister moves) and proved that any two planar diagrams generate isotopic knots (links) if and only if there exists a finite chain of Reidemeister moves from one of them to the other. In addition, he tabulated knot isotopy classes up to complexity seven, inclusively.

Since that time, to prove the invariance for some function on knots, one usually checks its invariance under Reidemeister moves.

Among books describing the state of knot theory at that time, we would like to point out those by Ashley [Ash], Crowell and Fox [CF] and that by Burde and Zieschang [BZ].

The next stage of development of knot theory was the discovery of the Conway polynomial [Con]. This discovery is based on so–called *skein relations*. These relations are purely combinatorial and based on the notion of the planar diagram. The Alexander polynomial [Ale1, Tor] can also be interpreted in terms of skein relations. Moreover, Alexander knew about this. However, Conway was the first to show that skein relations can be used axiomatically for defining a knot invariant.

This discovery stimulated further beautiful work presenting polynomials based on skein relations. By using these polynomials, some old problems were solved, e.g. Tait's problem [Tai].

Among the other skein polynomials, we would like to emphasise the

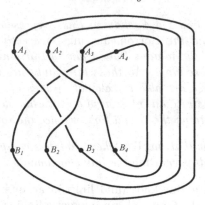

FIGURE 1.5: Closure of a braid

HOMFLY-PT polynomial and the Kauffman polynomial; HOMFLY is the abbreviation of the first letters of the authors: Hoste, Ocneanu, Millett, Freyd, Lickorish and Yetter (see [HOMFLY]). This polynomial was also discovered by Przytycki and Traczyk [PT].

The most powerful of the skein polynomials is the Jones polynomial of two variables; each of those named above can be obtained from it by a variable change. About Jones polynomials of one and two variables one can see [Jon1, Jon2].

The planar diagram approach for coding links is not the only possibility. Besides knot theory one should point out another theory — the theory of braids. Braids were proposed by the German mathematician Emil Artin, who gave initial definitions and proved basic theorems on the subject [Art1]; English translation [Art2]. There are four classical definitions of the braid group. Braids are closely connected with polynomials without multiple roots, discriminant theory, representation theory, etc. By an *n-strand braid* we mean a set of n ascending simple non-intersecting piecewise-linear curves (*strands*), connecting points $A_1, \ldots, A_n$ on a line with points $B_1, \ldots, B_n$ on a parallel line. Analogously to the case of knots, one can describe braids by their *planar diagrams*; the equivalence of braids is defined as an isotopy of strands preserving all strands ascending. *The product* of two braids a and b is obtained by juxtaposing one braid under the other and rescaling the height coordinate.

It is easy to see that by *closing* the braid in the most natural way (i.e., by connecting A_i with $B_i, i = 1, \ldots, n$; see Fig. 1.5) we obtain a link diagram.

In the present book we describe the three important theorems on braids and links. The Alexander theorem states that each link isotopy class can be obtained as a closure of braids. The Artin theorem [Art1] gives a presentation of the braid group by generators and relations. The Markov theorem [Mar, Mar'] gives a list of sufficient relations transforming one braid to the other in the case when their closures represent isotopic knots. We also present algorithms for braid recognition: a simple algorithm described in [GM] and that of Dehornoy.

The great advantage of braid theory (unlike knot theory) is that braids form a group. This simplifies some problems (i.e., reduces the word recognition problem to the trivial word recognition problem).

We would like to recommend the following monographs on those parts of knot theory described above: Louis Kauffman's two books [Kau2, Kau4], and those of Adams [Ada], Kawauchi [Kaw] and Atiyah [Ati2].

Suppose we have a knot and we want to change its isotopy class, by changing the smooth map of the circle in $\mathbb{R}^3$. By definition, it is impossible to do this without intersection. Thus, the most important moment of this map is the intersection moment. If there exists only a finite number of transversal intersections, one can speak about *a singular knot*. The space of all singular knots (that contains the space of all knots as a subspace) is called *the discriminant space*.

By studying the properties of the discriminant sets, V.A. Vassiliev proposed the notion of finite type invariants, later known as *Vassiliev's invariants*. Initially, Vassiliev's knot invariants required a complicated and non-trivial mathematical approach. However, a purely combinatorial interpretation of them was found. In the present book, we shall give a proof of the fact that the Vassiliev knot invariants are stronger than all the polynomial invariants named above.

The initial proof of the existence of the Vassiliev knot invariants is given in [Vas1]; the structure of these invariants was obtained by M.L. Kontsevich by means of his remarkable integral construction now known as *the Kontsevich integral*. The work of Kontsevich is published in [Kon1, Kon2]. We would also recommend the profound and detailed description of the Kontsevich integral in [BN1, BN3]. There one can also find good points of view for the connection between knots, Vassiliev's invariants, and the representation of Lie algebras.

The calculation of the Kontsevich integral had been very difficult before the work by Le and Murakami [LM] appeared. In this work, they present explicit techniques for the Kontsevich integral calculations. This technique is, however, very difficult.

One should also mention the outstanding work by Khovanov [Kho1] from 1997 who gave a generalisation of the Jones polynomial in terms of homologies of some formal algebraic complex. A new knot invariant appeared after so much had been discovered.

Here we would like to mention another way of representing knots and links. This is based on the notion of *d-diagram* (see [Man5]). A *d*–diagram can be seen as a circle with two families of non–intersecting chords where no point can be the end of two different chords. This theory has its origin in atoms and Hamiltonian systems. However, the *d*-diagram theory allows us to represent all links by using words in a finite alphabet, see [Man5]. Here we have an advantage in comparison with, say, encoding knots by braids: in the latter

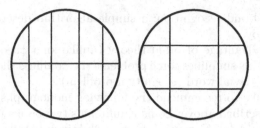

FIGURE 1.6: Left trefoil;   right trefoil.

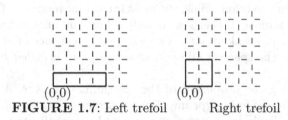

FIGURE 1.7: Left trefoil   Right trefoil

case one requires an infinite number of letters. This encoding can be easily generalised for cases of braids, singular knots and certain other purposes.

By using d–diagrams, one can represent all links as loops on checked paper; these loops should lie inside the first quadrant and come from the origin of coordinates.

The simplest d–diagrams corresponding to the left and right trefoil knots are shown in Fig. 1.6.

Their bracket structures look like

$$(((([))))]$$

for the left one and

$$(([[))]]$$

for the right one.

Example 1.3. *The left trefoil knot can be represented as the rectangle 1×4. The square 2×2 represents the right trefoil knot (see Fig 1.7).*

Actually, the theory of atoms first developed for the classification of Hamiltonian systems can be useful in many areas of geometry and topology, e.g. for coding 3-manifolds.

In the last few years, one of the most important branches of knot theory is the theory of virtual knots. A virtual knot is a combinatorial notion proposed by Louis H. Kauffman in 1996 (see [Kau5]). This notion comes from a generalisation of classical knot diagram with generalised Reidemeister moves. The theory can be interpreted as a "projection" of knot theory from knots in some 3-manifolds. This theory is developed very rapidly. There are many common approaches coming from classical knot theory, and there are "purely virtual invariants" (see [Man4]).

Chapter 2

Reidemeister moves. Knot arithmetics

In the present chapter, we shall discuss knots, their planar diagrams and the semigroup structure on knots; the latter is isomorphic to that on natural numbers with respect to multiplication. This allows us to investigate *knot arithmetics* and to establish some properties of multiplication and decomposition of knots.

As we have shown in the previous chapter, all links can be encoded by their regular planar diagrams. Obviously, while deforming a link, its planar projection might pass through some singular states. These singular states give the motivation for writing down the list of simplest moves for planar diagrams.

2.1 Polygonal links and Reidemeister moves

Because each knot is a smooth embedding of S^1 in $\mathbb{R}^3$, it can be arbitrarily closely approximated by an embedding of a closed broken line in $\mathbb{R}^3$. Here we mean a good approximation such that after a very small smoothing (in the neighbourhood of all vertices) we obtain a knot from the same isotopy class. However, generally this might not be the case.

Definition 2.1. An embedding of a disjoint union of n closed broken lines in $\mathbb{R}^3$ is called a *polygonal n–component link*. A *polygonal knot* is a polygonal 1–component link.

Definition 2.2. A link is called *tame* if it is isotopic to a polygonal link and *wild* otherwise.

Remark 2.1. *The difference between tame and wild knots is of great importance. To date, the serious systematic study of wild knots has not been started. We shall deal only with tame knots, unless otherwise specified.*

All C^1–smooth knots are tame; for a proof see [CF]. In the sequel, all knots are taken to be smooth, hence, tame.

In higher dimensions there is a difference between piecewise linear and smooth knots.

Definition 2.3. Two polygonal links are *isotopic* if one of them can be transformed to the other by means of an iterated sequence of *elementary isotopies*

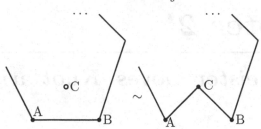

FIGURE 2.1: Elementary isotopy

and inverse transformation. Here the elementary isotopy is a replacement of
an edge AB with two edges AC and BC provided that the triangle ABC has
no intersection points with other edges of the link; see Fig. 2.1.

It can be proved that the isotopy of smooth links corresponds to that of
polygonal links; the proof is technically complicated. It can be found in [CF].
When necessary, we shall use either the smooth or polygonal approach for
representing links.

Like smooth links, polygonal links admit *planar diagrams* with overcrossings and undercrossings; having such a diagram one can restore the link up to
isotopy.

Exercise 2.1. *Show that all polygonal links with less than six edges are trivial.*

Exercise 2.2. *Draw a polygonal trefoil knot with six edges.*

Definition 2.4. By a *planar isotopy* of a smooth link planar diagram we
mean a diffeomorphism of the plane onto itself not changing the combinatorial
structure of the diagram.

Remark 2.2. *The polygonal knot planar diagram is defined analogously to
the smooth diagram. However bivalent vertices give us redundant information.
Thus, dividing an edge into two edges by an additional vertex we obtain a
diagram that is planar–isotopic to the initial one.*

Obviously, planar isotopy is an isotopy; i.e., it does not change the link
isotopy type in $\mathbb{R}^3$.

Theorem 2.1 (Reidemeister [Reid]). *Two diagrams D_1 and D_2 of smooth
links generate isotopic links if and only if D_1 can be transformed into D_2 by
using a finite sequence of planar isotopies and the three Reidemeister moves
$\Omega_1, \Omega_2, \Omega_3$, shown in Fig. 2.2.*

One can prove this fact by using the notion of codimension; it involves
some complicated technicalities. Here we give another proof for the case of
polygonal links.

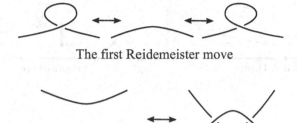

The first Reidemeister move

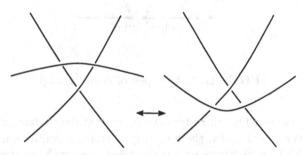

The second Reidemeister move

The third Reidemeister move

FIGURE 2.2: Reidemeister moves $\Omega_1, \Omega_2, \Omega_3$

Proof. One implication is evident: one should just check that moves $\Omega_1, \Omega_2, \Omega_3$ do not change the link isotopy class.

Now, let us prove the inverse statement of the theorem. Let D_1, D_2 be two planar diagrams of the two isotopic polygonal links K_1 and K_2. By definition, the isotopy between K_1 and K_2 consists of a finite number of elementary isotopies looking like $[AB] \to [AC] \bigcup [CB]$. Here the "link before" is reconstructed to the "link after". Without loss of generality, let us assume that for each step for the triangle ABC, the edges $[DA]$ and $[BE]$, coming from the ends of $[AB]$, do not intersect the interior of ABC. Otherwise, we can obtain it by using Ω_1.

Let L_0 be the projection of the "link before" on the plane ABC. Let us split the intersection components of ABC and L_0 into two sets: upper and lower according to the location of edges of the link K_0 with respect to plane ABC.

Let us tile ABC into small triangles of the four types in such a way that edges of small triangles do not contain vertices of L_0. Each first-type triangle contains only one crossing of L_0; here edges of L_0 intersect two sides of the triangle. The second-type triangle contains the only vertex of L_0 and parts

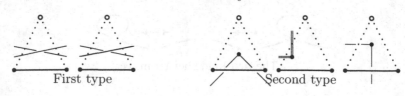

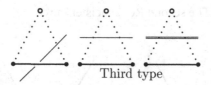

FIGURE 2.3: Types of small triangles

of outgoing edges. The third-type triangle contains a part of one edge of L_0 and no vertices. Finally, the fourth-type triangle contains neither vertices nor edges. All triangle types except the fourth (empty) one are illustrated in Fig. 2.3.

Such a triangulation of ABC can be constructed as follows. First, we cut all vertices and all crossings by triangles of the first and the second type, respectively. Then we tile the remaining part of ABC and obtain triangles of the last two types.

The plan of the proof is now the following. Instead of performing the elementary isotopy to ABC, we perform step–by–step elementary isotopies for small triangles, composing ABC. It is clear that these elementary moves (isotopies) can be represented as combinations of Reidemeister moves and planar isotopies.

More precisely, the first-type triangle generates a combination of the second and third Reidemeister moves, the second and the third-type triangles generate Ω_2 or planar isotopy. The fourth-type triangle generates planar isotopy. □

Exercise 2.3. *Show that two variants of the third Reidemeister moves, shown in Fig. 2.2, are not independent. More precisely, each of them can be obtained from the other and the second Reidemeister move.*

The above reasonings show that, in a general position, links have projections with many intersection points, which are simple and transverse. However, "the opposite case" is also interesting.

Difficult exercise 2.1 (H. Brunn). *Prove that for each link $L \subset L'$, there*

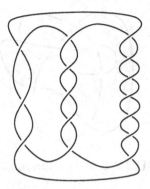

FIGURE 2.4: Non-invertible knot

exists a link L' and a plane P such that the projection of L' on P has the only intersection point, which is transverse for all branches.

Definition 2.5. A knot is called *invertible* if it is isotopic to the knot obtained from the initial one by the orientation change.

Remark 2.3. *Do not confuse knot invertibility and the existence of the inverse knot (in the sense of concatenation, see Def. 2.7).*

Exercise 2.4. *By using Reidemeister moves, show that each of the the two trefoil knots is invertible.*

The existence of non–invertible knots had been an open problem for a long time. This problem was solved positively in 1964 by Trotter [Tro]. The simplest non-invertible knot is 8_{17}, see the knot table [BNk, HTW2].

An example of a non-invertible knot is shown in Fig. 2.4.

Definition 2.6. A link diagram is called *alternating* if while moving along each component, one passes overcrossings and undercrossings alternately.

The simplest non–alternating knot (i.e. a knot not having alternating diagrams) is shown in Fig. 2.5.

2.2 Independence of Reidemeister moves

Let us show that each of the three moves $\Omega_1, \Omega_2, \Omega_3$ is necessary for establishing knot isotopy; i.e. that for each move there exist two diagrams of the same classical link which cannot be transformed to each other by using the two remaining Reidemeister moves (without the chosen one).

FIGURE 2.5: Non-alternating knot

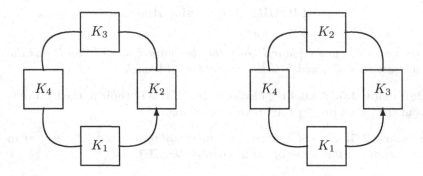

FIGURE 2.6: Diagrams L and M

Example 2.1. *The first Reidemeister move is the only move that changes the parity of the number of crossings. Thus, each unknot diagram with an odd number of crossings cannot be transformed to the diagram without crossings by using only* Ω_2, Ω_3.

Example 2.2. *Let* K_1, K_2, K_3, K_4 *be some diagrams of different prime (non-trivial) knots. Consider the diagrams* $L = K_1 \# K_2 \# K_3 \# K_4$ *and* $M = K_1 \# K_3 \# K_2 \# K_4$, *shown in Fig. 2.6.a,b. These diagrams represent the same knot.*

Let us show that there is no isotopy transformation from L *to* M *involving all Reidemeister moves but* Ω_2. *Actually, consider the subdiagrams* $K_i, i = 1, 2, 3, 4$ *inside* L. *Their order is such that between the knots* K_1 *and* K_3 *there are trivial knots at both sides. Let us show that this property remains true under* Ω_1, Ω_3. *Actually, while performing these moves, different* K_i *do not meet; thus one can always indicate each of these knots on the diagram. During the isotopy their order remains the same. However, in* M, *the knots* K_1 *and* K_3 *are adjacent. Thus* L *cannot be transformed to* M *by using only* Ω_1, Ω_3 *and planar isotopy.*

Example 2.3. *Now, let us consider the shadow of the standard Borromean*

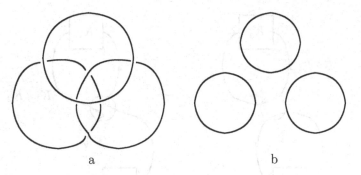

a   b

FIGURE 2.7: Diagrams L_1 and L_2

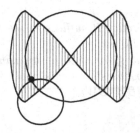

FIGURE 2.8: The domain U is shaded.

rings diagram and construct a link diagram as shown in Fig. 2.7.a. Denote this diagram by L_1.

Since the link represented by L_1 *is trivial, it has a diagram* L_2, *shown in Fig. 2.7.b.*

Let us show that the diagram L_1 *cannot be transformed to* L_2 *by using a sequence of* Ω_1 *and* Ω_2. *Actually, let us consider an arbitrary planar diagram of the three-component unlink and assign a certain element of* $\mathbb{Z}_2$ *to it as follows. Define the domain* U, *restricted by one link component. Let us fix this component* l. *It tiles the plane (sphere) into cells which admit a checkboard colouring. Let us use the colouring where the cell containing the infinite point is white. Denote by* $U(l)$ *the set of all black cells; see Fig. 2.8.*

Let us now consider crossings of the two components different from l, *select those lying inside* U *and calculate the parity of their number.*

Let us do the same for the second and the third component. Thus, we get three elements of $\mathbb{Z}_2$. *It is easy to see that for* L_1 *all these three numbers are equal to one and for* L_2 *they all are equal to zero.*

Then this (non-ordered) triple of numbers is invariant under Ω_1, Ω_2.

For Ω_1 *this statement is evident. In the case of* Ω_2 *the move is applied*

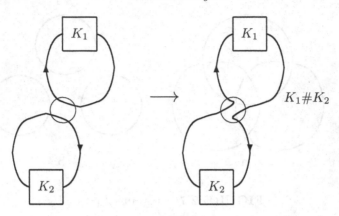

FIGURE 2.9: Connected sum of knots

to arcs of two different circles. It suffices to see that either both crossings lie outside $U(l)$ or they both lie inside l.

Thus, L_1 cannot be transformed to L_2 by using only Ω_1, Ω_2.

Let us note that analogous statements hold place in ornament theory, see Chapter 6 of [Vas3] and [Bjö, BW].

2.3   Knot arithmetics and Seifert surfaces

Let us now discuss the algebraic structure on the set of knot isotopy classes. Let K_1 and K_2 be two oriented knots.

Definition 2.7. By a *composition, concatenation* or *connected sum* of knots K_1 and K_2 is meant the oriented knot obtained by attaching the knot K_2 to the knot K_1 with respect to the orientation of both knots; see Fig. 2.9. Notation: $K_1 \# K_2$.

Exercise 2.5. *Show that the concatenation operation is well defined; i.e., the isotopy class of $K_1 \# K_2$ does not depend on the two places of attachment.*

Exercise 2.6. *Show that the concatenation operation is commutative: for any K_1, K_2 the knots $K_1 \# K_2$ and $K_2 \# K_1$ are isotopic.*

The exercise above is intuitively clear. However, the desired isotopy can be performed for the case of knots with fixed ends (or *long knots*).

Definition 2.8. Analogously, one defines the *disconnected sum*: one just takes two diagrams and situates them inside two non–intersecting domains on the plane.

Notation: $K_1 \sqcup K_2$.

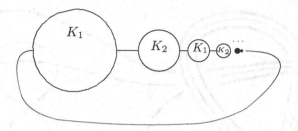

FIGURE 2.10: A wild knot

Theorem 2.2. *Let K_1 be a non-trivial knot. Then for each knot K_2, the knot $K_1 \# K_2$ is not trivial either.*

We shall give here three proofs of the theorem. One of them is a bit "speculative" but very clear. Another one is based on a beautiful idea that belongs to J.H. Conway. The last proof is a corollary from a stronger statement.

Let us begin now with the "speculative" one.

Proof. Suppose K_1 is not an unknot and $K_1 \# K_2$ is. Consider the sequence of knots

$$K_1 \# K_2, (K_1 \# K_2) \#(K_1 \# K_2), \ldots,$$

where the knot K_1 lies inside the ball with radius 1, the knot K_2 lies inside the ball of radius $\frac{1}{2}$, the knot K_3 lies inside the ball of radius $\frac{1}{4}$, and so on.

Thus, one can place the infinite series on a finite interval; see Fig. 2.10.

Thus we obtain a knot that will be, possibly, wild. Denote this knot by a. Since the knot $K_1 \# K_2$ is trivial, the knot a is trivial as well. On the other hand, we have the following decomposition: $a = K_1 \#(K_2 \# K_1)\#(K_2 \# K_1) \ldots$ Since the concatenation is commutative, the knot $K_2 \# K_1$ is trivial. Thus, the trivial knot a is isotopic to K_1. This contradiction completes the proof. $\quad\square$

There is a beautiful and simple proof of this fact proposed by J. H. Conway.

Proof. Let us look at Fig. 2.11.

Definition 2.9. By a *standardly embedded* (in $\mathbb{R}^3$) handlebody S_g for a natural number g we mean the small tubular neighbourhood of a graph lying in $\mathbb{R}^2 \subset \mathbb{R}^3$; see Fig. 2.12.

By a *standardly embedded* handle surface we mean the boundary of a standardly embedded handlebody.

Here we see that the connected sum $K_1 \# K_2$ has a tubular neighbourhood T isomorphic to the natural tubular neighbourhood of K_2. This neighbourhood is such that the intersection of each meridional disc D with $K_1 \# K_2$ is not trivial (homologically).

But for the unknot U the only possible neighbourhood T' with the property described above is a standardly embedded torus.

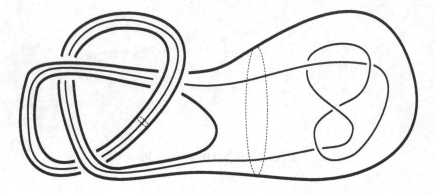

FIGURE 2.11: The product of non-trivial knots is not trivial

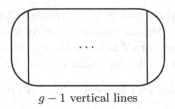

$g - 1$ vertical lines

FIGURE 2.12: A graph on the plane

The knot K_1 is not trivial and hence "thick" knots T' and T are not isotopic. Thus, $K_1 \# K_2$ is not a trivial knot.   □

Definition 2.10. A knot K is said to be *prime* if for any knots L, M, such that $K = L \# M$, one of the knots L, M is trivial. All other knots are said to be *composite*.

Definition 2.11. If for some knots K, L, M the statement $K = L \# M$ holds then one says that the knots L, M *divide* the knot K.

Thus, we have proved that all elements of the knot semigroup except the unknot have no inverse elements. Let us study another property of this semigroup.

Exercise 2.7. *Show that each knot (link) isotopy class can be represented as a curve (several curves) on some handled body standardly embedded in* $\mathbb{R}^3$.

Now let us give the definition of the Seifert surface first introduced by Seifert in [Sei1], see also [Sei2, Pra2, CF].

Definition 2.12. Let L be an oriented link. A *Seifert surface* for the link L is a closed compact oriented two–dimensional surface in $\mathbb{R}^3$, whose boundary is the link L and the orientation of the link L is induced by the orientation of the surface.

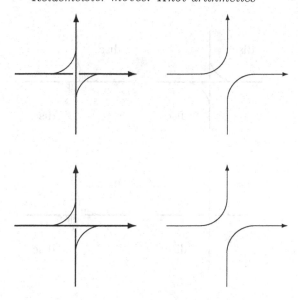

FIGURE 2.13: Smoothing the diagram crossings

Theorem 2.3. *For each link in* $\mathbb{R}^3$, *there exists a Seifert surface of it.*

Proof. Consider a planar diagram D of the link L. Let us smooth link crossings as shown below.

After such a smoothing, we obtain a set of closed nonintersecting simple curves on the plane.

Definition 2.13. These curves are called *Seifert circles*.

Let us attach discs to these circles. Though the interiors of these circles on the plane might contain one another, discs in 3–space can be attached without intersections.

In the neighbourhood of each crossing, two discs meet each other. Let us choose two closed intervals on the boundary of these discs and connect them by a twisted band; see Fig. 2.14. The boundaries of this band are two branches of the link incident to the chosen crossing. The two positions (upper and lower) in Fig. 2.14 show different ways of twisting (in one case the vertical line lies over the horizontal line, in the other case — vice versa).

Thus we obtain some surface that might not be connected. Connecting different components of this surface by thin tubes, we easily obtain a connected surface with the same boundary.

It remains to prove now that the obtained surface is orientable. Actually, consider the plane P of the link projection. Choose a positive frame of reference on P. This generates an orientation for any discs attached to a Seifert

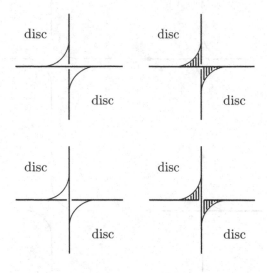

FIGURE 2.14: Attaching a band to discs

circle. Herewith, for two Seifert surfaces adjacent to the same vertex, these orientations are opposite (in the sense of the Seifert surface) since there is a twisted band between them.

It remains to show that for each sequence $C_1, \ldots C_n = C_1$ of Seifert circles, where any two adjacent circles C_i, C_{i+1} have a common vertex, the number n is odd. In other words, while passing from a circle to itself one should pass an even number of twistings. The latter follows from the fact that for a polygon with an odd number of sides one cannot choose an orientation of sides in such a way that any two adjacent sides have opposite orientations. □

Theorem 2.4. *The parity of the number of Seifert surfaces for the diagram of a k–component link with n crossings coincides with the parity of $n - k$.*

Proof. Let L be a diagram of a k–component link with n vertices. Consider a Seifert surface $S(L)$. Let us construct a cell decomposition of it. First let us choose the one–dimensional frame as follows: vertices of the diagram correspond to crossings of L (two vertices near each crossing), and edges correspond to edges of the diagram (two); one edge is associated with each crossing that connects the two vertices; see Fig.2.15. The number of cells of such a decomposition equals the number of Seifert surfaces. Now, let us attach discs to the boundary components (components of the initial link). Thus we obtain a closed oriented manifold. Its Euler characteristic should be even. It equals $2n - 3n + S + k$, where n is the number of crossings of the diagram L, and

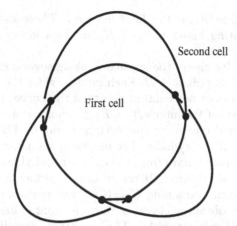

Second cell

First cell

FIGURE 2.15: Cell decomposition of a Seifert surface

S is the number of Seifert surfaces of this diagram. Taking into account that the number $2n - 3n + S + k$ is even, we obtain the claim of the theorem.

$\square$

The Seifert surface for the knot K is a compact 2–surface whose boundary is K. By attaching a disc to this surface we get a sphere with some handles.

Definition 2.14. A knot K is said to have *genus g* if g is the minimal number of handles for Seifert surfaces corresponding to K.

The notion of knot genus was also introduced by Seifert in [Sei1].

Remark 2.4. *Actually, the calculation problem for the knot genus is very complicated but it was solved by Haken. For this purpose, he developed the theory of* normal surfaces — *surfaces in 3–manifolds lying normally with respect to a cell decomposition, see [Hak]. The trivial knot (knot of genus 0) recognition problem is a partial case of this problem.*

Lemma 2.1. *The function g is additive; i.e. for any knots K_1, K_2 the equality $g(K_1) + g(K_2) = g(K_1 \# K_2)$ holds.*

Proof. First, let us show that $g(K_1 \# K_2) \leq g(K_1) + g(K_2)$.

Consider the Seifert surfaces F_1 and F_2 of minimal genii for the knots K_1 and K_2. Without loss of generality, we can assume that these surfaces do not intersect each other. Let us connect the two surfaces by a band with respect to their orientation.

Thus we obtain a Seifert surface for the knot $K_1 \# K_2$ of genus $g(K_1) + g(K_2)$. Thus

$$g(K_1 \# K_2) \leq g(K_1) + g(K_2).$$

Now, let us show that $g(K_1 \# K_2) \geq g(K_1) + g(K_2)$. Consider a minimal genus Seifert surface F for the knot $K_1 \# K_2$. There exists a (topological) sphere S^2, separating knots K_1 and K_2 in the connected sum $K_1 \# K_2$ at some points A, B.

The sphere S^2 intersects the surface F along several closed simple curves and a curve ended at points A, B. Each circle divides the sphere S^2 into two parts; one of them does not contain points of the curve AB. Without loss of generality, assume that the intersection $F \cap S^2$ consists of several closed simple curves $a_1, \ldots, a_k$ and one curve connecting A and B. The neighbourhood of each a_i in F looks like a cylinder. Let us delete a small cylindrical part that contains the circle (our curve) from the cylinder and glue the remaining parts of the cylinder by small discs. If the obtained surface is not connected, let us take the part of it containing $K_1 \# K_2$. After performing such operations to each circle, we obtain a closed surface F' containing the knot $K_1 \# K_2$ and intersecting S^2 only along AB. The operations described before can only decrease the number of handles. Thus, $g(F') \leq g(F)$. Because F has minimal genus, we conclude that $g(F') = g(F) = g(K_1 \# K_2)$.

The sphere S^2 divides the surface F' into surfaces that can be treated as Seifert surfaces for K_1 and K_2. Thus,

$$g(K_1) + g(K_2) \leq g(F') = g(K_1 \# K_2)$$

Taking into account that we have proved the "$\geq$"–inequality, we conclude that the genus is additive.   □

As a corollary, one can conclude that any non-trivial knot has no inverse, since the unknot has genus zero and the others have genus greater than zero. This is the third proof of the non-invertibility of non-trivial knots.

Exercise 2.8. *Show that the trefoil has genus one.*

Thus, each knot can be decomposed in no more than a finite number of prime knots. To clarify the situation about knot arithmetics, it remains to prove one more lemma, the lemma on unique decomposition.

Lemma 2.2. *Let L and M be knots, and let K be a prime knot dividing $L \# M$. Then either K divides L or K divides M.*

Proof. Consider the knot $L \# M$ together with some plane p, intersecting it at two points A and B and separating the knot L from the knot M. If we go from considering $\mathbb{R}^3$ to $S^3 = \mathbb{R}^3 \cup \infty$, the plane p turns into the sphere P.

Since $L \# M$ is divisible by K, there exists (topological) two-dimensional sphere S^2 intersecting the knot $L \# M$ at two points.

If this sphere does not intersect the sphere P, then the problem is solved. Without loss of generality, suppose that the sphere S^2 intersects the sphere P by some disjoint non-self-intersecting curves (circles). Each of these circles on the sphere P either separates the points A and B, or does not. The circles that do not separate the points A and B are easily destroyed by a small modification of the sphere: you need to start with the circles that do not contain other curves inside themselves.

When you destroy the curves c_i that separate the points A and B, you need use the simplicity of the knot K. The P sphere breaks the sphere $S^3 = \mathbb{R}^3 \cup \infty$ into two three-dimensional balls. Let B_K be the ball containing the knot K. Consider the connected component of $S^2 \cap B_K$ containing c_i. It may be either a two-dimensional disk or a cylinder. In the case of a cylinder, the curve c_i is destroyed by a small perturbation of the sphere S_2 together with another curve (the other edge of the same cylinder). In the first case the disc D^2 under consideration intersects the knot $L\#M$ exactly in one point. From the simplicity of the knot K it follows that in B_K it is entirely located on one side of the disk D^2. Therefore, we can destroy the disk D^2, and the curve together with it .

From the above, we conclude that if the connected sum $L\#M$ is divisible by K, then one of the links L or M is divisible by K.

$\square$

Remark 2.5. *This basic statement of knot arithmetics belongs to Schubert, see [Sch].*

Thus we have:

1. Knot isotopy classes form a commutative semigroup related to the concatenation operation; the unit element of the semigroup is the unknot.

2. Each non-trivial knot has no inverse element;

3. Prime decomposition is unique up to permutation;

4. The number of different prime knots is denumerable.

Exercise 2.9. *Prove the last statement.*

Hence the number of smooth knot isotopy classes is denumerable, and we get the following:

Theorem 2.5. *The semigroup of knot isotopy classes with respect to concatenation is isomorphic to the semigroup of natural numbers with respect to multiplication. Here prime knots correspond to prime numbers.*

However, the isomorphism described above is not canonical, hence there is no canonical linear order on the set of all knots (prime knots); i.e., it is impossible to say that "the (prime knot) right trefoil corresponds to the prime number three or to the prime number 2017". To do this, one should be able to recognise knots, which that is quite a difficult problem.

There exists only one semigroup defined by the properties described above (up to isomorphism).

In [Man5], we propose a purely algebraic description for this semigroup; i.e., we give an explicit isomorphism between this geometric semigroup and some algebraically constructed "bracket semigroup".

Chapter 3

Links in 2–surfaces in $\mathbb{R}^3$. Simplest link invariants

Knots (links) embedded in $\mathbb{R}^3$ can be considered as curves (families of curves) in 2–surfaces, where the latter surfaces are standardly embedded in $\mathbb{R}^3$. In this chapter we shall prove that all knots and links can be obtained in this manner.

In the present chapter, we shall describe some series of knots and links, e.g., torus links that can be represented as links lying on the torus standardly embedded in $\mathbb{R}^3$.

We shall also present some invariants of knots and links and demonstrate that the trefoil knot is not trivial.

3.1 Knots in 2–surfaces. The classification of torus knots

Consider a handle surface S_g standardly embedded in $\mathbb{R}^3$ and a curve (knot) K in it. We are going to discuss the following question: which knot isotopy classes can appear for a fixed g?

First, let us note that for $g = 0$ there exists only one knot embeddable in S^2, namely, the unknot.

Exercise 3.1. *Show, that for each link isotopy class $L \subset \mathbb{R}^3$ there exists a representative link L' (of the class L) lying in some handle surface standardly embedded in $\mathbb{R}^3$.*

The case $g = 1$ (torus, *torus knots*) gives us some interesting information.

Consider the torus as a Cartesian product $S^1 \times S^1$ with coordinates $\phi, \psi \in [0, 2\pi]$, where 2π is identified with 0. In Fig. 3.1, the torus is illustrated as a square with opposite sides identified.

Let us embed this torus standardly in $\mathbb{R}^3$; more precisely:

$$(\phi, \psi) \longrightarrow ((R + r \cos \psi) \cos \phi, (R + r \cos \psi) \sin \phi, r \sin \psi)$$

Here R is the outer radius of the torus, r is the small radius ($r < R$), ϕ is *the longitude*, and ψ is *the meridian* coordinates, the direction of the longitude and the meridian seen in Fig. 3.2.

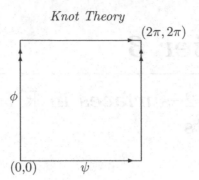

FIGURE 3.1: Torus

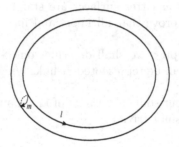

FIGURE 3.2: Longitude and meridian of the torus

For the classification of torus knots we shall need the classification of isotopy classes of non-intersecting curves in T^2: obviously, two curves isotopic in T^2 are isotopic in $\mathbb{R}^3$.

Without loss of generality, we can assume that the considered closed curve passes through the point $(0,0) = (2\pi, 2\pi)$. It can intersect the edges of the square several times. Without loss of generality, assume all these intersections to be transverse. Let us calculate separately the algebraic number of intersections with horizontal edges and that of intersections with vertical edges. Here passing through the right edge or through the upper edge is said to be positive; that through the left or the lower is negative.

Thus, for each curve of such type we obtain a pair of integer numbers.

Exercise 3.2. *Show that if both numbers are equal to zero then the knot is trivial.*

Remark 3.1. *In the sequel, we shall consider only those knots for which at least one of these numbers is not equal to zero.*

The following fact is left for the reader as an exercise.

Exercise 3.3. *For a non-self-intersecting curve these numbers are coprime.*

So, each torus knot passes p times the longitude of the torus, and q times its

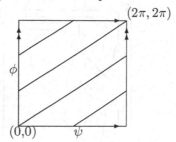

FIGURE 3.3: Trefoil on the torus

meridian, where $GCD(p, q) = 1$. It is easy to see that for any coprime p and q such a curve exists: one can just take the geodesic line $\{q\phi - p\psi = 0 (mod\ 2\pi)\}$. Let us denote this torus knot by $T(p, q)$.

Exercise 3.4. *Show that curves with the same coprime parameters* p, q *are isotopic on the torus.*

So, in order to classify torus knots, one should consider pairs of coprime numbers p, q and see which of them can be isotopic in the ambient space $\mathbb{R}^3$.

The simplest case is when either p or q equals one.

Exercise 3.5. *Show that for* $p = 1$, q *is arbitrary or* $q = 1, p$ *is arbitrary we get the unknot.*

The next simplest example of a pair of coprime numbers is $p = 3, q = 2$ (or $p = 2, q = 3$). In each of these cases we obtain the trefoil knot.

Let us prove the following important result.

Theorem 3.1. *For any coprime integers* p *and* q*, the torus knots* (p, q) *and* (q, p) *are isotopic.*

Proof. Let us take S^3 as the ambient space for knots (instead of $\mathbb{R}^3$). As we know, it does not affect isotopy.

It is well known that S^3 can be represented as a union of two full tori attached to each other according to the following boundary diffeomorphism. This diffeomorphism maps the longitude of the first torus to the meridian of the second one, and vice versa. More precisely, $S^3 = \{z, w \in \mathbb{C} \mid |z|^2 + |w|^2 = 1\}$. The two tori (each of them is one half of the sphere) are given by the inequalities $|z|^2 \geq |w|^2$ and $|z|^2 \leq |w|^2$; their common boundary is defined by the equation $|z| = |w| = \sqrt{\frac{1}{2}}$. It is easy to see that the circles defined as

$$|w| = \sqrt{\frac{1}{2}}, z \text{ — fixed with absolute value } \sqrt{\frac{1}{2}}$$

and

$$|z| = \sqrt{\frac{1}{2}}, w \text{ — fixed with absolute value } \sqrt{\frac{1}{2}}$$

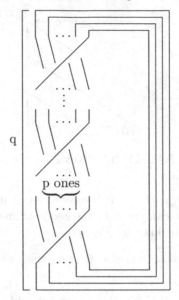

FIGURE 3.4: Torus p, q–knot diagram

are the longitude and the meridian of the boundary torus.

Thus, the (p, q) torus knot in one full torus is just the (q, p) torus knot in the other one. Thus, mapping one full torus to the other one, we obtain an isotopy of (p, q) and (q, p) torus knots.   □

Exercise 3.6. *Express this homotopy of full tori as a continuous process in* S^3.

Torus knots of type (p, q) can be represented by the following series of planar diagrams; see Fig. 3.4.

Remark 3.2. *Figure 3.4 demonstrates a way of coding a knot (link) as a (p–strand) braid closure. We shall speak about braids later in the book.*

Analogously to the case of torus knots, one can define *torus links* which are links embedded into the torus standardly embedded in $\mathbb{R}^3$.

We know the construction of torus knots. So, in order to draw a torus link one should take a torus knot $K \subset T$ (one can assume that it is represented by a straight-linear curve defined by the equation $q\phi - p\psi = 0(mod\ 2\pi)$) and add to the torus T some closed non–intersecting simple curves; each curve should be non-intersecting and should not intersect K. Thus, these curves should be embedded in $T \backslash K$; i.e. in the open cylinder.

Each closed curve on the cylinder is either contractible or passes the longitude of the cylinder once; see Fig. 3.5.

So, each curve in $T \backslash K$ is either contractible inside $T \backslash K$, or "parallel" to K

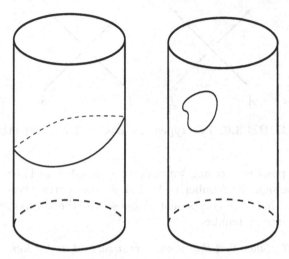

FIGURE 3.5: Possible simple curves on the cylinder

inside T; i.e. isotopic to the curve given by the equation $q\phi - p\psi = \varepsilon(mod\ 2\pi)$ inside $T\backslash K$.

Thus, the following theorem holds.

Theorem 3.2. *Each torus link is isotopic to the disconnected sum of a trivial link and a link that is represented by a set of parallel torus knots of the same type* (p,q).

3.2   The linking coefficient

From now on, we shall construct some invariants of links. As we know from the first chapter, a link invariant is a function defined on links that is invariant under isotopies. One can consider separately invariants of knots and links (oriented or non-oriented).

We shall represent links by using their planar diagrams. According to the Reidemeister theorem, in order to prove the invariance of some function on links, it is sufficient to check this invariance under the three Reidemeister moves.

First, let us consider the simplest integer–valued invariant of two–component links.

Let L be a link consisting of two oriented components A and B and let L' be a planar diagram of L. Consider those crossings of the diagram L' where the component A goes over the component B. There are two possible types of such crossings with respect to the orientation; see Fig. 3.6.

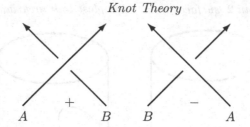

FIGURE 3.6: Two types of crossings for an oriented link

For each positive crossing we assign the number $(+1)$, for each negative crossing we assign the number (-1). Let us summarise these numbers along all crossings where the component A goes over the component B. Thus we obtain some integer number.

Exercise 3.7. *Show that this number is invariant under Reidemeister moves.*

Thus, we have an oriented link invariant.

Definition 3.1. The obtained invariant is called the *linking coefficient* .

Remark 3.3. *This invariant was first invented by Gauss [Gau]. He calculated it by means of his famous formula. This formula is named the Gauss electromagnetic formula in honour of him. The linking coefficient can be generalised for the case of p– and q–dimensional manifolds in $\mathbb{R}^{p+q+1}$.*

The formula for the parametrised curves γ_1 and γ_2 with radius–vectors $r_1(t), r_2(t)$ is given by the following formula

$$lk(\gamma_1, \gamma_2) = \frac{1}{4\pi} \int_{\gamma_1} \int_{\gamma_2} \frac{(r_1 - r_2, dr_1, dr_2)}{|r_1 - r_2|^3}.$$

The proof of this fact is left for the reader as an exercise.

Hint 3.1. *Prove that this function is the degree of a map from the torus generated by two link components to a sphere and, thus, it is invariant; then prove that it is proportional to the linking coefficient and find the coefficient of proportionality.*

The linking coefficient allows us to distinguish some two–component links.

Example 3.1. *Let us consider the trivial two–component link and enumerate its components in an arbitrary way. Obviously, their linking coefficient is zero. For the Hopf link, the linking coefficient equals ± 1 depending on the orientation of the components. Hence, the Hopf link is not trivial.*

Exercise 3.8. *Show that the linking coefficient of two "parallel" torus knots of type $(1, n)$ equals n.*

Example 3.2. *For any two components of the Borromean rings, the linking coefficient equals zero; each component of this link is a trivial knot. However, the Borromean rings are not isotopic to the trivial three–component link. This will be shown later in the text.*

For one–component link diagrams (knot diagrams), one can define the *self–linking coefficient*. To do this, one should take an oriented knot diagram K and draw a parallel copy K' of it on the plane. After this, one takes the linking coefficient of K and K'. It is easy to check that this is invariant under Ω_2 and Ω_2, but not Ω_1: adding a loop changes the value by ± 1.

There exists another approach to the link coefficients, namely that involving Seifert surfaces.

Definition 3.2. Let F be a Seifert surface of an oriented knot J. Assume that an oriented link K intersects F transversely in finitely many points. With each intersection point, we associate a number $\varepsilon_i = \pm 1$ according to the following rule.

Let us define the orientation of F, assuming the reference point $\{e_1, e_2\}$ positive, where e_1 is the speed vector of J and e_2 is the interior normal vector (it is perpendicular to e_1 and directed inside F). Let e_3 be the speed vector of the curve K at a point $a \in K \cap F$, and $\{c'_1, c'_2\}$ be a positive frame of reference on the surface F at a. Then, $\varepsilon_i = +1$ if the orientation $\{e'_1, e'_2, e_3\}$ coincides with the orientation of the ambient space $\mathbb{R}^3$; otherwise, let us set $\varepsilon_i = -1$.

The sum of all signs ε_i is called *the linking coefficient of J and K*.

Exercise 3.9. *Prove that the linking coefficient (in the latter sense) is well defined (does not depend on the choice of the Seifert surface) and coincides with the initial definition.*

Hint 3.2. *Use the planar projections and investigate the behaviour of two knot projections in neighbourhoods of crossings.*

3.3 The Arf invariant

We have constructed a simple invariant of two–component links, the link coefficient. For unoriented links this construction allows us to define an invariant with coefficients from $\mathbb{Z}_2$.

It turns out that there exists a knot invariant valued in $\mathbb{Z}_2$ that is closely connected with the link coefficient, namely, the so–called *Arf invariant*. There are many ways to define this invariant; here we follow [Kau1] and [Ada].

The Arf invariant comes from Seifert surfaces. Namely, let K be a knot and l be a band that is a part of the Seifert surface of the knot K, see the upper part of Fig. 3.7.

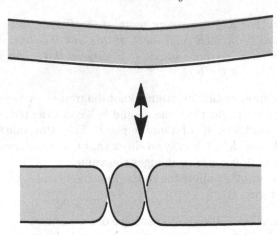

FIGURE 3.7: Simple twisting

Let us transform the knot by twisting this band by two full turns; see Fig. 3.8.

We obtain a knot K'.

Definition 3.3. Let us say that K and K' are *Arf equivalent.*

Definition 3.4. The *Arf invariant* is the complete invariant of the Arf equivalent classes.

Now, let us denote the Arf invariant by A and decree $A(\bigcirc) = 0$ and $A(\mathcal{Q}) = A(\mathcal{Q}) = 1$ where $\bigcirc$ denotes the unknot.

It turns out that the Arf invariant defined above has only two values, one and zero!

Theorem 3.3. *Each knot is either Arf equivalent to the unknot or to a trefoil (both trefoils are Arf equivalent).*

The main idea for proving this theorem (for a rigorous proof see, e.g. in [Ada]) is the following. First, let us mention that each Seifert surface can be thought of as a disc with several bands attached to its boundary. Each band can be twisted many times, and bands can be knotted. This observation is left to the reader as an exercise.

The number of half–turn twists for each band can be taken to be zero or one according to the Arf equivalence. Besides, the Arf equivalence allows us to change the disposition of bands in 3–space, for instance, to erase knottedness. Namely, a simple observation shows that passing one band through the other is also an Arf equivalence.

Thus we obtain a Seifert surface of a very simple type: it looks like a disc together with some bands attached to its boundary according to some "chord diagram" law; some of these bands are half–turn twisted; the others

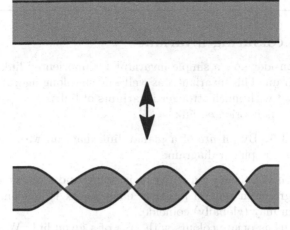

FIGURE 3.8: Double twisting

are not twisted. After this, one can perform some reducing operations, which will transform our knot into the connected sum of some trefoils. Finally, the only thing to do is the following exercise.

Exercise 3.10. *Show that the connected sum of the trefoil with another trefoil is Arf equivalent to the unknot. It does not matter which trefoils you take (right or left).*

In the sequel, all calculations concerning the Arf invariant will be performed modulo $\mathbb{Z}_2$.

It turns out that the Arf invariant is very easy to calculate.

Let ⊗ and ⊗ be two knot diagrams that differ in a small neighbourhood. Then, one of the diagrams)(, ⌣ is a knot diagram, and the other one is a link diagram. Suppose that)(is a link diagram, consisting of the two link components l_1 and l_2. Then the following theorem holds.

Theorem 3.4. *Under conditions described above,*

$$a(\otimes) - a(\otimes) \equiv lk(l_1, l_2) \ (mod \mathbb{Z}_2).$$

This relation in fact allows us to calculate the value of the Arf invariant for different knots. Later, we shall show how to transform any knot diagram to an unknot diagram only by switching some crossing types. Thus will work in many other situations while calculating some polynomials.

Furthermore, it looks similar to the so–called *skein relations* that will be defined later. The skein relations lead to much stronger knot invariants.

3.4   The colouring invariant

Let us consider now a simple invariant for unoriented links starting from the link diagram. This invariant was well known a long ago. It turns out that it is connected with much stronger invariants of links.

Consider a non–oriented link.

Definition 3.5. By an *arc* of a planar link diagram we mean a connected component of the planar diagram.

Thus, each arc always goes "over"; it starts and stops at undercrossings. For link diagrams in general position, each vertex is incident to three arcs. Some of them may (globally) coincide.

Now, let us associate colours with arcs of a given link. We shall use three colours: red, blue, white.

Definition 3.6. A colouring of a link diagram is said to be *proper* if for each crossing of the diagram, the three arcs incident to it have either all three different colours or one and the same colour.

Theorem 3.5. *The number of proper colourings is an invariant of link isotopy types.*

Proof. Let us prove the invariance in the following way. Consider a Reidemeister move and two diagrams L, L' obtained one from the other by using this move. Then we present a one–to–one correspondence of proper colourings for L and L'.

In the case of a Ω_1–move the situation is clear: the two edges of L' corresponding to one "broken" edge of L should have one colour because we have the situation where two of three edges meeting at a crossing are the same (and hence, there is no possibility to use three colours).

Thus, the desired one–to–one correspondence is evident. Analogously, there are "one colour" cases of the second and third Reidemeister moves. The invariance under second Reidemeister move is shown here: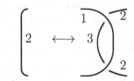

In Fig. 3.9, we give the corresponding colourings for edges taking part in the third Reidemeister moves. For the corresponding colouring all edges outside the area should be coloured identically. Here we show some examples of such colourings and their one–to–one correspondences. The other possibilities can be obtained from these by some permutation of colours (here colours are marked by numbers 1, 2, 3).

Thus, each proper colouring of the initial diagram uniquely corresponds to a colouring of the diagram obtained from the initial one by applying a Reidemeister move. Thus, the number of proper colourings is invariant under Reidemeister moves.

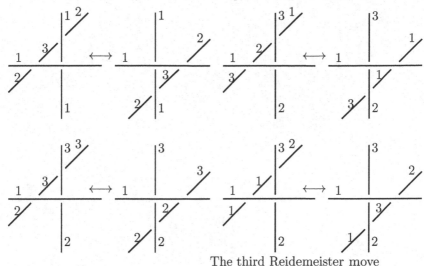

The third Reidemeister move

FIGURE 3.9: Invariance of proper colouring number

□

Let us now call the number of proper colourings the *colouring invariant*. **Notation:** for a link L we have $CI(L)$.

Obviously, for the unknot, we have $CI(\bigcirc) = 3$; for the k–component unlink, we have $CI = 3^k$.

Exercise 3.11. *Calculate the colouring invariant for the unknot and that for the (right) trefoil knot and show that the trefoil knot is not trivial.*

Exercise 3.12. *Calculate the colouring invariant for the Whitehead link and show that it is not trivial.*

Exercise 3.13. *Calculate the values of the colouring invariant for the Borromean rings. Show that they are not trivial.*

However, if we observe this invariant on the figure eight knot, we see that each colouring is monochrome. Thus, our invariant does not distinguish between the figure eight knot and the unknot. This encourages us to seek stronger knot and link invariants.

Chapter 4

Fundamental group. The knot group

4.1 Digression. Examples of unknotting

Let us now discuss a sympathetic example (or problem) concerning unknot diagrams and ways of unknotting them.

First, consider an arbitrary diagram of the trivial knot. Let us try to unknot it by using Reidemeister moves. In the "good" case this can be done only by decreasing the number of vertices (we mean the "decreasing" version of the first two Reidemeister moves and the third move).

However, this is not always so. More precisely, there exists an unknot diagram with a non-empty set of vertices such that we can apply neither decreasing versions of the first or second Reidemeister move nor the third Reidemeister move to this diagram. Thus, in order to transform this diagram to the unknot, one should first increase the number of vertices of it. We are going to construct such diagrams.

Example 4.1. *Consider the knot diagram shown in Fig. 4.1. Obviously, this is a diagram of the unknot because it can be obtained from the stretched unknotted circle by using a sequence of Ω_2's.*

Example 4.2. *Now let us consider the knot diagram shown in Fig. 4.2. This*

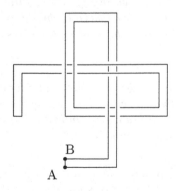

FIGURE 4.1: First unknot diagram

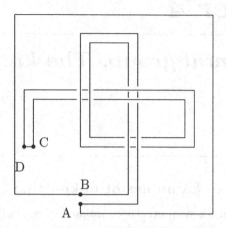

FIGURE 4.2: Second unknot diagram

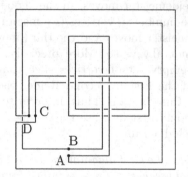

FIGURE 4.3: Third unknot diagram

can be obtained from that shown in Fig. 4.1 by throwing the arc AB over infinity (we think of the plane as compactified by the point at infinity). Obviously, this throwing does not change the knot isotopy class. Hence, we see the unknot in Fig. 4.2.

Example 4.3. *Let us consider now the knot diagram shown in Fig. 4.3. This can be obtained from that shown in Fig. 4.2 by applying the second Reidemeister move to CD and throwing it over infinity. Hence, the knot diagram shown in Fig. 4.3 also represents the unknot.*

Theorem 4.1. *Each Reidemeister move that can be applied to the unknot diagram shown in Fig. 4.3 increases the number of vertices of the diagram.*

Proof. Consider the knot diagram shown in Fig. 4.3 as a four–valent graph

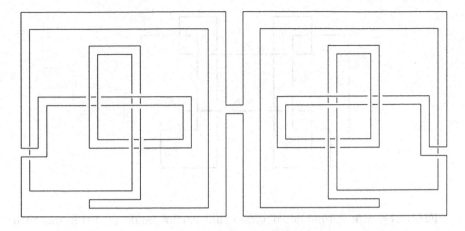

FIGURE 4.4: Fourth unknot diagram

with over– and undercrossings. It is sufficient to check that in order to perform the first or the second decreasing Reidemeister move, the shadow must contain loops or bigons. Additionally, in order to perform the third Reidemeister move, the shadow must contain a triangle such that one edge of it has no over-crossings. It is easy to see that the knot diagram shown in Fig. 4.3 has neither such loops, nor bigons, nor "good" triangles. Thus, each of the Reidemeister moves applicable to this diagram will increase the number of vertices. This completes the proof. □

However, one can consider not only planar knot (link) diagrams, but also *spherical* diagrams. Namely, one can think of the sphere as the plane compact-ified by the point at infinity. Without loss of generality, one can assume that the shadow of the link does not contain this point. In this case, there appears one more "elementary isotopy", when some edge of the shadow passes through the infinity. This operation is called the *infinity change.*

It is intuitively clear that the infinity change is indeed an isotopy. Actually, it can be represented as a sequence of Reidemeister moves.

Exercise 4.1. *Prove this fact directly.*

Thus, the knot shown in Fig. 4.3 can be unknotted only by using the second decreasing Reidemeister move, after the preliminary infinity change (throwing an arc over infinity) in the very beginning.

Actually, the knot shown in Fig. 4.4 cannot be unknotted only by non-decreasing Reidemeister moves even if we admit the infinity change (the infi-nite cell is no longer a bigon).

Now, if we consider the knot diagram from Fig 4.3 as a spherical diagram, we shall see the bigon, containing the infinite point. Thus, after the infinity

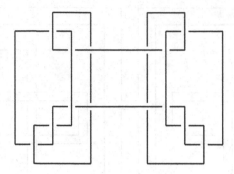

FIGURE 4.5: An unknot diagram that cannot be decreased in one turn

change, one gets this bigon inside a compact domain. Then one can easily untangle the diagram by using only decreasing moves.

In Fig. 4.4, we illustrate an example of a knot diagram having no loops, no bigons and no "good" triangles with one edge forming two overcrossings. These properties remain true even if considering this as a spherical diagram. Thus, in order to untangle the knot shown in Fig. 4.4, one should first perform some increasing moves.

Another example of such an unknot diagram that cannot be decreased in one turn is shown in Fig. 4.5. This example was invented by the student I.M. Nikonov who attended the author's lecture course.

4.2 Fundamental group. Basic definitions and examples

Let us now start the main part of the present chapter. We shall describe the notion of fundamental group for arbitrary topological spaces and show how to calculate it for link complements. We are going to introduce some presentations of this group.

The topological theory of the fundamental group can be read in, e.g., [Fom, FFG, Vas3, CF]. The theory of three-manifolds can be read in [Mat5].

Consider a topological space X and a point $x_0 \in X$. Fix a point a on the circle S^1. Consider the set of continuous mappings $f : S^1 \to X$ such that $f(a) = x_0$. The set of homotopy classes of such mappings admits a group structure. Indeed, the multiplication of two such mappings can be represented by concatenating their paths. The inverse element is obtained by passing over the initial path in the inverse order. Obviously, these operations are well defined up to homotopy.

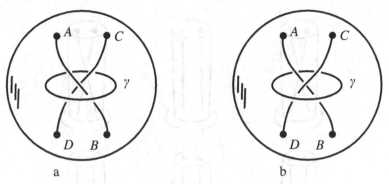

a   b

FIGURE 4.6: Changing the crossing

Definition 4.1. The obtained group is called the *fundamental group* of the space X; it is denoted by $\pi_1(X, x_0)$.

Exercise 4.2. *Show that for the case of connected X the group $\pi_1(X, x_0)$ does not depend on the choice of x_0; i.e., $\pi_1(X, x_0) \cong \pi_1(X, x_1)$.*

Remark 4.1. *There is no canonical way to define the isomorphism for fundamental groups with different initial points.*

The fundamental group is a topological space homotopy invariant.

Now let K be a link in $\mathbb{R}^3$.

Let $M_K = \mathbb{R}^3 \backslash K$ be the complement to K. It is obvious that while performing a smooth isotopy of K in $\mathbb{R}^3$ the complement always stays isotopic to itself. Hence, the fundamental group of the complement is an invariant of link isotopy classes.

In [GL] it is shown that the complement to the knot (more precisely, to its small tubular neighbourhood in $\mathbb{R}^3$) is a complete invariant of the knot up to amphicheirality.

However the analogous statement for links is not true. Before constructing a counterexample, let us prove the following lemma.

Lemma 4.1. *Let $D^3 \subset \mathbb{R}^3$ be a ball and $T \subset D^3$ be the full torus γ; see Fig. 4.6. There exists a homeomorphism of $\mathbb{R}^3 \backslash T$ onto itself, mapping the curves AB and CD (as they are shown in Fig. 4.6.a) to AB and CD (Fig. 4.6.b) that is constant inside the ball D^3.*

Proof. In order to have a more intuitive outlook, let us imagine that the interior diameter of T is very big (so that the "interior" boundary of it represents a high cylinder) in comparison with the exterior one. Thus, we have a deep hole surrounded by the boundary of the full torus; see Fig. 4.7.

Let us consider the ball D^3 and cut a circle from the plane, as shown in Fig. 4.7.a. Let us rotate the part of this cut (that is a circle with two marked points) in the direction indicated by arrows. This operation is possible

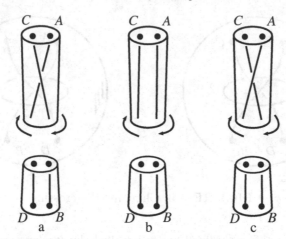

FIGURE 4.7: Rotating upper part of the cylinder

since the full torus T is deleted from D^3. Performing the 180–degree turn, we obtain the configuration shown in Fig. 4.7.b. Then, let us make one more turnover. Thus we obtain the embedding represented in Fig. 4.7.c. Thus, each point of the cut returns to the initial position. So, both copies of the cut can be glued together in such a way that the total space remains the same. In this way, we obtain a homeomorphism of the manifold $D^3 \backslash T$ onto itself. This homeomorphism can be extended to a homeomorphism of $\mathbb{R}^3 \backslash T$ onto itself, identical outside D^3. The latter homeomorphism realises the crossing change.   □

Exercise 4.3. *Show that links L_1 and L_2 (Fig. 4.8.a and 4.8.b) are not isotopic, but their complements are homeomorphic.*

This gives us an example of non–isotopic links with isomorphic fundamental groups.

Difficult exercise 4.1. *Find two non-isotopic (and not mirror) knots with isotopic fundamental groups.*

Definition 4.2. The link (knot) complement fundamental group is also called the *link (knot) group.*

Exercise 4.4. *Show that the fundamental group of the circle is isomorphic to the fundamental group of the complement to the unknot. Show that they are both isomorphic to $\mathbb{Z}$.*

Exercise 4.5. *Show that the fundamental group of the complement to the trivial n–component link is isomorphic to the free group in n generators.*

The link complement fundamental group is a very strong invariant. For instance, it recognises trivial links among links with the same number of components. This result follows from Dehn's theorem.

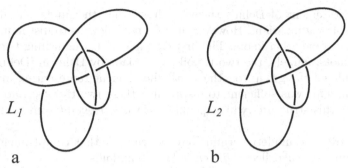

FIGURE 4.8: Non-isotopic links with isomorphic complements

Theorem 4.2 (Dehn). *An m–component link L is trivial if and only if $\pi_1(\mathbb{R}^3 \backslash L)$ is isomorphic to the free group in m generators.*

Thus, Dehn's theorem reduces the trivial link recognition problem to the free group recognition problem (for some class of groups). In the general case, the free group recognition problem is undecidable. For more details see [Bir1, Bir2] and [BZ, ECH..., MKS].

Dehn's theorem follows from the following lemma.

Lemma 4.2. *Let M be a 3–manifold with boundary and let γ be a closed curve on its boundary ∂M. Then if there exists an immersed 2–disc $D \to M$, such that $\partial D = \gamma$ then there exists an **embedded** disc $D' \subset M$ with the same boundary $\partial D' = \gamma$.*

This lemma was first proved by Dehn [Dehn1], but this proof contained lacunas. The rigorous proof was first found by Papakyriakopoulos [Pap]. This proof used the beautiful techniques of towers of 2–folded coverings.

Now the Dehn theorem (for the case of knots) is proved as follows. Having a knot $K \subset \mathbb{R}^3$ with its tubular neighbourhood $N(K)$, let us consider $\pi_1(\mathbb{R}^3 \backslash N(K), A)$ where $A \in T(K) = \partial N(K)$. Obviously, each closed loop which can be isotoped to a loop lying on $T(K)$ then can be represented via the longitude and meridian of $T(K)$, which are non-intersecting closed curves. Obviously, the *meridian* μ; i.e. the simple curve on $T(K)$ that lies in a small neighbourhood of some point on K and has linking coefficient with K equal to one, cannot be contracted to zero. Let λ be the *longitude*; i.e. a simple curve in T "parallel to K" and having linking coefficient zero with K. The curves μ and λ generate the fundamental group of the torus T.

Suppose the group of the knot K is isomorphic to $\mathbb{Z}$. Obviously, this group contains $\{\mu\} = \mathbb{Z}$. Besides, no power of λ can be equal to a non-trivial exponent of μ (because of linking coefficients). Thus, the curve λ is isotopic to zero in $\pi_1(\mathbb{R}^3 \backslash N(K))$. Thence, there is a singular disc bounded by λ. By Dehn's lemma, there is a disc embedded in $\mathbb{R}^3 \backslash N(K)$ bounded by λ. Contracting $N(K)$ to K, we obtain a disc bounded by K. Thus, K is the unknot. This completes the proof of Dehn's theorem.

The statement of Dehn's theorem shows that the fundamental group is rather a strong invariant. However, it does not allow us to distinguish mirror knots and some other knots. The first example of distinguishing two different mirror knots, namely, the two trefoils, was made by Dehn in [Dehn2]. There he considered the group together with the element representing an oriented meridian. This was sufficient to distinguish these two trefoils. Later, we will return to this structure (while speaking of the *peripheral system* of the knot complement).

According to modern terminology, we can say that complements to non-trivial links are so–called *sufficiently large* manifolds,

Definition 4.3. A manifold M is called *sufficiently large* if one can embed a handlebody (not the sphere) in M in such a way that the image map for the fundamental group has no kernel.

These manifolds are classified by S.V. Matveev; however the algorithm is quite formal and cannot be performed practically (e.g. by means of a computer program).

In [Mat1], he constructed the full invariant of knots, the knot *quandle (distributive groupoid)*. We shall consider this invariant later.

Exercise 4.6. *Show that $\pi_1(A_1 \vee A_2)$ (of the union of spaces A_1 and A_2 with one identified point) is isomorphic to the free product of $\pi_1(A_1)$ and $\pi_1(A_2)$ in the case when both A_1 and A_2 are pathwise connected.*

Let us demonstrate two ways of calculating a representation for the fundamental group of a knot complement. The first of them is more common; it can be used in many other situations.

Let X be a topological space that admits a decomposition $X = X_1 \cup X_2$, where each of the sets X_1, X_2, $X_0 = X_1 \cap X_2$ is open, pathwise–connected and non-empty. Choose a point $A \in X_0$. Suppose fundamental groups $\pi_1(X_1, A)$ and $\pi_1(X_2, A)$ have presentations $\langle a_1, \dots | f_1 = e, \dots \rangle$ and $\langle b_1, \dots | g_1 = e, \dots \rangle$, respectively. Suppose that the generators $c_1, c_2 \dots$ of $\pi_1(X_0, A)$ (which are elements of both groups $\pi_1(X_1, A)$ and $\pi_1(X_2, A)$) are represented as $c_i = c_i(a_1, \dots)$ and as $c_i = c_i(b_1, \dots)$ in the terms of $\pi_1(X_1, A)$ and $\pi_1(X_2, A)$, respectively.

Then the following theorem holds.

Theorem 4.3. *(The van Kampen theorem) The group $\pi_1(X, A)$ admits a presentation*

$$\langle a_i, b_i, | f_i = e, g_i = e, c_i(a) = c_i(b) \rangle.$$

In the case of CW–complices, the proof of the theorem is evident. For more details in the general case, see, e.g. [CF].

Corollary 4.1. *If both X_1 and X_2 described above are simply connected then X is simply connected as well.*

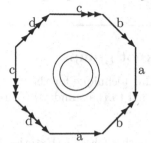

FIGURE 4.9: Gluing a handlebody

As an example, let us show how to calculate fundamental groups of 2–manifolds.

Theorem 4.4. *The fundamental group of the connected oriented 2–surface of genus g ($g > 0$) without boundary has a presentation*

$$\langle a_1, b_1, \ldots, a_g, b_g | a_1 b_1 a_1^{-1} b_1^{-1} \ldots a_g b_g a_g^{-1} b_g^{-1} = e \rangle.$$

Proof. Consider this handle surface as a $4g$–gon with some pairs of edges glued together; see Fig. 4.9.

Let us divide this manifold into two parts: one of them is located inside the big circle in Fig. 4.9; the other one is outside the small circle (it contains all edges and all gluings are performed for this part).

The first part is simply connected; the second one is contractible to the union of $2g$ circles and hence its fundamental group is isomorphic to the free group with generators[1] $a_1, b_1, \ldots, a_g, b_g$.

The intersection of the two areas described above is isotopic to the circle; hence its fundamental group is isomorphic to $\mathbb{Z}$. Thus, the only relation we have to add that deals with the generator of this $\mathbb{Z}$ is going to be e.

Applying the van Kampen theorem we get that

$$\pi_1(S_g) = \langle a_1, b_1, \ldots, a_g, b_g | a_1 b_1 a_1^{-1} b_1^{-1} \ldots a_g b_g a_g^{-1} b_g^{-1} = e \rangle.$$

$\square$

For each link L, there exists a handle surface S_g standardly embedded in $\mathbb{R}^3$ and a link $L' \subset S_g \subset \mathbb{R}^3$ isotopic to L in $\mathbb{R}^3$. Thus, one can calculate the fundamental group of the complement to L by using the van Kampen theorem: we divide the complement to L' into two parts lying on different sides of S_g.

---

[1]Here each letter means an oriented edge of the octagon; this edge is closed since all vertices are contracted to one point.

4.3   Calculating knot groups

Let us demonstrate this technique for the case of torus knots. Let K be a (p,q)–torus knot, embedded to the standard torus in $\mathbb{R}^3$.

Now, consider $\mathbb{R}^3 \backslash K$ as

(Interior full torus without K) $\cup$ (Exterior full torus without K).

Since deleting some set on the boundary does not change the fundamental group (of the full torus), we conclude that both fundamental groups are isomorphic to $\mathbb{Z}$ (one of them has the generator a, the other one has the generator b; they correspond to the longitude and meridian of the torus). The intersection of these parts is homeomorphic to the cylinder; the fundamental group of this cylinder is isomorphic to $\mathbb{Z}$. The only generator of this group can be expressed as a^p on one hand, and as b^q on the other hand. This implies the following theorem.

Theorem 4.5. *The fundamental group of the complement to the (p,q)–torus knot has a presentation with two generators a and b and one relation $a^p = b^q$.*

It is obvious that the fundamental group does not distinguish a knot (not necessarily torus) and its mirror image.

As they should be, fundamental groups of isotopic knots are isomorphic. Thus, $T(p,q)$ has the same group as $T(q,p)$, and the group of the knot $T(1,n)$ is isomorphic to $\mathbb{Z}$. Torus knots of types (p,q) and $(p,-q)$ are mirror images of each other, thus their groups coincide.

In all the other cases the fundamental group distinguishes torus knots.

Now, we present another way of calculating the fundamental group for arbitrary links. Consider a link L given by some planar diagram $\bar{L}$. Consider a point x "hanging" over this plane. Let us classify isotopy classes of loops outgoing from this point. It is easy to see (the proof is left to the reader) that one can choose generators in the following way. All generators are classes of loops outgoing from x and hooking the arcs of $\bar{L}$. Let us decree that the loop corresponding to an oriented edge is a loop turning according to the right–hand screw rule; see Fig. 4.10.

Now, let us find the system of relations for this group.

It is easy to see the geometrical connection between loops hooking adjacent edges (i.e., edges separated by an overcrossing edge). Actually, we have $b = cac^{-1}$, where c separates a and b; see Fig. 4.11.

Let us show that all relations in the fundamental group of the complement arise from these relations.

Actually, let us consider the projection of a loop on the plane of $\bar{L}$ and some isotopy of this loop. While transforming the loop, its written form in terms of generators changes only when the projection passes through crossings

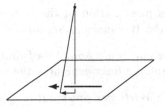

FIGURE 4.10: Loops corresponding to edges

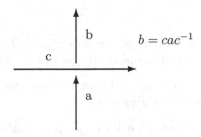

FIGURE 4.11: Relation for a crossing

of the link. Such an isotopy is shown in Fig. 4.12. During the isotopy process, the arc connecting P and Q passes under the crossing.

Obviously, the loop shown on the left hand (Fig. 4.12) generates the element $c^{-1}bc$, that on the right hand is just a.

Thus, our presentation of the fundamental group of the link complement is constructed as follows: arcs correspond to the generators and the generating relations come from crossings: we take $cac^{-1} = b$ for adjacent edges a and b, separated by c, when the edge b lies on the left hand from c with respect to the orientation of c.

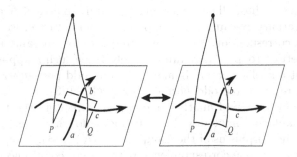

FIGURE 4.12: Isotopy generating relation

Definition 4.4. This presentation of the fundamental group for the knot complement is called *the Wirtinger presentation.*

Exercise 4.7. *Find a purely algebraic proof that the Wirtinger presentation for two diagrams of isotopic links generates the same group.*

Exercise 4.8. *Find a Wirtinger presentation for the trefoil knot and prove that the two groups presented as $\langle a, b | aba = bab \rangle$ and $\langle c, d | c^3 = d^2 \rangle$ are isomorphic.*

Remark 4.2. *The group with presentation $\langle a, b | aba = bab \rangle$ will appear again in knot theory. It is isomorphic to the* three–strand braid group.

It turns out that not only mirror (or equivalent) knots may have isomorphic groups.

Exercise 4.9. *Show that for the two trefoils $T_1 = $ and $T_2 = $, the fundamental groups of complements for $T_1 \# T_1$ and $T_1 \# T_2$ are isomorphic.*

Exercise 4.10. *Calculate a Wirtinger presentation for the figure eight knot (for the simplest planar diagrams).*

Exercise 4.11. *Calculate a Wirtinger presentation for the Borromean rings.*

Now let us prove the following theorem.

Theorem 4.6. *For each knot K, the number $CI(K) + 3$ is equal to the number of homomorphisms of $\pi_1(\mathbb{R}^3 \backslash K)$ to the symmetric group S_3.*

Proof. Consider a knot K and an arbitrary planar diagram of it. In order to construct a homeomorphism from $\pi_1(\mathbb{R}^3 \backslash K)$ to S_3, we should first find images of all elements corresponding to arcs of K.

Suppose that there exists at least one such element mapped to an even permutation. Consider the arc s, corresponding to this element. Then any arc s' having a common crossing A with s and separated from s by some overcrossing arc at A, should be mapped to some even permutation. Since K is a knot, we can pass from s to any other arc by means of "passing through undercrossings". Thus, all elements corresponding to arcs of K are mapped to even permutations. Because the group A_3 is commutative, we conclude that all elements corresponding to arcs are mapped to the same element of A_3 (even symmetric group). There are precisely three such mappings.

If at least one element–arc is mapped to an odd permutation then so are all arcs. There are three odd permutations: $(12), (23), (31)$. If we conjugate one of them by means of another one, we get precisely the third one. This operation is well coordinated with the proper colouring rule.

So, all homomorphisms of the group $\pi_1(\mathbb{R}^3 \backslash K)$ to S_3, except three "even" ones, are in one–to–one correspondence with proper colourings of the selected planar diagram of K.   □

Thus we obtain the following statement.

Corollary 4.2. *The colouring invariant does not distinguish mirror knots.*

This statement does not obviously follow from the definition of the colouring invariant.

Chapter 5

The knot quandle and the Conway algebra

5.1 Introduction

The aim of the present chapter is to describe the universal knot invariant discovered independently by S.V. Matveev [Mat2] and D. Joyce [Joy]. In Matveev's work and in other works by Russian authors, this invariant is usually called the *distributive groupoid*; in Western literature it is usually called *quandle*.[1] This invariant is a complete one;[2] however, it is barely recognisable. In the present chapter, we shall construct some series of "weaker" invariants coming from the knot quandle; the series of invariants to be constructed are easier to calculate and to compare. We shall tell about so-called Conway algebras, describing them according to [PT]. Both these directions, the knot quandle and the Conway algebras, allow us to construct various knot invariants.

First, let us return to the simplest knot invariant; i.e., to the colouring invariant. Why is it possible to construct an invariant function by so simple means?

Even the fact that this invariant is connected with maps from the knot group to the symmetric group S_3 does not tell us very much: an analogous construction with a greater number of colours does not work.

Let us now try to use a greater palette of colours. Let Γ be an arbitrary finite set (here the finiteness will be used in order to be able to *count the number* of colourings); all elements of Γ are to be called *colours*.

Suppose the set Γ is equipped with a binary operation $\alpha : \Gamma \times \Gamma \to \Gamma$; this operation will be denoted like this: $a \circ b \equiv \alpha(a, b)$.

Definition 5.1. By a *proper colouring* of a diagram D of an oriented link K we mean a way of associating some colour with each arc of D in such a way that for each overcrossing arc (that has colour b), undercrossing arc lying on the left hand (colour a) and undercrossing lying on the right hand (colour c), the relation $a \circ b = c$ holds; see Fig. 5.1.

---

[1]There are some other names for this and similar objects, e.g., crystal and rack.

[2]In a slightly weaker sense.

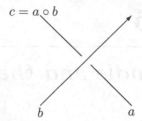

FIGURE 5.1: The rule of colourings

Which should be the conditions for ∘ for the number of proper colourings to be invariant under Reidemeister moves?

It is easy to show that the invariance of such a colouring function under Ω_1 implies the idempotence relation $a \circ a = a$ for all the elements $a \in \Gamma$ that can be associated to arcs and play the role of colour. However, in order to simplify the situation we shall not restrict ourselves only to this case, and require that $\forall a \in \Gamma : a \circ a = a$.

Analogously, the invariance under Ω_2 requires the left invertibility of the operation ∘: for any a and b from Γ, the equation $x \circ a = b$ should have only the solution $x \in \Gamma$ (in the case of the three–colour palette, the inverse operation for ∘ and the operation ∘ itself coincide).

Finally, the invariance under Ω_3 implies *right self distributivity* of the operation ∘, which means that $\forall a, b, c \in \Gamma$ the equation $(a \circ b) \circ c = (a \circ c) \circ (b \circ c)$ holds. In the sequel, each set with an operation ∘ satisfying the three properties described above, is called a *quandle*.

Each quandle generates a rule for proper colouring of link diagrams described above.

Thus, we conclude the following

Proposition 5.1. *The number of proper colourings by elements of any quandle is a link invariant.*

In any quandle, the inverse operation for ∘ is denoted by $/$. More precisely, the element b/a is defined to be the unique solution to the equation $x \circ a = b$.

Exercise 5.1. *Show that each quandle Γ (with operation ∘) is a quandle with respect to the operation $/$.*

Furthermore, prove the following identities for Γ: $(a \circ b)/c = (a/c) \circ (b/c)$, $(a/b) \circ c = (a \circ c)/(b \circ c)$.

There is a common way for constructing quandles by using their presentations by generators and relations.

Let A be an alphabet consisting of *letters*. A *word* in the alphabet A is an arbitrary finite sequence of elements of A and symbols $(,), \circ, /$. Now, let us define inductively the set $D(A)$ *of admissible words* according to the following rules:

1. For each $a \in A$, the word consisting of only the letter a is admissible.

2. If two words W_1, W_2 are admissible then the words $(W_1) \circ (W_2)$ and $(W_1)/(W_2)$ are admissible as well.

3. There are no other admissible words except for those obtained inductively by rules 1 and 2.

Sometimes we shall omit brackets when the situation is clear from the context. Thus, e.g. for letters a_1, a_2 we write the word $a_1 \circ a_2$ instead of $(a_1) \circ (a_2)$.

Let R be a set of *relations*; i.e., identities of type $r_\alpha = s_\alpha$, where $r_\alpha, s_\alpha \in D(A)$ and α runs over some set X of indices. Let us introduce the equivalence relation for $D(A)$, supposing $W_1 \equiv W_2$ if and only if there exists a finite chain of transformations starting from W_1 and finishing at W_2 according to the rules 1-5 described below:

1. $x \circ x \Longleftrightarrow x$;

2. $(x \circ y)/y \Longleftrightarrow x$;

3. $(x/y) \circ y \Longleftrightarrow x$;

4. $(x \circ y) \circ z \Longleftrightarrow (x \circ z) \circ (y \circ z)$;

5. $r_i \Longleftrightarrow s_i$.

The set of equivalence classes is denoted by $\Gamma\langle A|R\rangle$. It is easy to check that it is a quandle with respect to the operation $\circ$.

Remark 5.1. *There is an analogous construction of biquandles also known as rack [FR] which instead of edges deals with half-edges. By a half-edge we mean the result of breaking each edge at every overpass.*

Hence, for each crossing, we have four elements of the biquandle, say, two incoming and two outgoing.

Then one can define the colourings of the two outgoing edges in terms of colourings of the two incoming edges. The invariance under Reidemeister moves will lead us to some conditions, we are not going to write directly.

Similarly to the quandle, for biquandles we have a fundamental biquandle and its realisations.

Quandles are partial cases of biquandles where the outgoing edge forming an overpass has the same colour as the incoming edge forming the overpass.

Let us give one more example of a finite quandle (denoted by G_4). Denote by G_4 the set of four different elements a_1, a_2, a_3, a_4.

Remark 5.2. *This example comes from the universal construction of quandles from groups. Later, this construction will be discussed in detail.*

Let us define the operation ∘ according to the rule:

∘	a_1	a_2	a_3	a_4
a_1	a_1	a_3	a_4	a_2
a_2	a_4	a_2	a_1	a_3
a_3	a_2	a_4	a_3	a_1
a_4	a_3	a_1	a_2	a_4

Here the result $a_i \circ a_j$ occupies the position number j in the i-th line.

Exercise 5.2. *Check the quandle axioms for the operation described above.*

It is easy to see that the figure eight knot admits non–trivial colourings by elements of G_4. The same can be said about the trefoil knot.

Remark 5.3. *Note that the quandle G_4 is not especially connected with any knot; it is only used for construction of knot invariants.*

From Proposition 5.1 we see that quandles are useful for constructing knot invariants. It turns out that one can associate with each knot (link) a quandle, that is the universal (almost[3] complete) knot invariant. For the sake of simplicity, we shall describe this invariant for the case of a knot. The *universal knot quandle* can be described in two ways: geometrically and algebraically.

5.2 Geometric and algebraic definitions of the knot quandle

5.2.1 Geometric description of the quandle

Let K be an oriented knot in $\mathbb{R}^3$, and let $N(K)$ be its small tubular neighbourhood. Let $E(K) = \overline{(\mathbb{R}^3 \backslash N(K))}$ be the complement to this neighbourhood. Fix a base point x_K on $E(K)$. Denote by Γ_K the set of homotopy classes of paths in the space $E(K)$ with fixed initial point at x_K and endpoint on $\partial N(K)$ (these conditions must be preserved during the homotopy). Note that the orientations of $\mathbb{R}^3$ and K define the orientation of the tubular neighbourhood of the knot (right screw rule). Let m_b be the oriented meridian hooking an arc b. Define $a \circ b = [bm_b b^{-1} a]$, where for $x \in \Gamma_K$ the letter x means a representative path, and square brackets denote the class that contains the path $[x]$; see Fig. 5.2.

The quandle axioms can also be checked straightforwardly. Also, one can easily check that the groupoids corresponding to different points x_K are isomorphic. This statement is left for the reader as an exercise.

---

[3]Later, we shall comment on the incompleteness of the quandle in the proper sense.

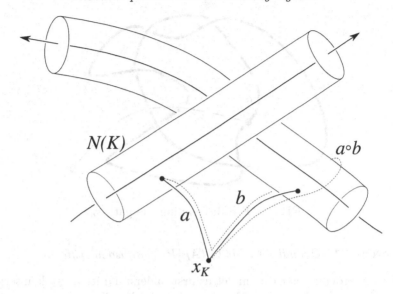

FIGURE 5.2: Intuitive description of the quandle operation.

Here, we should emphasise that we first define our map from barely the set of elements of the quandle to the set of elements of the group. Then it will naturally yield a quandle operation on the group; this quandle operation will then lead us to further examples.

There is a natural map from the knot quandle $\Gamma(K)$ to the group $\pi_1(\mathbb{R}^3 \backslash E(K))$. Let us fix a point x outside the tubular neighbourhood. Now, with each element γ of the quandle (path from x to $\partial E(K)$) we associate the loop $\gamma m \gamma^{-1}$, where m is the meridian at the point x.

This interpretation shows that *the fundamental group can be constructed by the quandle*: all meridians can play the role of generators for the fundamental groups, and all relations of type $a \circ b = c$ have to be replaced with $bab^{-1} = c$.

Besides, the fundamental group has an obvious action on the quandle: for each loop g and element of the quandle γ, the path $g\gamma$ is again an element of the quandle.

5.2.2 Algebraic description of the quandle

Let D be a diagram of an oriented knot K. Denote the set of arcs of D by A_D. Let P be a crossing incident to two undercrossing arcs a and c and an overcrossing arc b. Let us write down the relation: $a \circ b = c$, where a is the arc lying on the left hand with respect to b and c is the arc lying on the right hand with respect to b. Denote the set of all relations for all crossings by R_D. Now, consider the quandle $\Gamma\langle A_D | R_D \rangle$, defined by generators A_D and relations R_D.

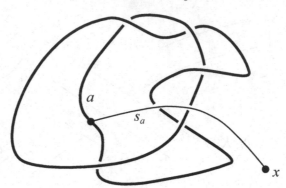

FIGURE 5.3: Defining the path s_a.

Theorem 5.1. *Quandles Γ_K and $\Gamma\langle A_D|R_D\rangle$ are isomorphic.*

Before proving the theorem, let us first understand its possible interpretations. On one hand, the theorem shows how to describe generators and relations for the geometrical quandle Γ_K. On the other hand, it demonstrates the independence $\Gamma\langle A_D|R_D\rangle$ of the choice of concrete knot diagram. Now, Proposition 5.1 (concerning colouring number) evidently follows from this theorem as a corollary because any proper colouring of a knot diagram by elements of Γ is a presentation of $\Gamma\langle A_D|R_D\rangle$ to Γ.

Proof of Theorem 5.1. With each arc a of the projection D, we associate the path s_a in $E(K)$ in such a way that

1. the path s_a connects the base point with a point of the part of the torus ∂N_K corresponding to the arc a;

2. at all points where the projection of s_a intersects that of D, the path s_a goes over the knot; see Fig. 5.3.

Obviously, these conditions are sufficient for the definition of the homotopy class of s_a.

Consequently, to each generator of $\Gamma = \langle A_D|R_D\rangle$, there corresponds an element of the quandle Γ_K. Thus we have defined the homomorphism $\phi : \Gamma\langle A_D|R_D\rangle \to \Gamma_K$. In order to define the inverse homomorphism $\psi : \Gamma_K \to \Gamma\langle A_D|R_D\rangle$, let us fix $s \in \Gamma_K$. Then, the path representing s is constructed in such a way that the projection of the path intersects D transversely and contains no diagram crossing.

Denote by $a_n, a_{n-1}, \ldots, a_1$ those arcs of D going over the path s. Denote by a_0 the arc corresponding to the end of s. Now, for each $s \in \Gamma_K$, let us assign the element $((\ldots(a_0\varepsilon_1a_1)\varepsilon_2\ldots a_{n-1})\varepsilon_na_n$ of the quandle $\Gamma\langle A_D|R_D\rangle$, where ε_i means / if s goes under a_i from the left to the right, or $\circ$, otherwise; see Fig. 5.4.

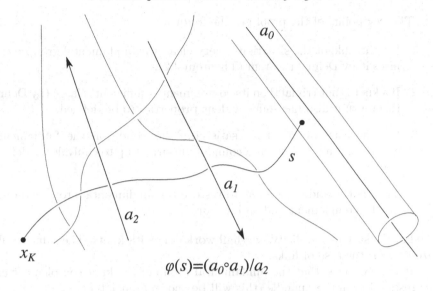

$$\varphi(s)=(a_0 \circ a_1)/a_2$$

FIGURE 5.4: Constructing the map $\psi : \Gamma_K \to \Gamma \langle A_D, R_D \rangle$.

It is easy to check that this map is well defined (i.e., it does not depend on the choice of representative s for the element of Γ_K) and that maps ϕ and ψ are inverse to each other. This completes the proof. □

The quandle corresponding to the knot, is a complete invariant. However, it is difficult to recognise quandles by their presentation. This problem is extremely difficult. But it is possible to simplify this invariant making it weaker but more recognizable.

5.3 Completeness of the quandle

Roughly speaking, the quandle is a complete knot invariant because it contains the information about the fundamental group and "a bit more". To prove this fact about the completeness of the quandle, we shall use one very strong result by Waldhausen [Wal] concerning three–dimensional topological surgery.

By Matveev, two (non–isotopic) knots are equivalent if one can be obtained from the other by changing both the orientation of the ambient space and that of the knot. In this sense, the two trefoils are equivalent and have isomorphic quandles.

Here by *complete* we mean that the quandle distinguishes knots up to equivalence defined above.

The key points of the proof are the following:

1. For the unknot the situation is very clear: the fundamental group recognises it by Dehn's theorem (Theorem 4.2).

2. If a knot is not trivial then its complement is sufficiently large (by Dehn's theorem); it also has some evident properties to be defined.

3. For the class of manifolds satisfying this condition the fundamental group "plus a bit more" is a complete invariant up to equivalence defined above.

4. The knot quandle allows us to restore the fundamental group structure for the complement and "a bit more".

For the sake of simplicity, we shall work only with knots. The same results are true for the case of links.

It is easy to see that the fundamental group of the knot complement can be restored from the quandle (this will be shown a bit later).

Let us first introduce some definitions.

Definition 5.2. A surface F in a manifold M is *compressible* in either of the following cases:

1. There is a non–contractible simple closed curve k in the interior of F and a disc D in M (whose interior lies in the interior of M) such that $D \cap F = \partial D = k$.

2. There is a ball E in M such that $E \cap F = \partial E$.

Otherwise the surface is called *incompressible*

Definition 5.3. A 3–manifold M is called *irreducible* if any sphere $S^2 \subset M$ is compressible.

A 3–manifold M with boundary is called *boundary–irreducible* if its boundary ∂M is incompressible.

Let K be an oriented knot. Consider the fundamental group π of $\mathbb{R}^3 \setminus N(K)$ where N is a tubular neighbourhood of K. Obviously, $\partial N = T$ is a torus that has an oriented meridian m (a curve that has linking coefficient 1 with K).

If K is not trivial then the fundamental group $\pi(T)$ is embedded in π. This result follows from Dehn's theorem.

Definition 5.4. For a non-trivial knot K, the embedded system $m \in \pi(T) \subset \pi$ is called a *peripherical system* of K.

The Waldhausen theorem[4] says the following:

---

[4]We use the formulation taken from [Mat2]

Theorem 5.2. *[Wal] Let M, N be irreducible and boundary–irreducible 3–manifolds. Let M be sufficiently large and let $\psi : \pi_1(N) \to \pi_1(M)$ be an isomorphism preserving the peripherical system. Then there exists a homeomorphism $f : N \to M$, inducing ψ.*

We are going to prove that the knot quandle is a complete knot invariant.

Now let K_1, K_2 be two knots. Suppose that ϕ is an isomorphism $\Gamma(K_1) \to \Gamma(K_2)$ of the quandles. Denote the complements to tubular neighbourhoods of K_1, K_2 by E_{K_1}, E_{K_2}, respectively.

Note that if K_1, K_2 are not trivial then the manifolds E_{K_1}, E_{K_2} are boundary–irreducible, sufficiently large and irreducible (by Dehn's lemma).

Now, let us suppose that one of the two knots (say, K_1) is trivial. Then $\pi(K_1)$ is isomorphic to $\mathbb{Z}$. Since the knot group can be restored from the quandle, we have $\pi(K_2)$ is also $\mathbb{Z}$. Thus, K_2 is trivial.

Now consider the case when K_1, K_2 are non-trivial.

In this case, we know that the knot group can be restored from the quandle; besides, the meridian can also be obtained from the knot quandle: it can be chosen to be the image of any element of the quandle under the natural morphism.

Now, let us prove that the normaliser of the meridian (as an element of the quandle representing the path from x to x_K) in the fundamental group consists precisely of the fundamental group of the tubular neighbourhood of the knot $\pi_1(T_2) = \mathbb{Z}^2$. Indeed, each element of the group $\pi_1(T_2)$ is a path looking like ana^{-1} where a represents the meridian in the quandle (path from the initial point to the point on T^2), and n is a loop on the torus T^2. So, we have: $ana^{-1} \cdot a = an$ which is homotopic to a in the quandle.

Now, suppose that for some g we have: ga is homotopic to a. Then, there exists a path on the torus drawn by the endpoint while performing this homotopy. Denote this path by x. So, we have: $gaxa^{-1} = e$, so g is homotopic to $ax^{-1}a^{-1}$ that belongs to the fundamental group of T_2.

The next step of the proof is the following. The quandle knows the peripheral system. Let K_1, K_2 be two non-trivial knots with the same peripheral structure. Consider an isomorphism of the knot groups. By the Waldhausen theorem, it generates a homeomorphism h between E_{K_1} and E_{K_2} and maps the meridian of the first one to a meridian of the second one. Thus, we have the same information on how to attach full tori N_1 and N_2 to E_{K_1} and E_{K_2} in order to obtain $\mathbb{R}^3$. Having a full torus N_i, $i = 1, 2$, in $\mathbb{R}^3$, we can contract its meridian to a point; hence, we get a curve λ_i which will be exactly the knot K_i. So, they are obviously isotopic. To perform all this, we must fix the orientation of E_{K_1} and E_{K_2}. Then we shall be able to choose the orientation of the meridian. If we choose the opposite orientation of them both, we shall obtain an equivalent knot.

However, if the orientation of the meridian is fixed, the knot can be uniquely restored.

5.4   Special realisations of the quandle: colouring invariant, fundamental group, Alexander polynomial

One can easily define the *free quandle* with generators $a_1, \ldots, a_k$. Namely, one takes into account idempotence, the existence of a unique solution $x \circ b = c$ and self–distributivity. No more conditions will be given for this quandle.

Let us give two more examples.

Example 5.1. *(see [Prz]) Consider the free group in an infinite number of generators. Define the quandle operation on it according to the rules:* $a \circ b = bab^{-1}, a/b = b^{-1}ab$. *All axioms can be checked straightforwardly. In this case, we have obtained a natural morphism of the free quandle to the free group. As in the case of quandles, free groups can be transformed to arbitrary groups; we only have to describe the quandle relations in terms of groups.*

It can be easily checked that the image of the knot quandle is the fundamental group of the knot complement (to do this, we should just compare the presentation of the quandle and the fundamental group corresponding to a knot diagram).

Other examples of quandle structure which can be obtained from the group operation are

$$a \circ b = ba^{-1}b$$

and (for a fixed n)

$$a \circ b = b^{-n}ab^n.$$

Example 5.2. *One can consider maps of the quandle to the free module over Laurent polynomial ring (with respect to a variable t) as well. To do this, one should decree*

$$a \circ b = ta + (1 - t)b, \quad a/b = \frac{1}{t}a + \left(1 - \frac{1}{t}\right)b.$$

In this case, we obtain the quotient ring that has a quadratic matrix of linear relations for the generators $a_1, a_2, \ldots$. This matrix is called the Alexander matrix *of the knot diagram. The* Alexander polynomial *of the knot can be defined as follows: we set one variable a_i to be equal to 0 and then we solve the system of n equations for $n - 1$ variables. Finally, we obtain some relation for the ring elements: $f(t) = 0$, where f is a Laurent polynomial. The function f (defined up to multiplication by $\pm t^k$), is called the* Alexander polynomial. *It can be calculated by taking any minor of the Alexander matrix of $(n - 1)$–th order.*

5.5 The Conway algebra and polynomial invariants

Now we shall describe the construction that allows us to look in the same way at different polynomial invariants of knots: those by Jones, Conway, and HOMFLY-PT. In particular, we shall prove the invariance and uniqueness of the HOMFLY-PT polynomial. Note that HOMFLY is not a surname, but an abbreviation of the first letters of six surnames, [HOMFLY]: Hoste, Ocneanu, Millett, Freyd, Lickorish, and Yetter. This polynomial was later rediscovered by Przytycki and Traczyk [PT]. We shall use the approach proposed in the article [PT], the book [Prz] etc.

The main idea is the following. Unlike the previous approach, where we associated a special algebraic object to **each link**, here we construct some algebraic object and assign some element of this algebraic object to each link. This is going to be the link invariant.

Caveat. We are now going to introduce the two operations, ∘ and /. They have another sense and other properties than those described above.

Let A be an algebra with two binary operations ∘ and / such that the following properties hold: for all $a, b \in A$ we have $(a \circ b)/b = a$, $(a/b) \circ b = a$. For each link (diagram) L, let us construct the element $W(L)$ of A as follows. Denote by a_n the element of A corresponding to the n–component trivial link.

Let us also require the following algebraic equation for any *Conway triple* (i.e. three diagrams coinciding outside a small circle and looking like ⨂, ⨂, ⟩⟨ inside this circle; such diagrams are called a *Conway triple*):

$$W(\text{⨂}) = W(\text{⨂}) \circ W(\text{⟩⟨}). \tag{5.1}$$

The uniqueness of the inverse element means that we must require the existence of the inverse function /, such that $W(\text{⨂})/W(\text{⟩⟨}) = W(\text{⨂})$. So, W is going to be a map from the set of all links to the algebraic object to be constructed. Later we shall see that some partial cases of the equation (5.1) coincide with some skein relations. Let us now take into account the following circumstance: because each link can be transformed to the trivial (ascending) one (Exercise 1.2) by switching some crossing types, the value of the function W on any m–component link with n crossings can be described only by using the value of W for the trivial m–component link and the value of W for some links with fewer crossings (see [Prz]).

Indeed, they can also be represented by a_i and values of W on links with less than $n - 1$ crossings. Consequently, for each link L with arbitrary number of crossings, the value of $W(L)$ can be expressed somehow (possibly, not uniquely) in $a_i, i = 1, 2, \ldots$, by using ∘ and /.

At the present moment, we do not know whether the function W is well defined and if so, whether it is a link invariant. Let us try to look at the algebra A and find the restrictions for the uniqueness of the definition.

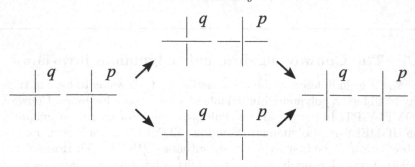

FIGURE 5.5: Two ways of resolving two crossings

Consider the trivial n–component link diagrams with only one crossing at one (twisted) circle. The value of W on this trivial link should be equal to a_n. Depending on the crossing type, the relation (5.1) leads to

$$a_n = a_n \circ a_{n+1} \tag{5.2}$$

and

$$a_n = a_n/a_{n+1}. \tag{5.3}$$

These two relations should hold for arbitrary $n \geq 1$.

There is one more argument that one may call "the switching order". Consider the diagram of L and choose two (say, positive) crossings p, q of it; see Fig. 5.5.

Denote by $L_{\alpha\beta}$ for $\alpha, \beta \in \{+, -, 0\}$ the link diagram coinciding with L outside small neighbourhoods of p, q and having type α at p and type β at q.

Let us consider the relation (5.1) at p and then at q.

We get: $W(L_{++}) = W(L_{-+}) \circ W(L_{0+}) = (W(L_{--}) \circ W(L_{-0})) \circ (W(L_{0-}) \circ W(L_{00}))$. See [Zhi]

Now, let us consider the same relation for q and later, for p (the other order). We have: $W(L_{++}) = W(L_{+-}) \circ W(L_{+0}) = (W(L_{--}) \circ W(L_{0-})) \circ (W(L_{-0}) \circ W(L_{00}))$.

Comparing the obtained equalities, we get:

$$(a \circ b) \circ (c \circ d) = (a \circ c) \circ (b \circ d), \tag{5.4}$$

where $a = W(L_{--}), b = W(L_{0-}), c = W(L_{-0}), d = W(L_{00})$.

We shall require the equation (5.4) for arbitrary a, b, c, d, which are going to be the elements of the algebra to be constructed.

In the case when both p and q are negative, we get the analogous equation

$$(a/b)/(c/d) = (a/c)/(b/d). \tag{5.5}$$

Analogously, if one crossing is positive, and the other is not, we get the equation

$$(a/b) \circ (c/d) = (a \circ c)/(b \circ d). \tag{5.6}$$

Thus, we have found some necessary conditions for $W(L)$ to be well defined. Let us show that these conditions are sufficient.

Definition 5.5. An algebra A with two operations $\circ$ and $/$ (inverse to each other) and a fixed sequence a_n of elements is called a *Conway algebra* if the conditions (5.2)–(5.6) hold.

We shall need later the following simple fact.

Lemma 5.1. *There exists a unique monomorphism ϕ of the universal Conway algebra A_U (see page 72) such that $\phi(a_n) = a_{n+1}$.*

Theorem 5.3. *For each Conway algebra, there exists a unique function $W(L)$ on link diagrams that has value a_n on the n–component unlink diagrams and satisfies (5.1). This function is an invariant of oriented links.*

Proof. First, let us show that this invariant is well defined on diagrams with numbered components. We shall use induction on the number of crossings.

Let C_k be the class of links having diagrams with no more than k crossings.

The main induction hypothesis. There exists a well-defined function $W(L)$ on C_k which is invariant under those Reidemeister moves, which do not let the diagram leave the set C_k and satisfy the relation (5.1) for all Conway triples with all elements from C_k. Note that if $L \sqcup \bigcirc$ is any diagram which can be obtained from a diagram L by adding a small circle bounding an empty disc and having no crossings, then $W(L \sqcup \bigcirc) = \phi(W(L))$

The induction base ($k = 0$) is trivial since in this case the class C_k consists only of unlinks and no Reidemeister moves can be performed. In order to perform the induction step, let us first choose a canonical way of associating $W(L)$ with a link $L \in C_{k+1}$ by ordering components and choosing a base point on each of them.

After this, we shall prove the independence of $W(L)$ from the choice of base points, its invariance under Reidemeister moves, and, finally, independence of the order of components.

The construction of W. Let us enumerate all components of L. Fix a point b_1 on the first component, b_2 on the second one, and so forth; base points should not coincide with crossings. Let us now describe how to construct the element $W_b(L)$ (here the index b means the ordered set $b_1, b_2, \ldots$ of base points). Let us walk along the link components according to the orientation. First, take the point b_1 and pass the first component until b_1, then take b_2 and pass the second component, and so forth. A crossing p is called *good* if it is first passed under, and then over. All other points are said to be *bad*.

Now we are going to use the (second) induction on the number of bad points. The induction step is obvious: if all points are good then the link is trivial (the diagram is ascending). In this case, we set by definition $W_b(L) = a_n$, where n is the number of link components.

Suppose we have defined W_b for all links with $k + 1$ crossings and no more than m bad points. Let L be a link diagram with $k + 1$ crossings and $m + 1$ bad points.

Let us fix the first bad point (in accordance with the chosen circuit). Without loss of generality, suppose that this crossing is positive. Let us apply the relation (5.1) to this crossing. Thus we obtain two links: L_- that has one bad point less than the diagram L (consequently, $W_b(L_-)$ is well defined by the induction hypothesis) and L_0 that belongs to C_k (and $W(L_0)$ (consequently, $W_b(L_0)$) is also well defined). By definition, let us put for L

$$W_b(L) = W_b(L_-) \circ W(L_0).$$

Thus, we have defined W_b for C_{k+1}.

Now, let us prove that the function W_b satisfies the relation

$$W_b(\text{⨉}) = W_b(\text{⨉}) \circ W(\text{)(})$$

for each crossing q. We have already used this relation **to define** the module. However, we have not yet proved that it does not yield to a contradiction. Thus, the relation (5.1) deserves proving. Suppose the point q is bad. If it is the first bad point according to our circuit, the desired equality holds by construction. Now, let us apply the induction method (the third induction) on the number N of this bad point. Suppose that if $N < m$ then the Conway relation holds. Let p be the first bad (say, positive) crossing. By using the definition of W_b, the induction hypothesis and the main induction hypothesis, (5.4), and again, the induction hypothesis and the main induction hypothesis, we obtain:

$$W_b(L_{++}) = W_b(L_{-+}) \circ W(L_{0+}) = (W_b(L_{--}) \circ W(L_{-0})) \circ (W(L_{0-}) \circ W(L_{00}))$$

$$= (W_b(L_{--}) \circ W(L_{0-})) \circ (W(L_{-0}) \circ W(L_{00})) = W_b(L_{+-}) \circ W(L_{+0}).$$

Here we use the following notation: the first index of $L_{\varepsilon_1 \varepsilon_2}$ is related to the point p and the second is related to the point q. Thus, the obtained formula is just what we wanted.

If q is a good point (say, positive) for L_+, then it is bad for L_-. As we have already proved, the identity $W_b(L_-) = W_b(L_+)/W(L_0)$ holds for this point, and the desired equality $W_b(L_+) = W_b(L_-) \circ W(L_0)$ is just its corollary.

Now, let us prove that the function W_b does not depend on the choice of base points (the order of components remains fixed). It is sufficient to consider only the case when one base point (say, b_k) passes through one crossing (denote it by q) to a position b'_k; see Fig. 5.6.

Suppose the crossing q is positive. If q is good for both base points $b = (b_1, \ldots, b_n)$ and $b' = (b_1, \ldots, b_{k-1}, b'_k, b_{k+1}, \ldots, b_n)$ then the equality $W_b(L) = W_{b'}(L)$ holds just by definition. If q is bad in both cases then we have $W_b(L_+) = W_b(L_-) \circ W(L_0), W_{b'}(L_+) = W_{b'}(L_-) \circ W(L_0)$. Since q is good for L_- for both choices of base points, we see that $W_b(L_-) = W_{b'}(L_-)$ and hence $W_b(L_+) = W_{b'}(L_+)$. It remains only to consider the case when q is

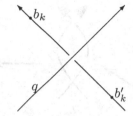

FIGURE 5.6: Changing the starting point

good for b and bad for b' (or vice versa). This happens only in the case when both parts of the link passing through q lie on the same component.

Now we can assume that L has no more bad points (either with respect to b or with respect to b'); in this case they all can be transformed to good ones by using the Conway relation (that preserves the equality $W_b(L) = W_{b'}(L)$).

So, the link L has no bad points with respect to b, hence, $W_b(L) = a_n$. The link L has the only bad point q with respect to b' (by definition, $W_{b'}(L) = W_{b'}(L_-) \circ W(L_0)$). Now, it remains to note that the links L_- and L_0 have no bad points either. Thus, they are trivial. Besides, $W(L_-) = a_n, W(L_0) = a_{n+1}$. Taking into account $a_n = a_n \circ a_{n+1}$, we get $W_b(L) = W_{b'}(L)$.

Let us prove now that the function $W_b(L)$ is invariant under Reidemeister moves that preserve the link in the class C_{k+1}.

The main idea is the following: suppose we perform some Reidemeister move inside the area U. Thus, we have two pictures inside U: the picture before and the picture after. Outside U, both diagrams have the same crossings and the same shadow. If we can arrange all other crossing types in order to obtain an ascending diagram (with respect to enumerated components) with each of the two fixed crossings inside U, then the invariance is trivial: we just express our diagrams *in the same way* via diagrams of unlinks. These unlink diagrams differ only inside U. In the other case, we need to perform the relation (5.1) and then consider the case described above.

Let us be more detailed. First, let us consider the move Ω_1. By moving the base point, the point added or deleted by Ω_1 can be thought to be good. In this case, the existence of the loop does not affect the result of the calculation of W_b.

For the same reason, the value of W_b does not change under Ω_2 if both points appearing (disappearing) while performing the move are good. If both points are bad and cannot be made good by moving base points (this may happen when two different components are involved in Ω_2), one should apply the Conway relation at both points and note that $L_{-0} \cong L_{0-}$ and $W(L_{-0}) = W(L_{0-})$; see Fig. 5.7.

Because $W_b(L_{-+}) = W_b(L_{--}) \circ W(L_{-0}) = (W_b(L_{+-})/W(L_{0-})) \circ W(L_{-0}) = W_b(L_{+-})$, then the procedure of making all bad points good does not affect the behaviour of W_b under Ω_2.

Finally, let us consider the case of Ω_3. Let us assume that all base points

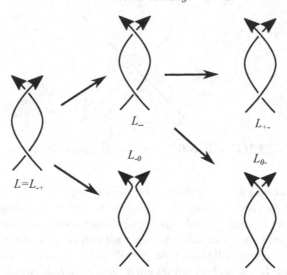

FIGURE 5.7: Different ways of resolving two crossings

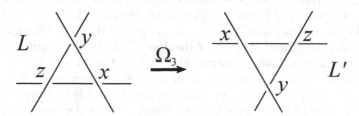

FIGURE 5.8: Third Reidemeister move

are outside the area of Ω_3. Denote the crossings by x (upper and lower arcs), y (upper and middle) and z (lower and middle); see Fig. 5.8

Denote the link obtained after performing Ω_3 by L'. Let us note that x cannot be the only bad point or the only good point from the set $\{x, y, z\}$. If, for instance, x is good, and y, z are bad then the branch xz is passed before xy with respect to the orientation, the branch xy is before yz, and yz is before xz, which leads to a contradiction.

Thus, if x is good then so is one of y, z. Suppose, y. In this case moving the branch xy over the crossing z does not change the invariant W_b by construction. If the crossing x is bad then one of y, z (say, z) is bad as well. Let us write the Conway relation for the crossings z, x. We get:

$$W_b(L_{++}) = W_b(L_{-+}) \circ W(L_{0+}) = (W_b(L_{--}) \circ W(L_{-0})) \circ W(L_{0+}),$$

$$W_b(L'_{++}) = W_b(L'_{-+}) \circ W(L'_{0+}) = (W_b(L'_{--}) \circ W(L'_{-0})) \circ W(L'_{0+}).$$

The first index for $L..$ is related to z and the second one is related to x. Since for the link L_{--} the points x, z are good, then $W_b(L_{--}) = W_b(L'_{--})$ (this case has already been considered). Diagrams L_{-0} and L'_{-0} coincide, and L'_{0+} can be obtained from L_{0+} by applying the move Ω_2 twice, which does not change the value of W_b. Consequently, $W_b(L_{++}) = W_b(L'_{++})$.

Thus, we have constructed a function W that is an invariant of links with marked components.

Let us show that it is a link invariant (no marked point information is needed).

In order to do this, let us introduce the crossing switch operation Ω_0. Denote the versions of Ω_1, Ω_2 decreasing the number of crossings by Ω_1^-, Ω_2^-.

We shall need the following lemma.

Lemma 5.2. *Each link diagram can be transformed to the unlink diagram without crossing by using $\Omega_0, \Omega_1^-, \Omega_2^-, \Omega_3$ and removing components without crossings.*

The necessity of removing trivial components in the reduction process was pointed out by Carlo Petronio.

We shall prove this lemma later.

Note that the equation $W_b(L) = W_{b'}(L)$ is invariant under Ω_0. As we proved earlier, it is also invariant under $\Omega_{1,2,3}$. Given a diagram D, removing a component without crossings corresponds to the equality $W_\beta(D) = \phi(W_\beta(D'))$ where $\beta = b, b'$, D' is the diagram with the component removed and ϕ is the map of Lemma 5.1.

Since for the standard diagram of the unlink we have $\phi^k(W_b(L)) = \phi^k(W_{b'}(L))$ for any orders of components with base points, where k is the number of removed trivial components. Since ϕ is a monomorphism, $W_b(L) = W_{b'}(L)$ holds in the general case. □

Below, we give a prove of the auxiliary lemma we have used.

Proof of Lemma 5.2. Let L be a link diagram. A branch l of this diagram is called a *loop* if it starts and ends at the same crossing. A loop is called *simple* if it has no self-intersection and if in the domain bounded by it there are no other loops. We say that two arcs $l_1, l_2 \supset L$ bound a *bigon* if they have no selfintersections, have common initial and final points and no other intersections. A bigon is called *simple* if it does not contain smaller bigons and loops inside. In Fig. 5.9 we show that if L has a simple bigon then the number of its intersection points can be decreased by making this bigon empty (using Ω_0, Ω_3 and removing trivial components) and deleting it by means of Ω_2^-.

In the case of a simple loop, we can do the same and finally delete it by Ω_1^-.

The only thing to note is that if a link diagram has at least one crossing then it contains either a simple bigon or a simple loop. This is left to the reader as a simple exercise. □

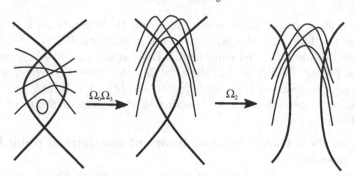

FIGURE 5.9: Removing arcs from a bigon

Let us show now that Jones, Conway and HOMFLY-PT polynomials are just special cases of the invariant W for some special algebras A.

We have shown that for each Conway algebra A there exists an A–valued invariant of oriented links $W(\cdot) \in A$. Among all such algebras there exists the universal algebra A_U. It is generated by $a_n, n \geq 1$ and has no other relations except for (5.2)–(5.6).

The universal link invariant corresponding to the universal algebra is the strongest one among all those obtained in this way. However, it has a significant disadvantage: it is difficult to recognise two different presentations of an element in the algebra A.

We are going to show how to construct a family of Conway algebras. The invariants to be constructed are more convenient than the universal one: they are easier to recognise.

Let A be an arbitrary commutative ring with a unit element, $a_1 \in A$ and α, β be some invertible elements of A. Let us define $\circ, /$ as follows:

$$x \circ y = \alpha x + \beta y \tag{5.7}$$

and

$$x/y = \alpha^{-1}x - \alpha^{-1}\beta y, \tag{5.8}$$

where

$$a_n = (\beta^{-1}(1 - \alpha))^{n-1} a_1, n \geq 1. \tag{5.9}$$

Then the following proposition holds.

Proposition 5.2. *For any choice of invertible elements α, β and element a_1, the ring A endowed with operations $\circ, /$ defined above and with elements a_n, see (5.9), is a Conway algebra.*

The proof follows straightforwardly from the axioms.

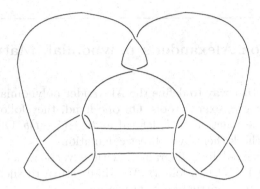

FIGURE 5.10: The Kinoshita–Terasaka knot

5.6   Realisations of the Conway algebra. The Conway–Alexander, Jones, HOMFLY-PT and Kauffman polynomials

Let us give here some examples of simple invariants that originate from the Conway algebra.

Example 5.3. *Let A be the ring of polynomials of variable x with integer coefficients. Let $\alpha = 1, \beta = x, a_1 = 1$. Then $W(L)$ coincides with the Conway polynomial (also called the* Conway *potential function).*

The Conway polynomial was the first among the polynomials satisfying the Conway relation. It was proposed in the pioneering work by Conway [Con]. All other polynomials and modifications appeared much later.

In Fig. 5.10, we present the celebrated Kinoshita–Terasaka knot. This knot is not trivial. However, it can be easily checked that this knot has Conway polynomial equal to one. Thus, the Conway (consequently, Alexander) polynomial does not always distinguish the unknot.

This knot has the non-trivial Jones polynomial. It is not yet known whether the Jones polynomial always distinguishes the unknot. We will touch on this problem later.

Example 5.4. *Let A be the ring of Laurent polynomials in $\sqrt{q}$, where $\alpha = q^2, \beta = q(\sqrt{q} - \frac{1}{\sqrt{q}}), a_1 = 1$. In this case, $W(L)$ coincides with the Jones polynomial of q.*

Example 5.5. *Let A be the integer coefficient Laurent polynomial ring of the variables l, m. Let $\alpha = -\frac{m}{l}, \beta = \frac{1}{l}, a_1 = 1$. Then the obtained invariant $\mathcal{P}(l, m)$ coincides with the HOMFLY-PT polynomial.*

Exercise 5.3. *Write down the skein relations for these polynomials.*

5.7 More on Alexander's polynomial. Matrix representation

There is another way to define the Alexander polynomial more exactly.

We shall not give exact proofs. On one hand, they follow from "quandle properties" of the function $a \circ b$ defined as $ta + (1 - t)b$. On the other hand, the reader can check them by a direct calculation.

Given an oriented link diagram L with n vertices, let us construct the *Alexander matrix* $M(L)$ as follows. We shall return to such matrices in the future when studying virtual knot invariants.

Let us enumerate all crossings by natural numbers $1, \ldots, n$. In the general position, there exists precisely one arc outgoing from each crossing (if there are no separated cyclic arcs). It is easy to see that each knot isotopy class has such a diagram. So, we can enumerate outgoing arcs by integers from 1 to n, correspondingly. Now, we construct an incidence matrix, where a crossing corresponds to a row, and an arc corresponds to a column.

Suppose that no crossing is incident twice to one and the same arc (no loops). Then, each crossing (number i) is incident precisely to three arcs: passing through this crossing (number j), incoming (number k) and outgoing (number i).

In this case, the i–th row of the Alexander matrix consists of the three elements at places i, j, k. If the i–th crossing is positive, then $m_{ii} = 1, m_{ik} = -t, m_{ij} = t - 1$. Otherwise we set $m_{ii} = t, m_{ik} = -1, m_{ij} = 1 - t$. See [CF].

Obviously, this matrix has determinant zero, because the sum of elements in each row equals zero.

Define the algebraic complement to m_{ij} by Δ_{ij}.

Then the following theorem holds.

Theorem 5.4. *All Δ_{ij} coincide up to multiplication by $\pm t^k$.*

Denote Δ_{ij} by $\Delta(L)$.

Theorem 5.5. *The function Δ defined on links (and normed properly) satisfies the following skein relation:*

$$\Delta(\overcrossing) - \Delta(\undercrossing) = (t^{1/2} - t^{-1/2})\Delta(\smoothing).$$

It is easy to check that for the unknot $\bigcirc$, we have $\Delta(\bigcirc) = 1$.

Thus, we can conclude that the polynomial Δ coincides with the Conway polynomial up to the variable change $x = t^{1/2} - t^{-1/2}$. So, it is a well–defined link invariant.

Remark 5.4. *This way derives Alexander–like polynomials from groups that can be found in, e.g., [CF, Cro].*

The Conway polynomial [Con] is obtained from the Alexander polynomial just by a variable change: $x = (t^{\frac{1}{2}} - t^{-\frac{1}{2}})$. Thus, the polynomial (denoted by C) satisfies the skein relation $C(\times) - C(\times) = x \cdot C(\,)(\,)$. Conway first proved that this relation (together with $C(\bigcirc) = 1$) can be axiomatic for defining a knot invariant. The approach by Przytycki and Traczyk described in this chapter was a generalisation of the Conway approach.

Chapter 6

Kauffman's approach to Jones polynomial

In the present chapter, we shall describe another approach for constructing invariant polynomials from link diagrams. It was proposed by Kauffman. It is quite expressive and allows us to construct polynomials (known as Kauffman polynomials in one and two variables).

The first polynomial to be constructed coincides with the Jones polynomial up to a suitable variable change. So, one can also speak about the Jones–Kauffman polynomial or the Jones polynomial in Kauffman's form.

The second polynomial includes some more sophisticated techniques in comparison with the first one. So, the second Kauffman polynomial is stronger than the first one. It is "in the general position" with the Jones two–variable polynomial that will be discussed later in the book.

6.1 State models in physics and Kauffman's bracket

First, let us seek an invariant polynomial for unoriented links. Let $\bar{L}$ be an unoriented link diagram having n crossings. Each crossing of $\bar{L}$ can be "smoothed" in two ways.

These ways $L \to L_A$ and $L \to L_B$ are shown in Fig. 6.1.

Now, let us try to construct some function (later, it will be called the *Kauff-*

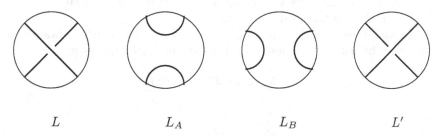

$$L \qquad\qquad L_A \qquad\qquad L_B \qquad\qquad L'$$

FIGURE 6.1: Two ways of smoothing the diagram

man bracket) in the three variables a, b, c satisfying the following axiomatic relations.

$$\langle L \rangle = a\langle L_A \rangle + b\langle L_B \rangle \tag{6.1}$$

$$\langle L \sqcup \bigcirc \rangle = c\langle L \rangle \tag{6.2}$$

$$\langle \bigcirc \rangle = 1. \tag{6.3}$$

Here we consider arbitrary diagrams $L = $ ⊗, $L' = $ ⊗, $L_A = $ ⊗ and $L_B = $ ⊃⊂ which coincide outside a small circle; inside this circle, the diagrams differ as shown in Fig. 6.1.

Herewith, $\bigcirc$ denotes the unknot, and $\sqcup$ denotes the disconnected sum.

Let us try to find the conditions for a, b, c in order to obtain a polynomial invariant under Reidemeiser moves.

First, let us test the invariance of the function to be constructed under the second Reidemeister move. By using (6.1) and (6.2), we obtain the following properties of the hypothetical function:

$$\langle \otimes \rangle = a\langle \otimes \rangle + b\langle \otimes \rangle = (a^2 + b^2)\langle \otimes \rangle + ab\langle \otimes \rangle + ab\langle \rangle\langle\rangle$$

$$= (a^2 + b^2 + abc)\langle \otimes \rangle + ab\langle \rangle\langle\rangle$$

Thus,

$$\langle \otimes \rangle = (a^2 + b^2 + abc)\langle \otimes \rangle + ab\langle \rangle\langle\rangle.$$

This equality should hold for all such triples looking like ⊗, ⊗, ⊃⊂ inside some small circle and coinciding outside it.

Thus, the Ω_2–invariance of the polynomial to be constructed should imply the following relations: $ab = 1$ and $a^2 + b^2 + abc = 0$.

Let us decree $b = a^{-1}$ and $c = -a^2 - a^{-2}$.

Thus, if the Ω_2–invariant link polynomial from $\mathbb{Z}[a, a^{-1}]$ exists, then it satisfies the properties described above and it is unique.

It turns out that here the Ω_2–invariance implies Ω_3–invariance.

Let us discuss this in more detail.

Consider the two diagrams ⊗ and ⊗ where one can be obtained from the other by using Ω_3. Smoothing them at one vertex we have:

$$\langle \otimes \rangle = a\langle \otimes \rangle + a^{-1}\langle \otimes \rangle$$

and

$$\langle \otimes \rangle = a\langle \otimes \rangle + a^{-1}\langle \otimes \rangle.$$

Let us compare the second parts of these equalities. We have: ⊗ ≡ ⊗. Furthermore, after applying Ω_2 twice, we obtain:

$$\overset{\frown}{\asymp} = \overset{\searrow}{\asymp} = \overset{\smile}{\asymp}.$$

Thus we have shown the invariance of the bracket under Ω_3. Now, one should mention that Ω_3 does not change the writhe number (see page 80).

This implies the invariance of the "hypothetical" bracket polynomial $\langle L \rangle$ under Ω_3.

However, while studying the invariance of the hypothetical polynomial under the first Reidemeister move, we meet the following unpleasant circumstance: addition (removal) of a curl multiplies the polynomial by $-a^3$ or by $-a^{-3}$. In fact, by applying (6.1) to the vertex incident to a curl, we obtain a sum of brackets of two diagrams. One of them gives us $p(-a^2 - a^{-2})\langle L \rangle$, the other one gives $p^{-1}\langle L \rangle$, where p is equal to $a^{\pm 1}$. The sign $\pm$ depends on the type of curl twisting.

Taking the sum of these two values, we get $(-a^{\pm 3})\langle L \rangle$.

Thus we have proved some properties that the Kauffman bracket **should** satisfy; i.e., we have deduced these properties from the axioms. However, we have not yet shown the main thing; i.e., the existence of such a polynomial. Let us prove that it exists.

Theorem 6.1. *There exists a unique function on link isotopy classes valued in $\mathbb{Z}[a, a^{-1}]$ satisfying relations (6.1)–(6.3) and invariant under Ω_2, Ω_3.*

Proof. Consider an unoriented diagram $\bar{L}$ of a link L that has n crossings. Let us enumerate all crossings of $\bar{L}$ by integers from 1 to n.

As before, we can smooth each crossing of the diagram in one of two ways, $A : \text{⧖} \to \text{)(}$ or $B : \text{⧖} \to \text{⌣⌢}$.

Definition 6.1. By a *state* of a crossing we mean one of the two possible ways of smoothing for it. By a *state* of the diagram L we mean the n states of crossings, one smoothing for each vertex.

Thus, the diagram $\bar{L}$ has 2^n possible states. Choose a state s of the diagram $\bar{L}$. Obviously, these smoothings turn $\bar{L}$ into a set of non-intersecting curves on the plane.

Let $\alpha(s)$ and $\beta(s)$ be the numbers of crossings in states A and B, respectively. Let $\gamma(s)$ be the number of circles of the diagram $\bar{L}$ in the state s.

If we "smooth" all crossings of the diagram $\bar{L}$ by means of (6.1), and then apply the relations (6.2) and (6.3) for calculating the bracket polynomials for the obtained diagrams, we get

$$\langle \bar{L} \rangle = \sum_s a^{\alpha(s) - \beta(s)}(-a^2 - a^{-2})^{\gamma(s)-1}, \tag{6.4}$$

where the sum is taken over all states s of the diagram $\bar{L}$.

Thus we have shown the *uniqueness* of the polynomial satisfying (6.1)–(6.3).

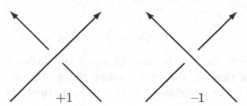

FIGURE 6.2: Local write numbers

It remains to show now that the formula (6.4) shows not only the uniqueness of $\langle \bar{L} \rangle$, but its *existence* too.

Actually, we can just *define* the polynomial $\langle \bar{L} \rangle$ as in (6.4).

Being defined by (6.4), the bracket polynomial evidently satisfies the conditions (6.2) and (6.3).

The invariance of the constructed polynomial under Ω_2, Ω_3 can be deduced straightforwardly from the definition. The theorem is proved. $\qquad \square$

6.2 Kauffman's form of Jones polynomial and skein relations

Thus, we have well–defined the bracket polynomial (*Kauffman's bracket*). This polynomial is defined on unoriented link diagrams and is Ω_2– and Ω_3–invariant. It turns out that it can be transformed into an invariant polynomial of oriented links; i.e., a function on oriented link diagrams invariant under all Reidemeister moves.

Consider an oriented diagram of a link L. Let us define an integer number $w(L)$ as follows. With each crossing of L we associate $+1$ or -1 as shown in Fig. 6.2. This number is called the *local writhe number*. Taking the sum of these numbers at all vertices, we get the *writhe number* $w(L)$.

It is easy to see that this number is invariant under Ω_2, Ω_3, but not invariant under Ω_1: under this move the writhe number is changed by ± 1. This circumstance allows us to normalise the bracket. Thus, we can define the **polynomial invariant** of links like this:

$$X(L) = (-a)^{-3w(L)} \langle |L| \rangle, \qquad (6.5)$$

where L is an oriented link diagram, and $|L|$ is the non–oriented diagram obtained from L by "forgetting" the orientation.

Definition 6.2. Let us call the invariant polynomial X according to (6.5) *the Kauffman polynomial*.

It turns out that the Kauffman polynomial satisfies a certain skein relation.

Actually, let $L_+ = $, $L_- = $ and $L_0 = $ be a Conway triple. Without loss of generality, we can assume that $w(L_+) = 1, w(L_-) = -1, w(L_0) = 0$. Consider the non–oriented diagrams $K_+ = $ | | $= $, $K_- = $ | | $= $, $K_A = $ | | $= $, and $K_B = $, that is K_B is the diagram where the corresponding vertex of L_+ is smoothed in the way B. From (6.1) and (6.5) we conclude that

$$X(\text{}) = (-a)^{-3}(a\langle K_A\rangle + a^{-1}\langle K_B\rangle) = -a^{-2}\langle K_A\rangle - a^{-4}\langle K_B\rangle \tag{6.6}$$

and, analogously,

$$X(\text{}) = -a^2\langle K_A\rangle - a^4\langle K_B\rangle \tag{6.7}$$

$$X(\text{}) = \langle K_A\rangle. \tag{6.8}$$

In order to eliminate K_B from (6.6) and (6.7) let us multiply (6.6) by a^4 and (6.7) by $(-a)^{-4}$ and take their sum. Thus, we get

$$a^4 X(\text{}) - a^{-4}X(\text{}) = (a^{-2} - a^2)X(\text{}). \tag{6.9}$$

This is the desired skein relation for the Kauffman polynomial. Now, it is evident that for each link, the value of the Kauffman polynomial contains only even degrees of a.

After the change of variables $q = a^{-4}$, we obtain another invariant polynomial called *the Jones polynomial*, originally invented by Jones [Jon1]. Obviously, it satisfies the following skein relation:

$$q^{-1}V(\text{}) - qV(\text{}) = (q^{\frac{1}{2}} - q^{-\frac{1}{2}})V(\text{}). \tag{6.10}$$

Exercise 6.1. *Prove that the Jones polynomial never equals zero.*

By definition, the value of the Jones polynomial on the n–component unlink is equal to $(-q^{\frac{1}{2}} - q^{-\frac{1}{2}})^{n-1}$

The Jones polynomial allows us to distinguish some mirror knots. For example, we can prove that the right trefoil knot is not isotopic to the left trefoil knot. Namely, from (6.10) we have:

$$V(\text{}) = q^2 \cdot V(\bigcirc) + q(q^{\frac{1}{2}} - q^{-\frac{1}{2}})V(\text{}) =$$

(here the Conway relation is applied to the upper left crossing)

$$q^2 + (q^{\frac{3}{2}} - q^{\frac{1}{2}})V(\text{}).$$

Taking into account

$$V(\text{}) = q^2(-q^{-\frac{1}{2}} - q^{\frac{1}{2}}) + q(q^{\frac{1}{2}} - q^{-\frac{1}{2}}) = -q^{\frac{5}{2}} - q^{\frac{1}{2}},$$

we see that

$$V(\text{⬡}) = q^2 + q(q^{\frac{1}{2}} - q^{-\frac{1}{2}})(-q^{\frac{5}{2}} - q^{\frac{1}{2}}) = -q^4 + q^3 + q.$$

Analogously, $V(\text{⬡}) = -q^{-4} + q^{-3} + q^{-1}$.

Thus, $\text{⬡} \neq \text{⬡}$.

Later, we shall present the perfect Jones proof of the invariance for a stronger two–variable polynomial, called *the Jones polynomial in two variables*.

6.3   Kauffman's two–variable polynomial

The Kauffman approach also allows us to construct a strong two–variable polynomial. The stronger one does not, however, satisfy any skein relation.

To construct it, one should consider a more complicated relation than that for the usual Kauffman polynomial. Namely, we first take four diagrams $L = \text{⬡}, L' = \text{⬡}, L_A = \text{⬡}, L_B = \text{⬡}$ of unoriented links that differ from each other only inside a small circle.

Then we construct the polynomial $K(z, a)$ satisfying the following axioms:

$$D(L) - D(L') = z(D(L_A) - D(L_B)); \tag{6.11}$$

$$D(\bigcirc) = \left(1 + \frac{a - a^{-1}}{z}\right); \tag{6.12}$$

$$D(X\#P) = aD(X), D(X\#Q) = a^{-1}D(X), \tag{6.13}$$

where

$$P = \text{⬡} \quad \text{and} \quad Q = \text{⬡}$$

are the two loops.

The polynomial D (like Kauffman's bracket) is invariant under Ω_2, Ω_3.

As in the case of the one–variable Kauffman polynomial, we normalise this function by using $w(L)$, namely for an oriented diagram L of a link, we set $Y(L) = a^{-w(|L|)}D(|L|)$, where $|L|$ is the unoriented diagram obtained from L by forgetting the orientation.

The obtained polynomial is called *the two–variable Kauffman polynomial.*

After this, one can show the existence, uniqueness and invariance under Reidemeister moves of the polynomial defined axiomatically as above.

For uniqueness, the proof is pretty simple. First we define its values on unlinks.

Later, in order to define the value of the polynomial somehow, we use induction on the number of classical crossings. The induction basis is trivial. The induction step is made by using the relation (6.11): we can switch all crossing types and express the desired value by the value on the unlink and links with smaller number of crossings.

The existence proof can be found in Kauffman's original book [Kau4].

Just recently, by using the Kauffman approach of statistical sums, B. Bollobás, L. Pebody, and D.Weinreich [BPW] found a beautiful explicit formula for calculating the HOMFLY-PT polynomial. This formula is even easier than that for the Kauffman polynomial in two variables.

Chapter 7

Properties of Jones polynomials. Khovanov's complex

7.1 Simplest properties

In this chapter we shall describe some properties of Jones polynomials and ways that this polynomial can be applied for solving some problems in knot theory.

After this, we shall formulate three celebrated conjectures concerning link diagrams which are related to properties of the Jones polynomial. Finally, we shall present a very sophisticated generalisation of the Jones polynomial, the Khovanov complex.

First, let us prove some properties of the Jones polynomial and deduce some corollaries from them.

Theorem 7.1. *1. The value of the Jones polynomial $V(L)$ is invariant under orientation change for the link diagram.*

2. The values of the Jones polynomial on mirror knots differ according to the variable change $q \to q^{-1}$.

Proof. The first statement is evident; it is left for the reader as a simple exercise (use induction on the number of crossings). The proof of the second statement also involves induction on the number of crossings. The induction basis is evident. Let us prove the induction step. To do it, let $L_+ = \vcenter{\hbox{$\times$}}, L_- = \vcenter{\hbox{$\times$}}, L_0 = \vcenter{\hbox{$)($}}$ be a Conway triple and L'_+, L'_-, L'_0 be the three diagrams obtained from the first ones by switching all crossings. Obviously, $\{L'_-, L'_+, L'_0\}$ is also a Conway triple.

By the induction hypothesis, we can assume that $V(L_0)$ is obtained from $V(L'_0)$ by the change of variables $q \to q^{-1}$. Let us now apply the relation (6.10). We get:

$$V(L_+) = q^2 V(L_-) + q(q^{\frac{1}{2}} - q^{-\frac{1}{2}})V(L_0),$$

$$V(L'_+) = q^{-2}V(L'_-) - q^{-1}(q^{\frac{1}{2}} - q^{-\frac{1}{2}})V(L'_0) = q^{-2}V(L'_-) + q^{-1}(q^{-\frac{1}{2}} - q^{\frac{1}{2}})V(L'_0).$$

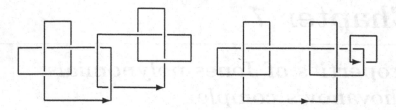

FIGURE 7.1: Two non-isotopic links with the same Jones polynomial

Now we can see that if the statement of the theorem holds for L_0 and L_-, then it holds for L_+ as well. This statement holds for L_0 by the induction conjecture. Thus, if it holds for some diagram, then it holds for any diagram with the same shadow.

Since the claim of the theorem is true for each unlink (hence, for each diagram of the unlink with arbitrary shadow), we can conclude that it is true for each diagram with any given shadow. Taking into account that unlinks can have arbitrary shadows, we obtain the desired result.   □

Theorem 7.2. *For arbitrary oriented links K_1 and K_2 the following holds:*

$$V(K_1 \# K_2) = V(K_1) \cdot V(K_2).$$

Proof. We shall prove this fact for the Kauffman polynomial.

It suffices to note that the Kauffman bracket is multiplicative with respect to the connected sum operation and $w(L)$ is additive.

We just note that the Kauffman bracket is multiplicative with respect to the connected sum operation (it follows from the definition of the bracket (6.4)) and $w(L)$ is additive with respect to the connected sum.

□

Remark 7.1. *This property is proved for* **any arbitrary** *connected sum of two links.*

Analogously, one can prove the following.

Theorem 7.3. *For arbitrary oriented knots K_1 and K_2 the following equality holds*

$$V(K_1 \sqcup K_2) = -(q^{-\frac{1}{2}} + q^{\frac{1}{2}})V(K_1) \cdot V(K_2).$$

Example 7.1. *Consider the two links shown in Fig. 7.1. According to Remark 7.1, the Jones polynomials for these links coincide. However, these links are not isotopic since their components are not so.*

This example shows that the Jones polynomial does not always distinguish

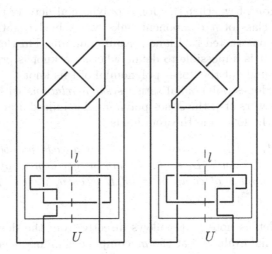

FIGURE 7.2: Left diagram L; right diagram L'

between different links. The reason is that the connected sum is not well defined. It turns out that the Jones polynomial is not a complete knot invariant. To show this, let us do the following.

Let L be a link diagram on the plane P. Suppose there exists a domain $U \subset P$ that is symmetric with respect to a line l in such a way that ∂U is a rectangle whose sides are parallel to coordinate axes. Suppose U intersects the edges of L only transversely at four points: two of theme lie on the upper side of the rectangle and the other two are on the lower sides. Suppose that these points are symmetric with respect to l. Let L' be the diagram that coincides with L outside U, and with reflection of L at l inside U; see Fig.7.2.

Lemma 7.1. *In this case $V(L) = V(L')$.*

Proof. Reflecting the parts of the diagram at l, we do not change the signs $\varepsilon = \pm 1$ of the crossings. Thus $w(L) = w(L')$. We only have to show that the Kauffman bracket does not change either. With each state of L, we can naturally associate a state of L'. So, it remains to see that after smoothing all crossings of L and L' (according to corresponding states), the number of circles is the same. The latter is evident. $\qquad\square$

Example 7.2. *It can be shown that the knots shown in Fig. 7.2 are not isotopic. Thus, the Jones polynomial is not a complete knot invariant.*

Remark 7.2. *Now it is an open problem whether the Jones polynomial distinguishes the unknot; i.e., is it true that $V(K) = 1$ implies the triviality of K, see the list of unsolved problems in Appendix D. This problem is equivalent to the faithfulness problem of the Burau representation for 4–strand braids (this was proved by Bigelow [Big2]). This problem will be formulated later.*

The question of whether the Jones polynomial detects the n–component unlink in the class of n–component links was solved negatively; a series of examples are constructed by Elahou, Kauffman, and Thistletwaite in [EKT].

As we see, it is impossible to define whether a knot is prime, having only the information about the Jones polynomial of this knot.

Hence the Jones polynomial satisfies *skein relations* with only integer or half–integer powers of q; the Jones polynomial is indeed a polynomial in $q^{\pm \frac{1}{2}}$. Furthermore, the following theorem holds.

Theorem 7.4. *(a) If an oriented link L has an odd number of components (e.g. it is a knot) then $V(L)$ contains only integer degrees of q.*

(b) If the number of components of L is even, then $V(L)$ contains only summands like $q^{\frac{(2k-1)}{2}}, k \in \mathbb{Z}$.

Proof. First, let us note (this follows directly from the definition) that the Jones polynomial evaluated at the m–component unlink is equal to

$$(-q^{-\frac{1}{2}} - q^{\frac{1}{2}})^{m-1}.$$

Consequently, the claim of the Theorem is true for unlinks.

Then, we can calculate the values of the Jones polynomial for arbitrary links by using the skein relation (6.10) for the Jones polynomial, knowing its values for unlinks.

We shall use induction on the number of crossings. The induction basis is evident. Suppose the induction hypothesis is true for less than n crossings. Let L be a diagram with n crossings. We can transform it to a diagram of the unlink by using the skein relation (6.10). Hence the statement is trivial for the unlink, but we have to check whether it remains true while switching crossing types.

Thus, given three diagrams L_+, L_-, L_0 (the first two having n vertices, the last one having $n-1$ vertices), the claim of the theorem is true for L_0 and for one of L_+, L_-. Without loss of generality, assume that this is L_-.

To complete the proof, we only have to see that:

1) The number of link components of L_+ and L_- coincide and have the same parity;

2) The number of link components of L_+ and L_0 have different parity. $\square$

Exercise 7.1. *Prove analogously that the value of the Conway polynomial on odd–component links (e.g. on knots) contains only terms of even degree and that on even–component links contains only monomials of odd degree.*

7.2 Tait's first conjecture and Kauffman–Murasugi's theorem

About 100 years ago, the famous English physicist and knot tabulator P.G. Tait formulated three very interesting conjectures. They had been unsolved for many years. Two of them were solved positively; the third one was solved negatively.

Remark 7.3. *All these conjectures are formulated for links with connected shadow.*

Definition 7.1. The *length* of a (Laurent) polynomial P is the difference between its leading degree and the lowest degree.

Notation: Span(P).

Tait's first conjecture (1898) states the following. If a link L with connected shadow has an alternating n–crossing diagram $\bar{L}$ without "splitting" points (i.e. points that split the diagram into two parts), then there is no diagram of L with less than n crossings.

This problem was solved independently by Murasugi [Mur1, Mur2], Kauffman, and Thistletwaite in 1987.

Theorem 7.5 (The Kauffman–Murasugi theorem). *The length of the Jones polynomial for a link with connected shadow is less than n or equal to n . The equality holds only for alternating diagrams without splitting points and connected sums of them.*

In Chapter 16, we shall give a proof of the Kauffman–Murasugi theorem based on the notion of the atom, and give some generalisations of it.

The first Tait conjecture follows immediately from the Kauffman–Murasugi theorem: if we assume the contrary (i.e. diagrams L, L' represent the same link, L is alternating with n crossings and without splitting points and L' has strictly less than n crossings), we obtain a contradiction when comparing the lengths of polynomials: $spanV(L) = n > spanV(L')$.

7.3 Menasco–Thistletwaite theorem and the classification of alternating links

The Murasugi theorem was a great step in the classification of alternating links. The classification problem for alternating (prime) links is reduced to the case of links with the same number of vertices.

The final step was made by William Menasco and Morwen Thistletwaite,

FIGURE 7.3: The flype move

[MT], when they proved the second Tait conjecture. This conjecture (known as the Tait flyping conjecture) was stated a very long ago and solved only in 1993.

In the present section, we consider diagrams up to infinity change.

In fact, Menasco and Thistletwaite proved the following theorem.

Theorem 7.6. *Any two diagrams of the same alternating knot can be obtained from each other by using a sequence of* flypes; *i.e. moves shown in Fig. 7.3.*

Obviously, the Kauffman–Murasugi theorem together with the Menasco-Thistletwaite theorem gives a solution for the alternating link classification problem. The algorithm is the following. First, one can consider prime alternating diagrams not having splitting points. Now, consider two alternating link diagrams L, L' without splitting points. Let us see whether they have the same number of crossings. If not, they are not isotopic. If yes, suppose the number of crossings equals n. Then, try to apply all possible flype moves to L in order to obtain L'. We shall start within finite type because the number of link diagrams with n crossings is finite.

For more details, we refer the reader to the original work [MT], where he can find beautiful rigorous proofs based on some inductions and three–dimensional imagination.

7.4   The third Tait conjecture

Any knot whose minimal diagram (with respect to the number of crossings) is odd is not amphicheiral.

The conjecture runs as follows.

If a knot has a minimal **alternating** diagram (without break points) then its Jones polynomial cannot be symmetric with respect to $q^{\frac{1}{2}} \to q^{\frac{1}{2}}$. Thus, it is not amphicheiral. So, the counterexample should not be an alternating knot.

This conjecture was disproved by Thistletwaite in 1998, see [HTW1].

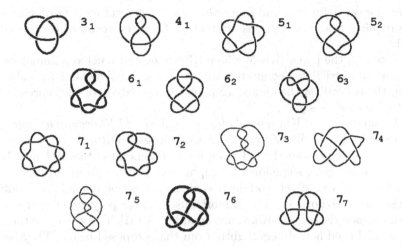

FIGURE 7.4: Prime knots with less than eight crossings

7.5 A knot table

Above we give a table of "prime knots diagrams" with less than or equal to seven crossings (up to mirror symmetry). It turns out that the Jones polynomial distinguishes all these diagrams, so, we can construct the simplest table of prime knots. See Fig. 7.4.

7.6 Khovanov's categorification of the Jones polynomial

We are now going to present a very interesting generalisation of the Jones polynomial of one variable, due to Khovanov, see [Kho1, Kho2]. Our description is close to that of the article [BN5] by Bar–Natan. In this article, Bar–Natan gives a clear explanation of Khovanov's theory, calculates various examples and shows that the homologies of the Khovanov complex constitute a strictly stronger invariant than the Jones polynomial itself.

In [BN6] an interesting construction was suggested. This construction describes a *topological* Khovanov complex, a formal chain complex, in which linear combinations of labeled sets of circles in the plane play the role of chains, where linear combinations of cobordisms are differentials. For invariance, some relations originating from the topology of two-dimensional complexes are imposed on such cobordism complexes. The general (algebraic) Khovanov complex is obtained from the geometrical one by "substitution" of concrete graded spaces for sets of circles, and concrete maps for elementary cobordisms; and it also requires that the relations originating from the relations on cobordisms hold.

Note also the paper [Kho5] where Khovanov constructed a homology theory for coloured links, connected with the Jones polynomial for cables, by using the cobordism theory and combinatorial analysis of the representation of the Lie algebra $\mathfrak{sl}_2$.

A counterpart of Khovanov homology called *odd Khovanov homology* was proposed by P.Ozsváth, Z.Szabó and J.Rasmussen [ORS].

Khovanov proposed the following idea: to generalise the notion of Kauffman's bracket using some formal complices and their cohomologies.

First, we give a slight modification of the Jones polynomial and Kauffman bracket due to Khovanov. The (unnormalised) Jones polynomial is the graded Euler characteristic of the Khovanov complex [Kho1]. The version of the Jones polynomial used here differs slightly from that proposed below. They become the same after a suitable variable change.

The axioms for the Kauffman bracket will be the following:

1. The Kauffman bracket of the empty set (zero–component link) equals 1.

2. $\langle L \sqcup \bigcirc \rangle = (q + q^{-1})\langle L \rangle$.

3. For any three diagrams $L = \langle\!\!\!\!\times\!\!\!\!\rangle, L_A = \rangle\!\!\langle, L_B = \asymp$ of unoriented links, we have
$$\langle L \rangle = \langle L_A \rangle - q\langle L_B \rangle.$$

Denote the state A of a vertex to be the 0–smoothing, and the state B to be the 1–smoothing. If the vertices are numbered then each way of smoothing for all crossings of the diagram is thought to be a vertex of the n–dimensional cube $\{0,1\}^{\mathcal{X}}$, where $\mathcal{X}$ is the set of vertices of the diagram.

Let the diagram L have n_+ positive crossings and n_- negative crossings; denote the sum $n_+ + n_-$ by n (that is the total number of crossings).

Denote the unnormalised Jones polynomial by

$$\hat{J}(L) = (-1)^{n_-} q^{n_+ + 2n_-} \langle L \rangle.$$

Let the Jones polynomial (denoted now by J, according to [BN5]) be defined as follows:

$$J(L) = \frac{\hat{J}(L)}{q + q^{-1}}.$$

Thus,

$$J(L) = (-1)^{n_-} q^{n_+ - 2n_-} \sum_s -q^{\beta(s)}(q + q^{-1})^{\gamma(s)-1}.$$

This normalised polynomial J differs from the Jones–Kauffman polynomial by a simple variable change: $a = \sqrt{(-q^{-1})}$. Namely,

$$(-a)^{-3(n_+ - n_-)} \sum_s a^{\alpha(s) - \beta(s)} (-a^2 - a^{-2})^{\gamma(s) - 1}$$

$$= (-1)^n a^{-3(n_+ - n_-)} \sum_s a^{-2\beta(s) + n} \cdot (q + q^{-1})^{\gamma(s) - 1}$$

$$= (-1)^n a^{4n_- - 2n_+} \sum_s (-q)^{\beta(s)} (q + q^{-1})^{\gamma(s) - 1}$$

$$= (-1)^{n_-} q^{n_+ - 2n_-} \sum_s (-q)^{\beta(s)} (q + q^{-1})^{\gamma(s) - 1}.$$

Khovanov's categorification idea is to replace polynomials by graded vector spaces with some "graded dimension". This makes the Jones polynomial a homological object. On the other hand, the graded dimension allows us to consider the invariant to be constructed as a polynomial in two variables.

We shall construct a "Khovanov bracket" (unnormalised complex that plays the same role for the Khovanov complex as the Kauffman bracket for the Jones polynomial). This will be denoted by double square brackets.

Let us start with the basic definitions and introduce the notation (which will differ from that introduced above!)

Let a linear space M (or a free module M over a ring $\mathcal{R}$) have a preferred *quantum* grading q. Then one has the following decomposition $M = \bigoplus_i M_i$, where M_i is the homogeneous component of grading i. By the *graded dimension* of the space M we mean the polynomial $\operatorname{qdim} M = \sum_i q^i \dim M_i$.

For such complexes there are naturally defined operations of the *height shift* $\mathcal{C} \mapsto \mathcal{C}[k]$ and the *grading shift* $\mathcal{C} \mapsto \mathcal{C}\{l\}$ defined according to the following rules: $(\mathcal{C}[k])^{i,j} = \mathcal{C}^{i-k,j}$; $(\mathcal{C}\{l\})^{i,j} = \mathcal{C}^{i,j-l}$. In the first case, together with chains, all differentials are shifted accordingly (i.e. the differential ∂_i, which was acting from $\mathcal{C}^{i,*}$ to $\mathcal{C}^{i+1,*}$, will now act in the same way from $\mathcal{C}^{i-k,*}$ to $\mathcal{C}^{i+1-k,*}$). By the *graded Euler characteristic* of the complex $\mathcal{C}^{i,j}$ we mean the alternating sum of the graded dimensions of the chain spaces, or, which is the same, the graded dimension of the homology groups. For chain spaces, we have:

$$\chi_q(\mathcal{C}^{i,j}) = \sum_i (-1)^i \operatorname{qdim} \mathcal{C}^i = \sum_{i,j} (-1)^i q^j \dim \mathcal{C}^{i,j}.$$

For such complexes, for every bigraded dimension (i, j) there is the (co)homology group $H^{ij}(\mathcal{C})$ which is defined as the quotient module of the corresponding module of cycles by the submodule of boundaries.

Definition 7.2. Two graded (respectively, bigraded) complexes $\mathcal{C}$ and $\mathcal{C}'$ are called *quasiisotopic*, if there exist two bigrading preserving maps $f: \mathcal{C} \to \mathcal{C}'$, $g: \mathcal{C}' \to \mathcal{C}$ together with a map u decreasing the height by one and preserving the second grading if such exists, such that $f \circ g = \operatorname{Id}_{\mathcal{C}'}$, and

$g \circ f - \mathrm{Id}_{\mathcal{C}} = d \circ u + u \circ d$. Here, $\mathrm{Id}_{\mathcal{C}}: \mathcal{C} \to \mathcal{C}$, $\mathrm{Id}_{\mathcal{C}'}: \mathcal{C}' \to \mathcal{C}'$ denote the corresponding identity maps.

Homology groups of quasiisomorphic complexes are isomorphic.

Let L, n and $n_{\pm}$ be defined as before. Let $\mathcal{X}$ be the set of all crossings of L. Let V be the graded vector space generated by two basis elements $v_{\pm}$ of degrees ± 1, respectively. Thus, $\mathrm{qdim}V = q + q^{-1}$.

Definition 7.3. By *bifurcation cube* we understand the cube $\{0,1\}^{\mathcal{X}}$ where each vertex is assigned the number of circles (as in the state cube), and each edge indicates which circles bifurcate when passing from a state to an adjacent one. The *height* of a state (a vertex of the cube) is the number of B-smoothings.

We orient the edges of the cube as the sum of coordinates increases (i.e. from an A-smoothing to a B-smoothing).

With every vertex α of the bifurcation cube $\{0,1\}^{\mathcal{X}}$ we associate the graded vector space $V_{\alpha}(L) = V^{\otimes k}\{r\}$, where k (formerly γ) is the number of circles in the smoothing of L corresponding to α and r is the height $|\alpha| = \sum_i \alpha_i$ of α (so that $\mathrm{qdim}V_{\alpha}(L)$ is the polynomial that appears at the vertex α in the cube). Now, let the r-th chain group $[[L]]^r$ be the direct sum of all vector spaces at height r, that is $\oplus_{\alpha:\ |\alpha|=r}V_{\alpha}(L)$.

Let us forget for a moment that $[[L]]$ is not endowed with a differential, and hence, is not a complex. Set $\mathcal{C} := [[L]][-n_-]\{n_+ - 2n_-\}$.

Remark 7.4. *It is easy to show that for a complex C the graded dimension $\chi_q(C)$ equals the alternating sum of the graded dimensions of its chain groups. This is quite analogous to the case of the usual Euler characteristics.*

Thus, we can calculate the graded Euler characteristic of C (taking into account only its graded chains); the differential will be introduced later.

Theorem 7.7. *The graded Euler characteristic of $C(L)$ is the unnormalised Jones polynomial $\hat{J}$ of L.*

Proof. This theorem is almost trivial. One should just take the alternating sum of graded dimensions of chain groups and mention that $\mathrm{qdim}(V^{\oplus n}) = n\,\mathrm{qdim}(V)$. The remaining part follows straightforwardly. $\square$

Now, let us prove that the Khovanov complex is indeed a complex. So, let us introduce the differentials for it. First, we set all $[[L]]^r$ to be the direct sums of the vector spaces appearing in the vertices of the cube with precisely r coordinates equal to 1.

The edges of the cube $\{0,1\}^{\mathcal{X}}$ can be labelled by sequences in $\{0,1,*\}$ of length n having precisely one $*$. This means that the edge connects two vertices, obtained from this sequence by replacing $*$ with one or zero.

Definition 7.4. The *height* $|\xi|$ of the edge ξ is defined to be the height of its tail (the end having lower height).

Thus, if the maps for the edges are called d_ξ, then we get $d^r = \sum_{\{|\xi|=r\}}(-1)^\xi d_\xi$.

Definition 7.5. The cube with partial differentials d_ξ going along edges in the coordinate increasing direction is called *commutative*, if each two-dimensional face of this cube is a commutative diagram and *anticommutative*, if each two-dimensional face is an anticommutative diagram.

Now, we have to explain the sign $(-1)^\xi$ and to define the edge maps d_ξ. Indeed, in order to get a "good" differential operator d, such that $d \circ d = 0$, it suffices to show that all square faces of the cube anticommute.

This can be done in the following way. First, we make all faces commutative, and then we multiply each d_ξ by $(-1)^\xi = (-1)^{\sum_{i<j} \xi_i}$, where j is the position of $*$ in ξ.

Exercise 7.2. *Show that such coefficients really make any commutative cube skew–commutative.*

Thus, we should find maps that can make our cube commutative.

Each edge represents some switch of the state for our diagram at some vertex. So, this means either dividing one cycle into two cycles, or joining two cycles together. In these cases, we shall use the comultiplication Δ and multiplication m maps defined as follows.

The map m:

$$\begin{cases} v_+ \otimes v_- \mapsto v_-, v_+ \otimes v_+ \mapsto v_+, \\ v_- \otimes v_+ \mapsto v_-, v_- \otimes v_- \mapsto 0 \end{cases} \tag{7.1}$$

The map Δ :

$$\begin{cases} v_+ \mapsto v_+ \otimes v_- + v_- \otimes v_+ \\ v_- \mapsto v_- \otimes v_-. \end{cases} \tag{7.2}$$

Because of the degree shifts, our maps m and Δ are chosen to have degree (-1).

Now, the only thing to check is that the faces of our cube for d_ξ (without ± 1 coefficients) commute. This follows from a routine verification.

The most interesting fact here is the invariance of *all homologies* of the Khovanov complex under all Reidemeister moves. Let us speak about this in more detail.

For a link diagram L, denote by $Kh(L)$ the expression

$$\sum_r t^r \text{qdim} \mathcal{H}^r(L).$$

Remark 7.5. *When we wish to emphasise the field* **F**, *we write* $Kh_\mathbf{F}(L)$.

Theorem 7.8 (the main theorem). *The graded dimensions of the homology groups* $\mathcal{H}^r(L)$ *are links invariants, hence* $Kh(L)$ *is a link invariant polynomial (of the variables* t, q*) that gives the unnormalised Jones' polynomial being evaluated at* $t = -1$.

Proof. We shall restrict ourselves only to three versions of the Reidemeister moves (one of Ω_1, one of Ω_2, and one of Ω_3). The other cases can be reduced to those we are going to consider.

In the case of the Kauffman bracket and the Jones polynomial, the invariance can be proved by reducing the Kauffman bracket of the "complicated case" of the move by using the rule ($\langle L \rangle = \langle L_A \rangle - q\langle L_B \rangle$). Here we will do almost the same, but since we deal with complices and homologies rather than with polynomials, we must interpret it in another language. Namely, we are going to use the following "cancellation principle".

Let $\mathcal{C}$ be a chain complex and let $\mathcal{C}' \subset \mathcal{C}$ be a subchain complex of $\mathcal{C}$. Then the following two statements hold.

Lemma 7.2 (Cancellation principle). *1. If $\mathcal{C}'$ is acyclic then $H(\mathcal{C}) = H(\mathcal{C}/\mathcal{C}')$.*

2. If $\mathcal{C}/\mathcal{C}'$ is acyclic (has no homology) then $H(\mathcal{C}) = H(\mathcal{C}')$.

Both statements follow straightforwardly from the following exact sequence:

$$\cdots \to H^r(\mathcal{C}') \to H^r(\mathcal{C}) \to H^r(\mathcal{C}/\mathcal{C}') \to \cdots$$

associated with the short exact sequence

$$0 \to \mathcal{C}' \to \mathcal{C} \to \mathcal{C}/\mathcal{C}' \to 0.$$

Now, let us prove the invariance of $Kh(\cdot)$ under the three Reidemeister moves.

Invariance under Ω_1.

Consider the three diagrams ⬭, ⬭, and ⬭.
While computing $\mathcal{H}(P)$, we encounter the complex

$$\mathcal{C} = [[\,⬭\,]] = \left([[\,⬭\,]] \xrightarrow{m} [[\,⬭\,]]\{1\} \right).$$

This means that the total n–dimensional cube for ⬭ is divided into two $(n-1)$-dimensional cubes, corresponding to the two smoothed diagrams (one of them is shifted); the differentials between these two cubes are all represented via m by definition.

As we can easily see, all chains in ⬭ where the small circle ∘ is v_+, "kill" all cycles in ⬭ according to our differential, because v_+ plays the role of the unit element in V with respect to the multiplication m. Thus, the only homologies we can have lie in $[[\,⬭\,]]$ when the small circle is marked by v_-. It is easy to see, that after the necessary normalisation, these homologies precisely coincide with those of $[[\,⬭\,]]$.

The case of the other curl ⬭ can be considered analogously.

In the case of Ω_2, we shall consider the only case. In this case, the $[[\,⨂\,]]$ will be represented in the terms of brackets of ⟨⟩, ⟩⟨, ≍, ⋈ and differentials between them:

$$
\mathcal{C} = \quad
\begin{array}{ccc}
[[\,\rangle\langle\,]]\{1\} & \rightarrow & [[\,\smile\frown\,]]\{2\} \\
\uparrow & & m\uparrow \\
[[\,\asymp\,]] & \overset{\Delta}{\rightarrow} & [[\,\circleddash\,]]\{1\}
\end{array}
$$

Thus, we have four cubes of codimension two and we know what the differentials in these small cubes look like: we may catch the cohomology elements in terms of these differentials. So, we only have to check whether they really represent cohomologies in the big cube.

The lower–left part of the diagram contains the diagram ⋈ (more precisely, all states corresponding to this local state).

Observation 1. It is easy to see that the members of this state cannot be cohomologies of the complex: their differentials have non-trivial projection to ≍$\{1\}$.

Observation 2. All members corresponding to $[[\,\rangle\langle\,\{1\}\,]]$ are not boundaries of members corresponding to $[[\,\bowtie\,]]$: the differential of each member from $[[\,\bowtie\,]]$ also has an impact on $[[\,\asymp\,]]\{1\}$.

Observation 3. The complex $[[\,\asymp\,]]\{1\}_{v_+} \overset{m}{\rightarrow} [[\,\smile\frown\,]]\{2\}$ is acyclic.

Observation 4. Each boundary element x in $[[\,\smile\frown\,]]\{2\}$ coming from an element $z \in [[\,\rangle\langle\,]]\{1\}$ has a unique compensating element in $y \in [[\,\asymp\,]]\{1\}$ such that $\partial y = \partial z = x$. This follows from observation 3. Thus, there exists a y in this complex such that $\partial y = x$.

Taking into account observations 2 and 4 we conclude that **all cohomologies containing elements from** ⟩⟨ **are in one–to–one correspondence with homologies of the complex** $C[[\,\bowtie\,]]$.

It is easy to check that the complex $\mathcal{C}$ has no other homologies (this follows from observations 1 and 3; the proof is left for the reader).

This results in the invariants of homologies up to height and degree shifts. Taking into account the normalisation constants, we obtain the invariance of the Khovanov complex under the second Reidemeister move Ω_2.

The invariance proof for the other cases of Ω_2 is quite analogous to the case considered above. The direct calculation via Ω_2 does not work, thus we have to use the cancellation method described above.

In the case of the third Reidemeister move Ω_3 the situation is more difficult than the similar one for the case of the Kauffman polynomial.

In this case we have the following local pictures; see Fig. 7.5.

Let us recall the invariance proof for the Jones one–variable polynomial under Ω_3. First, we smooth one crossing and then we see that this invariance follows from the invariance under Ω_2. We are going to do something similar: we consider our three–dimensional cubes and take their top layers that differ by a move Ω_2 (bottom layers of these cubes coincide).

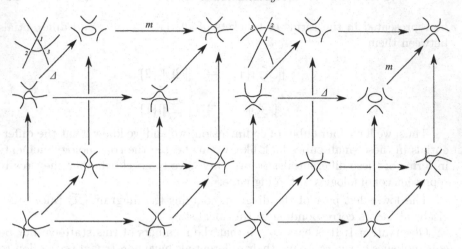

FIGURE 7.5: Behaviour of Khovanov's complex under Ω_3

If we consider the situation that occurs while performing the move Ω_2, we have the following complex.

The initial complex $\mathcal{C}$ looks like

$$
\begin{array}{ccc}
[[\;]]\{1\} & \xrightarrow{\;m\;} & [[\;]]\{2\} \\
\Delta \uparrow & & \uparrow \\
[[\;]] & \longrightarrow & [[\;]]\{1\}
\end{array}
.
$$

This complex contains the subcomplex $\mathcal{C}'$ that looks as follows

$$
\mathcal{C}' = \quad
\begin{array}{ccc}
[[\;]]_{v_+}\{1\} & \longrightarrow & [[\;]]\{2\} \\
\uparrow & & m \uparrow \\
0 & \longrightarrow & 0
\end{array}
$$

The acyclicity of the complex $\mathcal{C}'$ is obvious.

After factorising the complex $\mathcal{C}$ by $\mathcal{C}'$, we obtain the complex

$$
\begin{array}{ccc}
[[\;]]\{1\}/_{v_+=0} & \xrightarrow{\;m\;} & 0 \\
\Delta \uparrow & & \uparrow \\
[[\;]] & \longrightarrow & [[\;]]\{1\}
\end{array}
.
$$

Now, if we consider the special case of the top layer shown in Fig. 7.5, we see that the complex $\mathcal{C}'$ contains a subcomplex

$$
\mathcal{C}''' = \quad
\begin{array}{ccc}
& \beta \longrightarrow & 0 \\
\Delta \uparrow & \overset{\tau = d_{*0}\Delta^{-1}}{\searrow} & \uparrow \\
\alpha & \xrightarrow{d_{*0}} & \tau\beta,
\end{array}
$$

which is acyclic because Δ is an isomorphic map.

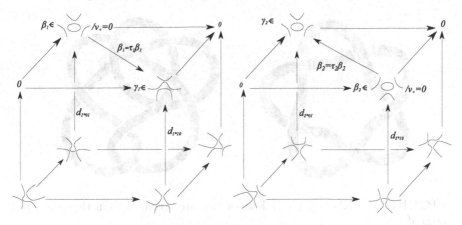

FIGURE 7.6: Invariance under Ω_3

Remark 7.6. *Here the arrow τ is not a differential. In the sequel, the diagonal arrow like $\beta = \tau\beta$ means that we identify two elements of the cube (arrows do not represent differentials).*

After this, we see that

$$(C/C')/C''' = \begin{array}{ccc} \beta & \longrightarrow & 0 \\ \uparrow & \searrow & \uparrow \\ 0 & \longrightarrow & \gamma. \end{array}$$

By the cancellation principle, we can perform this operation (factorising by C' and C''' defined for the top layers of the 3–cube) for the two cubes shown in Fig. 7.5 (only to the top layers of them). The resulting cubes are shown in Fig. 7.6.

Now, these two complices really are isomorphic via the map $\mathfrak{Y}$ which keeps the bottom layers shown in Fig. 7.6. in their place and transposes the top layers by mapping the pair (β_1, γ_1) to the pair β_2, γ_2.

The fact that $\mathfrak{Y}$ is really an isomorphism of spaces is obvious. To show that it is really an isomorphism of complices, we need to know that it commutes with the edge maps. In this case, only the vertical edges require a proof. The proof of this fact, namely that $\tau_1 \circ d_{1*01} = d_{2*01}$ and $d_{1*10} = \tau_2 \circ d_{2*10}$, is left to the reader as an exercise.

$\square$

Definition 7.6. Let us call by the *height* $h(\mathrm{Kh}(K))$ of the Khovanov polynomial of a link K the difference between the leading and lowest non-zero quantum gradings of non-zero terms of Khovanov polynomial of K.

The height of the Khovanov polynomial justifies the estimates coming from the span of the Kauffman bracket polynomial. The latter is responsible for non-cancellability of the leading and lowest terms in the decomposition (6.4);

FIGURE 7.7: Khovanov's $\mathbb{Q}$–homologies are stronger than the Jones polynomial

at the same time chains of the Khovanov complex are in natural one-to-one correspondence with monomials of the bracket multiplied by $(-a^2 - a^{-2})$.

By construction it is clear that

$$h(\mathrm{Kh}(K)) - 2 \geqslant \frac{\mathrm{span}\langle K \rangle}{2}.$$

As we have said before, the Khovanov polynomial (with rational homologies) is strictly stronger than the Jones polynomial. The example of two knots for which the Jones polynomial coincides and Khovanov's homologies do not, is shown in Fig. 7.7.

Exercise 7.3. *Perform the calculation check for this example.*

7.6.1 The two phenomenological conjectures

Obviously, the Khovanov complex (respectively, invariant polynomial) can be considered over an arbitrary field. We are interested in the two cases: $\mathbb{Q}$ and $\mathbb{Z}_2$.

Notation: $Kh_{\mathbb{Q}}, Kh_{\mathbb{Z}_2}$

Below we give the two phenomenological conjectures from [BN5]. They belong to Bar–Natan, Khovanov, and Garoufalidis.

Conjecture 7.1. *For any prime knot L there exist an even $s = s(L)$ and a polynomial $Kh'(L)$ in $t^{\pm 1}, q^{pm1}$ with only non-negative coefficients such that*

$$Kh_{\mathbb{Q}}(L) = q^{s-1}(1 + q^2 + (1 + tq^4)Kh'(L))$$

$$Kh_{\mathbb{Z}_2}(L) = q^{s-1}(1 + q^2 + (1 + tq^2)Kh'(L)).$$

Conjecture 7.2. *For the case of a prime alternating knot L, the number $s(L)$*

equals the signature of L, and the polynomial $Kh'(L)$ contains only powers of (tq^2).

These two conjectures were checked by Bar–Natan for knots with a reasonably small number of crossings (seven for $\mathbb{Q}$ and eleven for $\mathbb{Z}_2$).

It is easy to see that for the case of alternating prime knots, these two conjectures imply that the Khovanov polynomial is defined by the Jones polynomial.

A further phenomenological conjecture is presented in Garoufalidis' work [Garo]. All further information concerning these conjectures can be found in Bar–Natan's homepage [BNh].

The conjecture concerning alternating diagrams was solved positively by E.S. Lee, see [Lee1].

It is worth mentioning that Khovanov's homologies are functorial. This magnificent result is due to Magnus Jacobssen, see [Jac].

We also recommend to read the paper by O.Ya. Viro [Vir1] where a new "simple" approach to Khovanov's homologies is proposed.

7.6.2 Spanning tree for Khovanov complex

We shall describe a slightly different approach to calculating (more precisely, to estimating) the Khovanov homology, thanks to which some properties of the Khovanov homology became clearer.

Let us formulate the lemma from the theory of algebraic complexes, we shall follow S.Wehrli [Weh].

Lemma 7.3. *Let C_0 and C_1 be graded complexes and $C_i = A_i \oplus B_i$, where the complexes B_i have zero homology. Let $w: C_0 \to C_1$ be a map of chains preserving the grading, and let $w_{AA}: A_0 \to A_1$ be a "part" of the map w; i.e. the composition of the map w with the evident projection and embedding. Let A be a cone of the map w_{AA}, C be a cone of the map w, and B be contractible complex of type $B_0 \oplus B_1[1]$. Then the complexes C and $A \oplus B$ have the same homology.*

The proof of this theorem is purely algebraic, it does not concern the "internal" structure of differentials in the complexes A_i and B_i. The lemma is a key point in the proof of Theorem 7.9 about the spanning tree for the Khovanov complex.

The main idea of constructing the spanning tree leading to the proof of the theorem is the same as the Thistlethwaite idea which he used for constructing the spanning tree of the Kauffman bracket polynomial: It is necessary to take the bifurcation cube and split it into small subcubes corresponding to states from the set V_1 of the states with one circle.After that we have to consider the Khovanov homology for each of these subcubes; i.e. the copies of the homology groups of the unknot and apply Lemma 7.3 to them repeatedly. We should apply this lemma at each splitting of the cube into two parts.

Let us describe this construction in more detail. We shall consider a non-normalised Khovanov complex of a link In what follows we should take the "common normalizing factor" out; i.e. shift the height and the grading.

Let K be a link diagram. Let us consider its non-normalised bifurcation cube $[[K]]$ with the differential ∂. Enumerate all crossings of K and we shall split the cube $[[K]]$ successively into cubes according to Thistlethwaite's scheme. Namely, in the first step we investigate whether the first crossing is splitting (we call a crossing *splitting*, if under deleting the corresponding vertex from the diagram it becomes not connected) and, if it is not splitting, we pass to considering two cubes obtained from $[[K]]$ by fixing the first coordinate. These two cubes represent non-normalised Khovanov complexes for the diagrams K_0 and K_1 obtained from K by smoothings of type A and B. The Khovanov complex (unnormalised) for K_i has some set of homologies; if we consider K_0 and K_1 as non-separated complexes but compound parts of the Khovanov complex corresponding to K, we get some new differentials corresponding to passing from K_0 to K_1. Lemma 7.3 asserts that the initial (non-normalised) Khovanov complex for the diagram K has the same homology as the complex made only from homology of the complexes K_0 and K_1 (and as well as some acyclic part).

Further, we apply the second step: we consider the complexes K_0 and K_1 (as consistent parts of the new complex the homology of which coincides with the Khovanov homology of the link K) and investigate whether the corresponding diagrams split in the second crossing. If some of them (say, K_0) do not split, then we reconstruct the complex K_0 and get the complex of type $(K_{00} \to K_{01}) \oplus \langle$acyclic part$\rangle$.

We continue the process until we reach a diagram with all crossings smoothed . Each of these diagrams represents the unknot; therefore, we conclude that the Khovanov homology can be calculated with the help of a complex consisting of the Khovanov homology of the unknot. In terms of formula it looks like the following.

Theorem 7.9. *The non-normalised Khovanov complex of a link diagram K is isomorphic to some complex whose chain group looks like*

$$\bigoplus_{s \in \mathcal{V}_1} \mathcal{A}[\beta(s) + w(K_s)]\{\beta(s) + 2w(K_s)\}, \tag{7.3}$$

where $\mathcal{A}$ is the homology group of the unknot.

Later on, we shall use also the phrase *Wehrli's complex*, by bearing in mind the complex which is quasiisotopic to the Khovanov complex, the existence of the latter is given by Theorem 7.9.

7.6.3 The Khovanov polynomial and Frobenius extensions

The Khovanov theory of knots described earlier in this chapter is not unique when considering what one can get with the help of the Kauffman model and the (anti)commutative state cube. The present section is devoted to a generalisation of the Khovanov theory which uses Frobenius extensions.

Frobenius extensions

Let $\mathcal{R}$, $\mathcal{A}$ be commutative rings, and let $\iota\colon \mathcal{R} \to \mathcal{A}$ be an embedding of the commutative rings such that $\iota(1) = 1$. The restriction functor taking $\mathcal{A}$-modules to $\mathcal{R}$-modules has right and left adjoint functors: the induction functor $\mathrm{Ind}(M) = \mathcal{A} \otimes_{\mathcal{R}} M$ and the coinduction functor $\mathrm{CoInd}(M) = \mathrm{Hom}_{\mathcal{R}}(\mathcal{A}, M)$. One says that ι is a *Frobenius mapping*, if the induction functor coincides with the coinduction functor. Equivalently: the embedding ι is *Frobenius* if the restriction functor has a 3-sided dual functor. In this case one says also that the ring $\mathcal{A}$ is a *Frobenius extension* over $\mathcal{R}$ by means of ι.

The following proposition takes place.

Proposition 7.1 ([Kad]). *The embedding ι is Frobenius if there exist a mapping $\mathcal{A}$-bimodules $\Delta\colon \mathcal{A} \to \mathcal{A} \otimes_{\mathcal{R}} \mathcal{A}$ and a mapping $\mathcal{R}$-modules $\varepsilon\colon \mathcal{A} \to \mathcal{R}$ such that Δ is a coassociative and commutative multiplication, herewith $(\varepsilon \otimes \mathrm{Id})\Delta = \mathrm{Id}$.*

A Frobenius extension with a choice ε and Δ is denoted by $\mathcal{F} = (\mathcal{R}, \mathcal{A}, \varepsilon, \Delta)$ and called a *Frobenius system*, [Kad].

Frobenius extensions are convenient for constructing the Khovanov homology theory for the following reasons. In the module $\mathcal{A}$ defined over the ring $\mathcal{R}$ there are two natural operations: multiplication and comultiplication, the operation Δ.

We are going to use these operations for constructing the Khovanov homology theory for links. Meanwhile we (for evident reasons) restrict ourselves only to the case of commutative rings; moreover, we forget the operator ε (this operator is used for defining invariants of cobordisms and proving functoriality). In other aspects we follow the paper [Kho2] by Khovanov.

Khovanov construction for Frobenius extensions

As it was described earlier in this chapter the standard Khovanov theory is constructed over some arbitrary ring $\mathcal{R}$ (for example, the ring $\mathbb{Z}$ or the field $\mathbb{Q}$, or the field $\mathbb{Z}_p$); herewith the homology of the unknot is a graded two-dimensional module $\mathcal{A}$ over this ring, generated by vectors v_+ and v_- having gradings $+1$ and -1, respectively. Two maps are defined on these vectors: the multiplication m and comultiplication Δ. If one shifts the gradings of vectors (this requires a slight change (renormalization) in the construction of the homology theory), then one can set $\deg v_+ = 0$, $\deg v_- = 2$. Then the element v_+ can be considered as a unit (let us denote it by 1, and denote v_- by X), and the multiplication and comultiplication defined earlier turn the module $\mathcal{A}$ into a Hopf algebra over $\mathcal{R}$, in which the multiplication is defined by rules $X^2 = 0$, and the comultiplication looks like $\Delta(1) = 1 \otimes X + X \otimes 1$, $\Delta(X) = X \otimes X$.

In [Kho2] Khovanov solved the following problem: How can one find a condition for a couple of linear spaces $(\mathcal{A}, \mathcal{R})$ to get a link homology theory, where $\mathcal{R}$ is the basic coefficient ring and $\mathcal{A}$ (some Hopf algebra over $\mathcal{R}$) is the homology of the unknot (the main building bricks)? That means that we consider the state cube, with each vertex associated with a tensor power of $\mathcal{A}$ (over $\mathcal{R}$), with exponent equal to the number of circles in the given state, and define partial differentials by means of multiplication and comultiplication, and then add signs on edges and normalise the whole construction by grading shifts.

Khovanov showed that the invariance under the first Reidemeister move requires that $\mathcal{A}$ is two-dimensional as an $\mathcal{R}$-module and gave necessary and sufficient conditions for the existence of such a link homology theory.

In the same paper [Kho2], it is shown that any such theory can be obtained by some operations (base change, twisting and duality) from the following solution:

1. $\mathcal{R} = \mathbb{Z}[h, t]$,

2. $\mathcal{A} = \mathcal{R}[X]/(X^2 - hX - t)$,

3. $\deg X = 2$, $\deg h = 2$, $\deg t = 4$,

4. $\Delta(1) = 1 \otimes X + X \otimes 1 - h1 \otimes 1$,

5. $\Delta(X) = X \otimes X + t1 \otimes 1$.

As we see, the multiplication in the algebra $\mathcal{A}$ preserves the grading, and the comultiplication raises it by two.

We omit normalisations regulating these gradings.

We call this construction the *universal $(\mathcal{R}, \mathcal{A})$-construction*. The corresponding homology of a (classical oriented) link K will be denoted by $\mathrm{Kh}_U(K)$.

Khovanov proved that all other cases followed from the universal $(\mathcal{R}, \mathcal{A})$-construction. First, he investigates Frobenius extensions for the invariance of the obtained homology theory under the first classical Reidemeister move Ω_1. This leads it to two-dimensional $\mathcal{A}$ as an $\mathcal{R}$-module.

Later, Khovanov considers the universal topological construction by Bar-Natan [BN6], and constructs a functor from the topological category of Bar-Natan to the category of Frobenius extensions of rank two. The constructed functor is neither injective nor surjective, but it enjoys all nice properties needed for the invariance under the Reidemeister moves.

Thus Khovanov shows that any rank two Frobenius extension as above defines an extraordinary link homology theory. He shows also that any such theory without loss of information can be reduced to the universal theory described above by some algebraic operations.

We shall not go into the details of Khovanov's and Bar-Natan's constructions. We shall just consider the universal $(\mathcal{R}, \mathcal{A})$-construction.

Also, note that Khovanov also studied functoriality of his new homology theory, for example, its "good behaviour" under cobordisms (projective functoriality). To this end, besides multiplication and comultiplication operations, he also defined the unit and counit map and their transformations; we shall not touch on this subject.

7.6.4   Minimal diagrams of links

In the classification and tabulation of knots the important step is to describe diagrams having a minimal number of crossings. One of the main achievements in the development of knot theory is the Kauffman–Murasugi–Thistlethwaite theorem (Theorem 7.5) and the classification of alternating links by Menasco and Thistlethwaite [MT] following from this theorem.

In this section we shall prove theorems establishing the minimality of virtual and classical diagrams, see also [JS, Man17]. The inequality span $\langle K \rangle \leqslant 4n + 2(\chi - 2)$ for a virtual diagram K with n classical crossings and the atom with the Euler characteristic χ allowed one to prove the minimality in those cases, when the Euler characteristic could not be increased. If the inequality turns into the equality, then to decrease the number of crossings we have to increase the Euler characteristic of the atom or, the same, to decrease its genus. It turns out that by using Khovanov homology one can get estimates on the atom genus, at the same time in some cases one can see that this genus cannot be decreased. In this case the previous arguments together with non-reducibility of the genus lead to the minimality of the diagram.

We shall first mention the spanning tree theorem for Khovanov homology, proved independently by S.Wehrli [Weh] and A.Champanerkar and J.Kofman [ChKo].

More precisely, in [Weh] it is shown that the Khovanov homology is isomorphic to the homology of a certain complex. Let $\mathcal{V}_1(K)$ be the set of single-circle states of the virtual diagram K. From this a generalisation of Theorem 7.9 follows.

Lemma 7.4. *The non-zero Khovanov homology* $\mathrm{Kh}(K)$ *can have the bigrading only of the form* $(C_1 + \beta - w, C_2 + \beta - 2w \pm 1)$, *where* w *belongs to some finite set of integers,* β *belongs to the set of values* $\beta(s)$ *over all states* $s \in \mathcal{V}_1(K)$, *and* C_1, C_2 *are constants.*

An important particular case of this lemma is the statement of the Khovanov homology thickness (thickness was first introduced by Shumakovitch [Shu2, Shu3]).

Consider a link diagram K and its Khovanov homology over a certain non-graded ring R. Denote by $t_{\max}$ and $t_{\min}$ the maximal and minimal values of $2x - y$ over all pairs x, y such that the homology group of K with the bigrading (x, y) is non-trivial.

Definition 7.7. *The* thickness (width) $T_R(K)$ *of the Khovanov complex is* $(t_{\max} - t_{\min})/2 + 1$.

Remark 7.7. *This quantity is an integer for all links.*

Later on, by a *diagonal* we call the set of pairs of integer numbers (x, y) for which the number $2x - y$ is constant. Among diagonals there are the extreme left and the extreme right, at which the number $2x - y$ is minimal and maximal, respectively. Thus, the thickness measures the number of diagonals between two extreme diagonals.

Definition 7.8. By *thickness (width)* $T(K)$ of the link diagram K we mean the maximum of all $T_R(K)$ over all rings R without additional grading.

From Lemma 7.4 and the definition of atom (see Definition 16.1), we get the following lemma.

Lemma 7.5. *For any diagram K (with a connected atom) of a link we have:* $T(K) \leqslant g(K) + 2$, *where $g(K)$ is the genus of the atom corresponding to K.*

Definition 7.9. Let us call a link diagram K *2-complete*, if $T(K) = g(K) + 2$.

Indeed, for an estimate of the number of diagonals of the Wehrli complex (see Theorem 7.9) it is necessary for us to estimate the range of numbers $\beta(s)$ over all states $s \in \mathcal{V}_1(L)$. It is easy to see that in the case of alternating link diagrams all these numbers equal each other (this leads to the presence of two diagonals $t_{\max}$ and $t_{\min}$ such that $t_{\max} = t_{\min} + 2$), in the case of atoms with genus one the numbers $\beta(s)$ can equal x, $x + 1$, $x + 2$ for some x; in the case of atoms with the Euler characteristic χ they can take values in an interval from some number x to $x + (2 - \chi)$.

Now we have the following

Theorem 7.10. *Let $T(K) = g + 2$, span $\langle K \rangle = s$. Then the number of crossings of any connected diagram equivalent to K cannot be smaller than $s/4 + g$.*

In particular, if a diagram with n crossings and the atom with genus g is 1-complete and 2-complete, then it is minimal.

The last assertion means that all diagrams for which two properties of "natural non-reducibility" hold (in the decomposition of the Kauffman bracket polynomial the leading and lowest terms are not equal to zero and in the Wehrli complex each of the two extreme diagonals has at least one non-trivial element of the Khovanov homology) are minimal.

Chapter 8

Lee-Rasmussen invariant, slice knots, and the genus conjecture

The aim of the present chapter is to discuss one of the first striking applications of the Khovanov homology, the *Rasmussen invariant*.

In this chapter we follow closely the paper [Ras2].

We shall first define the Lee homology, which has the same pattern as the Khovanov homology (however, with a different differential), and leads to just two non-trivial generators for the case of a knot (and 2^k generators for the case of an k-component link). The Lee theory is not *bigraded* but rather *graded* with respect to homological grading and *filtered* with respect to the quantum grading.

Then the story starts, and the Khovanov homology can be considered as the starting term of the *spectral sequence* (see [McC]) which *converges to the Lee homology*. The fact that it converges is purely abstract and the filtration of the two (or 2^k in the case of a link) surviving terms is the crux of the matter.

These two filtrations differ by 2 and their average is the value of the *Rasmussen invariant of the knot K*, $s(K)$. In general, the calculation of the Rasmussen invariant is a very complicated task; in some cases, however, it can be calculated by hand.

One of the nicest properties of the Rasmussen invariant is its nice behaviour under cobordisms. Here we have to make a digression about two categories of sliceness.

The slice genus estimate for virtual knots is an interesting question. The Lee-Rasmussen theory which is used to estimate the slice genus deals with a TQFT with a division by 0; hence, one requires Khovanov homology theory with coefficients in $\mathbb{Z}$ rather than in $\mathbb{Z}_2$.

There is one important class of virtual links where the whole contents of this chapter generalises straighforwardly *even virtual links* or *virtual links with orientable atoms*. In fact, the only thing one needs to construct the Lee-Rasmussen is the source-sink structure of the atom.

Thus, one can deal with cobordisms where each slice admits a source-sink structure ("atomic cobordisms") and then it is in fact the case that the "even" slice genus of an "even" virtual knot can not be larger than the slice genus of an arbitrary knot.

When we do not have the source-sink structure, we can not apply the Lee–Rasmussen theory directly.

There are the following ways to tackle this problem.

1. To use our Khovanov homology theory (refer to Chapter 7) and construct the corresponding Lee–Rasmussen complex. This is done by Dye, Kaestner, and Kauffman in [DKK].

2. Use satellites of knots and links or 2-coverings of knots and links in a way similar to that in Chapter 22.

   This is not yet found in literature.

3. An interesting recent approach with "doubled Khovanov homology" due to William Rushworth; we are not going to discuss it here.

Note that for odd virtual knots (and even for simpler objects, the free knots) there is an elementary approach to handle the sliceness problem (cobordisms of genus 0), see papers [Man25, FeMa1, FeMa2].

Assume we have a knot $K \in S^3$ and we want it to be the boundary of a surface $\Sigma \subset B^4 : \partial B^4 = S^3, \partial \Sigma = \Sigma \cap S^3 = K$. We want to deal with the four-ball genus (*slice genus*) $g_*(K)$ of the K as the minimal genus of such a surface spanning the knot, but here we have a caveat. First, we observe that it is not allowed to consider the problem just in the continuous category. Indeed, for each knot K, we can take the cone over it from the center of B^3. As K is homeomorphic to a circle, the cone over it $C(K)$ is always homeomorphic to a ball. Thus, in the continuous category the problem is trivial.

In the sequel, by g_* we mean the slice genus in the *smooth* category.

The main theorem we are going to prove here is the following

Theorem 8.1. *[Ras2]* $|s(K)| \leq 2g_*(K)$.

There is, however, a beautiful theorem due to Freedman [Fre] saying that

Theorem 8.2. *If a knot K has trivial Alexander polynomial, then it is slice in the locally flat category.*

We shall not prove Theorem 8.2; we just mention that is closely related to some *Casson handle techniques* which appeared in M.Freedman's proof of the Poincaré conjecture in dimension 4 in the continuous category: *a closed simply connected 4-manifold with trivial second homology group is homeomorphic to S^4*.

Theorems 8.1 and 8.2 gave rise to various examples of knots which are slice in continuous category but not in the smooth category: to find such knots, it suffices just to be able to compute Rasmussen's invariant and the Alexander polynomial; both tasks are combinatorial.

In fact, Rasmussen proved more in his paper.

Theorem 8.3. *The map s induces a homomorphism from $Conc(S^3)$ to $\mathbb{Z}$ where $Conc(S^3)$ denotes the concordance group of knots in S^3.*

For alternating knots, $s(K)$ does not provide any new information about the genus $g_*(K)$:

Theorem 8.4. *If K is an alternating knot, then $s(K)$ is equal to the classical knot signature $\sigma(K)$.*

There is, however a class of knots for which $s(K)$ gives a *sharp* information. We say that a knot is *positive* if it admits a planar diagram with all positive crossings.

Theorem 8.5. *If K is a positive knot then $s(K) = 2g_* = 2g_K$, where $g(K)$ is the ordinary genus of K.*

As a corollary, we get a Khovanov homology proof of the following results which was first proved by P. Kronheimer and T. Mrowka [KrMr1] using gauge theory:

Corollary 8.1. *(The formerly Milnor conjecture) The slice genus of the (p,q)-torus knot is equal to $\frac{(p-1)(q-1)}{2}$.*

The theorems above all hold with $2\tau(K)$ in place of $s(K)$ (where $\tau(K)$ is the invariant defined via Floer homology). Indeed, the equality $s(K) - 2\tau(K)$ holds in many cases.

Based on these observations, Rasmussen formulated the following

Corollary 8.2. *For any knot $K \subset S^3, s(K) = 2\tau(K)$.*

8.1 Khovanov homology and Lee homology

Given a link diagram L with crossings labeled by integers from 1 to k, we form the cube of possible resolutions of L. As before, with each vertex v of the cube $[0,1]^k$ we associate the planar diagram D_v obtained by resolving the i-th crossing of L according to the i-th coordinate of v. Then D_v is a collection of circles. Let e be an edge of the cube; the coordinates of its two ends differ by one component, say, l-th. We call the end which has a 0 in this component the *initial end,* and denote it by $v_e(0)$. The other end will be called the *terminal end* and denoted by $v_e(1)$. We assign to e the *cobordism* $S_e : D_{v_e(0)} \to D_{v_e(1)}$, which is a product cobordism except in a neighbourhood of the l-th crossing where it is the saddle cobordism between the 0-resolution and the 1-resolution. We are now going to construct the *Lee complex* in analogy to the *Khovanov complex* (see Chapter 7). We associate a 1+1-dimensional TQFT $\mathcal{A}$ to the cube of resolutions. In other words, one replaces each vertex v with a group $\mathcal{A}(D_v)$ and each edge e with a map $\mathcal{A}(S_e) : \mathcal{A}(D_{v_e(0)}) \to \mathcal{A}(D_{v_e(1)})$. The underlying groups of the Khovanov complex $CKh(L)$ are the direct sum of groups $\mathcal{A}(D_v)$

for all vertices v, and the differential on the summand $\mathcal{A}(D_v)$ is a sum of the maps of edges $\mathcal{A}(S_c)$ for all edges e which have v as their initial end. As usual, we have

$$d(x) = \sum_{i=1}^{c_0(v)} (-1)^{s(e_i)} \mathcal{A}(S_{e_i}).$$

The cobordisms S_c come in two forms: two circles can merge into one or one circle can split into two circles; the corresponding maps $m : V \otimes V \to V$ and $\Delta : V \to V \otimes V$ for Khovanov homology theory are given in the Chapter 7.

Once these maps m and Δ satisfy certain conditions, the differential of the corresponding chain complex is well defined and its homology groups (after a certain normalisation) are invariant under Reidemeister moves.

However, if we want to deal not only with *invariants of knots*, but also with *cobordisms*, we shall need to define two other maps ι and ϵ. Corresponding to the addition of a 0-handle (the birth of a circle in a diagram), there is a map $\iota : \mathbb{Q} \to V$, and corresponding to the addition of a two handle (the death of a circle) there is a map $\varepsilon : V \to \mathbb{Q}$. These maps are given by

$$\varepsilon(\mathbf{v}_-) = 1, \quad \varepsilon(\mathbf{v}_+) = 0, \quad \iota(1) = \mathbf{v}_+.$$

The map $\mathcal{A}$ is especially nice because it is a graded TQFT. Recall that the grading is defined on V by setting $p(\mathbf{v}_\pm) = \pm 1$ and extended to the tensor product accordingly. We know that if $\mathbf{v}$ is a homogeneous element of $V^{\otimes n}$, then $p(S_c(\mathbf{v})) = p(\mathbf{v}) - 1$. The quantum grading q on Khovanov homology is defined by normalising $q(\mathbf{v}) = p(\mathbf{v}) + gr(\mathbf{v}) + n_+ - n_-$, where $n_\pm$ are the number o the positive crossings and the number of negative crossings of the diagram L. Recall that $gr(v) = |v| - n_-$, where v is the number of 1's among the coordinates of the crossing v.

8.1.1   Lee's homology

In [Lee1], Lee considered a similar construction, but with another TQFT $\mathcal{A}'$ instead of $\mathcal{A}$. The underlying vector spaces for these two TQFT's are the same, but the maps $m' : V \otimes V \to V$ and $\Delta' : V \to V \otimes V$ induced by cobordisms are slightly different. They are given by:

$$m'(\mathbf{v}_+ \otimes \mathbf{v}_+) = m'(\mathbf{v}_- \otimes \mathbf{v}_-) = \mathbf{v}_+; \quad m'(\mathbf{v}_+ \otimes \mathbf{v}_-) = m'(\mathbf{v}_- \otimes \mathbf{v}_+) = \mathbf{v}_- \quad (8.1)$$

and

$$\Delta'(\mathbf{v}_+) = \mathbf{v}_+ \otimes \mathbf{v}_- + \mathbf{v}_- \otimes \mathbf{v}_+; \quad \Delta'(\mathbf{v}_-) = \mathbf{v}_- \otimes \mathbf{v}_- + \mathbf{v}_+ \otimes \mathbf{v}_+. \quad (8.2)$$

The maps ι and ε corresponding to the handles are the same as before. We denote the resulting complex by $CKh'(L)$ and its homology by $Kh'(L)$. Using the obvious identification between the underlying groups of $CKh(L)$

and $CKh'(L)$, we can define a q-grading for the latter group as well. It is clear from (8.1), (8.2) the grading does not behave quite well with respect to the differential d'. Indeed, $\Delta'(\mathbf{v}_-)$ is not even homogeneous. It is easy to see, however, that $\mathbf{v} \in CKh'(L)$ is a homogeneous element then the q-grading of every monomial in $d'(\mathbf{v})$ is *greater than or equal to* the q-grading of $\mathbf{v}$. In other words, the q-grading defines a filtration on the complex $CKh'(L)$. This fact leads to the following

Theorem 8.6. *There is a spectral sequence with E_2-term $Kh(L)$ which converges to $Kh'(L)$. The E_2 and higher terms of this spectral sequence are invariants of the link L.*

The first part of the theorem is immediate from the above observation. The filtration on CKh' gives rise to a spectral sequence converging to Kh'. The differential in its E_1 term is the part of d' which preserves (rather than raises) the q-grading. Comparing the differentials for the Khovanov complex and for the Lee complex, we see that the E_1 term is the complex CKh.

The invariance of the Lee homology under Reidemeister moves is proved along the same lines as that of the Khovanov homology. In [Lee1], Lee defines maps $\rho_i' : CKh(L) \to CKh'(\bar{L})$ which induce isomorphisms on homology. Later we shall show that these maps induce isomorphisms on E_2 terms of spectral sequences, thus completing the proof of Theorem 8.6.

8.1.2 Calculation of Kh'

The Khomology group $Kh'(L)$ is surprisingly simple [Lee1]. To see this, it suffices only to introduce the new basis $\{\mathbf{a}, \mathbf{b}\}$ for V, where $\mathbf{a} = \mathbf{v}_- + \mathbf{v}_+, \mathbf{b} = \mathbf{v}_- - \mathbf{v}_+$.

In this basis, the maps m' and Δ' look as follows:

$$m'(\mathbf{a} \otimes \mathbf{a}) = 2\mathbf{a}; \quad m'(\mathbf{a} \otimes \mathbf{b}) = m'(\mathbf{b} \otimes \mathbf{a}) = 0; \quad m'(\mathbf{b} \otimes \mathbf{b}) = -2\mathbf{b};$$

$$\Delta'(\mathbf{a}) = \mathbf{a} \otimes \mathbf{a}; \quad \Delta'(\mathbf{b}) = \mathbf{b} \otimes \mathbf{b}.$$

One can also easily check that

$$\epsilon'(\mathbf{a}) = \epsilon'(\mathbf{b}) = 1, \quad \iota(1) = \frac{(\mathbf{a} - \mathbf{b})}{2}.$$

Using this basis, she proves

Theorem 8.7. *[Lee1] For an l-component link L, the homology $Kh(L')$ has rank 2^l.*

Proof. We shall now construct a bijection between the set of all orientations of L and a set of generators of $Kh'(L)$ which we refer to as *canonical generators*

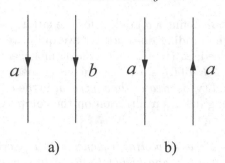

a)   b)

FIGURE 8.1: Local behaviour of the state s_o.

. This bijection can be described as follows. Given an orientation o of a link diagram L, let D_o be the corresponding oriented resolution. We shall label the circles in D_o with **a** and **b** according to the following rule. To each circle C we assign a mod 2 invariant, which is the mod 2 number of circles in D_o which separate it from infinity. In other words we can draw a ray starting from a point on this circle and count the number of intersections with other circles modulo 2. To this number, we add 1 if C has the counterclockwise orientation, and 0 if it has the clockwise orientation. Label C by **a** if the resulting invariant is 0 and by **b** if it is 1. We denote the resulting state by s_o.   □

Exercise 8.1. *Show that all canonical generators of a framed graph, which admits a source–sink structure, are induced by labeling its components with 0 and 1. In particular, a unicursal framed graph has two generators.*

  The name "canonical generator" is justified by the following result, whose proof will be given later.

Proposition 8.1. *Suppose L and $\bar{L}$ are related by the i-th Reidemeister move. Then an orientation o on L induces an orientation $\bar{o}$ of $\bar{L}$, and $\rho'_{i*}([s_o])$ is a nonzero multiple of $[s_o]$*

  We end this Section with an elementary but important observation.

Lemma 8.1. *(Coherent orientations) Suppose there is a region in the state diagram for s_o containing exactly two segments, as shown in Figure 8.1. Then either the orientations of the two are the same and the labels are different (like part a of the figure) or the orientations are different and the labels are the same (like part b).*

Proof. We consider three possible cases: either the two segments belong to the same circle in D_o, or they belong to two circles, one of which is contained inside the other, or they belong to two circles, neither of which is contained inside the other. In each case, it is easy to verify that the claim holds.   □

Corollary 8.3. *If two circles in the state diagram for s_o share a crossing, they have different labels.*

8.2 The Rasmussen invariant: Definition and basic properties of the invariant

Let K be a knot in S^3. By Theorems 8.6 and 8.7, we know that there is a spectral sequence associated to K which converges to $\mathbb{Q} \oplus \mathbb{Q}$. This spectral sequence is a relatively complicated object, but we can extract some simpler invariants of K from it. Let s_{max} and s_{min} (with $s_{max} \geq s_{min}$) be the q-gradings of the two surviving copies of $\mathbb{Q}$ which remain in the E_∞ term of the spectral sequence. Like all q-gradings for a knot, s_{max} and s_{min} are odd integers. Since the isomorphism type of the spectral sequence is an invariant of K, s_{max} and s_{min} are invariants

Before making this definition formal, we digress to establish some terminology related to filtrations. Suppose C is a chain complex. A *finite length filtration* of C is a sequence of subcomplexes

$$0 = C_n \subset C_{n-1} \subset C_{n-2} \subset \cdots \subset C_m = C.$$

To such a filtration, we associate a *grading* defined as follows: $x \in C$ has grading i if and only if $x \in C_i$ but $x \notin C_{i+1}$. If $f : C \to C'$ is a map between two filtered chain complexes, we say that f *respects the filtration* if $f(C_i) \subset C'_i$. More generally, we say that f is a *filtered map of degree* k if $f(C_i) \subset C'_{i+k}$.

A filtration $\{C_i\}$ on C induces a filtration $\{S_i\}$ on $H_*(C)$ defined as follows: a class $[x] \in H_*(C)$ is in S_i if and only if it has a representative which is an element of C_i. If $f : C \to C'$ is a filtered chain map of degree k, then it is easy to see that the induced map $f_* : H_*(C) \to H_*(C')$ is also filtered of degree k.

A finite length filtration $\{C_i\}$ on C induces a spectral sequence, which converges to the associated graded group of the induced filtration $\{S_i\}$. In other words, the group which survives at grading i in the spectral sequence is naturally identified with the group S_i/S_{i+1}.

Let us denote by s the grading on $Kh'(K)$ induced by the q-grading on $CKh'(K)$. Then the informal definition above is equivalent to

Definition 8.1. Set

$$s_{min}(K) = \min\{s(x) \mid x \in Kh'(K), x \neq 0\}$$
$$s_{max}(K) = \max\{s(x) \mid x \in Kh'(K), x \neq 0\}.$$

Since Kh of the unknot U has rank two and is supported in q-gradings ± 1, we have $s_{max}(U) = 1$, $s_{min}(U) = -1$.

Another proof that s_{max} and s_{min} are knot invariants could be given using

Proposition 8.2. *The maps ρ'_{i*} and $(\rho'_{i*})^{-1}$ both respect the induced filtration s on Kh'.*

We shall give the proof in Section 8.5.

8.2.1   The invariant s

Our first task in this section is to prove

Proposition 8.3.
$$s_{max}(K) = s_{min}(K) + 2$$

which justifies

Definition 8.2.

$$s(K) = s_{max}(K) - 1 = s_{min}(K) + 1$$

Since s_{max} and s_{min} are odd, $s(K)$ is always an even integer.

Before proving the proposition, we need some preliminary results.

Lemma 8.2. *Let n be the number of components of L. There is a direct sum decomposition $Kh'(L) \cong Kh'_o(L) \oplus Kh'_e(L)$, where $Kh'_o(L)$ is generated by all states with q-grading conguent to $2 + n \mod 4$, and $Kh'_e(L)$ is generated by all states with q-grading congruent to $n \mod 4$. If o is an orientation on L, then $\mathbf{s}_o + \mathbf{s}_{\bar\sigma}$ is contained in one of the two summands, and $\mathbf{s}_o - \mathbf{s}_{\bar\sigma}$ is contained in the other.*

Proof. Following Lee [Lee2], we write

$$m' = m + \Phi_m$$
$$\Delta' = \Delta + \Phi_\Delta$$

where m and Δ preserve the q-grading and Φ_m and Φ_Δ raise it by 4. This proves the first statement.

For the second statement, let $\iota \colon CKh'(L) \to CKh'(L)$ be the map which acts by the identity on CKh'_e and by multiplication by -1 on CKh'_o. We claim that $\iota(\mathbf{s}_o) = \pm\mathbf{s}_{\bar\sigma}$. To see this, we define a new grading on V with respect to which $\mathbf{v}_-$ has grading 0 and $\mathbf{v}_+$ has grading 2. Let $i \colon V \to V$ be given by $i(\mathbf{v}_-) = \mathbf{v}_-$, $i(\mathbf{v}_+) = -\mathbf{v}_+$, so that $i(\mathbf{a}) = \mathbf{b}$ and $i(\mathbf{b}) = \mathbf{a}$. Then the induced map $i^{\otimes n} \colon V^{\otimes n} \to V^{\otimes n}$ acts as the identity on elements whose new grading is congruent to 0 mod 4 and as multiplication by -1 on elements whose new grading is congruent to 2 mod 4. The new grading differs from the q-grading on D_o by an overall shift, so

$$\iota(\mathbf{s}_o) = \pm i^{\otimes n}(\mathbf{s}_o) = \pm\mathbf{s}_{\bar\sigma}$$

It follows that $\mathbf{s}_o + \iota(\mathbf{s}_o) = \mathbf{s}_o \pm \mathbf{s}_{\bar\sigma}$ is contained in one summand, while $\mathbf{s}_o - \iota(\mathbf{s}_o) = \mathbf{s}_o \mp \mathbf{s}_{\bar\sigma}$ is contained in the other. $\square$

Corollary 8.4.
$$s(\mathbf{s}_o) = s(\mathbf{s}_{\bar\sigma}) = s_{min}(K)$$

Corollary 8.5. $s_{max}(K) > s_{min}(K)$.

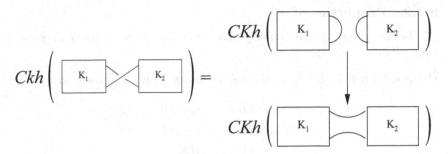

FIGURE 8.2: A short exact sequence for $CKh'(K_1 \# K_2)$.

Proof. Since $CKh'(K)$ decomposes as a direct sum, its affiliated spectral sequence decomposes too. The homology of each summand is $\mathbb{Q}$, so each must account for one of the surviving terms in the spectral sequence. The two summands are supported in different q-gradings, so the surviving terms must have different q-gradings as well. □

Lemma 8.3. *For knots K_1, K_2, there is a short exact sequence*

$$0 \to Kh'(K_1 \# K_2) \xrightarrow{p_*} Kh'(K_1) \otimes Kh'(K_2) \xrightarrow{\partial} Kh'(K_1 \# K_2) \to 0$$

The maps p_* and ∂ are filtered of degree -1.

Proof. Consider the diagram for $K_1 \# K_2$ shown in Figure 8.2. From it, we get a short exact sequence

$$0 \longrightarrow CKh'(D_1)\{1\} \longrightarrow CKh'(D_2) \xrightarrow{p} CKh'(D_3) \longrightarrow 0$$

where D_1 and D_2 are both diagrams for $K_1 \# K_2$, and D_3 is a diagram for the disjoint union $K_1 \coprod K_2$. Since $Kh'(K_1 \# K_2)$ has rank two and $Kh'(K_1 \coprod K_2) \cong Kh'(K_1) \otimes Kh'(K_2)$ has rank four, the resulting long exact sequence must split, giving the short exact sequence of the lemma. It is clear that the maps p_* and ∂ are filtered of some degree, which can be worked out by considering (for example) the case $K_1 = K_2 = U$. □

Proof. (of Proposition 8.3.) Consider the exact sequence of the previous lemma with $K_1 = K$ and K_2 the unknot. Denote the canonical generators of K by $\mathbf{s}_a$ and $\mathbf{s}_b$, according to their label near the connected sum point, and the canonical generators of U by $\mathbf{a}$ and $\mathbf{b}$. Without loss of generality, we may assume that $s(\mathbf{s}_a - \mathbf{s}_b) = s_{max}(K)$. From Figure 8.2, we see that $\partial((\mathbf{s}_a - \mathbf{s}_b) \otimes \mathbf{a}) = \mathbf{s}_a$. Since ∂ is a filtered map of degree -1, we conclude that

$$s((\mathbf{s}_a - \mathbf{s}_b) \otimes \mathbf{a}) \le s(\mathbf{s}_a) + 1$$
$$s_{max}(K) - 1 \le s_{min}(K) + 1$$

Since we already know that $s_{max}(K) \ne s_{min}(K)$, this gives the desired result. □

8.2.2   Properties of s

We check that s behaves nicely with respect to mirror image and connected sum.

Proposition 8.4. *Let $\overline{K}$ be the mirror image of K. Then we have*

$$s_{max}(\overline{K}) = -s_{min}(K)$$
$$s_{min}(\overline{K}) = -s_{max}(K)$$
$$s(\overline{K}) = -s(K)$$

Proof. Suppose that C is a filtered complex with filtration $C = C_0 \supset C_1 \supset \ldots \supset C_n = \{0\}$. Then the dual complex C^* has a filtration $\{0\} = C_0^* \subset C_{-1}^* \subset \ldots \subset C_{-n}^* = C^*$, where $C_{-i}^* = \{x \in C^* \mid \langle x, y \rangle = 0, \forall y \in C_i\}$.

To prove the proposition, we observe that the filtered complex $CKh'(\overline{K})$ is isomorphic to $(CKh'(K))^*$. Indeed, it is easy to see from equations (8.1),(8.2) that there is an isomorphism

$$r \colon (V, m', \Delta') \to (V^*, \Delta'^*, m'^*)$$

which sends $\mathbf{v}_\pm$ to $\mathbf{v}_\mp^*$. Then if $\mathbf{s}$ is a state of the diagram $\overline{K}$, we define $R(\mathbf{s})$ to be state of K obtained by applying r all the labels of $\mathbf{s}$. It is straightforward to check that the map $R : CKh'(\overline{K}) \to (CKh'(K))^*$ is the desired isomorphism. (Compare with Section 7.3 of [Kho1], where it is shown that $CKh(\overline{K}) \cong (CKh(K))^*$.)

We now appeal to the following general result, whose proof is left to the reader:

Lemma 8.4. *If C_1 and C_2 are dual filtered complexes over a field, then their associated spectral sequences E_n^1 and E_n^2 are dual, in the sense that $E_n^1 \cong (E_n^2)^*$.*

Thus if the two surviving generators in E_∞^1 have filtration gradings s_{min} and s_{max}, the surviving generators in E_∞^2 will have gradings $-s_{max}$ and $-s_{min}$.   □

Proposition 8.5.
$$s(K_1 \# K_2) = s(K_1) + s(K_2)$$

Proof. We use the short exact sequence of Lemma 8.3. Denote the canonical generators of K_i by $\mathbf{s}_a^i$ and $\mathbf{s}_b^i$, according to their label near the connected sum point. It is not difficult to see that $Kh'(K_1 \# K_2)$ has a canonical generator $\mathbf{s}_o$ which maps to $\mathbf{s}_a \otimes \mathbf{s}_b$ under p_*. Thus

$$s(\mathbf{s}_o) - 1 \le s(\mathbf{s}_a^1 \otimes \mathbf{s}_b^2)$$
$$s_{min}(K_1 \# K_2) - 1 \le s_{min}(K_1) + s_{min}(K_2)$$

Applying the same argument to $\overline{K}_1$ and $\overline{K}_2$, and using the fact that $s_{min}(K) = -s_{max}(K)$, we see that

$$s_{max}(K_1 \# K_2) + 1 \geq s_{max}(K_1) + s_{max}(K_2)$$
$$s_{min}(K_1 \# K_2) + 3 \geq s_{min}(K_1) + s_{min}(K_2) + 4$$

Thus

$$s_{min}(K_1 \# K_2) = s_{min}(K_1) + s_{min}(K_1) + 1$$
$$s_{max}(K_1 \# K_2) = s_{max}(K_1) + s_{max}(K_1) - 1.$$

This proves the claim.   □

8.3   Behaviour under cobordisms

Let L_0 and L_1 be two links in $\mathbb{R}^3$. An oriented cobordism from L_0 to L_1 is a smooth, oriented, compact, properly embedded surface $S \subset \mathbb{R}^3 \times [0,1]$ with $S \cap (\mathbb{R}^3 \times \{i\}) = L_i$. In this section, we define and study a map $\phi_S \colon Kh'(L_0) \to Kh'(L_1)$ induced by such a cobordism. Our construction follows Section 6.3 of [Kho1], where Khovanov describes a similar map for the homology theory Kh.

8.3.1   Elementary cobordisms

Following Khovanov, we decompose the cobordism S into a series of elementary cobordisms, each represented by a single move from one planar diagram to another. (See [CS] for a more detailed treatment of this material). For $i \in [0,1]$, let

$$L_i = S \cap (\mathbb{R}^3 \times \{i\})$$
$$S_i = S \cap (\mathbb{R}^3 \times [0,i]).$$

After a small isotopy of S, we can assume that L_i is a link in $\mathbb{R}^3$ for all but finitely many values of i. The orientation on S restricts to an orientation on S_i, which in turn determines an orientation on L_i. We denote this orientation by o_i. (Note that with this convention, o_0 is the inverse of the orientation induced on L_0 by S).

Next, we fix a projection $p : \mathbb{R}^3 \to \mathbb{R}^2$. After a further small isotopy of S, we can assume that p defines a regular projection of L_i for all but finitely many values of i, and that this set of special values is disjoint from the first set where L failed to be a link. The isotopy type of the oriented planar diagram L_i remains constant except when L passes through one of the two types of special

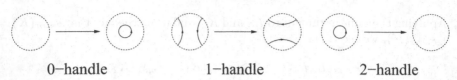

0–handle   1–handle   2–handle

FIGURE 8.3: Local pictures for Morse moves.

values, where it changes by some well-defined local move. Each of these moves corresponds to an elementary cobordism, so we can write the whole cobordism S as a composition of elementary cobordisms.

The necessary moves may be subdivided into two types: Reidemeister moves and Morse moves. There is one Reidemeister-type move for each of the ordinary Reidemeister moves, as well as one for each of their inverses. These moves do not change the topology of the surface S_i. The Morse moves correspond to the addition of a 0, 1 or 2-handle to S_i. They are illustrated in Figure 8.3.

8.3.2 Induced maps

Given a cobordism S from L_0 to L_1, we want to assign to it an induced map $\phi_S : Kh'(L_0) \to Kh'(L_1)$ which respects the filtration on Kh'. In addition, we would like this assignment to be functorial, in the sense that if S is the composition of two cobordisms S_1 and S_2, ϕ_S is the composition of ϕ_{S_1} and ϕ_{S_2}. Thus it suffices to consider the case when S is an elementary cobordism.

Suppose that S is an elementary cobordism corresponding to the i-th Reidemeister move or its inverse. Then we define ϕ_S to be ρ'_{i*} or its inverse. By Proposition 8.2, this is a filtered map of degree 0. If S is an elementary cobordism corresponding to a Morse move, then we take ϕ_S to be the map induced by $\psi : CKh'(L_0) \to CKh'(L_1)$, where ψ is the result of applying the TQFT $\mathcal{A}'$ to the corresponding map of cubes. In other words, if the move corresponds to the addition of a 0-handle or a 2-handle, we apply ι' or ϵ', respectively, to the summand at each vertex of the cube. If it corresponds to the addition of a 1-handle, we apply either m' or Δ', depending on whether the move results in a merge or a split at the vertex in question. It is easy to see that ϕ_S is a filtered map of degree 1 for a 0– or 2–handle addition and degree -1 for a 1–handle.

In general, given a cobordism S, we decompose it as a union of elementary cobordisms: $S = S_1 \cup S_2 \ldots \cup S_k$ and define the induced morphism $\phi_S : Kh'(L_0) \to Kh'(L_1)$ to be the composition $\phi_{S_k} \circ \ldots \circ \phi_{S_1}$, which is a filtered map of degree $\chi(S)$. We expect that the map ϕ_S will depend only on the isotopy class of S rel ∂S (*c.f* [Jac], where an analogous result is proved for the Khovanov homology), but since we do not need this fact, we will not pursue it here.

8.3.3   Canonical generators

The maps ϕ_S behave nicely with respect to canonical generators.

Proposition 8.6. *Suppose S is an oriented cobordism from L_0 to L_1 which is weakly connected, in the sense that every component of S has a boundary component in L_0. Then $\phi_S([\mathbf{s}_{o_0}])$ is a nonzero multiple of $[\mathbf{s}_{o_1}]$.*

Remark: Some sort of connectedness hypothesis is clearly necessary for the proposition to hold. For example, if we take S to be the union of a product cobordism and a trivially embedded sphere, the induced map on Kh' is the zero map.

Proof. In fact, we will prove a slightly stronger statement. Suppose i is a regular value for the cobordism S, so that L_i is a link. We divide the components of S_i into two sorts: those of the *first type*, which have a boundary component in L_0, and those of the *second type*, which do not. We say that an orientation o on S_i is *permissible* if it agrees with the orientation of S on components of the first type. (Here and in what follows, we use o_I to denote both a permissible orientation on S_i and the orientation it induces on L_i). We claim that

$$\phi_{S_i}([\mathbf{s}_{o_0}]) = \sum_I a_I [\mathbf{s}_{o_I}]$$

where $\{o_I\}$ runs over the set of permissible orientations on S_i and each coefficient a_I is nonzero. Note that the weak connectivity hypothesis implies that there is only one permissible orientation on S_1, so the proposition is implied by the claim.

To prove the claim, it suffices to check that if it holds for S_i, then it holds for $S_{i'}$ as well, where S_i' is the composition of S_i with a single elementary cobordism S_e. If this cobordism corresponds to a Reidemeister type move, this is a straightforward consequence of Proposition 8.1. Below, we check that it holds for each of the Morse-type moves as well.

0-*Handle Move:* In this case, $\phi_{S_e}(\mathbf{s}_{o_I}) = \mathbf{s}_{o_I} \otimes \frac{1}{2}(\mathbf{a} - \mathbf{b})$, where the second factor in the tensor product refers to the labels on the newly created circle. $S_{i'}$ has a new component of the second type — namely, the disk bounded by the new circle — and $\mathbf{s}_{o_I} \otimes \mathbf{a}$ and $\mathbf{s}_{o_I} \otimes \mathbf{b}$ are the canonical generators corresponding to the two possible orientations on $S_{i'}$ which agree with o_I on all components other than the new one.

1-*Handle Move:* Suppose that the orientation o_I is actually the orientation o_i induced by S_i. Then the two strands involved in the move have opposite orientations, so by Lemma 8.1, they must have the same label. Since

$$m'(\mathbf{a} \otimes \mathbf{a}) = 2\mathbf{a} \qquad\qquad \Delta'(\mathbf{a}) = \mathbf{a} \otimes \mathbf{a}$$
$$m'(\mathbf{b} \otimes \mathbf{b}) = -2\mathbf{b} \qquad\qquad \Delta'(\mathbf{b}) = \mathbf{b} \otimes \mathbf{b}$$

we see that $\phi_{S_e}(\mathbf{s}_{o_i})$ is a nonzero multiple of $\mathbf{s}_{o_{i'}}$.

More generally, the orientation o_I is either compatible with some orientation o_e on S_e, or it is not. In the former case, the two strands involved in the move point in opposite directions and have the same label, and $\phi_{S_e}(\mathbf{s}_{o_I})$ is a nonzero multiple of $\mathbf{s}_{o_I'}$ where o_I' is the orientation induced on $L_{i'}$ by o_e. In the latter case, the two strands point in the same direction and have different labels, so $\phi_{S_e}(\mathbf{s}_{o_I}) = 0$.

Now we consider what happens to the components of S_i during the move. If the move splits one component of L_i into two components of $L_{i'}$, then the number and type of components of S_i remain constant. In this case, the set of permissible orientations on S_i is naturally identified with the set of permissible orientations on $S_{i'}$. There is always an orientation on S_e compatible with o_I, and $\phi_{S_e}(\mathbf{s}_{o_I})$ is a nonzero multiple of $\mathbf{s}_{o_I'}$.

On the other hand, if the move merges two components of L_i into one component of $L_{i'}$, there are several possibilities to consider. If the merge involves only a single component of S_i, the situation is like the one above: there is always an orientation on S_e compatible with o_I, and $\phi_{S_e}(\mathbf{s}_{o_I})$ is a nonzero multiple of $\mathbf{s}_{o_I'}$. The same argument applies when S_e merges two components of S_i, both of which are of the first type.

Finally, suppose the merge joins two components of S_i, at least one of which is of the second type. Then the set of permissible orientations on $S_{i'}$ is only half as large as the set of permissible orientations on S_i. If o_I extends to a permissible orientation o_I' on $S_{i'}$, $\phi_{S_e}(\mathbf{s}_{o_I}) = \mathbf{s}_{o_I'}$, while if it does not, $\phi_{S_e}(\mathbf{s}_{o_I}) = 0$.

2-Handle Move: In this case, a permissible orientation o_I on S_i extends to a unique permissible orientation o_I' on $S_{i'}$. Since $\epsilon'(\mathbf{a}) = \epsilon'(\mathbf{b}) = 1$, $\phi_{S_e}(\mathbf{s}_{o_I}) = \mathbf{s}_{o_I'}$. To prove the claim, it suffices to show that two permissible orientations on S_i' cannot induce the same orientation on $L_{i'}$. But if this were the case, S_i would have a closed component, contradicting the hypothesis that S is weakly connected.

□

Corollary 8.6. *If S is a connected cobordism between knots K_0 and K_1, then ϕ_S is an isomorphism.*

Proof. Fix an orientation o on S. Then $\{\mathbf{s}_{o_0}, \mathbf{s}_{\overline{o}_0}\}$ is a basis for $Kh'(K_1)$. Its image under ϕ_S is $\{k_1 \mathbf{s}_{o_1}, k_2 \mathbf{s}_{\overline{o}_1}\}$ ($k_1, k_2 \neq 0$), which is a basis for $Kh'(K_2)$. □

8.3.4 The slice genus

We can now prove the first two theorems from the introduction.

Proof. (of Theorem 8.1.) Suppose $K \subset S^3$ bounds an oriented surface of genus g in B^4. Then there is an orientable connected cobordism of Euler characteristic $-2g$ between K and the unknot U in $\mathbb{R}^3 \times [0, 1]$. Let $x \in Kh'(K) - \{0\}$ be a class for which $s(x)$ is maximal. Then $\phi_S(x)$ is a nonzero element of $Kh'(U)$.

Now ϕ_S is a filtered map with filtered degree $-2g$, so

$$s(\phi_S(x)) \geq s(x) - 2g.$$

On the other hand, $s_{max}(U) = 1$, so

$$s(\phi_S(x)) \leq 1.$$

It follows that $s(x) \leq 2g + 1$, so $s_{max}(K) \leq 2g + 1$ and $s(K) \leq 2g$. To show that $s(K) \geq -2g$, we apply the same argument to $\overline{K}$ (which bounds a surface $\overline{S}$ of genus g) and use the fact that $s(\overline{K}) = -s(K)$. $\quad\square$

Proof. (of Theorem 8.3.) If K_1 and K_2 are concordant, then $K_1 \# \overline{K_2}$ is slice, so

$$0 = s(K_1 \# \overline{K_2}) = s(K_1) - s(K_2).$$

Thus s gives a well-defined map from $\mathrm{Conc}(S^3)$ to $\mathbb{Z}$. That this map is a homomorphism is immediate from Propositions 8.4 and 8.5. $\quad\square$

Corollary 8.7. *Suppose K_+ and K_- are knots that differ by a single crossing change — from a positive crossing in K_+ to a negative one in K_-. Then*

$$s(K_-) \leq s(K_+) \leq s(K_-) + 1$$

Proof. In [Liv], Livingston shows that this skein inequality holds for any knot invariant satisfying the properties of Theorems 8.1 and 8.3. $\quad\square$

8.4 Computations and relations with other invariants

Although the invariant $s(K)$ is algorithmically computable from a diagram of K, it is impossible to compute by hand for all but the smallest knots. In this section, we describe some techniques which enable us to efficiently compute s.

8.4.1 Using Kh

For many knots, it is a simple matter to compute $s(K)$ from the ordinary Khovanov homology $Kh(K)$. Although $Kh(K)$ is also hard to compute by hand, there are already a number of computer programs available for this purpose, including Bar-Natan's pioneering program [BNk] and a more recent, faster program written by Shumakovitch [Shu1], see also [KnotScape, KnotPlot].

In [BNk], Bar-Natan made the following observation (cf. Conjecture 7.1), based on his computations of Kh for knots with 10 and fewer crossings.

Conjecture 8.1. *(Bar-Natan) The graded Poincare polynomial $P_{Kh}(K)$ of $Kh(K)$ has the form*

$$P_{Kh}(K) = q^{s(K)}(q + q^{-1}) + (1 + tq^4)Q_{Kh}(K)$$

where $Q_{Kh}(K)$ is a polynomial with all positive coefficients.

In [Lee2], Lee showed that this conjecture holds whenever her spectral sequence for Kh' converges after the E_2 term. In this case, it is easy to see that the invariant $s(K)$ is equal to the exponent $s(K)$ which appears in Bar-Natan's conjecture.

To see how widely applicable this condition is, we introduce the notion of the homological *width* of a knot.

Definition 8.3. If K is a knot, let

$$\mu(K) = \{a - 2b \mid q^a t^b \text{ be a monomial in } P_{Kh}(K)\}.$$

The *width* $W(K)$ (cf. Definition 7.8) is one more than the difference between the maximum and minimum elements of $\mu(K)$.

In other words, $W(K)$ is the number of diagonals in the convex hull of the support of $Kh(K)$.

Proposition 8.7. *If $W(K) \leq 3$, then the spectral sequence for $Kh'(K)$ converges after the E_2 term, and Rasmussen's $s(K)$ is the same as Bar-Natan's.*

Proof. Suppose $W(K)$ has width ≤ 3. Then if x is an element of $Kh'(K)$ with q-grading a and homological grading b, the minimum possible q-grading of an element with homological grading $b-1$ is $a-6$. Since the differential d_n on the E_n term of the spectral sequence lowers the q-grading by $4(n-1)$, d_n must be trivial for all $n \geq 3$. □

Theorem 8.4 follows from this fact, since Lee has shown [Lee1] that if K is an alternating knot, then it has width two and Bar-Natan's s is equal to the knot signature $\sigma(K)$.

The proposition also applies to many non-alternating knots. Indeed, using Shumakovitch's tables and a computer, it is straightforward to check that there are only four knots with 13 or fewer crossings whose width is greater than three. Inspecting Kh of these four exceptions, one sees that in each case, the spectral sequence must converge after the E_2 term. Thus for all knots with 13 or fewer crossings, the value of $s(K)$ agrees with the value of Bar-Natan's s tabulated in [BNk] and [Shu1]. Below, we list those knots of 11 crossings or fewer for which $s(K) \neq \sigma(K)$. There are 22 such knots, and $|s(K)| > |\sigma(K)|$ (and thus provides a better bound on the slice genus) for precisely half of them.

K	$s(K)$	$\sigma(K)$	K	$s(K)$	$\sigma(K)$	K	$s(K)$	$\sigma(K)$
9_{42}	0	2	11_{n9}	6	4	11_{n70}	2	4
10_{132}	-2	0	11_{n12}	2	0	11_{n77}	8	6
10_{136}	0	2	11_{n19}	-2	-4	11_{n79}	0	2
10_{139}	8	6	11_{n20}	0	-2	11_{n92}	0	-2
10_{145}	-4	-2	11_{n24}	0	2	11_{n96}	0	2
10_{152}	-8	-6	11_{n31}	4	2	11_{n138}	0	2
10_{154}	6	4	11_{n38}	0	2	11_{n183}	6	4
10_{161}	-6	-4						

Knots with 10 or fewer crossings are labeled according to their numbering in Rolfsen, while those with 11 crossings use the *Knotscape* [KnotScape] numbering. The values of the signature are taken from [BNk]. All of the knots in the table have a homological width of 3, which raises the following question: if K has homological width 2 (*i.e.* is H-thin in the terminology of [Kho3]), must $s(K) = \sigma(K)$?

8.4.2 Positive knots

If K is a positive knot, $s(K)$ can be computed directly from the definition. To see this, consider a canonical generator $\mathbf{s}_o$ for a positive diagram of K. Since each crossing of K is positive, its oriented resolution is the 0-resolution. Thus the state $\mathbf{s}_o$ lives in the extreme corner of the cube of resolutions: it has homological grading 0, and there are no generators in $CKh'(K)$ with homological grading -1. It follows that the only class homologous to $\mathbf{s}_o$ is $\mathbf{s}_o$ itself, so

$$s_{min}(K) = s([\mathbf{s}_o]) = q(\mathbf{s}_o)$$

To compute $q(\mathbf{s}_o)$, we change back to the basis $\{\mathbf{v}_-, \mathbf{v}_+\}$. In the expansion of $\mathbf{s}_o$ with respect to this basis, there is a unique state with minimal q-grading, namely, the state in which every circle of the oriented resolution is labeled with a $\mathbf{v}_-$. If the positive diagram of K has n crossings, and its oriented resolution has k circles, then

$$q(\mathbf{s}_o) = p(\mathbf{s}_o) + gr(\mathbf{s}_o) + n_+ - n_-$$
$$= -k + 0 + n - 0$$

so

$$s(K) = -k + n + 1$$

On the other hand, Seifert's algorithm gives a Seifert surface S for K with Euler characteristic $k - n$, so

$$2g(K) \leq 2g(S) = n - k + 1 = s(K) \leq 2g_*(K)$$

Since $g_*(K) \leq g(K)$, the inequalities above must all be equalities. This completes the proof of Theorem 8.5.

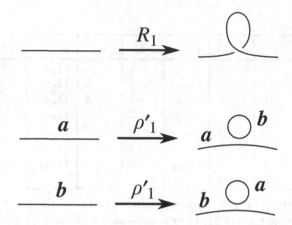

FIGURE 8.4: The Reidemeister I move and the map ρ_1'.

8.5 Reidemeister moves

In this section, we prove the results involving Reidemeister moves which were stated earlier.

Proof. (of Theorem 8.6.) The proof that the desired spectral sequence exists was sketched before. To prove its invariance, we use the following basic lemma, whose proof may be found in [McC], Proposition 3.2.

Lemma 8.5. *Suppose $F \colon C_1 \to C_2$ is a map of filtered complexes which respects the filtrations. Then F induces maps of spectral sequences $F_n \colon E_n^1 \to E_n^2$, and if F_n is an isomorphism, F_m is an isomorphism for all $m \geq n$.*

In Section 4 of [Lee2], Lee proves the invariance of Kh' by checking its invariance under the three Reidemeister moves. For each move, she exhibits a chain map between the complexes associated to the link diagram before and after the move. To prove the theorem, it suffices to check that these maps respect the q-filtration, and that they induce isomorphisms on the E_2 terms. The latter claim is straightforward, since in each case the induced maps on the E_1 terms are identical to the maps used in Section 5 of [Kho1] to prove invariance of Kh. Below, we sketch the proof of invariance for each move and explain why the maps in question respect the filtrations. For full details, we refer the reader to [Kho1] and [Lee2].

Reidemeister I Move: Let $\tilde{L}$ be the diagram L with an additional left-hand curl added in. Then $CKh'(\tilde{L})$ can be decomposed as a direct sum $X_1 \oplus X_2$, where X_2 is acyclic and X_1 is isomorphic to $CKh'(L)$ via the map $\rho_1' \colon CKh'(L) \to$

X_1 illustrated in Figure 8.4. In terms of the basis $\{\mathbf{v}_\pm\}$, we have

$$\rho_1'(\mathbf{v}_-) = \mathbf{v}_- \otimes \mathbf{v}_- - \mathbf{v}_+ \otimes \mathbf{v}_+$$
$$\rho_1'(\mathbf{v}_+) = \mathbf{v}_+ \otimes \mathbf{v}_- - \mathbf{v}_- \otimes \mathbf{v}_+$$

The corresponding map ρ_1 in [Kho1] is given by

$$\rho_1(\mathbf{v}_-) = \mathbf{v}_- \otimes \mathbf{v}_-$$
$$\rho_1(\mathbf{v}_+) = \mathbf{v}_+ \otimes \mathbf{v}_- - \mathbf{v}_- \otimes \mathbf{v}_+$$

so ρ_1' is filtration non-decreasing, and its induced map on E_1 terms is ρ_1.

Remark: There is another version of the first Reidemeister move, corresponding to the addition of a right-hand curl. Although it is not difficult to define an appropriate map ρ_1', for this move directly, for the sake of brevity we adopt the solution of [BNk] and [Lee2] and define it to be the composition of maps induced by an appropriate Reidemeister II move followed by a Reidemeister I move.

Reidemeister II Move: Let L and $\tilde{L}$ be as shown in Figure 8.5. In this case, $CKh'(\tilde{L})$ can be decomposed as a direct sum $X_1 \oplus X_2 \oplus X_3$, where X_2 and X_3 are acyclic and there is an isomorphism $\rho_2' : CKh'(L) \to X_1$, which is given by

$$\rho_2'(z) = (-1)^{\mathrm{gr}(z)}(z + \iota(d'_{01 \to 11}(z)))$$

The maps ι and $d'_{01 \to 11}$ are shown in the figure. The isomorphism ρ_2 in [Kho1] has the same form, but with $d_{01 \to 11}$ in place of $d'_{01 \to 11}$. Since $d - d'$ is strictly filtration increasing, it follows that ρ_2' is filtration non-decreasing, and its induced map on E_1 terms is ρ_2.

Reidemeister III Move: Let L and $\tilde{L}$ be as shown in Figure 8.6. Then there are direct sum decompositions

$$CKh'(L) \cong X_1 \oplus X_2 \oplus X_3$$
$$CKh'(\tilde{L}) \cong \tilde{X}_1 \oplus \tilde{X}_2 \oplus \tilde{X}_3$$

where $X_2, X_3, \tilde{X}_2,$ and $\tilde{X}_3$ are acyclic and there is an isomorphism $\rho_3' \colon X_1 \to \tilde{X}_1$. To describe X_1 and $\tilde{X}_1$, we first define maps

$$\beta' \colon CKh'(L_{100}) \to CKh'(L_{010}))$$
$$\tilde{\beta}' \colon CKh'(\tilde{L}_{010})) \to CKh'(\tilde{L}_{100})$$

by

$$\beta' = \iota \circ d'_{100 \to 110}$$
$$\tilde{\beta}' = \iota \circ d'_{010 \to 110}$$

Then

$$X_1 = \{x + \beta'(x) + y \mid x \in CKh'(L(*100)), y \in CKh'(L(*1))\}$$
$$\tilde{X}_1 = \{x + \tilde{\beta}'(x) + y \mid x \in CKh'(\tilde{L}(*010)), y \in CKh'(\tilde{L}(*1))\}$$

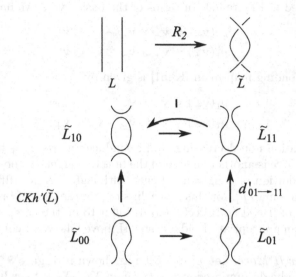

FIGURE 8.5: The Reidemeister II move and the maps ι and $d'_{01\to11}$.

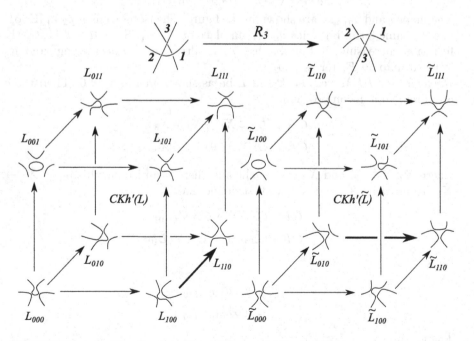

FIGURE 8.6: The Reidemeister III move. The relevant components of the differentials ($d'_{100\to110}$ and $d'_{010\to110}$) are marked in bold.

and

$$\rho_3'(x + \beta'(x) + y) = x + \tilde{\beta}'(x) + y.$$

The isomorphism ρ_3 in [Kho1] is defined similarly, except that it uses d instead of d' to define maps β and β'. Since d' does not increase the q-grading, we clearly have $q(\beta'(x)) \geq q(x)$. From this, it follows that ρ_3' does not decrease the q-grading. Since $d - d'$ strictly increases the q-grading, the map induced on E_1 terms by ρ_3' is equal to ρ_3. To finish the proof, we apply Lemma 8.5 three times: first to the inclusions $X_1 \hookrightarrow CKh'(L)$ and $\tilde{X}_1 \hookrightarrow CKh'(\tilde{L})$, and then to the map ρ_3'.

$\square$

Proof. (of Proposition 8.1.) We check the claim directly for each Reidemeister move:

Reidemeister I Move: In this case, it is easy to see that $\rho_1'(s_o) = s_{\tilde{o}}$.

Reidemeister II Move: Suppose that the two strands in L point in the same direction. Then by Lemma 8.1, they have different labels, so $d'_{01 \to 11}(s_o) = 0$. The oriented resolution of $\tilde{L}$ is contained in $CKh'(\tilde{L}(*01)) \cong CKh'(L)$, so $\rho_2'(s_o) = (-1)^0(s_{\tilde{o}}) = s_{\tilde{o}}$.

Now suppose the two strands point in different directions, so that they have the same label. Let us assume for the moment that this label is **a**. Then we define $s_{\tilde{ij}} \in Kh'(\tilde{L}_{ij})$ to be the state which is identical to s_o outside the area where the move takes place and has all components inside the area of the move labeled with an **a**. Then a direct computation shows that either

$$\rho_2'(s_o) = s_{\widetilde{01}} + \frac{1}{2}(s_{\widetilde{10}} - s_{\tilde{o}})$$
$$= -\frac{1}{2}(s_{\tilde{o}} + d'(s_{\widetilde{00}}))$$

if the two strands belong to the same component, or

$$\rho_2'(s_o) = s_{\widetilde{01}} + (s_{\widetilde{10}} - s_{\tilde{o}})$$
$$= -(s_{\tilde{o}} + d'(s_{\widetilde{00}}))$$

if they belong to different components. This proves the claim in the case where both strands are labeled with an **a**. We leave it to the reader to check that a similar argument applies when they are both labeled with a **b**.

Reidemeister III Move: Here there are three cases to consider. First, suppose that the two overlying strands in L are oriented as shown in Figure 8.7a. Then $s_o \in CKh'(L_1)$, and it is easy to see that $\rho_3'(s_o) = s_{\tilde{o}}$.

Next, suppose that the three strands are oriented as shown in Figure 8.7b. Then $s_o \in CKh'(L_{100})$ and $s_{\tilde{o}} \in CKh'(\tilde{L}_{010})$. Clearly $\beta'(s_o) = \tilde{\beta}'(s_{\tilde{o}}) = 0$, so $s_o \in X_1$ and $s_{\tilde{o}} \in \tilde{X}_1$. Again, it follows that $\rho_3'(s_o) = s_{\tilde{o}}$.

Finally, suppose the strands are oriented as shown in Figure 8.7c. In this case, the oriented resolution of L is in L_{010}, and the oriented resolution of $\tilde{L}$ is in $\tilde{L}_{100}$. Inside the region under consideration, s_o looks like the state of

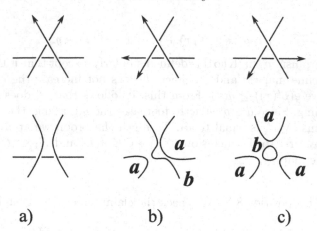

a)   b)   c)

FIGURE 8.7: Possible orientations for L and their respective canonical generators.

Figure 8.7c (perhaps with a's and b's inversed.) Our first step is to exhibit some $t \in X_1$ which is homologous to s_o. As before, we denote by s_{ijk} the unique state of L_{ijk} which is the same as s_o outside the area of the Reidemeister move and has all its components inside this area labeled by a's.

Assume for the moment that all three strands shown in L_{000} belong to different components. In this case, we can take

$$t = s_o - 2s_{100} - s_{010} - 2s_{001} = s_o - d'(s_{000}).$$

Indeed, $\beta'(-2s_{100}) = s_o - s_{010}$ and $s_{001} \in CKh'(L_1)$, so $t \in X_1$. Then

$$\rho_3'(t) = -2s_{\overline{010}} - 2\tilde{\beta}'(s_{\overline{010}}) - 2s_{\overline{001}}$$
$$= -2s_{\overline{010}} - 2s_{\overline{100}} + 2s_{\overline{o}} - 2s_{\overline{001}}$$
$$= 2s_{\overline{o}} - d'(s_{\overline{000}})$$

which proves the claim.

We leave it to the reader to check that a similar argument applies to each of the four other ways in which the segments outside the area of the move can be connected, as well as when the roles of a and b are inversed. In each case, it is not difficult to verify that $\rho_{3*}'([s_o])$ is one of $\pm[s_{\overline{o}}], \pm 2[s_{\overline{o}}]$, or $\pm\frac{1}{2}[s_{\overline{o}}]$.

$\square$

Proof. (of Proposition 8.2.) In the case of ρ_{1*}' and ρ_{2*}', the claim is immediate, since these maps are induced by filtered chain maps. For the others, we use the following

Lemma 8.6. *Suppose $f : C_1 \to C_2$ is a map of filtered chain complexes with the property that the induced map of spectral sequences $f_2 : E_1^2 \to E_2^2$ is an*

isomorphism. Then f_*^{-1} is a filtered map with respect to the induced filtrations on $H_*(C_1)$ and $H_*(C_2)$.

Proof. Since f_2 is an isomorphism, f_∞ (the induced map on the limiting term) is as well. It follows that f_* is an isomorphism. Suppose f_*^{-1} does not respect the filtration. Then there must be some $\mathbf{x} \in H_*(C_1)$ whose filtration is strictly increased by f_*. But this contradicts the fact that f_∞ is an isomorphism. $\square$

The remaining cases now follow easily from the results used in the proof of Theorem 8.6. Indeed, ρ_1' and ρ_2' both induce isomorphisms of E_2 terms, and $\rho_{3*}' = \iota_{1*} \circ \psi_* \circ \iota_{2*}^{-1}$, where ι_1, ι_2, and ψ all induce isomorphisms of E_2 terms. $\square$

Part II

Theory of braids

Part II

Theory of braids

Chapter 9

Braids, links and representations of braid groups

9.1 Four definitions of the braid group

In the previous chapters, we considered only one way of encoding links, namely, link planar diagrams.[1] In the present chapter, we are going to give an introduction to the theory of braids. On one hand, this theory gives us another point of view to knot theory. On the other hand, the theory of braids has some nice intrinsically interesting properties which are worth studying. Namely, the braid groups can be defined in many ways that lead to connections with different theories. Below, we are going to give some definitions of the braid groups and to discuss some of their properties.

9.1.1 Geometrical definition

Consider the lines $\{y = 0, z = 1\}$ and $\{y = 0, z = 0\}$ in $\mathbb{R}^3$ and choose m points on each of these lines having abscissas $1, \ldots m$.

Definition 9.1. An m–*strand braid* is a set of m non-intersecting smooth paths connecting the chosen points on the first line with the points on the second line (in arbitrary order), such that the projection of each of these paths to Oz represents a diffeomorphism.

These smooth paths are called *strands* of the braid.

An example of a braid is shown in Fig. 9.1.

It is natural to consider braids up to isotopy in $\mathbb{R}^3$.

Definition 9.2. Two braids B_0 and B_1 are *equal* if they are *isotopic*; i.e., if there exists a continuous family of braids $B_t, \{t \in \{0, 1\}\}$ of braids starting at B_0 and finishing at B_1.

Definition 9.3. The set of all m–strand braids generates a group. The operation in this group is just juxtaposing one braid under the other and rescaling the z–coordinate.

---

[1] Later, we shall also use the d–diagrams mentioned in the Introduction

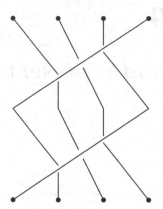

FIGURE 9.1: A braid

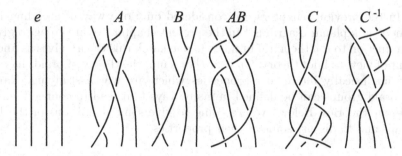

FIGURE 9.2: Unity. Operations in the braid group

The *unit element* or the *unity* of this group is the braid represented by all vertical parallel strands. The inverse element for a given braid is just its mirror image; see Fig. 9.2.

Exercise 9.1. *Check that the group structure on the set of braids is well defined.*

Definition 9.4. The *Artin m–strand braid group*[2] is the group of braids with the operation defined above.
  Notation: $Br(m)$.

One can consider only braids whose strands connect points with equal abscissas.

Definition 9.5. These braids are said to be *pure*. Pure braids form a subgroup of the braid group.
  Notation: $PB(m)$.

---

[2]In fact, there are other braid groups called Brieskorn braid groups. They are closely connected with Coxeter–Dynkin diagrams and symmetries. For more details see [Bri1, Bri2]

9.1.2 Topological definition

Definition 9.6. Given a topological space X, the *unordered m-configuration space for X* is the space (endowed with the natural topology) of all unordered sets of m pairwise different points of X.

Notation: $B(X, m)$.

Analogously, one can define the *m-ordered configuration space*.

Notation: $F(X, m)$.

Now, let $X = \mathbb{R}^2 = \mathbb{C}^1$.

Definition 9.7. The *m-strand braid group* is defined to be isomorphic to the fundamental group $\pi_1(B(X, m))$.

Definition 9.8. The group $\pi_1(F(X, m))$ is called *the pure m-strand braid group*.

9.1.3 Algebro–geometrical definition

Consider the set of all polynomials of degree m in one complex variable z with leading coefficient equal to one.

Obviously, this set (together with its intrinsic topological structure) is isomorphic to $\mathbb{C}^n$: its coefficients can be considered as its complex coordinates.

Now, delete the space Σ_m of all polynomials that have multiple roots (at least one). We obtain the set $\mathbb{C}^m \backslash \Sigma_m$.

Definition 9.9. The *m-strand braid group* is the group $\pi_1(\mathbb{C}^m \backslash \Sigma_m)$.

9.1.4 Algebraic definition

Definition 9.10. The *m-strand braid group* is the group given by the presentation with $(m-1)$ generators $\sigma_1, \ldots, \sigma_{m-1}$ and the following relations

$$\sigma_i \sigma_j = \sigma_j \sigma_i$$

for $|i - j| \geq 2$ and

$$\sigma_i \sigma_{i+1} \sigma_i = \sigma_{i+1} \sigma_i \sigma_{i+1}$$

for $1 \leq i \leq m - 2$.

These relations are called *Artin's relations*.

Definition 9.11. Words in the alphabet of σ's and σ^{-1}'s will be referred to as *braid words*.

9.1.5 Equivalence of the four definitions

Theorem 9.1. *The four definitions of the braid group $Br(m)$ given above are equivalent.*

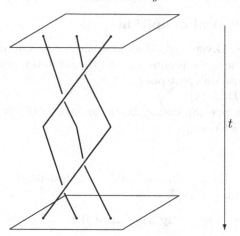

FIGURE 9.3: A braid in 3–space

Proof. The easiest part of the proof is to establish the equivalence of the topological and algebro–geometric definitions. Indeed, it is obvious that the two spaces, the space of polynomials of degree m without multiple roots with leading coefficient one and the unordered m–configuration space for $\mathbb{C}^1$, are homeomorphic. Thus, their fundamental groups are isomorphic.

Let us now show the equivalence of the geometrical and topological definitions. As we know, the fundamental group does not depend on the choice of the base point in the connected space. Thus, the base point A of the unordered m–configuration space can be chosen as the set of integer points $(1, 2, \ldots, m)$. Consider the space $\mathbb{R}^3$ as the product $\mathbb{C}^1 \times \mathbb{R}^1$.

With each closed loop, outgoing from A and lying in $B(\mathbb{C}^1, m)$, let us associate a set of lines in $\mathbb{R}^3$ as follows. Each of these (curvilinear) lines represents the motion of a point on the complex line $\mathbb{C}^1$ with respect to the time t, where t is the real coordinate; see Fig. 9.3.

Thus, with each topological braid we have uniquely associated a geometric braid. Obviously, with two homotopic (equal) topological braids we associate the same geometric braids.

So, it remains to show the equivalence of geometric and algebraic notions. In order to do this, let us introduce the notion of the *planar braid diagram*, analogous to the planar link diagram.

To see what this is, let us project a braid on the plane Oxz.

In the general case we obtain a diagram that can be described as follows.

Definition 9.12. A *braid planar diagram* (for the case of m strands) is a graph lying inside the rectangle $[1, m] \times [0, 1]$ endowed with the following structure and having the following properties:

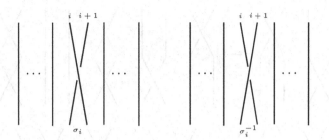

FIGURE 9.4: Generators of the braid group

1. Points $(i, 0)$ and $(i, 1), i = 1, \ldots, m$, are vertices of valency one; the other points of type $(x, 0)$ and $(x, 1)$ are not graph vertices.

2. All other graph vertices (crossings) have valency four; opposite edges at such vertices make angles π.

3. Unicursal curves; i.e., lines consisting of edges of the graph, passing from an edge to the opposite one, go from vertices with ordinate one and come to vertices with ordinate zero; they must be descending.

4. Each vertex of valency four is endowed with an over– and undercrossing structure.

Analogously to the planar isotopy of link diagrams, one defines the planar isotopy of braid diagrams.

Obviously, all isotopy classes of geometrical braids can be represented by their planar diagrams. Moreover, after a small perturbation, all crossings of the braid can be set to have different ordinates.

It is easy to see that each element of the geometrical braid group can be decomposed into a product of the following generators σ_i's: the element σ_i for $i = 1, \ldots, m - 1$ consists of $m - 2$ segments connecting $(k, 1)$ and $(k, 0), k \neq i, k \neq i + 1$, and two segments $(i, 0) - (i + 1, 1)$, $(i + 1, 0) - (i, 1)$, where the latter goes over the first one; see Fig. 9.4.

Different braid diagrams can generate the same braid. Thus we obtain some relations in $\sigma_1, \ldots, \sigma_m$.

Let us suppose that we have two equal geometrical braids B_1 and B_2. Let us represent the process of their isotopy in terms of their planar diagrams. Each interval of this isotopy either does not change the disposition of their vertex ordinates, or in this interval at least two crossings have (in a moment) the same ordinate; in the latter case the diagram becomes irregular.

We are interested in those moments where the algebraic description of our braid changes. We see that there are only three possible cases (all others can be reduced to these ones). In the first case (see Fig. 9.5.a) just one couple of crossings has the same ordinate. In the second case (see Fig. 9.5.b), two

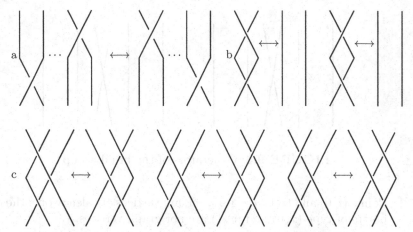

FIGURE 9.5: Intuitive expression of braid diagram isotopies

strands are tangent. In the third case (Fig. 9.5.c) we have a triple intersection point.

It is easy to see that the first case gives us the relation $\sigma_i \sigma_j = \sigma_j \sigma_i, |i-j| \geq 2$ (this relation is called *far commutativity*) or an equivalent relation $\sigma_i^{\pm 1} \sigma_j^{\pm 1} = \sigma_j^{\pm 1} \sigma_i^{\pm 1}, |i-j| \geq 2$, in the second case we get $aa^{-1} = 1$ (or $a^{-1}a = 1$), and in the third case we obtain one of the following three relations:

$$\sigma_i \sigma_{i+1} \sigma_i = \sigma_{i+1} \sigma_i \sigma_{i+1},$$

$$\sigma_i \sigma_{i+1} \sigma_i^{-1} = \sigma_{i+1}^{-1} \sigma_i \sigma_{i+1}, \quad \sigma_i^{-1} \sigma_{i+1} \sigma_i = \sigma_{i+1} \sigma_i \sigma_{i+1}^{-1}.$$

Obviously, each of the latter two relations can be reduced from the first one. This simple observation is left to the reader as an exercise. This completes the proof of the theorem.   □

In the m–strand braid group one can naturally define the subgroup $PB(m)$ of pure braids.

Exercise 9.2. *Show that $PB(m)$ is a normal subgroup in $Br(m)$, and the quotient group $Br(m)/PB(m)$ is isomorphic to the permutation group $S(m)$.*

9.1.6   The stable braid group

For natural numbers $m < n$, there exists the natural embedding $Br(m) \subset Br(n)$: a braid from $Br(m)$ can be treated as a braid from $Br(n)$ where the last $(n-m)$ strands are vertical and unlinked (separated) with the others.

Definition 9.13. The *stable braid group* Br is the limit of groups $Br(n)$ as $n \to \infty$ with respect to these embeddings.

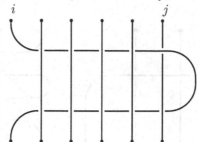

FIGURE 9.6: Generator b_{ij} of the pure braid group

The group Br has a presentation with generators $\sigma_1, \sigma_2, \ldots$, and the following relations $\sigma_i \sigma_j = \sigma_j \sigma_i$ for $|i - j| \geq 2, \sigma_i \sigma_{i+1} \sigma_i = \sigma_{i+1} \sigma_i \sigma_{i+1}$.

9.1.7 Pure braids

With each braid one can associate its permutation: this permutation takes an element k to m if the strand starting with the k–th upper point ends at the m–th lower point.

Definition 9.14. A braid is said to be *pure* (cf. Subsection 9.8) if its permutation is identical. Obviously, pure braids generate a subgroup $PB_n \subset B_n$.

There are other interpretations of PB_n. For instance, instead of the configuration space of unordered points of $\mathbb{R}^2$, one can consider the configuration space of ordered points.

The fundamental group of this space is obviously isomorphic to PB_n.

An interesting problem is to find an explicit finite presentation of the pure braid group on n strands.

Here we shall present some concrete generators (according to [Art2]). A presentation of this group can be found in e.g. [Maka].

Pure n–strand braids correspond to loops in the space of **ordered** point sets on the plane. They generate a finite–index subgroup in the braid group.

There exists an algebraic *Reidemeister–Schreier method* that allows us to construct a presentation of a finite–index subgroup having a presentation of a finitely defined group, see e.g [CF].

Here we give some generators of the pure braid group [Art1].

The following theorem holds.

Theorem 9.2. *The group $PB(m)$ is generated by braids*

$$b_{ij}, 1 \leq i < j \leq n \tag{9.1}$$

(see Fig. 9.6).

To prove the theorem, we shall use induction on the number n of strands.

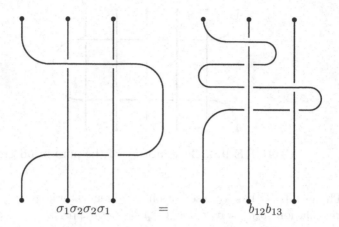

$$\sigma_1\sigma_2\sigma_2\sigma_1 \qquad = \qquad b_{12}b_{13}$$

FIGURE 9.7: Decomposing a pure braid

For $n = 2$ the statement is obvious: each 2–strand pure braid is some power of the braid $b_{12} = \sigma_1^2$.

Assume that the statement is proved for some n. Consider some pure $(n + 1)$–strand braid d_{n+1}. If we delete the first strand of it, we obtain some pure n–strand braid a_n. Now we can write $d_{n+1} = (d_{n+1}a_n^{-1})a_n$. By the induction hypothesis, the braid a_n can be decomposed in generators (9.1). The last n strands of $d_{n+1}a_n^{-1}$ are unlinked. Let us straighten them; i.e., make them vertical. In this case, the first strand is braided around them. Now, it is easy to see that this braid can be represented as a product of b_{1j}, b_{1j}^{-1}. This can be done as follows: every time when the first strand goes *under* some strand, it must be pulled back under all strands until the left margin, and after that returned to the previous place under all strands; see Fig. 9.7.

Thus, the product d_{n+1} can be decomposed in b_{ij}, b_{ij}^{-1}.

9.1.7.1   Pure braid groups and mapping classes

Now, let us give another description of the pure braid group (see, e.g., [PS]).

Denote by H_n the group of isotopy classes of homeomorphisms of the n–punctured disc on itself, identical on the boundary. It turns out that the pure braid group is closely connected with H_n. First, let us consider H_0.

Theorem 9.3 (Alexander's theorem on homeomorphism). *The group H_0 is isomorphic to the unity group; i.e., each homeomorphism of the disc that is identical on the boundary is homotopic to the identity map; moreover, such a homotopy can be found among those identical on the boundary.*

Proof. Consider the disc $|z| \leq 1$ in the complex space $\mathbb{C}^1$. Let h_0 be the

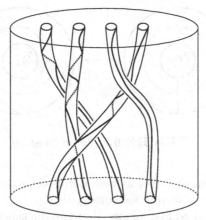

FIGURE 9.8: Thickened braid

identity map and h_1 be an arbitrary homeomorphism of the disc onto itself that is identical on the boundary. According to a known theorem, h_1 has a fixed point inside the circle. Without loss of generality, let us assume that this point coincides with the centre of the circle. Now, let us construct our homotopy. For $t \in (0,1)$, let us construct the homeomorphism h_t as follows: h_t is identical inside the ring $t \le |z| \le 1$. Inside the disc $|z| \le t$ we decree $h_t(z) = t(h_1(\frac{z}{t}))$. Obviously, h_t is a homotopy that satisfies the condition of the theorem. □

Let us consider now the group H_n. Let $g \in H_n$ be a homeomorphism of the n–punctured disc that is identical on the boundary of the punctured disc; i.e., on the boundary of the disc and on the boundaries of the holes. This homeomorphism can be extended to the homeomorphism of the entire disc, since g can be extended from the boundary of any hole to the hole itself (this can be done, say, by mapping the interior of the hole identically). Denote the obtained homeomorphism by h_1.

According to the previous theorem, there exists an isotopy h_t, connecting h_1 with $h_0 = id$. Fix the points $x_1, \ldots, x_n$ inside the holes. For each $i = 1, \ldots, n$ the set $(t, h_t(x_i)), 0 \le t \le 1$ is an arc connecting the points on the upper and lower base of the cylinder $I \times D$, where I is the interval of time and D is the disc. In Fig. 9.8 the spurs of fixed points are shown.

Thus we have obtained a pure braid. This braid "knows" a lot about h, but not all. Actually, during the isotopy one can watch the moving of the fixed point of the circle. The circle turns around and its spur represents a cylinder. To describe h up to isotopy, it is sufficient to consider all cylinders and to mark the spurs of fixed points on their boundaries. The pure braid with this additional information is called a *thickened braid*.

Obviously, to each homeomorphism of the disc with n punctures that is identical on the boundary there corresponds some thickened braid.

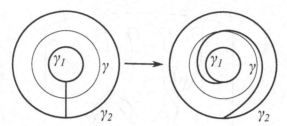

FIGURE 9.9: Dehn twisting

The inverse statement is true as well. Let us show that with each thickened braid one can associate such a homeomorphism.

Without loss of generality, assume that the thickened braid does not leave the cylinder. The bases of the cylinder are discs with n circular holes.

Let us lower the circle with holes in such a way that the points of the interior boundary move vertically downwards parallel to the axis of the cylinder, and the disc always stays planar. The boundaries of these holes move downwards and twist following the point of the strand. When the circle reaches the lower base, we obtain the required homeomorphism.

Let M^2 be an orientable 2–manifold (with or without boundary) and γ be a closed curve lying inside M^2. Consider a small neighbourhood U of the curve γ that is homeomorphic to $S^1 \times [0, 1]$. Let γ_1 and γ_2 be the boundaries of this neighbourhood.

Definition 9.15. The *Dehn twisting* of a 2–surface M^2 along a curve γ is the homeomorphism of M^2 onto itself, which is constant outside U that is represented by a full–turn twist of the curve γ_2 with curve γ_1 fixed inside U. It is shown in Fig. 9.9 how the Dehn twisting acts on curves connecting a point from γ_1 with a point from γ_2. In this figure, we show the image of the straight line connecting two circles.

A typical example of Dehn's twisting is the homeomorphism of a torus generated by twisting along the meridian; see Fig. 9.10.

Remark 9.1. *The group of thickened braids is an extension of the pure braid group PB_n. Moreover, it is easy to check that H_n is the direct sum of PB_n and $\mathbb{Z}^n$ (each group $\mathbb{Z}$ corresponds to twistings along boundaries of holes).*

Remark 9.2. *Actually, the braid group can be considered as the mapping class group of the punctured disc. Namely, we consider all homeomorphisms of the punctured disc P_n onto itself which are identical on the boundary and then factorise these homeomorphismsms by homeomorphisms isotopic to the identity.*

This approach is discussed in the book [Bir1]. This allows us to construct various representations of braid groups, and to solve some problems.

To clarify the situation completely, we have to prove the following theorem.

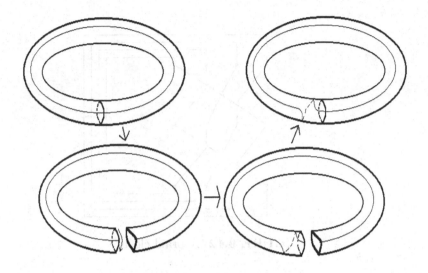

FIGURE 9.10: Twisting a torus along the meridian

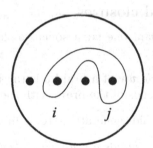

FIGURE 9.11: Twisting curve for the generator b_{ij}

Theorem 9.4. *The group H_n is generated by twistings along a finite number of closed curves in the circle.*

Proof. As it was proved before, the group H_n is isomorphic to the group of thickened braids. Suppose the homeomorphism $h \in H_n$ corresponds to thickened braid α'_n which is the sum $\alpha_n + \alpha$, where $\alpha_n \in PB_n$, and $a \in \mathbb{Z}^n$. The pure braid α_n can be represented in generators b_{ij} that correspond to twisting along curves going once around points i and j; see Fig. 9.11.

The thickened braid α is generated by the full–turn Dehn twisting along the curve going around the point a.

Summarising the facts described above, we obtain the claim of the theorem.

□

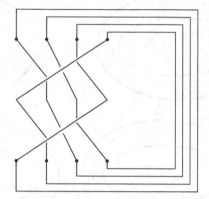

FIGURE 9.12: A braid closure

9.2   Links as braid closures

With each braid diagram, one can associate a planar knot (or link) diagram as follows.

Definition 9.16. The *closure* of a braid b is the link $Cl(b)$ obtained from b by connecting the lower ends of the braid with the upper ends; see Fig. 9.12.

Obviously, isotopic braids generate isotopic links.

Remark 9.3. *Closures of braids are usually taken to be oriented: all strands of the braid are oriented from the top to the bottom.*

Some links generate knots; the others generate links. In order to calculate the number of components of the corresponding link, one should take into account the following simple observation. In fact, there exists a simple natural epimorphism from the braid group onto the permutation group $\Sigma : Br(n) \to S_n$, defined by $\sigma_i \to s_i$, where s_i are natural generators of the permutation group.

Consider a braid B. Obviously, for all numbers p belonging to the same orbit of the natural permutation action (of $\Sigma(B)$) on the set $1, \ldots, n$, all upper vertices with abscissas $(p, 0)$ belong to the same link component.

Consequently, we obtain the following proposition.

Proposition 9.1. *The number of link components of the link of the closure $Cl(B)$ equals the number of orbits of action for $\Sigma(B)$.*

Exercise 9.3. *Construct braids whose closures represent both trefoils and the figure eight knot.*

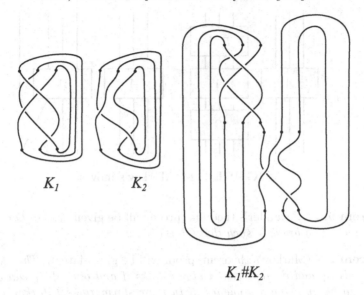

$$K_1 \qquad\qquad K_2$$

$$K_1 \# K_2$$

FIGURE 9.13: Representing the connected sum of braids

Obviously, non–isotopic braids might generate isotopic links. We will touch on this question later.

An interesting question is to define the minimal number of strands of a braid whose closure represents the given link isotopy class L. Denote this number by $Braid(L)$.

An interesting theorem on this theme belongs to Birman and Menasco.

Theorem 9.5. *For any knots K_1 and K_2, the following equation holds:*

$$Braid(K_1 \# K_2) = Braid(K_1) + Braid(K_2) - 1.$$

In Fig. 9.13 we show that if the knot K_1 can be represented by an n–strand braid, and K_2 can be represented by an m–strand braid, then $K_1 \# K_2$ can be represented by an $(n + m - 1)$-strand braid. This proves the inequality "$\leq$".

A systematic study of links via braid closures (including Markov's and Alexander's theorems, which will be discussed later) was done in the series of works by Birman [Bir3] and Birman and Menasco [BM1],[BM2],[BM3],[BM4],[BM5].

9.3   Braids and the Jones polynomial

First, let us formulate the celebrated Alexander and Markov theorems that we will use in this section.

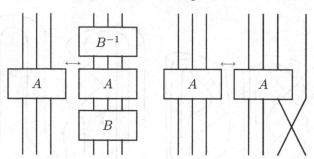

FIGURE 9.14: Markov's moves

Theorem 9.6 (Alexander's theorem, proof will be given later). *For each link L, there exists a braid B such that $Cl(B) = L$.*

Theorem 9.7 (Markov's theorem, proof will be given later). *The closures of two braids β_1 and β_2 represent isotopic links if and only if β_1 can be transformed to β_2 by using a sequence of two transformations (Markov's moves), shown in Fig. 9.14 (on the right, both types of the additional crossing are admissible).*

The main idea for constructing the Jones two–variable polynomial is the following. One can consider some functions looking like representations of braid groups and investigate their properties. It turns out that some of these functions (so–called Ocneanu's trace) have a good behaviour under Markov's moves. This idea was given to Jones by Joan Birman and led to the beautiful discovery of the Jones polynomial.

First, let us note that if we add the relations $\sigma_i^2 = 1$ to the standard presentation of the braid group B_n, we obtain the permutation group S_n.

Definition 9.17. The *Hecke algebra* $H(q, n)$ is the algebra generated by the following presentation:

$$\langle g_1, \ldots, g_{n-1} | g_i^2 = (q-1)g_i + q,$$

$$i = 1, \ldots, n-1, g_i g_{i+1} g_i = g_{i+1} g_i g_{i+1}, i = 1, 2, \ldots, n-2,$$

$$g_i g_j = g_j g_i, |i - j| \geq 2\rangle. \tag{9.2}$$

Remark 9.4. *Jones uses different (inverse) notation for the braid generators; namely, $\sigma_i = $ ⤫ and $\sigma_i^{-1} = $ ⤬.*

Let us now formulate the theorem on Ocneanu's trace.

Theorem 9.8 (Ocneanu, [HOMFLY]). *For each $z \in \mathbb{C}$ there exists a linear trace tr (that can be treated as a function in z, q) on $H(q, n)$ uniquely defined by the following axioms:*

1.

$$tr(ab) = tr(ba);$$

2.

$$tr(1) = 1;$$

3.

$$tr(xg_n) = ztr(x)$$

for any

$$x \in H(q, n).$$

Below, we present the proof of Jones [Jon1]. This proof differs slightly from the original Ocneanu proof and leads to the construction of the Jones polynomial.

Proof. The main idea of the proof is the following. If we add the new relations $g_i^2 = e$ to the braid group (with generators g_i instead of σ_i), we get exactly the permutation group. This group is finite and quite pleasant to work with.

Namely, all elements of the permutation group with generators $p_1, \ldots, p_{n-1}$, where p_i permutes the i-th and $(i+1)$-th elements, have a unique representation of the form:

$$\{(p_{i_1} p_{i_1-1} \cdots p_{i_1-k_1})(p_{i_2} \cdots p_{i_2-k_2}) \cdots (p_{i_j} \cdots p_{i_j-k_j})\}$$

for some

$$1 \leq i_1 < i_2 < \cdots < i_j \leq n-1.$$

But, in the Hecke algebra we have another quadratic relation instead of $g_i^2 = 1$:

$$g_i^2 = (q-1)g_i + q. \tag{9.3}$$

It turns out that this one is not worse than that of the symmetric group. Namely, each braid can be reduced to "basic" braids with some coefficients, for which we can easily define the Ocneanu trace.

This set of "basic" braids consists of just the same words as in the permutation group

$$\{(g_{i_1} g_{i_1-1} \cdots g_{i_1-k_1})(g_{i_2} \cdots g_{i_2-k_2}) \cdots (g_{i_p} \cdots g_{i_p-k_p})\} \tag{9.4}$$

for

$$1 \leq i_1 < i_2 < \cdots < i_p \leq n-1.$$

It is sufficient to prove that for any word W of type (9.4) and for any generator g_i, the words Wg_i and Wg_i^{-1} can be represented as linear combinations of words of type (9.4). Actually, taking into account the relation $g_i^2 = (q-1)g_i + q$, it is sufficient to consider only the first case.

Now, suppose W is decomposed as a product $W_1 \cdots W_k$, where the g_i's in each W_j have decreasing order, and the first letters of the W_i's have increasing order (according to (9.4)). After this, the additional generator g_i can be "taken through" W_k. Then we take it through $W_{k-1}, \ldots$ until the procedure stops.

For the sake of simplicity suppose the word W_k looks like

$$(g_{i_1} g_{i_1-1} \cdots g_{i_1-k_1}).$$

Then in the case $i = i_1 - k_1 - 1$, the new generator is just added to the word W_k, so there is nothing to prove.

If $i > i_1$, then the word $W g_i$ is already of type (9.4) because g_i can be treated as a new word W_{k+1}.

If $i < i_1 - k_1 - 1$, then we commute g_i with the word W_k and do not make any further changes with W_k: we shall work only with $W_1 \ldots W_{k-1} g_i$.

In the case when $i = i_1 - k_1$, the situation is again simple: at the end we obtain g_i^2 which can be transformed into a linear combination of g_i and e. So, the whole word will be $(q-1)W + qW_1 \ldots W_{k-1}W_k'$, where under W_k' we mean $g_{i_1} \ldots g_{i_1-k_1+1}$ (if $k_1 = 0$, this word is empty).

In the case when $i_1 - k_1 + 1 \leq i \leq i_1$, we have the following situation: we commute g_i with the last elements of W_k while possible, and then obtain the following subword in W_k: $g_i g_{i-1} g_i$, which equals $g_{i-1} g_i g_{i-1}$. Now, all letters in W_k before these subwords commute with g_{i-1}. So, we can take g_{i-1} to the left.

Thus, we have obtained a linear combination of words $W_1 \ldots W_{k-1} g_{i'} W_k$ and $W_1 \ldots W_{k-1} g_i W_k'$ for some i'. In all these cases i' is smaller than the index of the first letter in W_k.

The next step is just as the previous one: we take $g_{i'}$ through W_{k-1}. Then we perform the same with W_{k-2} and so on. The only thing we have to mention here is that the letter on the left side has an index always smaller than the initial letter of the last passed word W_j.

Thus we see that the dimension of $H(q,n)$ equals $n!$ (the number of permutations). The proof of the fact that the algebra does not collapse at all is well established. It can be found, e.g. in [Bou].

The construction above shows that for $(n+1)$–strand braids it is sufficient to consider only those generators containing the generator g_n once. Now, we are ready to define Ocneanu's trace explicitly (by using the induction method on the number of strands) by means of the following initial formulae:

$$tr(1) = 1$$

and

$$tr(xg_ny) = z \cdot tr(xy)$$

for all $x, y \in H(q,n)$.

The main problem is to prove the property $tr(ab) = tr(ba)$. By induction (on n), let us suppose that it is true for $x, y \in H(q,n)$.

Now, the only case that does not follow immediately from the definition is $tr(g_n x g_n y) = tr(x g_n y g_n)$. Namely, when we wish to show that for some element $A \in H(n+1)$ for any $B \in H(n+1)$ we have $tr(AB) = tr(BA)$ it is sufficient to check it only for $B = g_n$ and for $B \in H_n$. The latter follows from the definition.

It suffices to prove this for the case when one multiplicator is g_n and the other one lies in H_{n-1}. Obviously, it is sufficient to consider the following three cases:

1. $x, y \in H_{n-1}$;

2. one of x, y lies in H_{n-1}, the other equals $ag_{n-1}b$;

3. $x = ag_{n-1}b, y = cg_{n-1}d$, where $a, b, c, d \in H_{n-1}$.

The first case is trivial because the generator g_n commutes with all elements from H_{n-1}.

In the second case (we only consider the case when $y \in H_{n-1}$, so y commutes with g_n; the case $x \in H_{n-1}$ is completely analogous to the first one) we have:

$$tr(g_n a g_{n-1} b g_n y)$$

$$= tr(ag_n g_{n-1} g_n by) = tr(ag_{n-1} g_n g_{n-1} by)$$

$$= z \cdot tr(ag_{n-1}^2 by)$$

$$= (q-1)z \cdot tr(ag_{n-1} by) + qz \cdot tr(aby)$$

and

$$tr(ag_{n-1} b g_n y g_n)$$

$$= tr(ag_{n-1} b g_n^2 y) = (q-1)tr(ag_n b g_{n-1} y) + qtr(ag_{n-1} by)$$

$$= z(q-1)tr(ag_{n-1} by) + qz \cdot tr(aby).$$

Finally, in the third case we have

$$tr(g_n a g_{n-1} b g_n c g_{n-1} d)$$

$$= tr(ag_n g_{n-1} g_n bc g_{n-1} d) = tr(ag_{n-1} g_n g_{n-1} bc g_{n-1} d)$$

$$= z \cdot tr(ag_{n-1}^2 bc g_{n-1} d) =$$

$$= z(q-1)tr(ag_{n-1} bc g_{n-1} d) + zq \cdot tr(abc g_{n-1} d) =$$

$$z(q-1)tr(ag_{n-1}bcg_{n-1}d) + z^2 q \cdot tr(abcd)$$

and

$$tr(ag_{n-1}bg_n cg_{n-1}dg_n)$$

$$= tr(ag_{n-1}bcg_n g_{n-1}g_n d) = tr(ag_{n-1}bcg_{n-1}g_n g_{n-1}d)$$

$$= z \cdot tr(ag_{n-1}bcg_{n-1}^2 d) =$$

$$= z(q-1)tr(ag_{n-1}bcg_{n-1}d) + zq \cdot tr(ag_{n-1}bcd) =$$

$$= z(q-1)tr(ag_{n-1}bcg_{n-1}d) + z^2 q \cdot tr(abcd).$$

Thus, we have completed the induction step and defined correctly the Ocneanu trace.

This completes the proof of the theorem. $\qquad\square$

It follows directly from the proof that properties 1,2, and 3 allow us to calculate the trace for any given element of $H(q, n)$. Actually, first we transform this element to a combination of some "basic" braids. Then we use the formula $tr(xg_n y) = ztr(xy)$ and reduce our problem to the case of braids with a smaller number of strands. Then we just apply the induction method on the number of strands.

Exercise 9.4. *Let us calculate* $tr(g_1 g_2 g_3 g_2)$. *We have:*

$$tr(g_1 g_2 g_3 g_2)$$

$$= z \cdot tr(g_1 g_2^2) = z(q-1)tr(g_1 g_2) + zq \cdot tr(g_1)$$

$$= z^3(q-1) + z^2 q.$$

Let us consider oriented links as braid closures. In order to construct a link invariant, one should check the behaviour of some function defined on braids under the two Markov moves.

The function tr is perfectly invariant under the first Markov move (conjugation). Besides, it behaves quite well under the second move. The only thing to do now is the normalisation.

Let us normalise all g_i's in such a way that both types of the second Markov move affect the Ocneanu trace in the same way. To do this, let us introduce a variable Θ such that $tr(\Theta g_i) = tr((\Theta g_i)^{-1})$.

Simple calculations give us

$$\Theta^2 = \frac{\left(\frac{z-(q-1)}{q}\right)}{z} = \frac{z-q+1}{qz}.$$

Let us make a variable change. Namely, let $\lambda = \Theta^2$.
Now we are ready to define the invariant polynomial.

Definition 9.18. The *Jones two–variable polynomial* $X_L(q, \lambda)$ of an oriented link L is defined by

$$X_L(q, \lambda) = \left(-\frac{1 - \lambda q}{\sqrt{\lambda}(1 - q)}\right)^{n-1} (\sqrt{\lambda})^e tr(\pi(a)),$$

where $\alpha \in B_n$ is any braid whose closure is L, e is the exponent sum of α as a word on the σ_i's and π the presentation of B_n to $H_n : \pi(\sigma_i) = g_i$.

Notation. To denote the value of a polynomial on a link L, we put L in the lower index of the letter, denoting the polynomial. We do it for the sake of convenience because we are going to consider polynomials in some variables that will be put in brackets.

The invariance of X under conjugation is obvious (because of invariance of trace) and the invariance of X under the second Markov move follows straightforwardly from the properties of tr.

Theorem 9.9. *For any Conway triple, the Jones two–variable polynomial satisfies the following skein relation:*

$$\frac{1}{\sqrt{\lambda q}} \cdot X_{\diagdown} - \sqrt{\lambda q} X_{\diagup} = \frac{q-1}{\sqrt{q}} X_{)(}.$$

Proof. Indeed, consider three diagrams that differ at one crossing: $\diagdown$ has σ_i, $\diagup$ has σ_i^{-1} and $)($ has no crossing at all.
So, $X_{\diagdown} = \sqrt{\lambda} tr(g_i) M$, $X_{\diagup} = \frac{1}{\sqrt{\lambda}} tr(g_i^{-1}) M$, $X_{)(} = M$, where M is common for all three diagrams.
Then, writing down the Hecke algebra relation $g_i = (q - 1) + q g_i^{-1}$, we obtain the desired result. $\qquad\square$

Let us prove now that the HOMFLY-PT polynomial satisfies a certain skein relation. Indeed, let $t = \sqrt{\lambda}\sqrt{q}$, $x = (\sqrt{q} - \frac{1}{\sqrt{q}})$. Denote $X_L(q, \lambda)$ by $P_L(t, x)$. This is the famous HOMFLY-PT polynomial [HOMFLY].

Theorem 9.10. *The following skein relation holds:*

$$t^{-1} P_{\diagdown} - t P_{\diagup} = x P_{)(}.$$

The proof is quite analogous to that described above.

9.4   Representations of the braid groups

9.4.1   The Burau representation

The most natural way to seek a representation of the braid group is the following. One considers the braid group B_n ($\equiv Br(n)$) and tries to represent it by matrices $n \times n$. More precisely, one takes σ_i to some block–diagonal matrix with one 2×2 block lying in two rows $(i, i+1)$ and two columns $(i, i+1)$ and all the other (1×1) unit submatrices lying on the diagonal. Obviously this implies the commutation between images of σ_i, σ_j when $|i - j| \geq 2$. If one takes all σ_i having the same (2×2)–block (in different positions), then we shall only have to check the relation $\sigma_1 \sigma_2 \sigma_1 = \sigma_2 \sigma_1 \sigma_2$ for 3×3–matrices. Thus we can easily obtain the representation where the block matrix looks like

$$\begin{pmatrix} 1-t & t \\ 1 & 0 \end{pmatrix}.$$

This representation is called the *Burau representation* of the braid group. It was proposed by Burau, [Bura].

The faithfulness of this representation has been an open problem for a long time. In [Bir1], Joan Birman proved that it was faithful for the case of three strands.

In [Moo91], Moody found the first example of a kernel element for this representation.

To date, the problem is solved positively for $n \leq 3$ and negatively for $n \geq 5$; see, e.g. [Big1]. The case $n = 4$ still remains open. In [Big2], Stephen Bigelow shows that this problem is equivalent to the question of whether the Jones polynomial detects the unknot; i.e. Bigelow proved the following theorem.

Theorem 9.11. *[Big2] The Jones polynomial in one variable detects the unknot if and only if the Burau representation is faithful for $n = 4$.*

Besides this, Bigelow thinks that faithfulness of the Burau representation in this case seems to be beyond the reach of any known computer algorithm.

Here we shall demonstrate the proof for the case $n = 3$ and give a counterexample for $n = 5$, following Bigelow [Big1, Big2].

First, let us consider the following description of the braid group and the Burau representation. Denote by D_n the unit complex disc D with n punctures $x_1, \dots, x_n$ on the real line. The set of all automorphisms of D_n considered up to isotopy is precisely the braid group B_n. Let d_0 (the base point) be $-i$.

Now, the Burau representation can be treated as follows. The group $\pi_1(D_n)$ is a free group with n generators. For each loop $\gamma \in \pi_1(D_n)$ one can consider the number of full turns in γ in the counterclockwise direction, or, equivalently, the number of generators in any word representing γ. Thus, there exists a homomorphism $h : B_n \to \mathbb{Z}$. So, one can construct a cover $\tilde{B}_n \to B_n$ with the action of $\mathbb{Z}$ on it. This group has a generator $\langle q \rangle$. Let us consider the group $H_1(\tilde{B}_n)$. According to the arguments above, this group admits a module structure over $\mathbb{Z}[q, q^{-1}]$. Let $\tilde{d}_0$ be a preimage of d_0 under h.

The Burau representation is just obtained from module homomorphisms. More precisely, let $\bar{\beta}$ be a homomorphism representing the braid $\beta \in B_n$. The induced action of $\bar{\beta}$ on $\pi_1(D_n)$ satisfies $h \circ \bar{\beta} = h$. Thus, there exists a unique lift $\hat{\beta}$ of $\bar{\beta}$ such that the following diagram is commutative:

$$
\begin{array}{ccc}
(\tilde{D}_n, \tilde{d}_0) & \xrightarrow{\hat{\beta}} & (\tilde{D}_n, \tilde{d}_0) \\
\downarrow & & \downarrow \\
(D_n, d_0) & \xrightarrow{\bar{\beta}} & (D_n, d_0).
\end{array}
$$

Furthermore, $\hat{\beta}$ commutes with the action of q. Thus, $\hat{\beta}$ induces a $\mathbb{Z}[q^{\pm 1}]$–module homomorphism denoted by $\hat{\beta}_*$. Thus, we can define the Burau representation just as

$$
Burau(\beta) = \hat{\beta}_*.
$$

It is not difficult to check that this definition of the Burau representation coincides with the initial one.

Now let us prove that this representation is faithful in the case of three strands. First, let us introduce the notation of [Big2].

Definition 9.19. A *fork* is an embedded tree F in D with four vertices d_0, p_i, p_j, z such that:

1. the only puncture points of F are p_i, p_j;

2. F meets ∂D_n only at d_0;

3. all three edges of F have z as a vertex.

Definition 9.20. The edge of F containing d_0 is called the *handle* of F. The union of the other two edges forms one edge; let us call it the *tine edge* of F and denote it by $T(F)$. Let us orient $T(F)$ in such a way that the handle of F lies to the right of $T(F)$.

Definition 9.21. A *noodle* is an embedded oriented edge N in D_n such that

1. N goes from d_0 to another point on ∂D_n;

2. N meets ∂D_n only at endpoints;

3. a component of $D_n \backslash N$ contains precisely one puncture point.

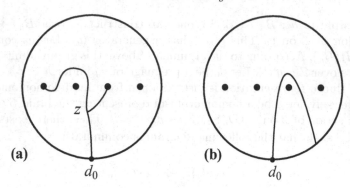

(a)   **(b)**

FIGURE 9.15: (a) A fork;   (b) a noodle.

Let F and N be a fork and a noodle, respectively. Let us define a pairing $\langle N, F \rangle$ in $\mathbb{Z}[q^{\pm 1}]$ as follows. Without loss of generality, let us assume that $T(F)$ intersects N transversely. Let $z_1, \ldots, z_k$ be the intersection points between $T(F)$ and N (with no order chosen). For each $i = 1, \ldots, k$, let γ_i be the arc in D_n which goes from d_0 to z_i along F and then goes back to d_0 along N. Let a_i be the integer such that $h(\gamma_i) = a_i$. Let ε_i be the sign of the intersection between N and F at z_i. Let

$$\langle N, F \rangle = \sum_{i=1}^{k} \varepsilon_i q^{a_i}. \tag{9.5}$$

One can easily check that this pairing is independent of the preliminary isotopy (which allows us to assume the transversality of $T(F)$ and N). Besides, this follows from the basic lemma.

The faithfulness proof follows from the two lemmas.

Lemma 9.1 (The basic lemma). *Let $\beta : D_n \to D_n$ represent an element of the kernel of the Burau representation. Then $\langle N, F \rangle = \langle N, \beta(F) \rangle$ for any noodle N and fork F.*

Lemma 9.2 (The key lemma). *In the case $n = 3$, the equality $\langle N, F \rangle = 0$ holds if and only if $T(F)$ is isotopic to an arc which is disjoint from N.*

Let us now deduce the faithfulness for the case of three strands from these two lemmas. Suppose β lies in the kernel of the Burau representation. We shall show that β represents the trivial braid.

Let N be a noodle. Take N to be a horizontal line through D_n such that the puncture points p_1 and p_2 lie above N and p_3 lies below N. Let F be a fork such that $T(F)$ is a straight line from p_1 to p_2 which does not intersect N. Then $\langle N, F \rangle = 0$. By the basic lemma we have $\langle N, \beta(F) \rangle = 0$. By the key lemma, $\beta(T(F))$ is isotopic to an arc which is disjoint from N. By applying isotopy to β, we can assume that $\beta(T(F)) = T(F)$.

Analogously, one can prove that each of the three edges of the triangle connecting p_1, p_2, p_3 is fixed by β. The only possibility for β to be non–trivial is to represent some whole twists of D, but one can easily check that this is not the case. Thus, the Burau representation is faithful for the case of three strands.

For those who are interested in details, we give the proofs of these two lemmas.

Proof of the basic lemma. The main idea is the following. We transform the definition (9.5) in such a way that it works with loops rather than with forks. Then the invariance for loops follows straightforwardly because our transformation lies in the kernel.

We can assume that the tine edges of both T and $\beta(T)$ intersect N transversely.

Let $\tilde{F}$ be the lift of F to $\tilde{D}_n$ with respect to h, containing $\tilde{d}_0$. Let $\tilde{T}(F)$ be the corresponding lift of $T(F)$. Then $\tilde{T}(F)$ intersects $q^a \tilde{N}$ transversely for any $a \in \mathbb{Z}$. Let $(q^a \tilde{N}, \tilde{T}(F))$ denote the algebraic intersection number of these two arcs. Then the following definition of pairing is equal to (9.5):

$$\langle N, F \rangle = \sum_{a \in \mathbb{Z}} (q^a \tilde{N}, \tilde{T}(F)) q^a. \tag{9.6}$$

Suppose now that $T(F)$ goes from p_i to p_j. Let $\nu(p_i)$ and $\nu(p_j)$ be small disjoint regular neighbourhoods of p_i and p_j, respectively. Let γ be a subarc of $T(F)$ which starts in $\nu(p_i)$ and ends in $\nu(p_j)$. Let δ_i be a loop in $\nu(p_i)$ with base point $\gamma(0)$ which goes counterclockwise around p_i. Denote by $T_2(F)$ the following loop:

$$T_2(F) = \gamma \delta_j \gamma^{-1} \delta_i^{-1}.$$

Let $\tilde{T}_2(F)$ be the lift of $T_2(F)$ which is equal to $(1 - q)\tilde{T}(F)$ outside a small neigbourhood of the punctures. Then the following definition (9.7) is equivalent to (9.5) and (9.6).

$$\langle N, F \rangle = \frac{1}{1 - q} \sum_{a \in \mathbb{Z}} (q^a \tilde{N}, \tilde{T}_2(F)) q^a. \tag{9.7}$$

Because β lies in the kernel of the Burau representation, the loops $\tilde{T}_2(F)$ and $\tilde{T}_2(\beta(F))$ represent the same element of the homology group $H_1(\tilde{D}_n)$. Thus, they have the same algebraic intersection number with any lift $q^a \tilde{N}$ of N. Thus, the claim of the lemma follows from (9.7). $\qquad \square$

Proof of the key lemma. The main idea is to find a fork whose tine edge intersects N at the minimal number of points and to show that if this number is not zero then $\langle N, F \rangle$ cannot be equal to zero.

First, let us recall that by definition (9.5): $\langle N, F \rangle = \sum_{i=1}^{k} \varepsilon_i q^{a_i}$.

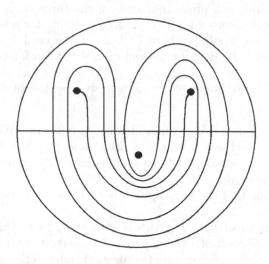

FIGURE 9.16: A tine edge and a noodle in D_3

By applying a homeomorphism to our picture, we can take N to be a horizontal straight line through D_3 with two punctures above it and one puncture point below it (the fork will therefore be twisted). Here we slightly change our convention that the punctures lie on the real line by a small deformation of $D_n \subset D$. Let D_n^+, D_n^- denote the upper and lower components of $D_n \backslash N$, respectively. Let us relabel puncture points in such a way that D_n^+ contains p_1, p_2 and D_n^- contains p_3. Now, let us consider the intersection of $T(F)$ with D_n^-. It consists of a disjoint collection of arcs having both endpoints on N (possibly, one arc can have p_3 as an endpoint). An arc $T(F) \cap D_n^-$ which has both endpoints on N must enclose p_3; otherwise one could remove it together with some intersection points. Thus $T(F) \cap D_n^-$ must consist of a collection of parallel arcs enclosing p_3, and, possibly, one arc with p_3 as an endpoint.

Similarly, each arc in $T(F) \cap D_n^+$ either encloses one of p_1, p_2 or has an endpoint at p_1 or p_2. There can be no arc in $T(F) \cap D_n^+$ which encloses both points; otherwise the outermost such arc together with the outermost arc of the lower part will form a closed loop.

Now, we are going to calculate carefully the intersection number and see that all summands evaluated at $q = -1$ have the same sign.

Indeed, let z_i and z_j be two points of intersection between $T(F)$ and N which are joined by an arc in $T(F) \cap D_n^+$ or $T(F) \cap D_n^-$. This arc, together with a subarc of N, encloses a puncture point. Thus, $a_j = a_i \pm 1$. Besides, the two signs of intersection are opposite: $\varepsilon_j = -\varepsilon_i$. So, $\varepsilon_j(-1)^{a_j} = \varepsilon_i(-1)^{a_i}$. Arguing as above, we prove that all summands for $\langle N, F \rangle$ evaluated at $q = -1$ have the same sign. Thus, $\langle N, F \rangle$ is not equal to zero.

$\square$

9.4.2 A counterexample

In fact, the proof of faithfulness does not work in the case of $n > 3$. Roughly speaking, the problem is that the key lemma uses the intrinsic properties of the three–strand braid groups.

The idea of the example is the following. Each curve α with endpoints at punctures generates an automorphism $\tilde{\tau}_\alpha$ of $H_1(D_n)$. This automorphism is generated by Dehn's half–turn twist about α (permutating the endpoints).

A direct calculation shows that if $\langle \alpha, \beta \rangle = 0$ then $\tilde{\tau}(\alpha)$ and $\tilde{\tau}(\beta)$ commute.

To show that the Burau representation is not faithful it suffices to provide an example of oriented embedded arcs $\alpha, \beta \in D_n$ with endpoints at punctures such that $\langle \alpha, \beta \rangle = 0$, but the corresponding braids τ_α and τ_β do not commute.

For $n \geq 6$, the simplest known example is the following. We set $\phi_1 = \sigma_1^2 \sigma_2^{-1} \sigma_5^{-2} \sigma_4$ and $\phi_2 = \sigma_1^{-1} \sigma_2 \sigma_5 \sigma_4^{-1}$. Take γ to be the simplest arc connecting x_3 with x_4 (so that the corresponding braid is σ_3).

After this, a straightforward check shows that for $\alpha = \phi_1(\gamma), \beta = \phi_2(\gamma)$ so that the corresponding braids are $\tau_\alpha = \phi_1 \sigma_3 \phi_1^{-1}$ and $\tau_\beta = \phi_2 \sigma_3 \phi_2^{-1}$, and we have $\langle \alpha, \beta \rangle = 0$.

It follows from a straightforward check that $\langle \alpha, \beta \rangle = 0$. So, the braid $\tau_\alpha \tau_\beta \tau_\alpha^{-1} \tau_\beta^{-1}$ belongs to the kernel of the Burau representation. To prove that this braid is not trivial, one can use an algorithm for braid recognition, say, one of those described later in the book.

This element has length 44 in the standard generators of the group $Br(6)$. In fact, an example of a kernel element for the Burau representation for five strands can be constructed by using similar (but a bit more complicated) techniques. This element has length 120.

9.5 The Krammer–Bigelow representation

While developing the idea that the Burau representation comes from some covering, Bigelow proposed a more sophisticated covering that leads to another representation. Bigelow proved its faithfulness by using the techniques of forks and noodles. We begin with the formal definition according to Krammer's work [Kra1, Kra2]. After that, we shall describe the main features of Bigelow's work [Big1].

9.5.1 Krammer's explicit formulae

Let n be a natural number and let R be a commutative ring with the unit element. Suppose that $q, t \in R$ are two invertible elements of this ring. Let V be the linear space over R of dimension $\frac{n(n-1)}{2}$ generated by elements $x_{i,j}, 1 \leq i < j \leq n$.

Let us define the action of the braid group $Br(n)$ on the space V according to the following rule:

$$\sigma_k(x_{i,j}) = \begin{cases} x_{i,j} & k < i-1 \text{ or } k > j; \\ x_{i-1,j} + (1-q)x_{i,j} & k = i-1; \\ tq(q-1)x_{i,i+1} + qx_{i+1,j} & k = i < j-1; \\ tq^2 x_{i,j} & k = i = j-1; \\ x_{i,j} + tq^{k-i}(q-1)^2 x_{k,k+1} & i < k < j-1; \\ x_{i,j-1} + tq^{j-i}(q-1)x_{j-1,j} & i < k = j-1; \\ (1-q)x_{i,j} + qx_{i,j+1} & k = j; \end{cases} \qquad (9.8)$$

where $\sigma_k, k = 1, \ldots, (n-1)$, are generators of the braid group.

It can be clearly checked that these formulae give us a representation of the braid group.

Denote the Krammer–Bigelow representation space of the braid group $Br(n)$ by L_n. Because the basis of L_n is a part of the basis for L_{n+1}, we have $L_n \subset L_{n+1}$.

Formula (9.8) implies that L_n is an invariant space under the action of the representation (9.8) of $Br(n+1)$ in L_{n+1}.

While passing from matrices of the representation for $Br(n)$ to those for $Br(n+1)$, the upper–left block stabilises. Thus, one may speak of the infinite–dimensional linear representation of the stable braid group.

We shall not give the (algebraic) proof of the faithfulness because it involves a lot of sophisticated constructions. For the details, see the original work by Krammer [Kra2].

9.5.2   Bigelow's construction and main ideas of the proof

Here we shall describe Bigelow's results following [Big1,PP,Tur2]. There he constructs a more sophisticated covering than that corresponding to the Burau representation. This covering gives a faithful representation that coincides with Krammer's.

In fact, the Bigelow representation deals with **the braid group itself as the mapping class group** rather than with any presentation of it. One should point out the work of Lawrence [Law] where he extended the idea of the Burau representation via coverings for the configuration spaces in D_n, and was able to obtain all of the so–called Temperley–Lieb representations. Just the representations of Lawrence were shown to be faithful (by Krammer and by Bigelow).

The Bigelow proof of the faithfulness is based on the ideas used in the proof of faithfulness of the Burau representation for three strands. Namely, we have the following three steps.

1. The basic lemma.

2. The key lemma.

3. Deducing the faithfulness from these lemmas.

Below, we shall modify the scalar product for curves defined above by using a more sophisticated bundle over a four–dimensional space together with action on its 2–homologies. As we have proved above, the basic lemma (new version) works for all n even for the Burau representation. For the remaining two steps we shall use the modified covering and scalar product.

Let D, D_n be as above, the punctured points are $-1 < x_1 \cdots < x_n < 1$ which we shall puncture.

Let C be the space of all unordered pairs of points in D_n. This space is obtained from $D_n \times D_n \backslash diagonal$ by the identification $\{x, y\} = \{y, x\}$ for any distinct points $x, y \in D_n$. It is clear that C is a connected non–compact four–dimensional manifold with boundary. It is endowed with a natural orientation induced by the counterclockwise orientation of D_n. Let $d = -i$ and let $d' = -ie^{\frac{\varepsilon \pi i}{2}}$ for small positive ε. We take $c_0 = \{d, d'\}$ as the base point of C.

A closed curve $\alpha : [0, 1] \to C$ can be written in the form $\{\alpha_1(s), \alpha_2(s)\}$, where $s \in [0, 1]$ and α_1, α_2 are arcs in D_n such that $\{\alpha_1(0), \alpha_2(0)\} = \{\alpha_1(1), \alpha_2(1)\}$. In this case, the arcs α_1, α_2 are either both loops or can be composed with each other. Thus, they form a closed oriented 1–manifold α in D_n. Let $a(\alpha) \in \mathbb{Z}$ be the total winding number of this manifold α around all the punctures of D_n.

Consider the map $s : [0, 1] \to S^1$ given by $s \mapsto \frac{\alpha_1(s) - \alpha_2(s)}{|\alpha_1(s) - \alpha_2(s)|}$ and the natural projection $S^1 \to \mathbb{R}P^1$. Thus, we obtain a loop in $\mathbb{R}P^1$. The corresponding element of $H_1(\mathbb{R}P^1)$ is denoted by $b(\alpha)$. The formula $\alpha \to q^{a(\alpha)} t^{b(\alpha)}$ thus defines a homomorphism ϕ from $H_1(C)$ to the free commutative group generated by q, t. Let $R = \mathbb{Z}[q^{\pm 1}, t^{\pm 1}]$ be the group ring of this group.

Let $\tilde{C} \to C$ be the regular covering corresponding to the kernel of ϕ. The generators q, t act on $\tilde{C}$ as commuting covering transformations. The homology group $H_2(\tilde{C}, \mathbb{Z})$ thus becomes an R–module.

Any homeomorphism h of D_n onto itself induces a homeomorphism $C \to C$ by $h(\{x, y\}) = \{h(x), h(y)\}$ (we preserve the notation h).

It is easy to check that $h(c_0) = c_0$ and the action of h on the homologies $H_1(c)$ commutes with ϕ. Thus, the homeomorphism $h : C \to C$ can be lifted uniquely to a map $\tilde{C} \to \tilde{C}$ that fixes the fibre over c_0 pointwise and commutes with the covering transformation. Consider the representation $B_n \to Aut(H_2(\tilde{C}))$ mapping the isotopy class of h to the R–linear automorphism $\tilde{h}_*$ of $H_2(C)$.

Theorem 9.12 ([Big1]). *The representation $B_n \to Aut(H_2(\tilde{C}))$ is faithful for all $n \geq 1$.*

The proof of this fact uses the techniques of noodles and forks.

We kindly ask the reader to be patient while reading all definitions.

For arcs $\alpha, \beta : [0, 1] \to D_n$ such that for all $s \in [0, 1]$ $\alpha(s) \neq \beta(s)$ denote by $\{\alpha, \beta\}$ the arc in C given by $\{\alpha, \beta\}(s) = \{\alpha(s), \beta(s)\}$. We fix a point $\tilde{c}_0 \in \tilde{C}$ lying over $c_0 \in C$.

We should be slightly more precise about the notion of a noodle. Namely, by a *noodle* we mean an embedded arc $N \subset D_n$ with endpoints d and d'. For any noodle N, the set $\Sigma_N = \{\{x, y\} \in C | x, y \in N, x \neq y\}$ is a surface in C containing c_0. This surface is homeomorphic to a triangle with one edge removed. We orient N from d to d' and orient Σ_N as follows: at a point $\{x, y\} = \{y, x\} \in \Sigma_N$ such that x is closer to d than y along N, the orientation of Σ_N is the product of orientations of N at x and at y in this order.

Let $\tilde{\Sigma}_N$ be the lift of Σ_N to $\tilde{C}$ containing the point c_0. The orientation of Σ_N naturally induces an orientation for $\tilde{\Sigma}_N$. Obviously, $\tilde{\Sigma}_N \cap \partial \tilde{C} = \partial \tilde{\Sigma}_N$

Having a fork $F = T \cup H$ with endpoints $d = d_0, x_i, x_j$ and vertex z, we can push it slightly (fixing x_i and x_j and moving d to d') and obtain a parallel copy F' with tine edges T' and handle H'. Denote $T' \cap H'$ by z'. We can assume that F' intersects F only in common vertices $\{x_i, x_j\}$ and in one more point $H \cap T'$ lying close to z and z'.

We shall use $\tilde{\Sigma}_N$ and $\tilde{\Sigma}_F$ to establish the duality between N and D. Without loss of generality we can assume that N intersects the tine edge T of F transversely at m points $z_1, \ldots, z_m$. We choose the parallel fork $F' = T' \cup H'$ as above in such a way that T' intersects N transversely in m points $z'_1, \ldots, z'_m$, where each pair z_i, z'_i is joined by a short arc in N that lies in the narrow strip bounded by $T \cup T'$ and meets no other z_i, z_j. Then, the surfaces Σ_F, Σ_N intersect transversely in m^2 points $\{z_i, z'_j\}$ for $i, j = 1, \ldots, m$. Thus, for any $a, b \in \mathbb{Z}$, the image $q^a t^b \tilde{\Sigma}_N$ under the covering transformation meets $\tilde{\Sigma}_F$ transversely.

Now, we are ready to define the pairing.

Definition 9.22. Consider the algebraic intersection number $q^a t^b \tilde{\Sigma}_N \cdot \tilde{\Sigma}_F \in \mathbb{Z}$ and set

$$\langle N, F \rangle = \sum_{a,b} \in \mathbb{Z}(q^a t^b \tilde{\Sigma}_N \cdot \tilde{\Sigma}_F) q^a t^b. \tag{9.9}$$

Now, we shall highlight the main ideas of proof of the main theorem.

1. First, one should check the invariance of pairing under homotopy. This follows from a routine verification.

2. It can be observed that the sign $\varepsilon_{i,j}$ of the intersection point $\{i, j\}$ is given by the formula $\varepsilon_{i,j} = -(-1)^{b_{i,i} + b_{j,j} + b_{i,i}}$.

3. The key lemma

   Lemma 9.3. *If a homeomorphism h of D_n onto itself is an element of the kernel of $B_n \rightarrow Aut(H_2(\tilde{C}))$ then for any noodle N and any fork F, we have $\langle N, h(F) \rangle = \langle N, F \rangle$.*

   The proof follows from a routine verification by rewriting the definition of the pairing.

4. The basic lemma:

Lemma 9.4. $\langle N, F \rangle = 0$ *if and only if the tine edge T can be isotoped of N.*

For the Burau representation we used the fact that we only have three points and thus we obtained that for a minimal state all intersections have the same sign for $q = -1$. Here we have another argument: two–dimensionality.

So, suppose $\langle N, F \rangle = 0$. Assume that the intersection between N and T cannot be removed and consider the minimal intersection. Let us use the *lexicographic ordering* of monomials $q^a t^b > q^c t^d$ if $a > c$ or $a = c, b > d$. The ordered pair $\{i, j\}$ is *maximal* if $q^{a_{i,j}} t^{b_{i,j}} \geq q^{a_{k,l}} t^{b_{k,l}}$ for any $k, l \in \{1, \ldots, m\}$.

Then the following statement holds.

5. If the pair (i, j) is maximal then $b_{i,i} = b_{j,j} = b_{j,i}$.

Thus, all entries of the maximal monomial, say, $q^a t^b$ in (9.9) have the same sign $-(-1)^b$. Thus $\langle N, F \rangle \neq 0$.

6. From the key lemma and the basic lemma we see that each element of the kernel of the representation preserves the lines connecting punctures. Thus, the only possibility for this kernel element is to be some power of the half–turn Dehn twist. However, such a twist is a multiplication by $q^{2n} t^2$.

So, we have constructed a faithful representation of the braid group with polynomials in two variables as coefficients. It follows from classical number theory theorems that there exists a pair of real numbers transforming this representation to a faithful representation with **real coefficients**. Thus, all braid groups are linear.

Chapter 10

Braids and links. Braid construction algorithms

In the present chapter, we shall present two algorithms for constructing a braid B by a given link L such that $Cl(B) = L$ and thus prove Alexander's theorem.

10.1 Alexander's theorem

Throughout the present section, we shall work only with oriented links. For any given unoriented link, we choose an arbitrary orientation of it.

Exercise 10.1. *Construct a braid whose closure represents the Borromean rings.*

The main statement of this chapter is Alexander's theorem that each link can be obtained as a braid closure. We shall give two proofs of this theorem: the original one by Alexander and the one by Vogel that realises a faster algorithm for constructing a corresponding braid.

Theorem 10.1. *(Alexander's theorem [Ale2]) Each link can be represented as the closure of a braid.*

Proof. We shall prove this theorem for the case of polygonal links.

Consider a diagram L of an oriented polygonal link and a point O on the plane P of the diagram (this point should not belong to edges and should not coincide with vertices of the diagram). We say that L is *braided around O* if each edge of L is visible from O as counterclockwise–oriented.

Definition 10.1. For any L and O, let us call edges visible as counterclockwise–oriented *positive*; the other ones will be *negative*.

If there exists a point O such that our link diagram is braided around O, then the statement of Alexander's theorem becomes quite clear: we just cut the diagram along a ray coming from O and "straighten the diagram"; see Fig. 10.1.

Thus, in order to prove the general case of the theorem, we shall reconstruct

163

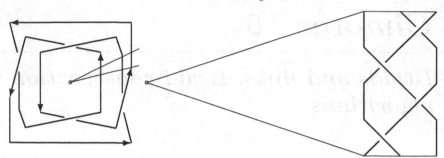

FIGURE 10.1: Constructing a braid by a braided link

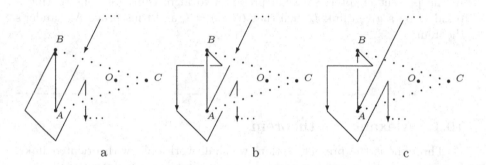

FIGURE 10.2: Alexander's trick

our arbitrary link diagram in order to obtain a diagram braided around some point O.

First, fix a point O. Now, we are going to use the *Alexander trick* as follows. Consider a negative edge AB of our polygonal link and find some point C on the projection plane P such that the triangle ABC contains O. Then we replace AB by AC and CB. Both edges will evidently be positive; see Fig. 10.2.

We shall use this operation till we get a diagram braided around O.

Let us describe this construction in more detail. In the case when the negative edge AB contains no crossings, the Alexander trick can be easily performed directly; see Fig. 10.2.a. Actually, one can divide the edge AB into two parts (edges) and then push them over O.

The same can be done in the case when AB contains the only crossing that is an overcrossing with respect to the other edge; see Fig. 10.2.b.

Finally, if AB contains the only crossing that is an undercrossing with respect to the other edge, then we can push it under, as shown in Fig. 10.2.c. □

Exercise 10.2. *By using the Alexander trick, construct a braid whose closure represents the connected sum of the two right trefoil knots.*

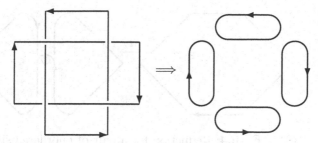

FIGURE 10.3: Smoothing of crossings and Seifert circles

The method of proof for Alexander's theorem described above certainly gives us a concrete algorithm for constructing a braid from a link. However, this algorithm is too slow. Below, we give a simpler algorithm for constructing braids by links.

10.2 Vogel's algorithm

Here we describe the algorithm proposed by Pierre Vogel [Vog1].
We start with a definition.

Definition 10.2. First we say that an oriented link diagram is *braided* if there exists a point on the plane of the diagram around which the link diagram is braided.

A braided link diagram can be easily represented as a closure of a braid.

Remark 10.1. *Obviously, the property of a diagram to be braided does not depend on the crossing structure. We may say that we shall work only with shadows of links. In the sequel, we shall never use this structure.*

Given an oriented closed diagram D of a link L, one can correctly define the operation of *crossing smoothing* for it. To do it, we just "smooth" the diagram at each vertex as shown in Fig. 10.3 and consider all Seifert circles of it. Denote this smoothing by σ.

Definition 10.3. Let us say that all Seifert circles of some planar diagram are *nested* if they all induce the same orientation of the plane and bound an enclosed disc system.

Obviously, if all Seifert circles of some planar link diagram are nested, then the corresponding diagram is braided. Moreover, in this case, the number of strands of the braid coincides with the number of Seifert circles.

Let us fix some link diagram D and consider now the shadow of D. This

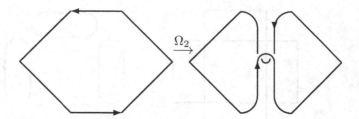

FIGURE 10.4: Reduction for a pair of unordered sides

shadow divides the sphere (one–point compactification of the plane) into 2–cells, called *sides*. The *interior* side is that containing infinity.

Definition 10.4. A side S is *unordered* if it has two edges a, b that belong to different Seifert circles A_1, A_2 and induce the same orientation of S, and *ordered* otherwise.

In the first case we say that the edges A_1, A_2 generate the unordered side.

One can apply the move Ω_2 to unordered sides, as shown in Fig. 10.4. In this case, the set of sides becomes "more ordered". More precisely, the following proposition holds.

Proposition 10.1. *If all edges of the side Σ belong to two Seifert circles then this side is ordered.*

Proof. Actually, consider the edges of this side. It is easy to see that all edges belonging to the same Seifert circle have the same orientation. Consider two adjacent edges belonging to different Seifert circles. They have different orientations. Thus, any two edges of the given diagram belonging to different Seifert circles must have different orientations. Hence, the side is ordered. $\square$

Proposition 10.2. *If a diagram D of the link L has no unordered sides, then it can be transformed to a braided diagram by using an infinity change.*

Proof. Suppose the diagram D has no unordered sides. Consider some side of the planar tiling generated by our link diagram. Any two adjacent edges of this side either have the same orientation (in this case, they belong to the same Seifert circle) or they have different orientations (and belong to different Seifert circles). If we consider the points of adjacent edges belonging to the same Seifert circle as the points of one "long" edge then we obtain some polygon M (or a whole Seifert circle). The edge orientations of the edges of M are alternating. Thus, the number of such edges is even (or equal to one when all edges belong to the same Seifert circle). Since this side is not unordered, all edges of it belong to no more than two different Seifert circles. Thus we conclude that each Seifert circle that defines some edge of the polygon M is adjacent either to one Seifert circle or to two Seifert circles (lying on different

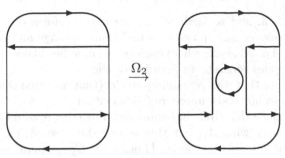

FIGURE 10.5: Eliminating an unordered side

sides of M). Otherwise, there would be an unordered side with edges belonging to three different Seifert circles.

The remaining part of the proof is left to the reader as a simple exercise.

□

The Vogel algorithm works as follows. First we eliminate all crossings by the rule: ⊗ →)(, ⊗ →)(. Then, by using Ω_2 we remove unordered sides. Finally, if Seifert circles are not nested, we change the infinity.

Let us describe this algorithm in more detail.

First, let us smooth all crossings of the diagram. Thus we obtain several Seifert circles. Denote the number of these circles by s. Some pairs of these circles might generate unordered sides. Let us construct a graph whose vertices are Seifert circles; two vertices should be connected by an edge if there exists a side (ordered or not) that is incident to the two circles. Let us remove from this graph a vertex, corresponding to some "interior" Seifert circle. We obtain some graph Γ_1. Let us change the notation for the remaining $s - 1$ circles and denote them by $A_1, A_2, \ldots, A_{s-1}$, in such a way that A_i and A_{i+1} contain edges that generate an unordered side. This means that our graph Γ_1 is connected. It is easy to see that in the disconnected case we should apply this algorithm to each connected component; it will work even faster.

We shall perform the following operation. Let us take the unordered side, generated by A_1 and A_2, and perform Ω_2 to it as described above. Instead of circles A_1 and A_2, we shall get two Seifert circles; one of them lies inside the other. Besides this, they do not generate an unordered side; see Fig. 10.5.

We have got two new circles; one of them lies inside the other. Denote the exterior circle by A_1 and the interior one by A_2. Because "the former A_2" generated an unordered side together with A_3 then the new circle A_1 also generates an unordered side together with A_3 (the latter stays the same).

Let us now perform Ω_2 on the circles A_1 and A_3 and change the notation again: the exterior circle will be A_1 and the interior one will be A_3, and so on. Finally (after $s - 2$ operations Ω_2), we obtain one interior circle A_1 that makes no unordered sides. Now, we shall not touch A_1, but perform the same procedure with the pairs (A_2, A_3), (A_2, A_4), and so on. Then we do the same for $A_3, A_i, i > 3$, and so forth. Thus, we have performed $\frac{(s-1)(s-2)}{2}$ second Reidemeister moves and (possibly) one infinity change and obtained the set of circles $A_1, A_2, \ldots, A_{s-1}$, where each next circle lies inside all previous ones and no two circles generate an unordered side.

Let us show that the remaining circle (that we "removed" in the very beginning) does not make unordered sides either.

Actually, since this circle has some exterior edge it could generate an unordered side only with A_1, but this is not the case. After this, we should change the infinity (if necessary). Thus, after C_{s-2}^2 operations (for the connected case; in the unconnected case we shall use even fewer operations) we obtain a braided diagram.

So, we have proved the following.

Theorem 10.2. *If the link diagram D has n crossings and s Seifert circles then*

1. *The Vogel algorithm requires no more than C_{s-2}^2 second Reidemeister moves.*

2. *The total number of strands of the obtained braid equals s and the number of crossings does not exceed $n + (s-1)(s-2)$.*

Below, we perform the Vogel algorithm for the knot named 5_2 according to the standard classification, given in the end of the book.

To do it, we perform the second Reidemeister move Ω_2 twice and then the infinity change; see Fig. 10.2. Thus, the two moves Ω_2 would be sufficient.

Finally, we get a braided diagram, see the lower part of Fig. 10.2.

Thus we can now construct braids corresponding to given links even faster than by using the Alexander trick.

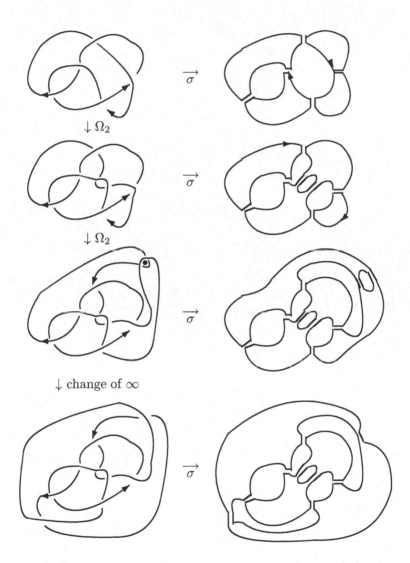

FIGURE 10.6: Planar knot diagrams and Seifert circles

Chapter 11

Algorithms of braid recognition

Until now, several braid recognition algorithms have been constructed. The first of them was contained in the original work of Artin [Art1] (and, in more detail, [Art2]). However, the approach proposed by Artin was not very clearly explained; both articles are quite difficult to read. There were other works on braid recognition (by Birman [Bir1], Garside [Gars], Thurston [Thu], et al.) Here we are going to describe a geometrically explicit algorithm (proving the completeness of a slightly modified Artin invariant according to [GM], see also [BZ]) and the algebraic algorithm by Dehornoy.

In [Gars], Garside proposed a method of normal forms; by using this method, he solved not only the word problem for the braid groups, but the conjugation problem as well; the conjugation problem is in fact more complicated. Unfortunately, we do not present here any solution of the conjugation problem. In this chapter, we shall also present a result by M. Berger concerning the minimal braid–word in $Br(3)$ representing the given braid isotopy class.

11.1 The curve algorithm for braids recognition

Below, we shall give a proof of the completeness of one concrete invariant for the braid group elements invented by Artin, see [GM].

11.1.1 Introduction

We are going to describe the construction of the above mentioned invariant for the classical braid group $Br(n)$ for arbitrary n.

The invariant to be constructed has a simple algebraic description as a map (non-homeomorphic) from the braid group $Br(n)$ to the n copies of the free group in n generators.

Several generalisations of this invariant, such as the spherical and cylindrical braid group invariant, are also complete. They will be described later in this chapter. The key point of such a completeness is that these invariants originate from several curves, and the braid can be uniquely restored from these curves.

Moreover, this aprroach finally led to the algorithmic recognition of virtual braids due to Oleg Chterental. We shall touch on virtual braids in Chapter 21.

11.1.2 Construction of the invariant

Let us begin with the definition of notions that we are going to use, and let us introduce the notation.

Definition 11.1. By an *admissible system of n curves* we mean a family of n non–intersecting non–self–intersecting curves in the upper half plane $\{y > 0\}$ of the plane Oxy such that each curve connects a point having ordinate zero with a point having ordinate one and the abscissas of all curve ends are integers from 1 to n. All points $(i, 1)$, where $i = 1, \ldots, n$, are called *upper points*, and all points $(i, 0), i = 1, \ldots, n$, are called *lower points*.

Definition 11.2. Two admissible systems of n curves A and A' are *equivalent* if there exists a homotopy between A and A' in the class of curves with fixed endpoints lying in the upper half plane such that no interior point of any curve can coincide with any upper or lower point during the homotopy.

Analogously, the equivalence is defined for one curve (possibly, self–intersecting) with fixed upper and lower points: during the homotopy in the upper half plane no interior point of the curve can coincide with an upper or lower point.

In the sequel, admissible systems will be considered up to equivalence.

Remark 11.1. *Note that curves may intersect during the homotopy.*

Remark 11.2. *In the sequel, the number of strands of a braid equals n, unless otherwise specified.*

Let β be a braid diagram on the plane, connecting the set of lower points $\{(1, 0), \ldots, (n, 0)\}$ with the set of upper points $\{(1, 1), \ldots, (n, 1)\}$. Consider the upper crossing C of the diagram β and push the lower branch along the upper braid to the upper point of it as shown in Figure 11.1.

Naturally, this move spoils the braid diagram: the result, shown in Figure 11.1.b is not a braid diagram. The advantage of this "diagram" is that we have a smaller number of crossings.

Now, let us do the same with the next crossing. Namely, let us push the lower branch along the upper branch to the end. If the upper branch is deformed during the first move, we push the lower branch along the deformed branch (see Fig. 11.2).

Reiterating this procedure for all crossings (until the lowest one), we get an admissible system of curves. Denote its equivalence class by $f(\beta)$.

Theorem 11.1. *The function f is a braid invariant; i.e., for two diagrams β, β' of the same braid we have $f(\beta) = f(\beta')$.*

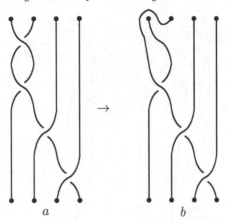

FIGURE 11.1: Pushing the upper crossing

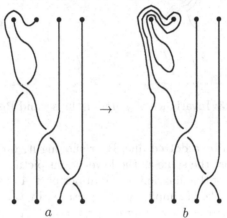

FIGURE 11.2: Pushing the next crossing

Proof. Having two braid diagrams, we can write the corresponding braid–words, and denote them by the same letters β, β'. We must prove that the admissible system of curves is invariant under braid isotopies.

The invariance under the commutation relations $\sigma_i \sigma_j = \sigma_j \sigma_i, |i - j| \geq 2$, is obvious: the order of pushing two "far" branches does not change the result.

The invariants under $\sigma_i \sigma_i^{-1} = e$ can be readily checked; see Fig. 11.3.

In the leftmost part of Fig. 11.3, the dotted line indicates the arbitrary behaviour for the upper part of the braid diagram. The rightmost part of Fig. 11.3 corresponds to the system of curves without $\sigma_i \sigma_i^{-1}$.

Finally, the invariance under the transformation $\sigma_i \sigma_{i+1} \sigma_i \to \sigma_{i+1} \sigma_i \sigma_{i+1}$ is shown in Fig. 11.4. In the upper part (over the horizontal line) we demonstrate the behaviour of $f(A\sigma_i \sigma_{i+1} \sigma_i)$, and in the lower part we show that of $f(A\sigma_{i+1} \sigma_i \sigma_{i+1})$ for an arbitrary braid A. In the middle–upper part, part of

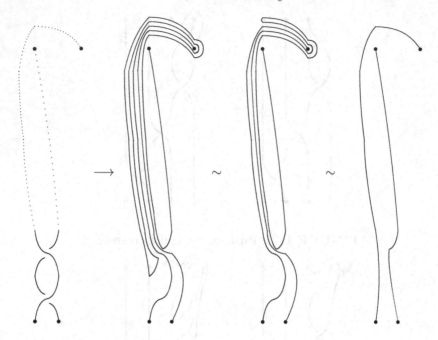

FIGURE 11.3: Invariance of f under the second Reidemeister move

the curve is shown by a dotted line. By removing it, we get the upper–right picture which is just the same as the lower–right picture.

The behaviour of the diagram in the upper part A of the braid diagram is arbitrary. For the sake of simplicity it is pictured by three straight lines.

Thus we have proved that $f(A\sigma_i\sigma_{i+1}\sigma_i) = f(A\sigma_{i+1}\sigma_i\sigma_{i+1})$.

This completes the proof of the theorem.   □

Now, let us prove the following lemma.

Lemma 11.1. *If for two braids a and b we have $f(a) = f(b)$ then for each braid c we obtain $f(ac) = f(bc)$.*

Proof. The claim $f(ac) = f(bc)$ follows directly from the construction. Indeed, we just need to attach the braid c to the admissible system of curves corresponding to a (or b) and then to push the crossings of c.   □

In fact, a much stronger statement holds.

Theorem 11.2 (The main theorem). *The function f is a complete invariant.*

To prove this statement, we shall use some auxiliary definitions and lemmas.

In order to prove the main theorem, we should be able to restore the braid from its admissible system of curves.

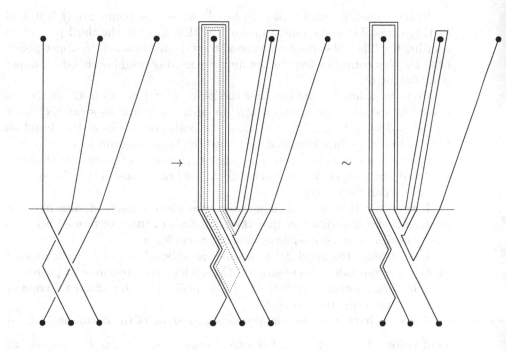

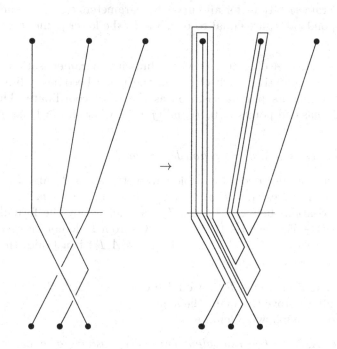

FIGURE 11.4: Invariance of f under the third Reidemeister move

In the sequel, we shall deal with braids whose end points are $(i, 0, 0)$ and $(j, 1, 1)$ with all strands coming upwards with respect to the third projection coordinates. They obviously correspond to standard braids with upper points $(j, 0, 1)$. This correspondence is obtained by moving neighbourhoods of upper points along Oy.

Consider a braid b and consider the plane $P = \{y = z\}$ in $Oxyz$. Let us place b in a small neighbourhood of P in such a way that its strands connect points $(i, 0, 0)$ and $(j, 1, 1), i, j = 1, \ldots, n$. Both projections of this braid on Oxy and Oxz are braid diagrams. Denote the braid diagram on Oxy by β.

The next step now is to transform the projection on Oxy without changing the braid isotopy type; we shall just deform the braid in a small neighbourhood of a plane parallel to Oxy.

It turns out that one can change abscissas and ordinates of some intervals of strands of b in such a way that the projection of the transformed braid on Oxy constitutes an admissible system of curves for β.

Indeed, since the braid lies in a small neighbourhood of P, each crossing on Oxy corresponds to a crossing on Oxz. Thus, the procedure of pushing a branch along another branch in the plane parallel to Oxy deletes a crossing on Oxy, preserving that on Oxz.

Thus, we have described the geometric meaning of the invariant f.

Definition 11.3. By an *admissible parametrisation* (in the sequel, all parametrisations are thought to be smooth) of an admissible system of curves we mean a set of parametrisations for all curves by parameters $t_1, \ldots, t_n$ such that at the upper points all t_i are equal to one, and at the lower points t_i are equal to zero.

Any admissible system A of n curves with admissible parametrisation T generates a braid representative: each curve on the plane becomes a braid strand when we consider its parametrisation as the third coordinate. The corresponding braid has end points $(i, 0, 0)$ and $(j, 1, 1)$, where $i, j = 1, \ldots, n$. Denote it by $g(A, T)$.

Lemma 11.2. *The result $g(A, T)$ does not depend on T.*

Proof. Indeed, let us consider two admissible parametrisations T_1 and T_2 of the same system A of curves. Let $T_i, i \in [1, 2]$, be a continuous family of admissible parametrisations between T_1 and T_2, say, defined by the formula $T_i = (i-1)T_1 + (2-i)T_2$. For each $i \in [1, 2]$, the curves from T_i do not intersect each other, and for each $i \in [1, 2]$ the set of curves $g(A, T_i)$ is a braid, thus $g(A, T_i)$ generates the desired braid isotopy. $\square$

Thus, the function $g(A) \equiv g(A, T)$ is well defined.

Now we are ready to prove the main theorem.

First, let us prove the following lemma.

Lemma 11.3. *Let A, A' be two equivalent admissible systems of n curves. Then $g(A) = g(A')$.*

Proof. Let $A_t, t \in [0,1]$, be a homotopy from A to A'. For each $t \in [0,1]$, A_t is a system of curves (possibly, not admissible). For each curve $\{a_{i,t}, i = 1, \ldots, n, t \in [0,1]\}$ choose points $X_{i,t}$ and $Y_{i,t}$, such that the interval from the upper point (upper interval) of the curve to $X_{i,t}$ and the interval from the lower point (lower interval) do not contain intersection points. Denote the remaining part of the curve (middle interval) between $X_{i,t}$ and $Y_{i,t}$ by $S_{i,t}$. Now, let us parametrise all curves for all t by parameters $\{s_{i,t} \in [0,1], i = 1, \ldots, n\}$ in the following way: for each t, the upper point of each curve has parameter $s = 1$, and the lower point has parameter $s = 0$. Besides, we require that for $i < j$ and for each $x \in S_{i,t}, y \in S_{j,t}$ we have $s_{i,t}(x) < s_{j,t}(y)$. This is possible because we can vary parametrisations of upper and lower intervals on $[0,1]$; for instance, we parametrise the middle interval of the j–th strand by a parameter on $[\frac{j}{n+2}, \frac{j+1}{n+2}]$.

It is obvious that for $t = 0$ and $t = 1$ these parametrisations are admissible for A and A'. For each $t \in [0,1]$ the parametrisation s generates a braid B_t in $\mathbb{R}^3$: we just take the parameter $s_{i,t}$ for the strand $a_{i,t}$ as the third coordinate. The strands do not intersect each other because parameters for different middle intervals cannot be equal to each other.

Thus the system of braids B_t induces a braid isotopy between B_0 and B_1. $\square$

So, the function g is well defined on equivalence classes of admissible systems of curves.

Now, to complete the proof of the main theorem, we need only to prove the following lemma.

Lemma 11.4. *For any braid b, we have $g(f(b)) = b$.*

Proof. Indeed, let us place b in a small neighbourhood of the "inclined plane" P in such a way that the ends of b are $(i, 0, 0)$ and $(j, 1, 1)$, $i, j = 1, \ldots, n$.

Consider $f(b)$ that lies in Oxy. It is an admissible system of curves for b. So, there exists an admissible parametrisation that restores b from $f(b)$. By Lemma 11.2, each admissible parametrisation of $f(b)$ generates b. So, $g(f(b)) = b$. $\square$

11.1.3 Algebraic description of the invariant

The general situation in the construction of a complete invariant is the following: one constructs a new object that is in one–to–one correspondence with the described object. However, the new object might also be badly recognisable.

Now, we shall describe our invariant algebraically. It turns out that the final result is very easy to recognise. Namely, the problem is reduced to the recognition problem of elements in a free group. So, there exists an injective map from the braid group to the (n copies of) the free group with n generators that is not homomorphic.

Each braid β generates a permutation. This permutation can be uniquely restored from any admissible system of curves corresponding to β. Indeed, for an admissible system A of curves, the corresponding permutation maps i to j, where j is the ordinate of the strand with the upper point $(i, 1)$. Denote this permutation by $p(A)$. It is obvious that $p(A)$ is invariant under equivalence of A.

Let n be an integer. Consider the free product G of n groups $\mathbb{Z}$ with generators $a_1, \ldots, a_n$. Denote by E_i the right residue classes in G by $\{a_i\}$; i.e., $g_1, g_2 \in G$ represent the same element of E_i if and only if $g_1 = a^k g_2$ for some k.

Definition 11.4. An *n–system* is a set of elements $e_1 \in E_1, \ldots, e_n \in E_n$.

Definition 11.5. An *ordered n–system* is an n–system together with a permutation from S_n.

Proposition 11.1. *There exists an injective map from equivalence classes of admissible systems of curves to ordered n–systems.*

Since the permutation for equivalent admissible systems of curves is the same, we can fix the permutation $s \in S_n$ and consider only equivalence classes of admissible systems of curves with permutation s (i.e., with all lower points fixed depending on the upper points in accordance with s). Thus we only have to show that there exists an injective map from the set of admissible systems of n curves with fixed lower points to n–systems.

To complete the proof of the proposition, it suffices to prove the following.

Lemma 11.5. *Equivalence classes of curves with fixed points $(i, 1)$ and $(j, 0)$ are in one–to–one correspondence with E_i.*

Proof. Denote $P \backslash \cup_{i=1, \ldots, n} (i, 1)$ by P_n. Obviously, $\pi_1(P_n) \cong G$. Consider a small circle C centered at $(i, 1)$ for some i with the lowest point X on it. Let ρ be a curve with endpoints $(i, 1)$ and $(j, 0)$. Without loss of generality, assume that ρ intersects C in a finite number of points. Let Q be the first such point that one meets while walking along ρ from $(i, 1)$ to $(j, 0)$. Thus we obtain a curve ρ' coming from C to $(j, 0)$. Now, let us construct an element of $\pi_1(P_n, X)$. First it comes from X to Q along C clockwise. Then it goes along ρ until $(j, 0)$. After this, it goes along Ox to the point $(i, 0)$. Then it goes vertically upwards till the intersection with C in X. Denote the constructed element by $W(\rho)$.

If we deform ρ outside C, we obtain a continuous deformation of the loop, thus $W(\rho)$ stays the same as the element of the fundamental group. The deformations of ρ inside C might change $W(\rho)$ by multiplying it by a_i on the left side. So, we have constructed a map from equivalence classes of curves with fixed points $(i, 1)$ and $(j, 0)$ to E_i.

The inverse map can be easily constructed as follows. Let W be an element of $\pi_1(P_n, X)$. Consider a loop L representing W. Now consider the curve L'

that first goes from $(i, 1)$ to X vertically, then goes along L', after this goes vertically downwards until $(i, 0)$ and finally, horizontally until $(j, 0)$. Obviously, $W(L') = W$. It is easy to see that for different representatives L of W we obtain the same L'. Besides, for $L_1 = a_i L_2$, the curves L'_1 and L'_2 are isotopic. This completes the proof of the lemma. $\qquad\qquad\square$

Thus, for a fixed permutation s, admissible systems of curves can be uniquely encoded by n–systems, which completes the proof of the theorem.

Now, we see that this invariant is a quite simple object: elements of E_i can easily be compared.

Exercise 11.1. *Implement a computer program realising this algorithm.*

Let us describe the algebraic construction of the invariant f in more detail.

Let β be a word–braid, written as a product of generators $\beta = \sigma_{i_1}^{\varepsilon_1} \dots \sigma_{i_k}^{\varepsilon_k}$, where each ε_j is either $+1$ or -1; $1 \le i_j \le n - 1$ and $\sigma_1, \dots, \sigma_{n-1}$ are the standard generators of the braid group $Br(n)$.

We are going to construct the n–system step–by–step while writing the word β. First, let us write n empty words (in the alphabet $a_1, \dots, a_n$). Let the first letter of β be σ_j. Then all words except for the word e_{j+1} should stay the same (i.e., empty), and the word e_{j+1} becomes a_j^{-1}. If the first crossing is negative; i.e., σ_j^{-1} then all words except e_j stay the same and e_j converts to a_{j+1}. While considering each next crossing, we do the following. Let the crossing be $\sigma_j^{\pm 1}$. Let p and q be the numbers of strands coming from the left side and from the right side respectively. If this crossing is positive; i.e., σ_j, then all words except e_q stay the same, and e_q becomes $e_q e_p^{-1} a_p^{-1} e_p$. If it is negative, then all crossings except e_p stay the same, and e_p becomes $e_p e_q^{-1} a_q e_q$. After processing all the crossings, we get the desired n–system.

Example 11.1. *For the trivial braid written as $\sigma_1 \sigma_2 \sigma_1 \sigma_2^{-1} \sigma_1^{-1} \sigma_2^{-1}$ the construction operation works as follows:*

$$(e, e, e) \to (e, a_1^{-1}, e) \to (e, a_1^{-1}, a_1^{-1}) \to (e, a_1^{-1}, b_1^{-1} a_1^{-1}) \to$$
$$\to (e, e, b_1^{-1} a_1^{-1}) \to (e, e, b_1^{-1}) \to (e, e, e).$$

A priori these words may be non-trivial; they must only represent trivial residue classes, say, (a_1, a_2^2, a_3^{-1}).

However, it is not the case.

Proposition 11.2. *For the trivial braid, the algebraic algorithm described above gives trivial words.*

Proof. Indeed, the algebraic number of occurencies of a_i in the word e_i equals zero. This can be easily proved by induction on the number of crossings. In the initial position all words are trivial. The induction step is obvious. Thus, the final word e_i equals a_i^p, where $p = 0$. $\qquad\qquad\square$

From this approach, one can easily obtain the well known invariant (action) as follows. Instead of a set of n words $e_1, \ldots, e_n$, one can consider the words $e_1 a_1 e_1^{-1}, \ldots, e_n a_n e_n^{-1}$. Since e_i's are defined up to a multiplication by a_i's on the left, the obtained elements are well defined in the free groups. Besides, these elements $\mathfrak{E}_i = e_i a_i e_i^{-1}$ are generators of the free group. This can be checked by a step–by–step confirmation. Thus, for each braid b we obtain a set $Q(b)$ of generators for the braid group. So, the braid b defines a transformation of the free group $\mathbb{Z}^{*n}$. It is easy to see that for two braids, the transformation corresponding to the product equals the composition of transformation. Thus, one can speak about the *action of the braid group on the free group*. Since f is a complete invariant, this action has an empty kernel.

Definition 11.6. This action is called *the Hurwitz action* of the braid group B_n on the free group $\mathbb{Z}^{*n}$.

11.2 *LD*–systems and the Dehornoy algorithm

Another algorithm for braid recognition is purely algebraic. It was proposed by French mathematician Patrick Dehornoy [Deh3]. The algorithm to be described is rather fast.

The idea to be used is very closed to that used in the distributive groupoid (quandle). We take some set (of colours) and associate colours with arcs of the braid from this set. Then we show how the braid can be reduced to the trivial braid and if it can not be reduced, why it is not trivial (because of some colour reasons). More precisely, for "good" colour systems (having structure similar to that of groupoids), each braid defines an operator on this colour system, and this operator can not be trivial for the case of a non-trivial braid.

Let us first remember that braids have a group structure. Thus, in order to compare some two braids a and b it is sufficient to check whether the braid ab^{-1} is trivial.

Let us start with the definition.

Given a braid written algebraically as a word W in the alphabet $\sigma_i^{\pm 1}$, $i = 1, 2 \ldots, n-1$.

Definition 11.7. We say that the word W is a 1–*positive* braid if it is equivalent to a word W', where the letter σ_1 occurs only in positive powers (and does not occur in negative powers). Analogously, one defines a 1–*negative* braid.

If a braid can be written by a braid–word W' without σ_1 and σ_1^{-1}, we say that this braid is 1–*neutral*.

Remark 11.3. *Since any braid can be encoded by different braid words, one cannot say a priori that the classes described above have no intersections. Later, we shall prove that it is not the case.*

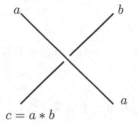

$$c = a * b$$

FIGURE 11.5: A relation

Theorem 11.3. *Each 1–positive (respectively, 1–negative) braid is not trivial.*

Remark 11.4. *Actually, a much stronger statement holds: each braid represented by a braid–word containing $\sigma_i{}^e$ but not $\sigma_i{}^{-e}$ for some $e = \pm 1$ is not trivial. We shall not prove this statement, see [Deh1, Deh3].*

The first aim of this section is to show that these three sets are in fact non-intersecting. We are going to present an algorithm that transforms each braid word to an equivalent one that is either unity or positive or negative (the set of 1–neutral braids is actually subdivided into more sets according to the next strands starting from the second one).

To prove Theorem 11.3, we shall need some auxiliary definitions and lemmas.

Consider a positive braid–word β and all *lower arcs* of it.

Definition 11.8. By an *lower arc* we mean a part of the braid diagram going from one overcrossing to the next one and passing only undercrossings. Lower arcs correspond to arcs of the mirror image of (upper) arcs.

We wish to label the braid diagram by elements from M in the following manner: we are going to associate with each arc some element of M in such a way that:

1. all lower arcs outgoing from upper ends of the braid are marked by variables which are allowed to have values in M;

2. each "lower" label is uniquely defined by all "upper" labels over it;

3. the operator f expressing lower labels by upper labels is invariant under isotopies of braids.

To set such a labeling, we have to consider the crossings of the diagram. Suppose a crossing is incident to lower arcs a, b, and c. Let us write down the following relation $c = a * b$ as shown in Fig. 11.5

Let us analyse the invariance of f under isotopies. Each elementary isotopy is associated with one of the following formulae:

$$\sigma_i \sigma_i^{-1} = e, i = 1, \ldots, n - 1,$$

(move Ω_3)

$$\sigma_i\sigma_{i+1}\sigma_i = \sigma_{i+1}\sigma_i\sigma_{i+1}, i = 1,\ldots n-2,$$

(far commutativity)

$$\sigma_i\sigma_j = \sigma_j\sigma_i, |i-j| \geq 2, 1 \leq i,j \leq n-1.$$

The relation $\sigma_i\sigma_i^{-1} = e$ will be considered later (now we consider only braid words with positive exponents of generators).

The function to be constructed is invariant under far commutativity by construction.

The move Ω_3 gives us the self–distributivity relation (in the case of a quandle we needed right self–distributivity):

$$a*(b*c) = (a*b)*(a*c). \tag{11.1}$$

Definition 11.9. A set M with a left self-distributive operation $*$ is called an *LD–system* or *LD–set*.

Remark 11.5. *It is obvious that the operation $*$ is not sufficient to define the operator f for arbitrary braid words: a letter σ_i^{-1} spoils the situation.*

In the sequel, we shall add two more operations $\vee$ and $\wedge$ on M as follows.

Obviously, the map f is invariant under the commutation relation (transposing σ_i and σ_j for $|i-j| \geq 2$).

It remains to check the invariance of the map f under Ω_3; i.e., under transformation $a_ia_{i+1}a_i \to a_{i+1}a_ia_{i+1}$. It is easy to see (Fig. 11.6) that the invariance under Ω_3 means the left distributivity operation.

Thus, having an *LD*–set M, we can label lower arcs of a positive braid–word by elements of M; the set of elements at lower points is uniquely defined by the set of elements at upper points.

Denote the latter by $p_i, 1 \leq i \leq n$; the elements at lower points will be denoted by q_i. For instance, for the trivial braid, we have $p_i = q_i$.

For an *LD*–set one can define a partial order relation $<$. Namely, for each $a, c \in M \quad a < c$ if $\exists b \in M : c = a*b$.

Definition 11.10. A partially ordered *LD*–set M is called *ordered* if $<$ is acyclic; i.e., there exists no sequence $a_1 < a_2 < \cdots < a_k < a_1$.

Example 11.2. *The sets $\mathbb{R}$ and $\mathbb{Q}$ admit some left-distributive (but not acyclic) operations:*

1. $a*b = max(a,b)$

2. $a*b = \frac{(a+b)}{2}$

3. $a*b = (a+1)$.

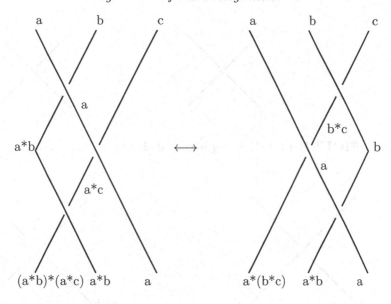

FIGURE 11.6: Invariance of the map f under Ω_3

Let us now give an example of an acyclic LD–system. To do this, we have to introduce more difficult structures, enclosing the LD–structure. Denote the semigroup of non-negative braids (or *positive braid monoid*) by $Br(n)^+$.

Remark 11.6. *This monoid played the key role in Garside's theory of normal form.*

Definition 11.11. Let $(M, \wedge)$ be a set endowed with a binary operation. Let us define the right action of the semigroup $Br(n)^+$ on $M^n = \underbrace{M \times \cdots \times M}_{n}$ inductively. First, for $\vec{a} = (a_1, \ldots, a_n)$ we set

$$(\vec{a})\varepsilon = \vec{a}, \quad (\vec{a})\sigma_i w = (a_1 \ldots, a_i \wedge a_{i+1}, a_i, a_{i+2}, \ldots, a_n)w, \qquad (11.2)$$

where ε is the unity element.

Let us now try to colour all braid diagrams (not only positive). In order to do this, we shall have to introduce two more operations. Let us change the notation: denote $*$ by $\wedge$ and introduce new operations $\vee$ and $\circ$. Then we can colour the braid diagram as shown in Fig. 11.7.

As before, we can express the labels of lower points by the labels of upper points; thus we define the operator f.

Lemma 11.6. *Let $(M, \wedge, \circ, \vee)$ be a system with three binary operations. Then, the operator f defined above is invariant under isotopies generated by the second Reidemeister move if and only if the following relations hold in M:*

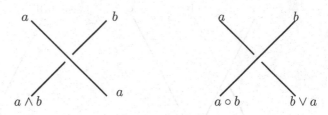

FIGURE 11.7: Defining lower labels at a negative crossing

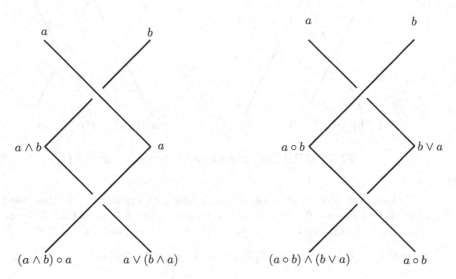

FIGURE 11.8: Invariance under Ω_2

$$\forall x, y \in M \qquad x \circ y = y, x \wedge (x \vee y) = x \vee (x \wedge y) = y.$$

Proof. Follows straightforwardly from Fig. 11.8.

$\square$

Definition 11.12. An *LD–system* M (with respect to wedge) endowed with an extra operation $\vee$ is said to be an *LD–quasigroup* if the relation

$$x \wedge (x \vee y) = x \vee (x \wedge y) = y$$

holds for arbitrary $x, y \in M$.

Remark 11.7. *Unlike distributive groupoid (quandle), LD–quasigroups do not require the idempotence relation.*

Remark 11.8. *The isotopy generated by Ω_2 preserves f, by definition. Thus,* *if M is an LD–quasigroup, then the operator f is invariant under all braid isotopies.*

The following proposition can be checked straightforwardly.

Proposition 11.3. *Let G be a group. Then the binary operations $x \wedge y = xyx^{-1}$, $x \vee y = x^{-1}yx$, $x \circ y = y$, define the LD–quasigroup structure on G.*

Thus, for a given group G we can use the system $(G, \wedge, \vee)$ (where $x \circ y = y \ \forall x, y$) for colouring braids. In particular, let FG_n be the free group generated by $\{x_1, \ldots, x_n\}$. For an n–strand braid word b, let us define $\tilde{b}$ as the braid word obtained from b by reversing the order of letters. Thus, $\sigma_1\sigma_2^{-1}\sigma_3\sigma_1^{-1} \mapsto \sigma_1^{-1}\sigma_3\sigma_2^{-1}\sigma_1$.

For given elements $x_1, \ldots, x_n$, define the elements $y_1, \ldots, y_n$ according to the rule: $(y_1, \ldots y_n) = (x_1, \ldots x_n)\tilde{b}$. Let $\phi(b)$ be an automorphism of FG_n, mapping all x_i's to y_i's. Then ϕ is an homomorphism of the braid group $Br(n)$ inside $Aut(FG_n)$ because $\phi(b^{-1})$ coincides with $\phi(b)^{-1}$ by construction.

Denote by FG_∞ the limit of embeddings $FG_1 \subset FG_2 \subset FG_3 \subset \ldots$. In this way, we obtain an homomorphism $\phi : B_\infty \to Aut(FG_\infty)$.

Denote $\phi(\sigma_i)$ by α_i.

By construction, we have:

$$\alpha_i(x_k) = \begin{cases} x_k & k < i \text{ or } k > i + 1 \\ x_i x_{i+1} x_i^{-1}, & k = i \\ x_i, & k = i + 1. \end{cases}$$

Remark 11.9. *This action coincides with the action of the generator σ_i on the colours of arcs; see Fig. 11.7, cf. the definition of the Hurwitz action.*

We have the operation sh on the generators: $sh(x_i) = x_{i+1}$. Now, let us define the shift sh on FG_∞, taking x_i to x_{i+1} for each i. Let us define the action of sh on $Aut(FG_\infty)$ according to the following rule. For $\varphi \in Aut(FG_\infty)$, let $sh(\varphi)(x_1) = x_1$ and $sh(\varphi)(x_{i+1}) = sh(\varphi(x_i))$ for $i \geq 1$.

Then the operation $\wedge$ on $Aut(FG_\infty)$ is defined by

$$\varphi \wedge \psi = \varphi \circ sh(\psi) \circ \alpha_1 \circ sh(\varphi^{-1}) \tag{11.3}$$

and ϕ is the homomorphism from $(B_\infty, \wedge)$ to $(Aut(FG_\infty), \wedge)$. We are going to prove that the operation defined by (11.3) is left self–distributive.

Note that α_1 commutes with the image of the automorphism sh^2. To complete the proof of self–distributivity of the operation $\wedge$, it remains to prove the following.

Proposition 11.4. *Let G be a group, a be a fixed element of G, and s be an automorphism[1] of G. Then the formula $x \wedge y = xs(y)as(x^{-1})$ defines a left*

---

[1] Later, this operation will play the role of shift in the free group.

self–distributive operation if and only if the element a commutes with images of the map s^2 and the following relation holds

$$as(a)a = s(a)as(a). \tag{11.4}$$

Moreover, in this case there is a homomorphism of LD-systems from B_∞ to G.

Proof. Since s is an endomorphism, we obtain

$$x \wedge (y \wedge z) = xs(y)s^2(z)s(a)s^2(y^{-1})as(x^{-1})$$

$$(x \wedge y) \wedge (x \wedge z) = xs(y)as^2(z)s(a)s^2(x^{-1})as^2(x)s(a^{-1})s^2(y^{-1})s(x^{-1}).$$

If the operation $\wedge$ is left self–distributive, then, taking $x = y = z = 1$, we get $s(a)a = as(a)as(a^{-1})$.

It is easy to check that the inverse statement holds as well.

Finally, formula (11.4) and the hypothesis that a commutes with $s^2(z)$ for all z implies the fact that the map f defined by $f(\sigma_i) = s^{i-1}(a)$ generates a homomorphism from B_∞ to G.   □

Thus, we have constructed a left self–distributive system on $Aut(FG_\infty)$ with operation $\wedge$.

Theorem 11.4. *This system $(Aut(FG_\infty), \wedge)$ is acyclic.*

The proof follows from two auxiliary lemmas on free reductions that appear while calculating $\alpha_i(x)$.

Let us denote the set of words in the alphabet $\{x_1^{\pm 1}, x_2^{\pm 1}, \cdots\}$ by W_∞.

We say that a word from W_∞ is *free reduced* if it does not contain the following subwords: $x_i x_i^{-1}$ and $x_i^{-1} x_i$. For each $w \in W_\infty$, let us denote by $red(w)$ the word obtained from w by means of consequent deleting of such subwords. Thus, we can identify the free group FG_∞ with the set of all free reduced words; this set is endowed with the operation $u \cdot v = red(uv)$.

Definition 11.13. For a letter x from the alphabet $\{x_1^{\pm 1}, x_2^{\pm 1}, \ldots\}$, denote by $E(x)$ the subset of FG_∞, containing all reduced words whose final letter is x.

Let us now investigate the image of the set $E(x_1^{-1})$ with respect to the action of $\alpha_i^{\pm 1}$.

Lemma 11.7. *Let ϕ be an arbitrary element from $Aut(FG_\infty)$. Then the automorphism $sh(\phi)$ maps the set $E(x_1^{-1})$ to itself.*

Proof. Let f be an automorphism of W_∞ mapping x_i to some reduced word $\phi(x_i)$ for each i. Let w be an arbitrary element from $E(x_1^{-1})$. Then $w = ux_1^{-1}$, where u is some reduced word that does not belong to $E(x_1)$. In this case we have $sh(f)(w) = sh(f)(u) \cdot sh(f)(x_1^{-1})$; i.e., $red(sh(f)(u)x_1^{-1})$.

Assume that the latter does not belong to $E(x_1^{-1})$. This means that the last letter x_1^{-1} is reduced with a letter x_1 that occurs in $sh(f)(u)$. But any letter x_1 in $sh(f)(u)$ can originate only from some x_1 in u. Thus, we should have a decomposition $u = u_1 x_1 u_2$ such that $red(sh(f)(u_2))$ is the empty word; i.e., $sh(f)(u_2) = 1$. Since $sh(f)$ is an automorphism of FG_∞ then $u_2 = 1$. This means $u_2 \in E(x_1)$. The contradiction to the initial hypothesis completes the proof. $\qquad\square$

Lemma 11.8. *The automorphism α_1 maps $E(x_1^{-1})$ to itself.*

Proof. Let w be an element from $E(x_1^{-1})$. By definition, $w = ux_1^{-1}$, where u is some reduced element that does not belong to $E(x_1)$, and $\alpha_1(w)$ is $red(\alpha_1(w))$. More precisely,

$$\alpha_1(w) = red(\alpha_1(u)x_1 x_2^{-1} x_1^{-1}).$$

Suppose that the word in the right-hand side part does not belong to the set $E(x_1^{-1})$. This means that the final letter x_1^{-1} is reduced by some x_1 in the end of $\alpha_1(u)$. This letter originates either from some x_2, or from some $x_1^{\pm 1}$ in the word $\alpha_1(u)$.

In the first case, let us write down the letter x_2 that takes part in this reduction, and represent u as $u_1 x_2 u_2$, where u_2 is a reduced word whose initial letter is not x_2^{-1}. Thence,

$$\alpha_1(w) = red(\alpha_1(u_1)x_1\alpha_1(u_2)x_1 x_2^{-1} x_1^{-1}),$$

and the hypothesis can be reformulated as follows: $red(\alpha_1(u_2)x_1 x_2^{-1})$ is the empty word (because $\alpha_1(u_2) = x_2 x_1^{-1}$).

Now, let α_1 be an automorphism, and $x_2 x_1^{-1}$ be the image of $x_2^{-1} x_1$ with respect to α_1. Thus, the only possible case is $u = x_2^{-1} x_1$. But, in this we assume the contradiction: u should not begin with x_2^{-1}.

In the second case, we analogously write $u = u_1 x_1^e u_2$ for $e = \pm 1$. In this case we obtain:

$$\alpha_1(w) = red(\alpha_1(u_1)x_1 x_2^e x_1^{-1}\alpha_1(u_2)x_1 x_2^{-1} x_1^{-1}).$$

Now our hypothesis is that $red(x_2^e x_1^{-1}\alpha_1(u_2)x_1 x_2^e)$ is the empty word.

In this case, we conclude that $\alpha_1(u_2) = x_1 x_2^{1-e} x_1^{-1}$. The latter word equals $\alpha_1(x_1^{1-e})$. Thus, u_2 should be equal to x_1^{1-e}. If $e = +1$ then the word u_2 is empty. If $e = -1$ then $u_2 = x_1^2$. In both cases, u_2 belongs to $E(x_1)$ which is a contradiction. $\qquad\square$

Lemma 11.9. *Suppose that ϕ is an automorphism of FG_∞, that can be expressed as a composition of images of sh and α_1, whence the latter takes place* *at least once. Then $\phi(x_1)$ belongs to $E(x_1^{-1})$. In particular, the automorphism ϕ is not identical.*

Proof. The condition of the lemma is that ϕ has the following representation:

$$\phi = sh(\phi_0) \circ \alpha_1 \circ sh(\phi_1) \circ \alpha_1 \circ \cdots \circ \alpha_1 \circ sh(\phi_p).$$

Then we have $sh(\phi_p)(x_1) = x_1$ and hence $\alpha_1(x_1) = x_1 x_2 x_1^{-1}$; i.e., $\alpha_1(x_1) \in E(x_1^{-1})$. In this case, each next map $sh(\phi_k)$ (as well as α_1) takes $E(x_1^{-1})$ to $E(x_1^{-1})$.

$\square$

Let us now prove Theorem 11.4. We have to show that in $Aut(FG_\infty)$ equalities like

$$\phi = (\dots(\phi \wedge \psi_1) \wedge \dots) \wedge \psi_p)$$

cannot hold.

By using the definition of $\wedge$, we get the representation of ϕ as

$$\phi = \phi \circ \text{some mess}$$

Here "some mess" contains α_1 and sh.

Thus, Id has a presentation by sh and α_1 where the latter occurs at least once. This is a contradiction to the last lemma.

Let us complete now the proof of Theorem 11.2. Let b be a 1–positive braid word. Consider the automorphism $\phi(b)$. It satisfies the conditions of Lemma 11.9. Thus, it is not identical. Consequently, the braid b is not trivial.

We can also present the following "intuitive" proof.

Suppose we have an LD–quasigroup Q that is an acyclic LD–system with the order operation $<$. Let us show that the existence of this quasigroup Q results in the claim of Theorem 11.2. Consider the elements $a_1, a_2, \dots, a_k$, corresponding to lower arcs, corresponding to leftmost crossings, and the elements $b_1, b_2, \dots$ on the right hand from the a_i's; see Fig. 11.9.

It is easy to see that $a_{i+1} = a_i \wedge b_i > a_i$. Thus, $a_k > a_1$. Because the operation $<$ is acyclic, we have: $a_k \neq a_1$.

However, for the trivial braid, the elements of the set M, corresponding to upper points coincide with those corresponding to lower points. Thus we obtain a contradiction which completes the proof of the theorem.

One can easily prove the following corollaries.

Corollary 11.1. *The braid that is inverse to the 1–positive braid is 1–negative; the inverse to a 1–neutral braid is 1–neutral.*

Corollary 11.2. *Each braid B belongs to no more than one of the three types: 1–positive, 1–negative, 1–neutral.*

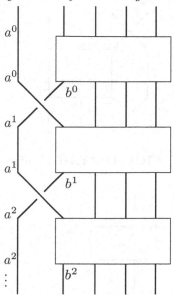

FIGURE 11.9: 1–positive braid is not trivial

Proof. Suppose that B is simultaneously 1–positive and 1–negative. Then there exists a 1–positive braid word B' representing the inverse braid for B. Consequently, the unit braid BB' is 1–positive. Thus we obtain a contradiction.

The other cases can be proved analogously. □

Corollary 11.3. *The toric (p, q)–braid is not trivial.*

As we know, no 1–positive or 1–negative braid–word can represent the trivial braid.

All 1–neutral braid words can be divided into 2–positive, 2–negative, and 2–neutral braids with respect to occurencies of σ_2 and σ_2^{-1}.

Analogously to Theorem 11.5, one can prove the following.

Proposition 11.5. *Each 2–positive or 2–negative braid word represents a non-trivial braid.*

Analogously to 1– and 2–positive (negative, neutral) braid words one can define k–positive (negative, neutral) braid words. Arguing as above, one proves that the first two types of braids do not contain the trivial braid.

Thus, according to our classification, there is only one n–strand $(n-1)$–neutral braid. Namely, it is the trivial braid.

Now, we have to show that all braids can be classified in this manner. To do this, we shall have to prove the following theorem.

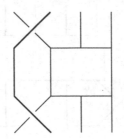

FIGURE 11.10: Handle

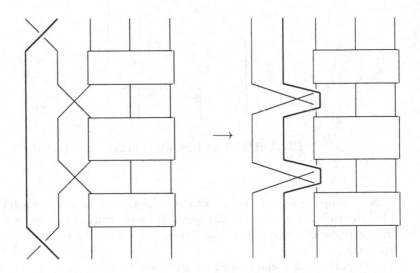

FIGURE 11.11: Handle reduction

Theorem 11.5. *Each braid is either 1–positive or 1–negative or 1–neutral.*

Let us first discuss this theorem for a while.

Suppose we have some braid word representing an n–strand braid K and we wish to use the relations

$\sigma_i \sigma_j = \sigma_j \sigma_i$ for $|i - j| \geq 2, \sigma_i \sigma_{i+1} \sigma_i = \sigma_{i+1} \sigma_i \sigma_{i+1}$ for removing either σ_1 or σ_1^{-1}.

Suppose that L is a subword of K looking like $\sigma_i{}^p w \sigma_i{}^{-p}$, where $p = \pm 1$, and the word w contains only generators $\sigma_j^{\pm 1}$ for $j > i$.

Definition 11.14. Such a word is called an i–*handle*.

Geometrically, a 1–handle is shown in Fig. 11.10.

For such a handle, one can perform a *reduction*; i.e., a move, pulling the first strand over the nearest crossings as shown in Fig. 11.11.

Proposition 11.6. *In this case, the braid word $\sigma_1^e v \sigma_1^{-e}$ becomes the word v' that is obtained by replacing all occurrences of $\sigma_2^{\pm 1}$ in v with $\sigma_2^{-e}\sigma_1^{\pm 1}\sigma_2^{e}$.*

Proof. Actually, let us consider the handle reduction shown in Fig. 11.11. All crossings lying on the right hand with respect to σ_2 will stay the same. The initial and the final crossings $\sigma_1^{\pm 1}$ will disappear. The crossings where the strand goes over are changed by the rule described in the formulation of the statement. □

Let us now consider the braid word K and consequently reduce all handles in it. If the process stops (i.e., we eliminate all handles) then the obtained braid word has either 1–positive or 1–negative or 1–neutral form.

Let us demonstrate now that this approach does not always work.

Convention. For the sake of convenience, let us write small Latin letters $a, b, c, \ldots$ instead of generators $\sigma_1, \sigma_2, \ldots$ and capital letters $A, B, C, \ldots$ instead of $\sigma_1^{-1}, \sigma_2^{-1}, \ldots$.

Example 11.3. *Consider the word $abcBA$. It is a 1–handle. After applying the handle reduction, we obtain the braid word $B(abcBA)b$ that contains the initial braid word (handle) as a subword. Thus, by applying handle reductions many times, we shall always have this handle and increase the length of the whole word.*

The matter is that this 1–handle encapsulates a 2–handle; after reducing the 1–handle, the 2–handle goes out and becomes a 1–handle.

For this braid word, one can first reduce the "interior" handle and then the "exterior" one. Then we get $a(bcB)A \to (aCbcA) \to CBabc$; thus, we conclude that this braid is 1–positive.

Fortunately, the existence of k–handles inside j handles ($k > j$) is the only obstruction for reducing the braid word to another word without 1–handles.

Let us prove the following lemma.

Lemma 11.10. *If a j–handle has no $(j + 1)$–handle inside (as a subword) then after reducing this j–handle no new handle appears.*

Proof. Without loss of generality, we can assume that $j = 1$. Let $u = \sigma_1^e v \sigma_1^{-e}$ be a 1–handle, $e = \pm 1$. Since v contains no 2–handles, we see that v contains either only positive exponents of σ_2 or only negative.

The same can be said about exponents of σ_1 in the word u' obtained from u by means of the handle reduction.

Consequently, the word u' contains no handles. □

Definition 11.15. A handle containing other handles as subwords is called a *nest*.

Definition 11.16. An i–handle reduction not containing $(i + 1)$–handles inside is called *proper*.

Now, let us describe the algorithm.

First, we reduce all interior handles (which are not nests); then we reduce the handles containing the handles that have already been reduced, and so on. Finally, we obtain a braid word that is either 1–positive or 1–negative or 1–neutral.

In the first two cases, everything is clear. In the third case, we forget about the first strand and repeat the same for all other strands.

Example 11.4. *Consider the braid word* $ABacBCBaCbaa$. *Let us transform it to an equivalent braid word without handles.*

ABacBCBaCbaa
bABcBCBaCbaa
bbABcbABCbABCbaa
bbABcbABCbAcBCaa
bbABcbABCbcbABCa
bbABcbABCbcbbABC

Here we underline the subword representing the handle to be reduced.
The obtained word is 1–*negative. Thus, the braid is not trivial.*

Thus, we have obtained a simple and effective algorithm for braid recognition.

Exercise 11.2. *Write a computer program realizing this algorithm.*

We have not yet proved that the Dehornoy algorithm stops within a finite number of steps (the number of letters in the braid word grows and the thus we cannot guarantee the finite time of work).

Exercise 11.3. *Proof that the Dehornoy algorithm works directly in the case of 3–strand braids.*

Below, we sketch the proof of the fact that the Dehornoy algorithm stops in a finite time. For a more detailed proof see [Deh3].

11.2.1   Why the Dehornoy algorithm stops

Definition 11.17. A braid word is called *positive* (resp., *negative*) if it contains only σ_i's (respectively, σ_i^{-1}'s).

It turns out that while performing the handle reductions $w_0 \to w_1 \to w_2 \to \ldots$, each word w_k can be represented by drawing a path in a special finite labelled graph; starting from the word $w = w_0$, only a finite number of such words may occur.

Let us be more detailed.

We should give some definitions.

Definition 11.18. By a *positive (negative) equivalence* of two words we mean a relation where only σ_i in positive powers (respectively, negative) occurs.

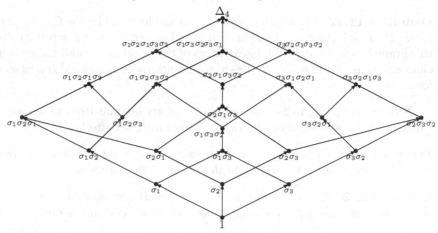

FIGURE 11.12: The Cayley graph for Δ_4

Example 11.5. $\sigma_1\sigma_2\sigma_1 \to \sigma_2\sigma_1\sigma_2$ *is a positive equivalence;*
$\sigma_1^{-1}\sigma_3^{-1} = \sigma_3^{-1}\sigma_1^{-1}$ *is a negative equivalence;*
$\sigma_1\sigma_2\sigma_1^{-1} = \sigma_2^{-1}\sigma_1\sigma_2$ *is neither a positive nor a negative equivalence.*

Definition 11.19. By a *word reversing* we mean a simple equivalence
$\sigma_i^{-1}\sigma_j \to \sigma_j\sigma_i^{-1}$ if $|i - j| \geq 2$ or $\sigma_i^{-1}\sigma_j \to \sigma_i\sigma_j\sigma_i^{-1}\sigma_j^{-1}$ if $|i - j| = 1$ (right reversing).
Analogously, one defines the left reversing.

It is easy to see that by reiterated left reversing, each braid word w can be transformed to a form $N_L D_R^{-1}$, where N_L and D_R are positive braid words. They are called the *left denominator* and *right denominator*.

Analogously, right–reversing can transform each braid word to a form $D_L^{-1}N_R$ of the *left denominator* D_L and right numerator N_R.

It is obvious that for each braid word u, $D_L(u)N_R(u) = N_L(u)D_R(u)$.

Definition 11.20. For a braid word w, the *absolute value* is $D_L(w)N_R(w)$.
    Notation: $|w|$.

Definition 11.21. For a positive braid X, the *Cayley graph* of X is the following graph: its vertices are the trivial braid e, the braid X and positive braids Y such that there exists a positive braid Z such that $YZ = X$. Two vertices P and Q are connected by an edge if there exists i such that $P = Q\sigma_i$ or $P = Q\sigma_i$. In this case, the edge is oriented from the "smaller" braid to the "bigger" braid.

Example 11.6. *Consider the positive braid* $\Delta_4 = \sigma_1\sigma_2\sigma_1\sigma_3\sigma_2\sigma_1$. *Then the Cayley graph for this braid is shown in Fig. 11.12.*

Definition 11.22. Let u be a positive braid and let $C(u)$ be the Cayley graph for u. Then each path (with the first and the last points chosen arbitrarily) on this graph can be expressed by generators of the braid group and their inverse elements; thus, each path generates a word. Such paths are called *traced words* for u.

Now, we are going to formulate some auxiliary lemmas from which we can conclude that the Dehornoy algorithm stops within a finite type.

Lemma 11.11. *Each proper reduction (without nested 2–handles) can be reduced to word–reversing, right equivalence, and left equivalence.*

Lemma 11.12. *For any braid word w, all words obtained from w by word reversing (right and left) and positive and negative equivalence are traced for $|w|$.*

From these two lemmas we deduce the following lemma.

Lemma 11.13. *Each word obtained from w by proper handle reductions is traced for $|w|$.*

Definition 11.23. Let w be a braid word. The *height* of w is the maximal number of letters σ_i occurring in a word containing no σ_i^{-1} and traced in the Cayley graph of $|w|$.

Assume that $w_0 = w, w_1, w_2, \ldots$ is a sequence of handle reductions from w. The first point is that the number of σ_1–handles on w_i is not larger than the number of σ_1–handles in w.

Definition 11.24. The p–th critical prefix $pref_p(w_k)$ is a braid represented by the prefix of w_k that ends with the first letter of the p–th σ_1–handle of w_k.

The point is now that for every p, the prefix is "not increasing" from w_k to w_{k+1} and actually "increasing" if this handle is reduced. More precisely, the following lemma holds.

Lemma 11.14. *Assume that the p–th σ_1–handle is reduced from w_k to w_{k+1}. Then there exists a braid word $u_{p,k}$ traced in the Cayley graph of w from $pref_p(w_k)$ to $pref_p(w_{k+1})$ that contains one letter σ_1^e and no letter σ_1^{-e}.*

Now, it is not difficult to prove the following lemma.

Lemma 11.15. *Assume that w is a braid word of length l and height h. Then the number of handle reductions from w is bounded by lh.*

This lemma shows that any braid word can be reduced to a 1–positive, 1–negative or 1–neutral word in finite time. Thus, the Dehornoy algorithm stops in finite time.

11.3 Minimal word problem for $Br(3)$

Among other problems arising in braid theory we would like to note the problem of finding the minimal braid word representing braids from a given class. This problem, of course, gives an algorithm for braid recognition since only the trivial braid has the braid word of length 0.

Here we give an algorithm solving this problem for the case of three strands. This algorithm is due to M. Berger, see [Ber].

For the braid group $Br(3)$ there exists a natural automorphism that transposes σ_1 and σ_2. Let T be a three–strand braid word; denote the braid obtained from T by applying this isomorphism by $\hat{T}$.

The *length* of a braid word is simply the number of characters in this word.

Definition 11.25. $\Delta = \sigma_1\sigma_2\sigma_1 = \sigma_2\sigma_1\sigma_2$

Note that Δ "almost commutes" with all braids. Namely, given a braid A, we have $A\Delta = \Delta\hat{A}$.

Definition 11.26. A *wrap* is any of the four words

$$\sigma_1\sigma_2, \quad \sigma_2\sigma_1, \quad \sigma_1^{-1}\sigma_2^{-1}, \quad \sigma_2^{-1}\sigma_1^{-1}.$$

The algorithm to be described consists of three steps.

The first step. Consider a braid word B_0. Let us search for any occurrence of Δ or Δ^{-1} and take them to the left by using $A\Delta = \Delta\hat{A}$.

We proceed until B_0 has been converted to a word having the form $\Delta^n B_1$, where B_1 is free of Δ's.

The second step clears away the wraps. Namely, finding a wrap in B_1, we replace it as follows:

$$\sigma_1\sigma_2 = \Delta\sigma_1^{-1},$$

$$\sigma_2\sigma_1 = \Delta\sigma_2^{-1},$$

$$\sigma_1^{-1}\sigma_2^{-1} = \Delta^{-1}\sigma_1,$$

$$\sigma_2^{-1}\sigma_1^{-1} = \Delta^{-1}\sigma_2.$$

Thus we obtain the form $B = \Delta^p B_2$ or $B = \Delta^p \hat{B}_2$, where

$$B_2 = \sigma_1^{q_1}\sigma_2^{-r_1}\sigma_1^{q_2}\sigma_2^{-r_2}\ldots\sigma_1^{q_m}\sigma_2^{-r_m},$$

where all q_i's and r_i's are some positive integers.

Finally, in the third step we partially reverse the second step. For definiteness, suppose that $p > 0$ and that $B = \Delta^p B_2$.

First, we take one Δ from the left and bring it to the right searching for σ_2^{-1}. We replace Δ with σ_2^{-1} by the wrap $\sigma_2\sigma_1$. Each time this is done, the length is reduced by 2.

Now we repeat this operation until there are no Δ's remaining (if $p \leq r$) or no σ_2^{-1} remaining ($p \geq r$). Denote the obtained braid word by B_{min}.

The main theorem of Berger's work is the following:

Theorem 11.6. *Given B, the word B_{min} has the minimum length among all braid words equivalent to B.*

11.4   Spherical, cylindrical, and other braids

Actually, the invariant described above (by means of admissible systems of curves) admits generalisations for cases of braids in different spaces.

11.4.1   Spherical braids

We recall that a *spherical* braid on n strands is an element of $\pi_1(X_n)$, where X_n is the configuration space of non-ordered n–point sets on the standard sphere S^2.

As in the case of ordinary braids, spherical braids admit a simple representation by n strands in the space[2] $S_z^2 \times I_t$ coming downwards with respect to the coordinate t (height) and connecting fixed points $A_i \times \{1\}$ and $A_j \times \{0\}$, where $A_1, \ldots, A_n$ are fixed points on S^2. Like ordinary braids, spherical braids are considered up to natural isotopy: we decree isotopic braids to be the same; spherical braids form a group. Denote it by $SB(n)$.

Without loss of generality one can assume that there exists a point $X \in S^2$ such that no strand of a given spherical braid contains $X \times t$ for any $t \in [0,1]$.

This means that each spherical braid comes from a (not necessarily unique) ordinary braid. More precisely, there exists a homomorphic map h from $Br(n)$ onto $SB(n)$ defined as follows: each braid b in $\mathbb{R}^3 = \mathbb{R}^2 \times \mathbb{R}$ generates a spherical braid b' simply by compactifying $\mathbb{R}^2$ by a point, thus by mapping $\mathbb{R}^3$ to $S^2 \times \mathbb{R}$. The homomorphic property of the braid group map follows in a straightforward way.

It is known (see, e.g. [Fra]) that the kernel of h is generated by the only element $\Sigma_n = (\sigma_1 \ldots \sigma_{n-1})^n$ for all $n \geq 2$, see [Fra].

Let us now prove this theorem explicitly. The proof that Σ_n commutes with the whole group $Br(n)$ is obvious. Actually, to the braid Σ_n, one can attach a band such that the first and the last strands are parts of the boundary of this band and all the other strands divide the band into smaller bands; see Fig. 11.13.

---

[2]Here I is a unit segment; z and t denote coordinates on S^2 and I, respectively.

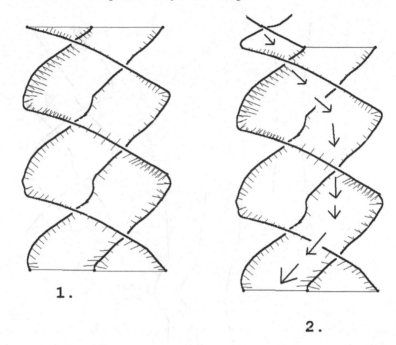

1.

2.

FIGURE 11.13: Σ_N commutes with generators

Now, each generator σ_i of the braid group can be taken along the corresponding smaller band from the top to the bottom, as shown in Fig. 11.13.

This means that Σ_n really lies in the centre of the braid group.

The remaining part of the proof can also be expressed in the language of bands. To do it, one should use induction on the number of strands (starting from three strands). Here we should slightly modify the induction basis: each **pure braid** that commutes with the whole braid group is a power of Σ_n. We do it in order to be able to start from the case of two strands. For two strands, the induction basis is evident.

Then, the induction step can be proved in the following manner: we take our n–strand braid that commutes with anything. It should be pure. Thus, we can consider n pure braids obtained from this one by deleting some strand (one of n). By the induction hypothesis, each of these braids should be Σ_{n-1}^k for some integer k. The remaining part of the proof is left to the reader.

Obviously, the invariant f (see page 172) distinguishes Σ_N and the trivial braid; thus it is not an invariant for $SB(n)$. Moreover, the described kernel coincides with the centre of $Br(n)$.

The main idea of the proof (see, e.g. [Fra]) is the following. Consider the trivial braid represented in the most natural way in $\mathbb{R}^3 \subset S^2 \times \mathbb{R}$. Let us attach a band to it in the simplest way. Now, while isotoping the braid in $SB(n)$, one can observe what can happen with the band. The only thing that

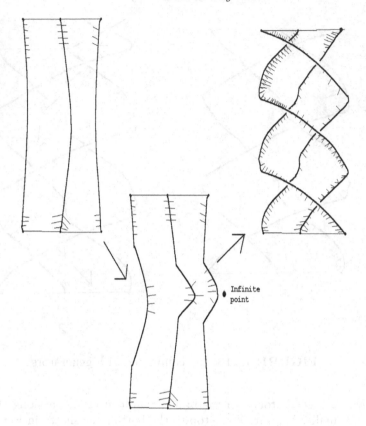

Infinite
point

FIGURE 11.14: The twist of the band

can happen is the twist of the band. This occurs when we pass through the compactification point $X \in S^2 = \mathbb{R}^2 \cup \{X\}$; see Fig. 11.14.

Now, it is evident that after a certain number of twists, our braid (in the sense of $Br(n)$) just becomes some power of Σ_n. Thus, we have proved that no other braids but powers of Σ_n lie in the kernel of the map $Br(n) \to SB(n)$. On the other hand, Σ_n really represents the trivial braid in $SB(n)$ for the same "twist" reasons.

The aim of this subsection is to correct the invariant f for the case of spherical braids.

We shall do this in the following way. We take a spherical braid b and its (infinitely many) preimages b_α with respect to h. Then we take their images $f(b_\alpha)$, which are, certainly, different. Thus the aim is to construct a map acting on $f(\cdot)$ that should bring all $f(b_\alpha)$ together. This is the way to construct a spherical braid invariant. We now construct a thin invariant that for any other braid b' and its preimages b'_α does not glue $f(b_\alpha)$ and $f(b'_\alpha)$. Thus, the invariant to construct must be complete.

Let us introduce the sets $E'_1, \ldots, E'_n$ by factorising E_i with respect to the relation $a_1 \ldots a_n = e$. Thus we get a map $\mathfrak{h} : (E_1, \ldots, E_n) \mapsto (E'_1, \ldots, E'_n)$.

Definition 11.27. A *spherical n–system* is a set of elements $e'_1 \in E'_1, \ldots, e'_n \in E'_n$. An *ordered spherical n–system* is a spherical n–system together with a permutation from S_n.

Now, let us define the map f_S from ordinary braids to spherical n–systems as follows. For each braid b the ordered n–system $f(b)$ consists of the permutation s corresponding to b and a set $e_i \in E_i, i = 1, \ldots, n$. Then the ordered spherical n–system $f_S(\beta)$ consists of the permutation s and the set $\mathfrak{h}(e_1), \ldots, \mathfrak{h}(e_n)$. Obviously, f is an ordinary braid invariant, and so is f_S.

Theorem 11.7. *The function f_S is a complete invariant of spherical braids; i.e., two braids $b, b' \in B_n$ generate the same spherical braid if and only if $f_S(b) = f_S(b')$.*

Proof. First, let us note that the statement of Lemma 11.1 is true for the invariant f_S as well. The proof is literally the same.

Thus, for any braid b we have $f_S(b) = f_S(\Sigma b)$, and, hence f_S is a braid invariant and Σ commutes with b, $f_S(b\Sigma) = f_S(\Sigma b) = f_S(b)$. Thus, if b and b' generate the same spherical braid then $f_S(b) = f_S(b')$. So, f_S is invariant.

Now, let us prove the inverse statement; i.e., that f_S is complete.

Indeed, we have to show that if β_1, β_2 are spherically equivalent braids, then $f_S(\beta_1) = f_S(\beta_2)$. By Lemma 11.1 for f_S we see that it suffices to show that f_S recognises the trivial spherical braid. Suppose β is an ordinary braid, and $h(\beta)$ is the spherical braid generated by β. Suppose that $f_S(h(\beta)) = e$. By definition, the value $f(\beta)$ is the following. The permutation of the braid is trivial and the n–system is $((a_1 \ldots a_n)^{k_1}, \ldots, (a_1 \ldots a_n)^{k_n})$ for some integer $k_1, \ldots, k_n$. Recall that the n–system comes from the admissible system of curves (non-intersecting). Thus we see that $k_1 = k_2 = \cdots = k_n$. Let $k = k_1 = \cdots = k_n$.

The only thing to check is that if the n–system is

$$\{((a_1 \ldots a_n)^k, \ldots, (a_1 \ldots a_n)^k)\}$$

and the permutation is trivial then $\beta = \Sigma^k$.

But β represents the trivial spherical braid; i.e., $h(\beta) = e$. This completes the proof.

$\square$

11.4.2 Cylindrical braids

Let C be the cylinder $S^1 \times I$.

Definition 11.28. A *cylindrical n–strand braid* is an element of $\pi_1(C_n)$, where C_n is the configuration space of non-ordered n–point sets on C. Cylindrical braids are considered up to natural isotopy. Like ordinary and spherical braids, cylindrical n–strand braids form a group. Denote this group by $CB(n)$.

The construction of the invariant for cylindrical braids is even simpler than that for spherical braids. This simplicity results from the structure of C, which is the product of the interval I and the circle.

A cylindrical n-strand braid can be considered as a set of n curves in $C \times I_t$ coming downwards from $t = 2$ to $t = 1$ in such a way that the ends of the curves generate the set $\{Y_i \times \{1\}, Y_j \times \{2\}, i, j = 1, \ldots, n, Y_i \in C\}$. The set $C = C \times I_t = S^1_\varphi \times I_s \times I_t$ can be considered in $R^3 = Oxyz$: the coordinate t corresponds to $z \in [1, 2]$, and $\varphi \in [0, 2\pi), s \in [1, 2]$ form a polar coordinate system of the plane Oxy.

For each curve in C we can consider its projection on the cylinder $S^1 \times I_t$. Thus, for a cylindrical braid β we have a system of curves on the cylinder $S^1 \times I_t$, with coordinate t decreasing from two to one. In a general position these curves have only double transversal crossing points, lying on different levels of t. For each crossing we must indicate which curve has the greater coordinate X (forms an overcrossing); the other curve forms an undercrossing.

Fix a point $x \in S^1$. Now, *a singular level* is a value t such that $S^1 \times \{t\}$ contains either a crossing or an intersection of a braid strand with the line $x \times \mathbb{R}$.

Let us require that no crossings lie in $x \times \mathbb{R}$; all intersections of strands with $x \times \mathbb{R}$ are transversal and each singular level contains either only one crossing or only one intersection point. Let us also require that no crossing lies on the intersection line.

Definition 11.29. Such a curve endowed with an undercrossing structure is called *a diagram of a cylindrical braid.*

Remark 11.10. *Obviously, all ordinary braids generate cylindrical braids by embedding of $\mathbb{R}^1$ in S^1 and $\mathbb{R}^2$ in $S^1 \times \mathbb{R}^1$. The inverse statement, however, is not true: if a strand represents a non-trivial element of $\pi_1(S \times \mathbb{R}^1)$, then the braid does not come from an ordinary braid. For instance $CB(1) \cong \mathbb{Z}$.*

Like the ordinary braid group, the cylindrical braid group $CB(n)$ has a simple presentation by generators and relations.

Indeed, let β be a braid diagram on the cylinder $S^1 \times \mathbb{R}_t$. Then, having a cylindrical braid diagram β, we can write a word as follows. Denote the set $I \times \mathbb{R} = (S^1 \backslash x) \times \mathbb{R}$ by T. Each non-singular level of the braid β consists of n points (coming from the strand). So, each crossing can be given a number $\sigma_i^{\pm 1}$ as in the case of ordinary braids, $i = 1, \ldots, n$.

For the intersection point we write τ if while walking along the strand downwards we intersect the rightmost boundary of T and return from the left side, and τ^{-1} otherwise.

Here τ represents an additional generator (with respect to the ordinary braid group generator).

Obviously, the elements $\sigma_1 \ldots, \sigma_{n-1}$ together with τ form a system of generators.

An example of a braid word obtained from a cylindrical braid diagram is shown in Fig. 11.15.

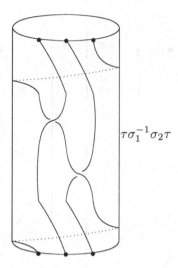

$$\tau\sigma_1^{-1}\sigma_2\tau$$

FIGURE 11.15: A cylindrical braid diagram and the corresponding word

As in the case of ordinary braids, the set of moves concerning cylindrical braids can be easily constructed. They are:

1. Moves of the diagram preserving the combinatorial structure of crossings (but, possibly, changing the height order of a crossing); see Fig. 11.16.

2. The second Reidemeister move; see Fig. 11.17.

3. The third Reidemeister move; see Fig. 11.18.

Besides the relations for the ordinary braid group $\{\sigma_i\sigma_j = \sigma_j\sigma_i, |i-j| > 1\}$ and $\{\sigma_i\sigma_{i+1}\sigma_i = \sigma_{i+1}\sigma_i\sigma_{i+1}\}$ we get more relations: $\tau\sigma_{i+1} = \sigma_i\tau$ for $i = 1,\ldots,n-2$ and $\sigma_{n-1}\tau^2 = \tau^2\sigma_1$.

The geometric meaning of the additional relations is as follows. The first series represents the change of crossing numeration under the action of τ: σ_i becomes σ_{i+1} when the rightmost strand appears on the left flank. The second additional series (of one relation) means that the rightmost crossing is moved by one full turn together with the two strands, generating it.

It can easily be checked that this system of relations is complete.

Remark 11.11. *It is easy to show that the additional relations place the element τ^n at the centre of the cylindrical braid group.*

Remark 11.12. *Both the second and the third Reidemeister moves for cylindrical braids are considered in a part of cylinder; i.e., they are just the same as in the case of an ordinary braid.*

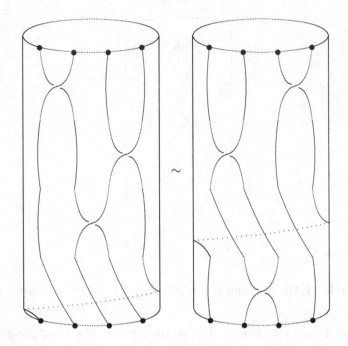

FIGURE 11.16: Transforming a cylindrical braid diagram

Now, having a diagram β of a cylindrical braid b, let us construct the invariant $f_C(b) \equiv f(\beta)$ (in the sequel, we prove that it is well defined).

Consider the cylinder $S^1_\varphi \times \mathbb{R}^1$.

Definition 11.30. An *admissible cylindrical system of n curves* is a family of n non–intersecting non–self-intersecting curves in the upper half–cylinder $S^1 \times \mathbb{R}_+$ such that each curve connects a point with abscissa zero with a point with abscissa one, such that the coordinate φ for all curve ends runs through the set

$$\left\{0, \frac{2\pi}{n}, \ldots, \frac{2(n-1)\pi}{n}\right\}$$

All points

$$\left(\frac{2\pi j}{n}, 1\right),$$

where $j = 1, \ldots, n$ are called *upper points*, and all points $(j, 0)$ are called *lower points*.

Denote

$$S^1 \backslash \cup_i \left\{\frac{2\pi i}{n}\right\}$$

by C_n.

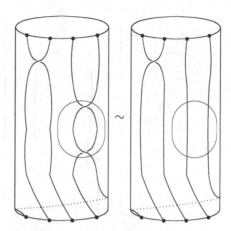

FIGURE 11.17: Applying the 2nd Reidemeister move to a cylindrical braid diagram

Consider the diagram β on the cylinder. Now let us resolve all crossings of β starting from the upper one as in the case of the ordinary braid.

Thus we obtain an admissible cylindrical system of curves.

The next definition is similar to that for the case of ordinary braids.

Definition 11.31. Two admissible cylindrical systems of curves A and A' are called *equivalent* if there exists a homotopy between A and A' in the class of curves with fixed end points, lying in the upper half–cylinder, such that no interior point of any curve can coincide with an upper point.

Having a diagram β of a braid b, we obtain an admissible cylindrical system $A(\beta)$ of curves, corresponding to it. We can take an equivalence class of $A(\beta)$. Denote it by $f_C(\beta)$.

Now, let us prove the following theorem.

Theorem 11.8. *1. Map f_C is a braid invariant; i.e., for different β_1, β_2 representing the same braid b we have $f_C(\beta_1) = f_C(\beta_2)$. In this case we shall write simply $f(b)$.*

2. The invariant f_C is complete; i.e., $f_C(b_1) = f_C(b_2)$ implies $b_1 = b_2$.

Proof. To prove the first part of the theorem, we only have to check the invariance of f_C under the second and the third Reidemeister moves. The proof is the same as in the case of ordinary braids, see Figs. 11.3 and Fig.11.4.

The proof of the second part is also analogous to the proof of completeness in the ordinary case. By an admissible cylindrical system of curves we restore a cylindrical braid (with lower points $r = 1, \varphi = \frac{2\pi j}{n}$ and upper points $r = 2, \varphi = \frac{2\pi k}{n}$) by parametrising each curve from the system from 1 to 2. Now,

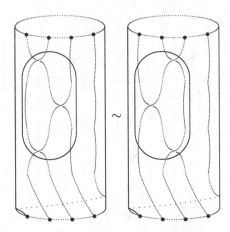

FIGURE 11.18: Applying the 3rd Reidemeister move to a cylindrical braid diagram

let us prove that equivalent admissible cylindrical systems of curves generate the same braid.

To do this, let us fix the permutation $s \in S_n$ (obviously, two admissible curves can be equivalent only if their permutations coincide). Then we choose two equivalent admissible systems A and A' of n curves and choose admissible parametrisations for them. The rest of the proof we leave for the reader as an exercise.   □

Like the ordinary braid invariant f, the invariant f_C is also easily recognisable. Indeed, instead of curves on P_n, we consider curves in C_n. So, our invariant can be completely encoded by the following object.

Let G_t be a free group with generators $a_1, \ldots, a_n, t$. Denote by G_i, $i = 1, \ldots, n$, the right residue class of G by a_i.

Definition 11.32. A *cylindrical n–family* is a set $g_1 \in G_1, \ldots, g_n \in G_n$ together with a permutation $s \in S_n$.

Obviously, values of the invariant f_C can be completely encoded by cylindrical n–families. The permutation is taken directly from the admissible system of curves, and elements g_i correspond to curve homotopy types in C_n with fixed points, where t stands for the element of C_n obtained by passing along the parallel of C_n.

Chapter 12

Markov's theorem. The Yang–Baxter equation

In his celebrated work [Mar], A.A. Markov has described the theorem about necessary and sufficient conditions for braids to represent isotopic links. However, his proof did not contain all rigorous details. He left this problem to N.M. Weinberg, who died soon after his first publication on the subject [Wei]. The first published rigorous proof belongs to Joan Birman, [Bir1]. The newest proofs of Markov's theorem can be found in [Tra] and in [BM6].

We shall describe the proof according Hugh Morton [Mor1], where a shorter (than Markov's one) proof is given.

After this, we shall give some precisions of Alexander's and Markov's theorems due to Makanin.

In the third part of the chapter, we shall discuss the Yang–Baxter equation which is closely connected with braid groups and knot invariants.

12.1 Markov's theorem after Morton

12.1.1 Formulation. Definitions. Threadings.

The Markov theorem gives an answer to the question of when the closures of two braids represent isotopic links. However, this answer does not lead to an algorithm; i.e., it gives only a list of moves, necessary and sufficient to establish such an isotopy, but does not say *how to use these moves* and *when to stop*.

In his work, Morton uses the original idea of *threading* — an alternative way of representing a link as a closure of a braid (besides those proposed by Alexander and Vogel).

Remark 12.1. *We shall consider each closure of braids as a set of curves inside the cylinder not intersecting its axis; the axis of the braid is the closure of the curve coinciding with the axis of the cylinder inside the cylinder.*

We start with the definitions.

Definition 12.1. Let K be an oriented link in $\mathbb{R}^3$. Let L be an unknotted

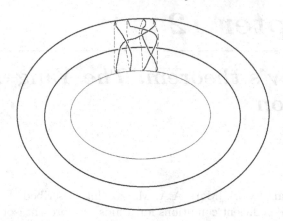

FIGURE 12.1: Representing a braid in a full torus

curve. We say that K is *braided with respect to* L or $K \cup L$ is a *braid–link*, if K and L represent the closure of some braid and the axis of this braid, respectively (i.e., K lies inside the full torus $S^1 \times D$, where the coordinate of S^1 is increasing, and L is the axis of the full torus); see Fig. 12.1.

Having a planar diagram of some braid closure, the corresponding braid–link can be obtained from it by threading this diagram by a circle; see Fig. 12.2.

Let K be a planar diagram of some oriented link. Consider some curve L on the projection plane P of the link K such that the curve L intersects the projection of the link K transversely and does not pass through crossings of K.

Definition 12.2. A *choice of overpasses* for a link diagram K is a union of two sets $S = \{s_1, \ldots, s_k\}$, $F = \{f_1, \ldots, f_k\}$ of points at the edges of K (points should not coincide with crossings) such that while passing along the orientation of K, the points from S alternate with points from F; besides, each interval $[s, f]$ does not contain undercrossings and each $[f, s]$ does not contain overcrossings; i.e. $[s, f]$ are arcs and $[f, s]$ are lower arcs.

Definition 12.3. We say that a curve L whose projection on the plane of the link K is a simple curve *threads* K *according to a given choice* (S, F) *of overpasses* if the interval of K goes over L when it starts in a domain containing points from S, and it passes under L if it starts in a domain containing points from S, as it is shown in Fig. 12.3.

Remark 12.2. *We do not require that this interval of L contain elements of the set S or F.*

For a given link diagram K and a curve L on the projection plane, such that L separates points from S from points from F then we can arrange over– and undercrossings at intersection points between K and L in such a way that the curve L' obtained from L threads the link K.

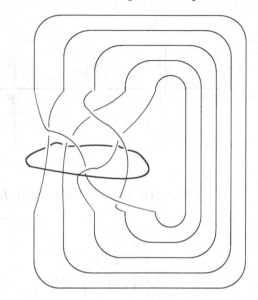

FIGURE 12.2: A threaded braid closure

Let us now prove the following theorem.

Theorem 12.1. *If L threads the link K then K is a braid with respect to L.*

Proof. Let us choose the overpasses (S, F) for the diagram of the link K (in an arbitrary way) and some curve L on the projection plane P of the diagram K such that L separates points from S from points belonging to F. Let us straighten the curve L in the plane P by a homeomorphism of P onto itself: we require that the transformed L is a straight line inside a domain D, containing K; L should be closed outside D (say, by a large half–circle). Consequently, points of S lie on one side of this line, and points of F lie on the other side. Such a transformation is shown in Fig. 12.4.

Without loss of generality, we can suppose that all under– and overcrossings of the diagram L lie in two planes parallel to P (just over and under the images of the corresponding projections).

Now, let us change the point of view and think of P as the plane Oxz and L as the axis Oz that is closed far away from the origin of coordinates.

Let us consider the line L (without its "infinite" circular part) as the axis of cylindrical coordinates. Then the plane P is divided into two half–planes; one of them is given by the equation $\{\theta = 0\}$ and the other one satisfies the equation $\{\theta = \pi\}$. Here the half–plane $z = 0, x > 0$ is thought to have coordinate $\theta = 0$; points over this half-plane are thought to have positive coordinates.

Let us construct a link isotopic to K as follows. Place all lower arcs of K (i.e., all intervals $[f, s]$) on the half planes $\{\theta = -\varepsilon\}$ and $\{\theta = \pi + \varepsilon\}$, and all

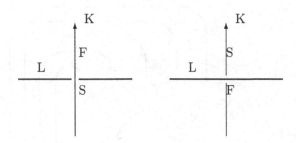

FIGURE 12.3: Crossings with L

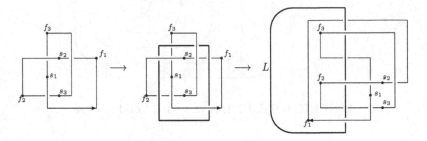

FIGURE 12.4: Straightening the curve L

arcs on the half–planes $\{\theta = \varepsilon\}$ and $\{\theta = \pi - \varepsilon\}$, where ε is small enough. Herewith, we shall add small intervals over all points belonging to S or F such that each interval is projected to one point on Oxz.

Let us represent the arcs where K intersects with L by vertical arcs.

Thus, we have made the polar coordinate θ of the link K to be always constant or increasing.

In Figs. 12.5 and 12.6 we show how to construct a knot with non-decreasing polar coordinate. This knot is isotopic to the knot shown in Fig. 12.4.

After a small deformation of the obtained link, we can make this coordinate strictly monotonic.

Thus, the transformed link (which we shall also denote by K) will represent a braid with respect to L. $\qquad\square$

It follows from Theorem 12.1 that if some link K is a braid with respect to an unknotted curve L then K is isotopic to a closure of some braid.

Theorem 12.2. *Each closure K of any braid B admits a threading by some curve L in such a way that K is a braid with respect to L.*

Proof. Let $D^2 \times I$ be a cylinder. Consider B as a braid connecting points lying on the upper base of the cylinder with points on the lower base of the same

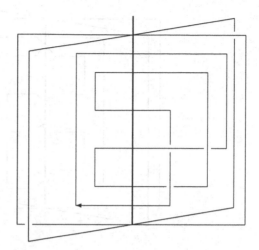

FIGURE 12.5: Link diagram intersecting L

cylinder. Now, let us close the braid as follows. Connect the lower points with the upper ones by lines, going horizontally along the bases and vertically at some discrete moments, as shown in Fig. 12.7.

Let us apply the isotopy that straightens the strands and changes homeo-morphically the upper base of the cylinder. Thus we obtain a link that admits a simple threading that can be constructed as follows. Let h_0 be the height level of the lower base and h_1 be the level of the upper base. Denote the set of lower ends of the braid B by A_1 and the set of upper ones by A_2.

One can assume that the levels h_0 and h_1 contain some additional sets of vertices B_1 and B_2 by means of which we are going to construct the closure of the braid. More precisely, the points from A_1 are connected by parallel lines with points from A_2, and points from B_1 are connected by parallel lines with points from B_2. Now, let us consider the circle lying on the plane at the level $h = \frac{h_1 + h_2}{2}$ and separating sets of lines $A_1 A_2$ and $B_1 B_2$.

Let us project the diagram on the base of the cylinder and take the set A_1 as S and A_2 as F.

It is easy to see that in this case the projection of the circle is really a threading of the link. □

Theorems 12.1 and 12.2 imply the Alexander theorem; the proofs of these theorems give us a concrete algorithm (different from Alexander's and Vogel's methods) to represent any link as a closure of a braid.

Let us now recall the main theorem of this chapter. It has already been formulated in Chapter 9.

Theorem 12.3. *The closures of braids A and B are isotopic if and only if B can be obtained from A by a sequence of the following moves (Markov moves):*

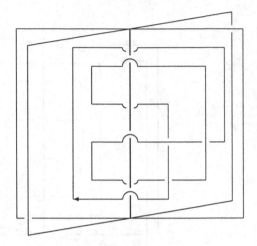

FIGURE 12.6: The polar coordinate is increasing while moving along the link

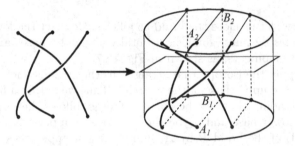

FIGURE 12.7: A braid and a braided link

1. *conjugation* $b \to a^{-1}ba$ *by an arbitrary braid* a *with the same number of strands as* b,

2. *the move* $b \to b\sigma_n^{\pm 1}$, *where* b *is a braid on* n *strands and the obtained braid has* $n + 1$ *strands,*

3. *the inverse transformation of 2.*

The necessity of these two moves is evident. The isotopies between corresponding pairs of braid closures are shown in Fig. 12.8.

In Fig. 12.8.b the first Reidemeister move comes into play. This move did not take part in braid isotopies, so this kind of knot isotopy appears here.

12.1.2 Markov's theorem and threadings

Let us now reformulate the difficult part of the Markov theorem.

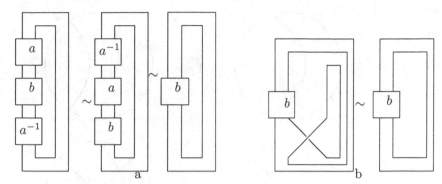

FIGURE 12.8: Two Markov moves represent isotopy

To do it, we shall need some definitions.

Definition 12.4. We say that two braided links $K \cup L$ and $K' \cup L'$ are *simply Markov equivalent* if there exists an isotopy of the second one, taking L' to L and K' to the link coinciding with K everywhere except one arc. The link K contains an arc α and K' contains a link α' with the same ends.

1. The polar coordinate is constant on the arc α and monotonically increasing on α'.

2. The arcs α, α' bound a disc intersecting L transversely at a unique point.

Definition 12.5. Two braided links are *Markov equivalent* if one of them can be transformed to the other by a sequence of isotopies and simple Markov equivalences.

Exercise 12.1. *Show that closures of two n–strand braids are isotopic in the class of closures of n–strand braids if and only if these two braids are conjugated.*

Lemma 12.1. *If links $K \cup L$ and $K' \cup L'$ are simply Markov equivalent then they represent threaded closures of braids, which are isotopic to some braids $\beta \in Br(n)$ and $\beta \sigma_n^{\pm 1} \in Br(n+1)$.*

Proof. Suppose the polar coordinate evaluated at points of the curve α equals θ_0. Consider the arc α_0. Without loss of generality, one can assume that the coordinate is almost everywhere equal to θ_0 and in some small neighbourhood of L' the arc α_0 makes a loop and this loop corresponds to the n-th (last) strand of the braid α. We can isotope the braids K and K' in the neighbourhood $\{\alpha = \theta_0 \pm \varepsilon\}$ in such a way that the final points of the arcs α and α_0 lie in a small neighbourhood of L'. The remaining part of the Lemma is now evident. $\square$

Lemma 12.1 together with Exercise 12.1 allows us to reformulate the difficult part of the Markov theorem as follows:

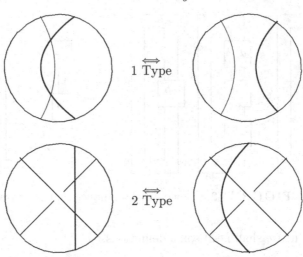

FIGURE 12.9: Types of transformations

Theorem 12.4. *Let β and γ be two braids whose closures B and Γ are isotopic as oriented links. Let us thread these closures and obtain some braided links B' and Γ'. Then B' and Γ' are Markov equivalent.*

To go further, we shall need some auxiliary lemmas and theorems.

Lemma 12.2. *Consider an oriented link diagram K on the plane P and fix the choice of overpasses (S, F). Then the threadings of K by different curves L and L' separating the sets S and F are Markov equivalent.*

Proof. The main idea of the proof is the following. First we consider the case when the curves L and L' are isotopic in the complement $P\backslash(S \cup F)$. In this case, one of them can be transformed to the other by means of moves in such a way that each of these moves is a Markov equivalence.

In the common case we shall use one extra move when two branches of the line L pass through some point from S (or F). Such a move is a Markov equivalence as well (this will be clear from the definition).

Let us give the proof in more detail.

The case a).

Suppose L and L' are isotopic in $P\backslash(S \cup F)$. Then $K \cup L$ and $K \cup L'$ can be obtained from each other by a sequence of transformations of the first and the second type, shown in Fig. 12.9.

The first type is represented either by the second Reidemeister move or by the "hooking" move, that adds two crossings in alternating order. It will be shown below that this move is a simple Markov equivalence.

The second type of transformation is an isotopy in all cases except that shown in Fig. 12.10.

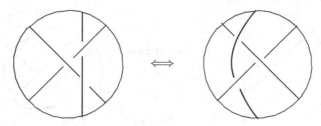

FIGURE 12.10: Nonisotopic transformation of type II

In this case, the first threading can be transformed to the second one by a sequence of moves of the first type and isotopies; see Fig. 12.11.

Here one should note that the passes of the diagram of K under the line L are alternating with passes of K over L while going along the link K. It remains to show that the two threadings obtained from each other by a transformation of the first type are simply Markov equivalent.

Thus, the part of the link K shown in Fig. 12.11 belongs either to an upper branch or to the lower branch. In the threading construction we can assume that both parts of the link K on the same side of the line L lie on one and the same level p_L (it might be either $\{\theta = \pi \pm \varepsilon\}$ or $\{\theta = \pm\varepsilon\}$ depending on the side of overcrossing).

Let us connect them by an arc as shown in Fig. 12.12.

Now, the arcs α and α' bound a disc. Thus we obtain a simple Markov equivalence of the two threadings.

The case b).

In the general case, note that if the curve M' of P that separates the set S from the set F can be isotoped to the curve M by means of moving the two arcs of the link K through some point of S or F (as shown in Fig. 12.13) then M and M' represent isotopic threadings.

Such an isotopy of the "curvilinear" line L (or, equivalently, motions of points from the sets S and F) is divided into several steps. Between these steps, we apply discrete moves changing the combinatorial type of the disposition for L with respect to the sets S and F. Thus, one can consider the discrete set of such dispositions, between two of each some elementary transformation takes place.

Without loss of generality, we may assume the following.

Let S and F consist of points $\{(-1, a_i)\}$ and $\{(1, a_i)\}$ for some $a_1, \dots a_k$, respectively. Let L be a part of Oy closed by a large half–circle in such a way that the interior of L contains F. We may assume that K is parallel to Ox near each of the points $s_1, \dots, s_k$.

Let L' be any simple closed curve separating S from F and restricting a domain that contains the set F. Without loss of generality, we can suppose that the curve L' intersects the rays $y = a_i, x < -1$ transversely; see Fig.12.14. Let us enumerate these intersections according to the decreasing of the ab-

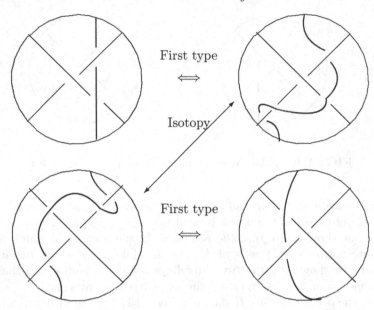

FIGURE 12.11: Transformation of type II reduces to transformations of type I

scissa. For each of the rays, the number of intersections is even. For each ray, let us group them pairwise: $(1,2),(3,4),\ldots$. To do this, we first isotope L' inside $P\backslash(S\bigcup F)$ in such a way that between each pair of points there are no crossings of the diagram K.

Now, let us move the curve line L' to the right in such a way that after performing the operation all points lie on the left side of the curve L. Let us divide such a transformation into stages when L' does not contain points from S and moments when L' does. In the first case, such a transformation is a Markov equivalence as in the case a). In the second case, let us assume that the intersection points of L' with each ray disappear pairwise; i.e., the curve L' consequently passes two crossings with the same point s_i. This move is a Markov equivalence as well; see Fig. 12.13.

Thus, the threading by means of L' is Markov equivalent to the threading by the curve lying on the right hand related from all s_i. This curve is isotopic to L inside the set $P\backslash(S\bigcup F)$; i.e., the threading by means of such curve is Markov equivalent to the threading by means of L, see a). Consequently, the threading by L' is Markov equivalent to the threading by L. This completes the proof of the lemma.

$\square$

Lemma 12.3. *Given a choice (S,F) of overpasses for a link diagram K, and*

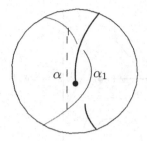

FIGURE 12.12: Transformation of type I is a simple Markov equivalence

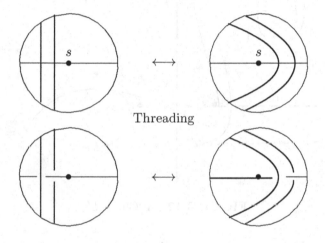

Threading

FIGURE 12.13: Moving two arcs

a point s in K, not belonging to F, then there exists a choice of overpasses $(\bar{S}, \bar{F})$ such that $s \in \bar{S}, S \subset \bar{S}, F \subset \bar{F}$.

Proof. The idea of the proof is pretty simple: we add elements of S or F where we want compensating them by corresponding elements of F or S. If s lies on an upper arc of (S, F) then one can choose f just before s with respect to the orientation of K; thus the interval $[f, s] \subset K$ contains no overcrossings. In the case when s lies on a lower arc, we can add f just after s; in this case $[s, f] \subset K$ contains no undercrossings. □

Theorem 12.5. *Each two threadings $K \cup L$ and $K \cup L'$ of the same diagram K are Markov equivalent.*

Proof. Let us choose some overpasses (S, F) for the threading $K \cup L$ and (S', F') for the threading $K \cup L'$. According to Lemma 12.3, there exists a choice of overpasses (S'', F'') such that $(S, F), (S', F') \subset (S'', F'')$, and the two threadings with this choice of overpasses, the first of which is Markov

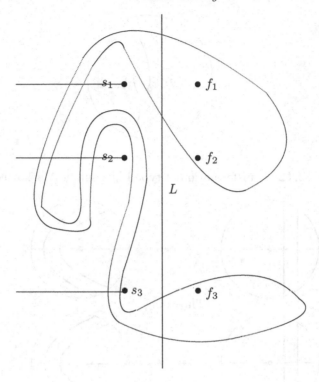

FIGURE 12.14: Curve L'

equivalent to $K \cup L$ and the second is Markov equivalent to $K \cup L'$. By Lemma 12.2, these two threadings are Markov equivalent. Thus, the initial two threadings are Markov equivalent and this completes the proof. □

Theorem 12.6. *Any two planar diagrams of isotopic links have Markov-equivalent threadings.*

Proof. To prove this theorem, we have to show how to construct Markov equivalent threadings for diagrams obtained from each other by using Reidemeister moves.

By Theorem 12.5, we can take any choice of overpasses for each of these diagrams. The idea is to be able to reconstruct the choice of overpasses together with L after each Reidemeister move.

Without loss of generality, for the first two Reidemeister moves we can choose the separating curve outside the small disc of the move (in the case of Ω_1 we choose one vertex s inside the disc; in the case of Ω_2 all vertices from (S, F) are outside the disc).

We have shown that the link diagrams obtained from each other by Ω_1 or Ω_2 obtain Markov equivalent threadings.

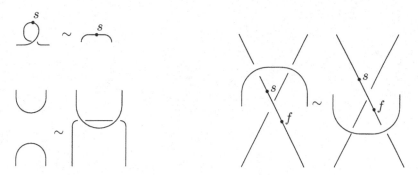

FIGURE 12.15: Choice of overpasses for Reidemeister moves

For Ω_3 we can take one vertex s and one vertex f inside the disc and all the other vertices outside the disc; see Fig. 12.15.

In Fig. 12.16 we show that the threading corresponding to diagrams obtained from each other by Ω_3 are isotopic and hence Markov equivalent (we show only one case, the other cases of Ω_3, with orientation and disposition of L and K in the left picture are quite analogous).

In the upper part of this figure, we show how the line L can be transformed with respect to this move Ω_3. In the lower part, we show how one concrete transformation is realised with over– and undercrossings between L and K.

□

Now, we are ready to prove the difficult part of the Markov theorem.

Proof of Theorem 12.4. Let $K \cup L$, $K' \cup L'$ be two braided links whence K and K' are isotopic as links.

By Theorem 12.2, the link $K \cup L$ is a threading of some diagram K and the link $K' \cup L'$ is a threading of some diagram K'. By Theorem 12.6, one can choose Markov–equivalent threadings for the first and the second diagram. By Theorem 12.5, the first one is Markov equivalent to $K \cup L$ and the second one is Markov equivalent to $K' \cup L'$.

Consequently, the threading $K \cup L$ is Markov equivalent to $K' \cup L'$, which completes the proof of Markov's theorem. □

Let us now present an example of how to use the Markov theorem.

As we have proved before, for each two coprime numbers p and q, the toric knots of types (p, q) and (q, p) are isotopic. Let us demonstrate the Markov moves for the braids, whose closures represent trefoils: $(2, 3)$, $(3, 2)$.

Example 12.1. *Actually, the first braid has two strands and is given by σ_1^{-3}; the second one (which has three strands) is given by $\sigma_1^{-1}\sigma_2^{-1}\sigma_1^{-1}\sigma_2^{-1}$. Let us write down a sequence of Markov moves transforming the first braid to the second one:*

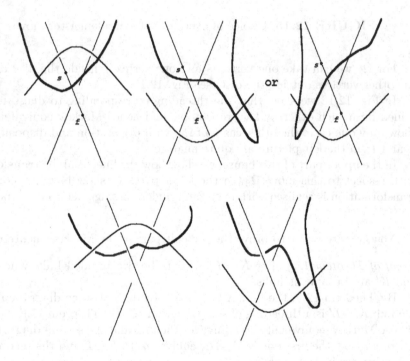

FIGURE 12.16: Isotopic threadings of diagrams differ by Ω_3

$$\sigma_1^{-3} \xrightarrow{2 \ move.} \sigma_1^{-3}\sigma_2^{-1} \xrightarrow{conj.} \sigma_1^{-2}\sigma_2^{-1}\sigma_1^{-1}$$

$$= \sigma_1^{-1}(\sigma_1^{-1}\sigma_2^{-1}\sigma_1^{-1}) \xrightarrow{braid \ isotopy} \sigma_1^{-1}\sigma_2^{-1}\sigma_1^{-1}\sigma_2^{-1}.$$

Exercise 12.2. *Perform the analogous calculation for the case of toric knots* $T(2, 2n+1)$ *and* $T(2n+1, 2)$ *and for the knots* $T(3, 4), T(4, 3)$.

12.2 Makanin's generalisations. Unary braids

In his work [Mak], G.S. Makanin proposed a nice refinement of the Alexander and Markov theorems: he proved that all knots (not braids) can be obtained as closures of so–called unary braids. Besides, he proved that for any two unary braids representing the same knot, there is a change of Markov moves from one to the other that lies in the class of unary braids. Furthermore, Makanin also gave some generating system of (harmonic) braids, such that their adjoint action has unary braids as an invariant set. For more details, see the original work [Mak].

Throughout this section, all knots are taken to be oriented.

Definition 12.6. An $(n+1)$–strand braid is called *unary* if the strand having ordinate one on the top has ordinate $(n+1)$ on the bottom and, after deleting the first strand, we obtain the trivial n–strand braid.

Obviously, unary braids generate only knots (not links).

Theorem 12.7 ([Mak]). *For each knot isotopy class K, there exists a unary braid B, such that $Cl(B)$ is isotopic to K.*

The first step here is to note that each knot K can be represented by a braid β with permutation $P = (1 \to 2 \to 3 \cdots \to n+1 \to 1)$.

The second step of the proof is to show that each braid (e.g. β with permutation P) is conjugated to some unary braid β'. Thus, by Markov's theorem, $Cl(\beta')$ is isotopic to K.

Here is the key lemma.

Lemma 12.4. *Let K be a braid from $Br(n+1)$ with permutation P. Then there exists a unary braid Y that is conjugated with K by means of a strand from $Br(n)$.*

This lemma follows from the construction of Artin [Art1] of so–called "reine Zöpfe". There is a nuance concerning mathematical terminology in German. "Reine Zöpfe" is literally "pure braids", but they have some other meaning in

German rather than in English. The word for "pure braids" used in German is "gefärbte Zöpfe" which literally means "coloured braids".

For more details concerning these notions and the proof of the lemma, see the original work of Makanin [Mak].

Theorem 12.8 ([Mak]). *Let K_1, K_2 be two isotopic knots. Let B, B' be two unary braids representing the knots K_1 and K_2, respectively. Then there exists a chain of unary braids $B = B_1, B_2, \ldots, B_k = B'$ such that each N_i is obtained from B_{i-1} by a Markov move (for all $i = 2, \ldots, k$).*

Proof. Our strategy is the following: first we find some chain "connecting" these braids by Markov's moves and then modify it by means of some additional Markov moves in order to obtain only unary braids.

Let K be a unary braid from $Br(p+1)$ and L be a unary braid from $Br(q+1)$. Thus, there exists a sequence of braids $K, Q_1, Q_2, \ldots, Q_t$ where each braid Q_{i+1} can be obtained from the previous one by one Markov move. Without loss of generality, we can assume that no two conjugations are performed one after the other. Besides, between any two Markov moves of second type we can place a conjugation by the unit braid. Thus, we might assume that each move $Q_{2j-1} \to Q_{2j}$ is a conjugation, and $Q_{2j} \to Q_{2j+1}$ is the second Markov move (either addition or removal of a strand). The other case can be considered analogously.

Furthermore, one can easily see that when an m–strand braid and a unary braid are conjugated, the braid for conjugation can be chosen from $Br(m-1)$. This follows from the Lemma 12.4.

Now, let us construct our chain. For each braid Q_i having n_i strands, let Y_i be a unary braid conjugated with the braid Q_i by a $(n_i - 1)$-strand braid.

Obviously, each Y_{2j-1} is conjugated with Y_{2j} by some braid from $Br(n_{2j})$. Besides, the braid $Y_{2j} \in Br(n_{2j})$ is conjugated with $Q_{2j} \in Br(n_{2j})$ by means of some braid δ from $Br(n_{2j} - 1)$. In the case when the transformation $Q_{2j} \to Q_{2j+1}$ adds a strand, we can perform the same operation for Y_{2j}. We obtain a braid Y_{2j}^*. Obviously, the braids Y_{2j} and Q_{2j+1} are conjugated by means of δ. So, Y_{2j}^* and Y_{2j+1} are conjugated.

In the other case when $Q_{2j} \to Q_{2j+1}$ deletes a strand, the inverse move $Q_{2j+1} \to Q_{2j}$ adds a loop. Arguing as above, we see that the braids Y_{2j} and Y_{2j+1} are connected by some Markov moves involving only unary braids.

Thus, we have constructed a chain of unary braids, connecting B with B', which completes the proof. $\square$

The Makanin work allows us to encode all knots by using words in some finite alphabet. Namely, in order to set an n–strand braid, we should just describe the behaviour of the first strand of it; i.e., we must indicate when it goes to the right (to the left) and when it forms an over (under)crossing. To do this, it would be sufficient to use four brackets: $\overset{+}{\to}, \to, \overset{+}{\leftarrow}$, and $\leftarrow$.

The only condition for such a word to give a braid is that the total number of $\rightarrow$ minus the total number of $\leftarrow$ never exceeds n in all initial subwords, and equals precisely n for the whole word.

Another approach with some detailed encoding of knots and links by words in a finite alphabet (bracket calculus) will be described later in Chapter 16.

12.3　The Yang–Baxter equation, braid groups and link invariants

The Yang–Baxter equation (YBE) was first developed by physicists. However, these equations turn out to be very convenient in many areas of mathematics. In particular, they are quite well suited for describing braids and their representations. In our book we shall not touch on the connection between the YBE and physics, for more details see in e.g. [Kau4].

Just after the revolutionary works by Jones [Jon1, Jon2], it became clear that the Jones polynomial and its generalisations (e.g. the HOMFLY-PT polynomial) are in some sense a vast family of knot invariants coming from quantum representations – the quantum invariants; for the details see [CES, Dri1, Dri2, Jon3, RT, Tur1].

Let V be the finite dimensional vector space with basis $e_1, \ldots, e_n$ over some field F. Let $R : V \otimes V \rightarrow V \otimes V$ be some endomorphism of $V \otimes V$. Consider the endomorphism

$$R_i : Id \otimes \cdots \otimes Id \otimes R \otimes Id \otimes \cdots \otimes Id : V^{\otimes n} \rightarrow V^{\otimes n},$$

where $V^{\otimes n}$ means the n–th tensor power of V, Id is the identity map, and R_i acts on the product of spaces $V \otimes V$, which have numbers $(i, i+1)$.

Definition 12.7. An operator R is said to be an R–*matrix* if it satisfies the following conditions:

$$R_i R_{i+1} R_i = R_{i+1} R_i R_{i+1}, \ i = 1, \ldots, n-1$$

$$R_i R_j = R_j R_i, \ |i - j| \geq 2.$$

The first of these conditions is called *the Yang–Baxter equation*; the second one is just far commutativity for $R_1, \ldots, R_{n-1}$.

The YBE look quite similar to Artin's relations for the braid group: they differ just by replacing σ with R, and we obtain one equation from the others.

Now, we are going to show the mathematical connection between the YBE, braid group representations and link invariants.

Having an R–matrix, one can construct a link invariant by using the following construction (proposed by V.G. Turaev in [Tur1]). Below, we just sketch this construction; for details see the original work.

First, we construct a representation of the braid group $Br(n)$ to the tensor power $V^{\otimes n}$ as $\rho(\sigma_i) = R_i$, where σ_i are standard generators of the braid groups.

After this, for any given link L we find a braid b with closure $Cl(b)$ that is isotopic to L.

One can set $T(L) = trace(\rho(b))$. In this case, $T(L)$ is invariant under braid isotopies and the first Markov move (conjugation). The latter follows from the simple fact that for any square matrices A and B of the same size, we have $trace(A) = trace(BAB^{-1})$.

In some cases (see [Tur1]), the function $T(L)$ is invariant under the second Reidemeister move as well. Thus, $T(L)$ gives a link invariant. As we have shown before, one can also introduce some specific traces (namely, the Ocneanu trace) instead of the ordinary trace. Such traces behave quite well under both Markov's moves and lead to the Jones polynomial in two variables.

Remark 12.3. *Numerical values of the Jones polynomial can be obtained by means of R–matrices.*

In quantum mechanics, one also considers the *quantum YBE* that looks like

$$R_{12}R_{13}R_{23} = R_{23}R_{13}R_{12}. \tag{12.1}$$

Here R_{ij} are obtained from some fixed matrix R generating an automorphism $R : V^{\otimes n^2} \to V^{\otimes n^2}$ (e.g. $R_{12} = R \otimes I_n$, where I_n is the $(n \times n)$ identity matrix). Each solution of (12.1) is called *quantum R–matrix*.

For the study of quantum invariants, we recommend the beautiful book by Ohtsuki, [Oht].

A review of classical R–matrices can be read in [Sem]. An interesting work concerning the problem of finding R–matrices was written by the well-known Dutch mathematician Michiel Hazewinkel [Haz].

Part III

Vassiliev's invariants. Atoms and d-diagrams

Chapter 13

Definitions and basic notions of Vassiliev invariant theory

The Vassiliev knot invariants were first proposed around 1989 by Victor A. Vassiliev [Vas1] while studying the topology of discriminant sets of smooth maps $S^1 \to \mathbb{R}^3$. A bit later, Mikhail N. Goussarov [Gus] independently found a combinatorial description of the same invariants.

13.1 Singular knots and the definition of finite type invariants

Throughout this part of the book, all knots are taken to be oriented, unless otherwise specified. Besides, we deal only with knots, not links. The analogous theory can be constructed straightforwardly for the case of links; the definitions are, however, a bit more complicated.

As we know, each knot can be transformed to the unknot by switching some crossings. This switch can be thought of as performed in $\mathbb{R}^3$.

Having a knot invariant f, one can consider its values on two knots that differ at only one crossing. Certainly, these two knots might not be isotopic; hence, these values might not coincide.

While switching the crossing continuously, the most interesting moment is the intersection moment: in this case we get what is called a *singular knot*. More precisely, a *singular knot* of degree n is an immersion of S^1 in $\mathbb{R}^3$ with only n simple transverse intersection points (i.e., points where two branches intersect transversely).

Singular knots are considered up to isotopy. The isotopy of singular knots is defined quite analogously to that for the case of classical knots. The set of singular knots of degree n (for $n = 0$ the set $\mathcal{X}_0$ consists of the classical knots) is denoted by $\mathcal{X}_n$. The set of all singular knots (including $\mathcal{X}_0$) is denoted by $\mathcal{X}$.

So, while switching a crossing of a classical knot, at some moment we get a singular knot of order one.

Then, we can define the *derivative f'* of the invariant f according to the following relation:

$$f'(\times) = f(\times) - f(\times). \tag{13.1}$$

This relation holds for all triples of diagrams that differ only outside a small domain (two of them represent classical knots and $\times$ represents the corresponding singular knot).

This relation is called *the Vassiliev relation.*

It is obvious that the invariant f' is a well-defined invariant of singular knots because with each singular knot and each vertex of it, we can associate the positive and the negative resolutions of it in $\mathbb{R}^3$. If we isotope the singular knot, the resolutions are "isotoped" together with it.

Having a knot invariant $f : \mathcal{X}_0 \to A$,[1] one can define all its derivatives of higher orders. To do this, one should take the same formula for two singular knots of order n and one singular knot of order $n+1$ (n singular vertices of each of them lie outside of the "visible" part of the diagram) and then apply the Vassiliev relation (13.1).

Thus, we define some invariant on the set $\mathcal{X}$. This invariant is called the *extension of f for singular knots.*

Notation: $f^{(n)}$.

Example 13.1. *Let us calculate the extension of the Jones polynomial evaluated on the simplest singular knot of order two. After applying the Vassiliev relation twice, we have:*

$$V''(\bigcirc) = V'(\bigcirc) - V'(\bigcirc) =$$

$$V(\bigcirc) - V(\bigcirc) - V(\bigcirc) + V(\bigcirc)$$

$$= V(\bigcirc) - V(\bigcirc) = q + q^3 - q^4 - 1$$

Definition 13.1. An invariant $f : \mathcal{X}_0 \to A$ is said to be a *(Vassiliev) invariant of order $\leq n$* if its extension for the set of all $(n+1)$–singular knots equals zero identically.

Denote by $\mathcal{V}_n$ the space of all Vassiliev knot invariants of order less than or equal to n.

---

[1] A can be a ring or a field; we shall usually deal with the cases of $\mathbb{Q}, \mathbb{R}$ and $\mathbb{C}$

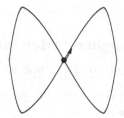

FIGURE 13.1: The simplest singular knot

Definition 13.2. A Vassiliev invariant of order (type) $\leq n$ is said to have order n if it is not an invariant of order less than or equal to $n - 1$.

13.2   Invariants of orders zero and one

The definition of the Vassiliev knot invariant shows us that if an invariant has degree zero then it has the same value on any two knots having diagrams with the same shadow that differ at precisely one crossing. Thus, it has the same value on all knots having the same shadow. Let K be a knot diagram, and S be the shadow of K. There is an unknot diagram with shadow S. So, the value of our invariant on K equals that evaluated on the unknot.

Thus, such an invariant is constant.

It turns out that the first order gives no new invariants (in comparison with 0–type invariants, which are constants).

Indeed, consider the simplest singular knot U shown in Fig. 13.1.

Let S be a shadow of a knot with a fixed vertex which is a singular point.

Exercise 13.1. *Prove that one can arrange all other crossing types for S to get a singular knot isotopic to U.*

It is easy to see that for each Vassiliev knot invariant I such that $I'' = 0$ we have $I'(U) = 0$. Indeed, $I'(U) = I(\infty) - I(\infty) = 0$.

Now, consider an invariant I of degree less than or equal to one. Let K be an oriented knot diagram. By switching some crossing types, the knot diagram K can be transformed to some unknot diagram. Thus, $I(K) = I(\bigcirc) + \sum \pm I'(K_i)$ where K_i are singular knots with one singular point. But, each K_i can be transformed to some diagram U by switching some crossing types. Thus, $I'(K_i) = I'(U) + \sum \pm I''(K_{ij})$, where K_{ij} are singular knots of second order. By definition, $I'' \equiv 0$, thus $I'(K_i) = 0$ and, consequently, $I(K) = I(\bigcirc)$. Thus, the invariant function I is a constant. So, there are no invariants of order one.

13.3   Examples of higher–order invariants

Consider the Conway polynomial C and its coefficients c_n.

Theorem 13.1. *For each natural n, the function c_n is a knot invariant of degree less than or equal to n.*

Proof. Indeed, we just have to compare the Vassiliev relation and the Conway skein relation:

$$c_n'(\text{\Large X}) = c_n(\text{\Large X}) - c_n(\text{\Large X}) = x \cdot c_n(\text{\Large)(}).$$

Thus we see that the first derivative of C is divisible by x; analogously, the n–th derivative of C is divisible by x^n. Thus, after $n + 1$ differentiations, c_n vanishes.   □

This gives us the first non-trivial example. The second coefficient c_2 of the Conway polynomial is the second-order invariant (one can easily check that it is not constant; namely, its value on the trefoil equals one).

However, this invariant does not distinguish the two trefoils because the Conway polynomial itself does not. In the next chapter, we shall show how an invariant of degree three can distinguish the two trefoils.

As will be shown in the future, all even coefficients of the Conway polynomial give us finite–order invariants of corresponding orders.

13.4   Symbols of Vassiliev's invariants coming from the Conway polynomial

As we have shown, each coefficient c_n of the Conway polynomial has order less than or equal to n.

Let v be a Vassiliev knot invariant of order n. By definition, $v^{(n+1)} = 0$. This means that if we take two singular knots K_1, K_2 of n–th order whose diagrams differ at only one crossing (one of them has the overcrossing and the other one has the undercrossing), then $v^{(n)}(K_1) = v^{(n)}(K_2)$. Thus, for singular knots of n–th order one can switch crossing types without changing the value of $v^{(n)}$. Hence, the value of $v^{(n)}$ does not depend on knottedness "that is generated" by classical crossings. It depends only on the order of passing singular points.

Definition 13.3. The function $v^{(n)}$ is called the *symbol* of v.

Definition 13.4. By a *chord diagram* we mean a finite cubic graph consisting of one oriented cycle (circle) and unoriented *chords* (edges connecting different points on this cycle). The *order* of a chord diagram is the number of its chords.

Remark 13.1. *Chord diagrams are considered up to natural graph isomorphism taking chords to chords, circle to the circle and preserving the orientation of the circle.*

Remark 13.2. *We shall never indicate the orientation of the circle on a chord diagram, always assuming that it is oriented counterclockwise.*

The above statements concerning singular knots can be put in formal diagrammatic language. Namely, with each singular knot one can associate a *chord diagram* that is obtained as follows. We think of a knot as the image of the standard oriented Euclidian S^1 in $\mathbb{R}^3$ and connect by chords the preimages of the same point in $\mathbb{R}^3$.

So, each invariant of order n generates a function on the set of chord diagrams with n chords. We can consider the formal linear space of chord diagrams with coefficients, say, in $\mathbb{Q}$, and then consider linear functions on this space generated by symbols of n–th order Vassiliev invariants (together with the constant zero function that has order zero).

Now, it is clear that the space $\mathcal{V}_n/\mathcal{V}_{n-1}$ is just the space of symbols that can be considered in the diagrammatic language.

We shall show that for even n, the coefficient c_n of the Conway polynomial has order precisely n. Moreover, we shall calculate its symbols, according to [CDL].

Consider a chord diagram D of order n. Let us "double" each chord and erase small arcs between the ends of parallel chords. The constructed object (oriented circle without $2n$ small arcs but with n pairs of parallel chords) admits a way of walking along itself. Indeed, starting from an arbitrary point of the circle, we reach the beginning of some chord (after which we can see a "deleted small arc"), then we turn to the chord and move along it. After the end of the chord we again move to the arc (that we have not deleted), and so on. Obviously, we shall finally return to the initial points. Here we have two possibilities.

In the first case we pass all the object completely; in the second case we pass only a part of the object.

By performing a small perturbation in $\mathbb{R}^3$ we can make all chords non-intersecting. In this case our object becomes a manifold $m(D)$. The first possibility described above corresponds to a connected manifold and the second one corresponds to a disconnected manifold

Proposition 13.1 ([CDL]). *The value of the n–th derivative of c_n on D equals one if $m(D)$ has only one connected component and zero, otherwise.*

Proof. Let L be a singular knot with chord diagram D. Let us resolve vertices of D by using the skein relation for the Conway polynomial and the Vassiliev relation:

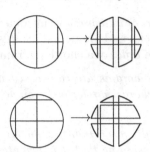

FIGURE 13.2: Calculation of the invariant c_4

$$C'(\!\!\includegraphics{x}\!\!) = x \cdot C(\!\!\includegraphics{}\!\!).$$

Applying this relation n times, we see that the value of the n-th derivative of the invariant C on L (on D) equals the value of C on the diagram obtained from D by resolving all singular crossings, multiplied by x^n. Herewith, the coefficient c_n of the n-th derivative of the Conway polynomial for the case of the singular knot is equal to the coefficient c_0 evaluated at the "resolved" diagram.

This value does not depend on crossing types: it equals one on the unknot and zero on the unlink with more than one component. That completes the proof.   □

It turns out that knots (as well as odd–component links) have only even–degree non-zero monomials of the Conway polynomial: $c_n \equiv 0$ for odd n.

This fact can be proved by using the previous proposition. Let D be a chord diagram of odd order n. Suppose that the curve $m(D)$ corresponding to D has precisely one connected component. Let us attach a disc to this closed curve. Thus we obtain an orientable (prove it!) 2–manifold with disc cut. Thus, the Euler characteristic of this manifold should be odd. On the other hand, the Euler characteristic equals $V - E + S = 2n - 3n + 1 = -n + 1$. Taking into account that n is odd, we obtain a contradiction that completes the proof.

Obviously, for even n, there exist chord diagrams, where c_n does not vanish.

Example 13.2. *The invariant c_4 evaluated at the diagram $\bigoplus$ (see Fig. 13.2, upper picture) is equal to zero; c_4 evaluated at $\bigotimes$ (see Fig.13.2, lower picture), is equal to one.*

Exercise 13.2. *Show that for each even n the value of the n–th derivative of the invariant c_n evaluated on the diagram with all chords pairwise intersecting is equal to one.*

This exercise shows the existence of Vassiliev invariants of arbitrary even orders.

Thus we have proved that the Conway polynomial is weaker than the Vassiliev knot invariants.

Thus, we can say the same about the Alexander polynomial that can be obtained from the Conway polynomial by a simple variable change.

13.5 Other polynomials and Vassiliev's invariants

If we try to apply formal differentiation to the coefficients of other polynomials, we might fail. Thus, for example, coefficients of the Jones polynomial themselves are not Vassiliev invariants. The main reason is that the Jones polynomial evaluated at some links might have negative powers of the variable q in such a way that after differentiation we shall still have negative degrees.

In [JP] the authors give a criterion to detect whether the derivatives of knot polynomials are Vassiliev invariants. They also show how to construct a polynomial invariant by a given Vassiliev invariant.

Although other polynomials can not be obtained from the Conway (Alexander) polynomial by means of a variable change, Vassiliev invariants are stronger than any of those polynomial invariants of knots (possibly, except for the Khovanov polynomial). The results described here first arose in the work by Birman and Lin [BL] (the preprint of this work appeared in 1991); see also [BN1, Kal].

First, let us consider the Jones polynomial. Recall that the Jones polynomial satisfies the following skein relation:

$$q^{-1}V(\rotatebox{0}{\includegraphics{}}) - qV(\rotatebox{0}{\includegraphics{}}) = (q^{\frac{1}{2}} - q^{-\frac{1}{2}})V(\rotatebox{0}{\includegraphics{}})$$

Now, perform the variable change $q = e^x$. We get:

$$e^{-x}V(\rotatebox{0}{\includegraphics{}}) - e^{x}V(\rotatebox{0}{\includegraphics{}}) = (e^{\frac{x}{2}} - e^{-\frac{x}{2}})V(\rotatebox{0}{\includegraphics{}}).$$

Now let us write down the formal Taylor series in x of the expression above and take all members divisible by x explicitly to the right part.

In the right part we get a sum divisible by x and in the left part we obtain the derivative of the Jones polynomial plus something divisible by x:

$$V(\rotatebox{0}{\includegraphics{}}) - V(\rotatebox{0}{\includegraphics{}}) = x\langle\text{some mess}\rangle$$

. Arguing as above, we see that after the second differentiation, only terms divisible by x^2 arise in the right part.

Consequently, after $(n + 1)$ differentiations, the n-th term of the series expressing the Jones polynomial in x, becomes zero. Thus, all terms of this series, are Vassiliev invariants. So, we obtain the following theorem.

Theorem 13.2. *The Jones polynomial in one variable and the Kauffman polynomial in one variable are weaker than Vassiliev invariants.*

One can do the same with the Jones polynomial (denoted by $\mathcal{X}$)) in two variables.

Let us write down the skein relation for it:

$$\frac{1}{\sqrt{\lambda}\sqrt{q}}\mathcal{X}(\mathbb{X}) - \sqrt{\lambda}\sqrt{q}\mathcal{X}(\mathbb{X}) = \frac{q-1}{\sqrt{q}}\mathcal{X}(\mathbb{)(})$$

and let us make the variable change $\sqrt{q} = e^x$, $\sqrt{\lambda} = e^y$ and write down the Taylor series in x and y.

In the right part we get something divisible by x and in the left part something divisible by xy plus the derivative of the Jones polynomial.

Finally, we have

$$\mathcal{X}(\mathbb{X}) - \mathcal{X}(\mathbb{X}) = x\langle\text{some mess}\rangle.$$

Thus, after $(n+1)$ differentiations, all terms of degree $\leq n$ in x, vanish. Consequently, we get the following theorem.

Theorem 13.3. *The Jones polynomial in two variables is weaker than Vassiliev invariants.*

Since the HOMFLY-PT polynomial is obtained from the Jones polynomial by a variable change, we see that the following theorem holds.

Theorem 13.4. *The HOMFLY-PT polynomial is weaker than Vassiliev invariants.*

The most difficult and interesting case is the Kauffman 2–variable polynomial because this polynomial does not satisfy any Conway relations. This polynomial can be expressed in the terms of functions z, a, and $\frac{a-a^{-1}}{z}$. In order to represent the Kauffman polynomial as a series of Vassiliev invariants, we have to represent all these functions as series of positive powers of two variables.

We recall that the Kauffman polynomial in two variables is given by the formula[2]

$$Y(L) = a^{-w(L)}D(L),$$

where D is a function on the chord diagram that satisfies the following relations:

---

[2]Here we denote the oriented and the unoriented diagrams by the same letter L.

$$D(L) - D(L') = z(D(L_A) - D(L_B));   \tag{13.2}$$

$$D(\bigcirc) = \left(1 + \frac{a - a^{-1}}{z}\right);   \tag{13.3}$$

$$D(X \# \text{⊙}) = aD(X), D(X \# \text{⊙}) = a^{-1}D(X),   \tag{13.4}$$

where the diagrams $L = \text{⊗}$, $L' = \text{⊗}$, $L_A = \text{⊗}$, $L_B = \text{)(}$ coincide outside a small neighbourhood of some vertex.

Let us rewrite (13.2) for Y. We get:

$$a^{-1}Y(\text{⊗}) - aY(\text{⊗}) = z(Y(\text{⊗}) - Y(\text{)(})) \cdot \langle\text{Power of } a\rangle.   \tag{13.5}$$

Let us perform the variable change: $p = \ln(\frac{a-1}{z})$. Then, in terms of z and p, one can express $z, a, \frac{a-a^{-1}}{z}$ by using only positive powers and series. Actually, we have:

$$z = z,$$

$$a = ze^p + 1 = z(1 + p + \ldots) + 1,$$

$$a^{-1} = 1 - z(1 + p + \ldots) + z^2(1 + p + \ldots)^2 + \ldots,$$

$$\frac{a - a^{-1}}{z} = a^{-1}(a + 1)e^p$$

Each of these right parts can evidently be represented as sequences of positive powers of p and z.

Thus, the value of the Kauffman polynomial in two variables on each knots is represented by positive powers of p and z. On the other hand, taking into account that $a = 1 + z\langle\text{some mess}_1\rangle$ and $a^{-1} = 1 + z\langle\text{some mess}_2\rangle$, we can deduce from (13.5) and (13.4) that

$$Y' = z\langle\text{some mess}\rangle.$$

Herewith, all terms of our double sequence having degree less than or equal to n in the variable z, vanish after the $(n + 1)$–th differentiation. Thus, all these terms are Vassiliev invariants.

Thus, we have proved the following theorem.

Theorem 13.5. *The Kauffman polynomial in two variables is weaker than Vassiliev invariants.*

Let us show how to calculate the derivative of products of two functions.

For any two functions f and g defined on knot diagrams one can formally define the derivatives f' and g' on diagrams of first-order singular knots just as we define the derivatives of the invariants. Analogously, one can define higher-order derivatives.

Consider the function $f \cdot g$ and consider a singular knot diagram K of order n. By a *splitting* is meant a choice of a subset of i singular vertices of n singular vertices belonging to K. Choose a splitting s. Let K_{1s} be the diagram obtained from K by resolving $(n-i)$ unselected vertices of s negatively, and let K_{2s} be the knot diagram obtained by resolving i selected vertices positively.

Lemma 13.1. *Let K be a chord diagram of degree n. Then the* Leibniz formula *holds:*

$$(fg)^{(n)}(K) = \sum_{i=0}^{n} \sum_{s} f^{(i)}(K_{1s}) g^{(n-i)}(K_{2s}).$$

Proof. We shall use induction on n.

First, let us establish the induction base (the case $n = 1$). Given a singular knot of order one, let us consider a diagram of it and the only singular vertex A of this diagram. Write down the Vassiliev relation for this vertex:

$$(fg)'(\text{⊗}) = f(\text{⊗})g(\text{⊗}) - f(\text{⊗})g(\text{⊗})$$
$$= g(\text{⊗})(f(\text{⊗}) - f(\text{⊗})) + f(\text{⊗})(g(\text{⊗}) - g(\text{⊗}))$$
$$= f'(\text{⊗})g(\text{⊗}) + g'(\text{⊗})f(\text{⊗}). \quad (13.6)$$

The equality (13.6) holds by definition of f' and g'. Thus, we have proved the claim of the theorem for $n = 1$. Note that we can apply the obtained formula for functions on *singular* (not ordinary) knots, when all singular points do not take part in the relation; i.e., lie outside the neighbourhood.

Now, for any given singular knot K of order n, let us fix a singular vertex A of the knot diagram K. The value of $(fg)^{(n)}$ on K equals the difference of $(fg)^{(n-1)}$ evaluated on two singular knots K^1 and K^2; these two diagrams of singular knots of order $n - 1$ are obtained by positive and negative resolution of A, respectively.

By the induction hypothesis, we have:

$$(fg)^{(n-1)}(K^i) = \sum_{i=0}^{n-1} \sum_{s} f^{(i)}(K_{1s}^i) g^{(n-1-i)}(K_{2s}^i), \quad (13.7)$$

where s runs over the set of all splittings of order $(n - 1)$.

We have:

$$(fg)^{(n)}(K) = (fg)^{(n-1)}(K^1) - (fg)^{(n-1)}(K^2)$$

$$= \sum_{i=0}^{n-1}\sum_{s} \left[f^{(i)}(K_{1s}^1)g^{(n-1-i)}(K_{2s}^1) - f^{(i)}(K_{1s}^2)g^{(n-1-i)}(K_{2s}^2) \right]$$

$$= \sum_{i=0}^{n-1}\sum_{s} \Big[f^{(i)}(K_{1s}^1)g^{(n-1-i)}(K_{2s}^1) - f^{(i)}(K_{1s}^2)g^{(n-1-i)}(K_{2s}^1)$$

$$+ f^{(i)}(K_{1s}^2)g^{(n-1-i)}(K_{2s}^1) - f^{(i)}(K_{1s}^2)g^{(n-1-i)}(K_{2s}^2) \Big]$$

$$= \sum_{i=0}^{n-1}\sum_{s} \left[f^{(i+1)}(K_{1s})g^{(n-1-i)}(K_{2s}^1) - f^{(i)}(K_{1s}^2)g^{(n-i)}(K_{2s}) \right]$$

$$= \sum_{i=0}^{n}\sum_{s} f^{(i)}(K_{1s})g^{(n-i)}(K_{2s}).$$

$$\square$$

Lemma 13.1 implies the following corollary.

Corollary 13.1. *Let f and g be two functions defined on the set of knot diagrams (not necessarily knot invariants) such that $f^{(n+1)} \equiv 0$, $g^{(k+1)} \equiv 0$. Then $(fg)^{(n+k+1)} \equiv 0$.*

In particular, the product of Vassiliev invariants of orders n and k is a Vassiliev invariant of order less than or equal to $(n + k)$.

13.6   An example of an infinite-order invariant

Until now, we have dealt only with invariants either having finite-order or invariants that can be reduced to finite order invariants. We have not yet given any proof that some knot invariant has infinite order.

Here we give an example of a knot invariant that has infinite order, [BL].

Definition 13.5. The *unknotting number $U(K)$* of an (oriented) link K is the minimal number $n \in \mathbb{Z}_+$ such that K can be transformed to the unlink by passing n times through singular links. In other words, n is the minimal number such that there exists a diagram of K that can be transformed to an unlink diagram by switching n crossings.

By definition, our invariant equals zero only for unlinks.

Theorem 13.6. *The invariant U has infinite order.*

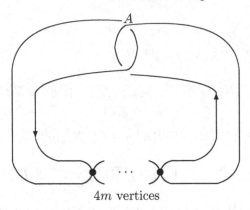

4m vertices

FIGURE 13.3: Singular knot, where $U^i \neq 0$.

Proof. Let us fix an arbitrary $i \in \mathbb{N}$. Now, we shall give an example of the singular knot for which $U^{(i)} \neq 0$. Fix an integer $m > 0$ and consider the knot K_{4m} with $4m$ singularity points which are shown in Fig. 13.3.

By definition of the derivative, the value of $U^{(4m)}$ on this knot is equal to the alternating sum of $2^{(4m)}$ summands; each of them is the value of U on a knot, obtained by somehow resolving all singular vertices of K_{4m}.

Note that for each such singular knot the value of U does not exceed one: by changing the crossing at the point A, we obtain the unknot. On the other hand, the knot obtained from K_{4m} by splitting all singular vertices is trivial if and only if the number of positive splittings equals the number of negative splittings (they are both equal to $2m$).

The case of q positive and $4m - q$ negative crossings generates the sign $(-1)^q$.

Thus we finally get that $U^{(4m)}(K_{4m})$ is equal to

$$U^{(4m)}(K_{4m}) = 2[C_{4m}^0 - C_{4m}^1 + \cdots - C_{4m}^{2m-1}].$$

This sum is, obviously, negative: $U^{(4m)}(K_{4m}) \neq 0$. So, for $m > \frac{i}{4}$, we get $U^{(i)} \neq 0$. Thus, the invariant U is not a finite type invariant of order less than or equal to i. Since i was chosen arbitrarily, the invariant U is not a finite type invariant.   □

Remark 13.3. *We do not claim that U cannot be represented via finite type invariants.*

The unknotting number is some "measure" of complexity for a knot. Thus, it would be natural to think that it is realised on minimal diagrams (i.e. the minimal diagram can be transformed to the unknot diagram by precisely n

switchings if the unknotting number is equal to n). However, this is not true. The first results in this directions were obtained by Bleiler and Nakanishi [Ble, Nak]. Later, an infinite series of knots with this property was constructed by D.J. Garity [Gari].

Chapter 14

The chord diagram algebra

14.1 Basic structures

In the present chapter, we shall study the algebraic structure that arises on the set of Vassiliev knot invariants.

In the previous chapter, we defined symbols of the Vassiliev knot invariants in the language of chord diagrams.

Now, the main question is: *Which functions on chord diagrams can play the role of symbols?*

The simplest observation leads to the following fact. If we have a chord diagram $C = $ ⊖ with a small solitary chord, then each symbol evaluated at this diagram equals zero. We have already discussed this in the language of singular knots.

This relation is called a $1T$–relation (or one–term relation).

One can easily prove the *generalised* $1T$–*relation* where we can take a diagram $C = $ ⊜ with a chord that does not intersect any other chord. Then, each symbol of a Vassiliev knot invariant evaluated at the diagram C equals zero. The proof is left to the reader.

There exists another relation, consisting of four terms, the so-called $4T$–relation. In fact, let us prove the following theorem.

Theorem 14.1 (The four–term relation). *For each symbol v^n of an invariant v of order n the following relation holds:*

$$v^n(\bigotimes) - v^n(\bigotimes) - v^n(\bigotimes) + v^n(\bigotimes) = 0.$$

This relation means that for any four diagrams having n chords, where $(n-2)$ chords (not shown in the Figure) are the same for all diagrams and the other two look as shown above, the above equality takes place.

Proof. Consider four singular knots S_1, S_2, S_3, S_4 of the order n, whose diagrams coincide outside some small circle, and their fragments s_1, s_2, s_3, s_4 inside this circle look like this:

$$s_1 = \bigotimes, s_2 = \bigotimes, s_3 = \bigotimes, s_4 = \bigotimes.$$

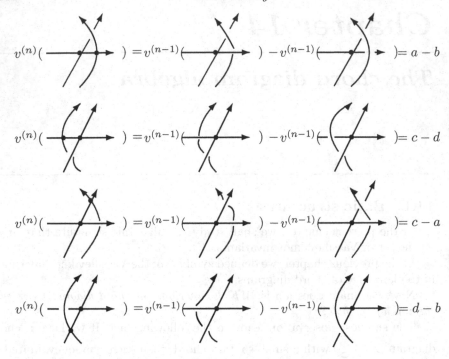

FIGURE 14.1: The same letters express $v^{(n-1)}$ for isotopic long knots

Consider an invariant v of order n and the values of its symbol on these four knots. Vassiliev's relation implies the relations shown in Fig. 14.1.

Obviously,

$$(a - b) - (c - d) + (c - a) - (d - b) = 0.$$

In order to get singular knots, one should close the fragments s_1, s_2, s_3, s_4. There are two possibilities to do this as shown in Fig. 14.2.

Thus, the diagrams S_1, S_2, S_3, S_4 satisfy the relation

$$v^{(n)}(S_1) - v^{(n)}(S_2) + v^{(n)}(S_3) - v^{(n)}(S_4) = 0. \qquad (14.1)$$

Each of the chord diagrams corresponding to S_1, S_2, S_3, S_4 has n chords; $(n-2)$ chords are the same for all diagrams, and only two chords are different for these diagrams.

Since the order of v equals n, the symbol of v is correctly defined on chord diagrams of order n. Thus, the value of $v^{(n)}$ on diagrams corresponding to singular knots S_1, S_2, S_3, S_4 equals the value on the singular knots themselves.

Taking into account the formulae obtained above, and the arbitrariness of the remaining $(n-2)$ singular vertices of the diagrams S_1, S_2, S_3, S_4, we obtain the statement of the theorem.

$\square$

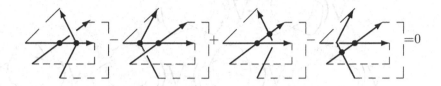

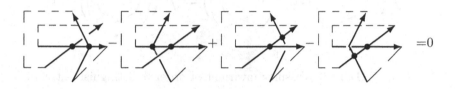

FIGURE 14.2: Closures of fragments imply $4T$–relation

Both $1T$– and $4T$–relations can be considered for chord diagrams and on the dual space of linear functions on chord diagrams (since these two dual spaces can obviously be identified). For the sake of simplicity, we shall apply the terms $1T$– and $4T$–relation to both cases.

Definition 14.1. Each linear function on chord diagrams of order n, satisfying these relations, is said to be a *weight system* (of order n).

Notation: Denote the space of all weight systems of order n, by $\mathcal{A}_n$ or by Δ_n.

In the last chapter, we considered invariants of orders less than or equal to two. The situation there is quite clear: there exists the unique non–trivial (modulo $1T$–relation) chord diagram that gives the invariant of order two. As for dimension three, there are two diagrams: ⊕ and ⊗. It turns out that they are linearly dependent. Namely, let us write the following $4T$–relation (here the fixed chord is represented by the dotted line):

$$\bigcirc - \bigcirc = \bigcirc - \bigcirc.$$

This means that ⊗ = 2⊕.

So, if there exists an invariant of order three, then its symbol is uniquely defined by a value on ⊗. Suppose we have such an invariant v and $V'''(⊗) = 1$. Let us show that this invariant distinguishes the two trefoils; see Fig. 14.3

The existence of this invariant will be proved later.

Let us consider the formal space Δ_4.

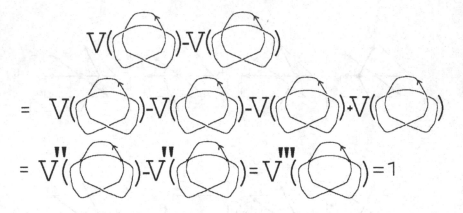

FIGURE 14.3: Vassiliev invariant of order 3 distinguishes trefoils

Exercise 14.1. *Prove the following relations:*

$$\otimes = \oplus + \oplus, \quad \oplus + \otimes = 2\otimes,$$

$$\oplus = \oplus + \otimes, \quad \oplus = \otimes + \oplus,$$

$$\oplus + \otimes = \otimes + \otimes.$$

Exercise 14.2. *Prove that $dim\Delta_4 = 3$ and that the following three diagrams can be chosen as a basis:*

$$\{\oplus, \oplus, \otimes\}.$$

It turns out that the chord diagrams factorised by the $4T$–relation (with or without the $1T$–relation) form an algebra. Namely, having two chord diagrams C_1 and C_2, one can break them at points $c_1 \in C_1$ and $c_2 \in C_2$ (which are not ends of chords) and then attach the broken diagrams together according to the orientation. Thus we get a chord diagram. The obtained diagram can be considered as the product $C_1 \cdot C_2$. Obviously, this way of defining the product depends on the choice of the base points c_1 and c_2; thus, different choices might generate different elements of $\mathcal{A}^c$. However, this is not the case since we have the $4T$–relation.

Theorem 14.2. *The product of chord diagrams in $\mathcal{A}^c$ is well defined; i.e., it does not depend on the choice of initial points.*

To prove this theorem, we should consider *arc diagrams* rather than chord diagrams.

Definition 14.2. By an *arc diagram* we mean a diagram consisting of one straight oriented line and several arcs connecting points of it in such a way that each arc connects two different points and each point on the line is incident to no more than one arc.

These diagrams are considered up to the natural equivalence; i.e., a mapping of the diagram, taking the line to the line (preserving the orientation of the line) and taking all arcs to arcs.

Obviously, by breaking one and the same chord diagram at different points, we obtain different arc diagrams.

Now, we can consider the 4T–relation for the case of the arc diagrams, namely the relation obtained from a 4T–relation by breaking all four circles at the same point (which is not a chord end).

The point is that the two arc diagrams A_1 and A_2 obtained from the same chord diagram D by breaking this diagram at different points are equivalent modulo 4T–relation. This will be sufficient for proving Theorem 14.2. Obviously, one can obtain A_2 from A_1 by "moving a chord end through infinity". Thus, it suffices to prove the following lemma.

Lemma 14.1 (Kontsevich). *Let A_1, A_2 be two arc diagrams that differ only at the chord: namely, the rightmost position of a chord end of A_2 corresponds to the leftmost position of the corresponding chord end of A_1; the other chord ends of A_1 and A_2 are on the same places. Then A_1 and A_2 are equivalent modulo the four–term relation.*

Proof. Suppose that each of the diagrams A_1 and A_2 have n arcs. Denote the common arc ends A_1 and A_2 by $X_1, X_2, \ldots, X_{2n-1}$ enumerated from the left to the right. They divide the line into $2n$ intervals $I_1, \ldots, I_{2n}$ (from the left to the right). Denote by D_j the arc diagram having the same "fixed" arc ends as A_1 and A_2 and one "mobile" arc end at I_j. Thus, $A_1 = D_1, A_2 = D_{2n}$. Suppose that the second end of the "mobile" arc is X_k. Then, obviously, $D_k = D_{k+1}$.

Now, consider the following expression

$$A_{2n} - A_1 = A_{2n} - A_{2n-1} + A_{2n-1} - A_{2n-2} + \ldots$$
$$+ A_{k+2} - A_{k+1} + A_k - A_{k-1} + \ldots A_2 - A_1.$$

Here we have $4n - 4$ summands. It is easy to see that they can be divided into $n - 1$ groups, each of which forms the 4T–relation concerning one immobile chord and the mobile chord.

Thus, $A_{2n} = A_1$. This completes the proof of the theorem. $\qquad\square$

14.2 Bialgebra structure of algebras $\mathcal{A}^c$ and $\mathcal{A}^t$. Chord diagrams and Feynman diagrams

The chord diagram algebra $\mathcal{A}^c$ has, however, very sophisticated structures. It is indeed a bialgebra. The coalgebra structure of $\mathcal{A}^c$ can be introduced as follows.

Let C be a chord diagram with n chords. Denote the set of all chords of the diagram C by $\mathcal{X}$. Let $\Delta(C)$ be

$$\sum_{s \in 2^{\mathcal{X}}} C_s \otimes C_{\mathcal{X} \setminus s},$$

where the sum is taken over all subsets s of $\mathcal{X}$, and C_y denotes the chord diagram consisting of all chords of C belonging to the set y. Now, let us extend the coproduct Δ linearly.

Now we should check that this operation is well defined. Namely, for each four diagrams $A = \bigotimes, B = \bigotimes, C = \bigcirc, D = \bigcirc$ such that $A - B + C - D = 0$ is the $4T$–relation, one must check that $\Delta(A) - \Delta(B) + \Delta(C) - \Delta(D) = 0$.

Actually, let A, B, C, D be four such diagrams (A differs from B only by a crossing of two chords, and D differs from C in the same way). Let us consider the comultiplication Δ. We see that when the two "principal" chords are in different parts of $\mathcal{X}$, then we have no difference between A, B as well as between C, D. Thus, such subsets of $\mathcal{X}$ give no impact. And when we take both chords into the same part for all A, B, C, D, we obtain just the $4T$–relation in one part and the same diagram at the other part. Thus, we have proved that Δ is well–defined.

Now, let us give the formal definition of the bialgebra.[1]

Definition 14.3. An algebra A with algebraic operation μ and unit map e and with coalgebraic operation Δ and counit map ε is called a bialgebra if

1. e is an algebra homomorphism;

2. ε is an algebra homomorphism;

3. Δ is an algebra homomorphism.

Definition 14.4. An element x of a bialgebra B is called *primitive* if $\Delta(x) = x \otimes 1 + 1 \otimes x$.

---

[1]In [Oni] this is also called a *Hopf algebra*. One usually requires more constructions for the algebra to be a Hopf algebra, see e.g. [Cas, MiMo]. However, the bialgebras of chord and Feynman diagrams that we are going to consider are indeed Hopf algebras: the antipode map is defined by induction on the number of chords. We shall not use the antipode and its properties.

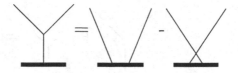

FIGURE 14.4: STU–relation

Obviously, for the case of $\mathcal{A}^c$ with natural e, ε and endowed with the product and coproduct Δ, e and ε are homomorphic. The map Δ is monomorphic: it has the empty kernel because for each $x \neq 0$, $\Delta(x)$ contains the summand $x \otimes 1$.

Thus, $\mathcal{A}^c$ is a bialgebra.

There is another interesting algebra $\mathcal{A}^t$ that is in fact isomorphic to $\mathcal{A}^c$.

Definition 14.5. A *Feynman diagram*[2] is a finite connected graph of valency three at each vertex with an oriented cycle (circle)[3] on it. All vertices not lying on the circle are called *interior* vertices. Those lying on the circle are *exterior vertices*. Each interior vertex should be endowed with a cyclic order of outgoing edges.

Remark 14.1. *Feynman diagrams on the plane are taken to have the counterclockwise orientation of the circle and counterclockwise cyclic order of outgoing edges at each interior vertex.*

Definition 14.6. The *degree* of a Feynman diagram is half the number of its vertices.

Obviously, all chord diagrams are Feynman diagrams; in this case the two definitions of the degree coincide.

Consider the formal linear space of all Feynman diagrams of degree n. Let us factorise this space by the STU–relation that is shown in Fig. 14.4.

Denote this space by $\mathcal{A}_n^t$.

Theorem 14.3. *There exists a natural isomorphism $f : \mathcal{A}_n^t \to \mathcal{A}_n^c$ which is identical on $\mathcal{A}_n^c$. Moreover, the STU–relation implies the following relations for $\mathcal{A}^t$:*

1. Antisymmetry; see Fig. 14.5.

2. IHX–relation; see Fig. 14.6.

Proof. First, let us prove that the algebras $\mathcal{A}^t$ and $\mathcal{A}^s$ are isomorphic. Obviously, the STU–relation implies the 4T–relation for the elements from $\mathcal{A}^c$. Let us construct now the isomorphism f. For all elements from $\mathcal{A}^c \subset \mathcal{A}^t$ we

---

[2]Also called *Chinese diagram* or *circular diagram*.

[3]This circle is also called the *Wilson loop*; we shall not use this term.

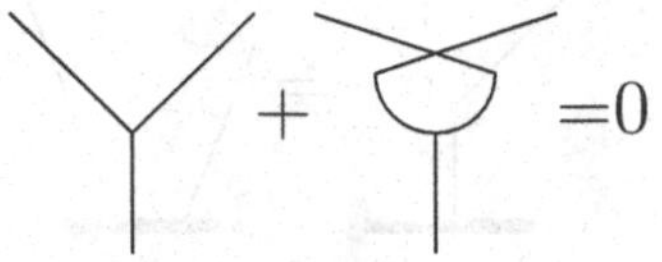

FIGURE 14.5: The antisymmetry relation

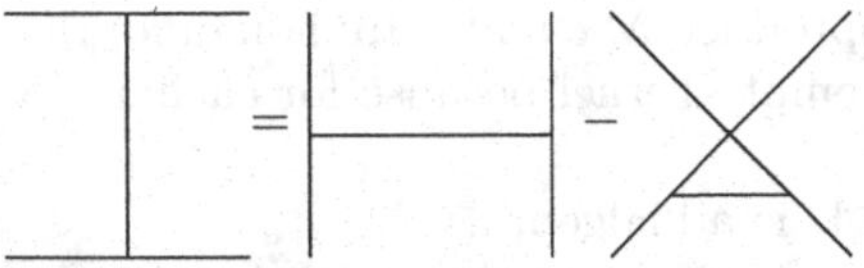

FIGURE 14.6: The IHX–relation

decree f to be the identity map. To define f on all Feynman diagrams, we shall use induction on the number x of interior vertices. For $x = 0$, there is nothing to prove.

Suppose f is well defined for all Feynman diagrams of degree d. Let K be a Feynman diagram of degree $d + 1$.

Obviously, there exists an interior vertex V of K that is adjacent to some exterior vertex by an edge v. Thus, we can apply the STU–relation to this vertex and obtain two diagrams of degree d. However, this operation is not well defined: it depends on the choice of such a vertex V and the edge v. Suppose there are two such pairs V, v and U, u, where $V \neq U, v \neq u$. In this case, we can prove that our operation is well–defined by applying the STU– relation twice; see Fig. 14.7.

In the case when $U = V$ and $u \neq v$, we can try to find another pair. Namely, the pair W, w, where W is a vertex adjacent to an exterior vertex and w is an edge connecting this vertex with the circle. Then we prove that the result for U, u equals that for W, w and then it equals that for U, v.

Finally, we should consider the case when $U = V$, $u \neq v$, and U is the only interior vertex adjacent to the circle. In this case, we are going to show that our diagram is equivalent to zero modulo the STU–relation.

In this case, we can indicate some domain containing all interior vertices except one. This domain has only one connection with exterior vertices , namely the connection via the vertex U and one of the chords u, v. In this case, the two possible splittings are equals because the product on Feynman diagrams is well defined. By the induction hypothesis, we see that the product of two Feynman diagrams of total degree d is well defined and commutative. By using this commutativity, we can move the vertex with the small domain from one

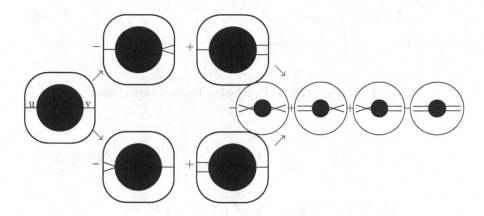

FIGURE 14.7: Applying the STU–relation twice

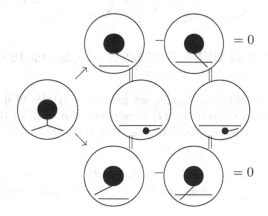

FIGURE 14.8: STU-reduction to zero diagram

point to the other one. Thus, we see that each of both splittings gives us zero. The concrete calculations are shown in Fig. 14.8.

Let us prove now that STU implies the antisymmetry relation.

Applying the STU–relations many times, one can reduce the antisymmetry relation to the case when all chords outgoing from the given interior points finish at exterior points. In this case, the antisymmetry relation follows straightforwardly; see Fig. 14.9.

The proof of the fact that the IHX–relation holds can be reduced to the case when one of the four vertices (say, lower left) is an exterior one. This can be done by taking the lower left vertex for all diagrams that have to satisfy the IHX–relation and then splitting all interior vertices between this vertex and the circle in the same manner for all diagrams. Then we repeat this procedure

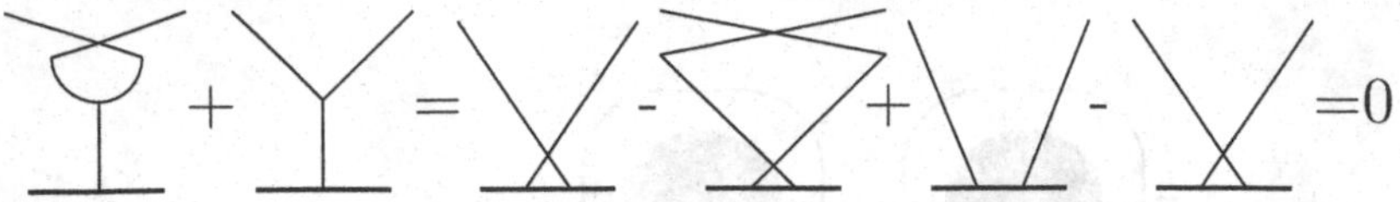

FIGURE 14.9: STU–relation implies antisymmetry

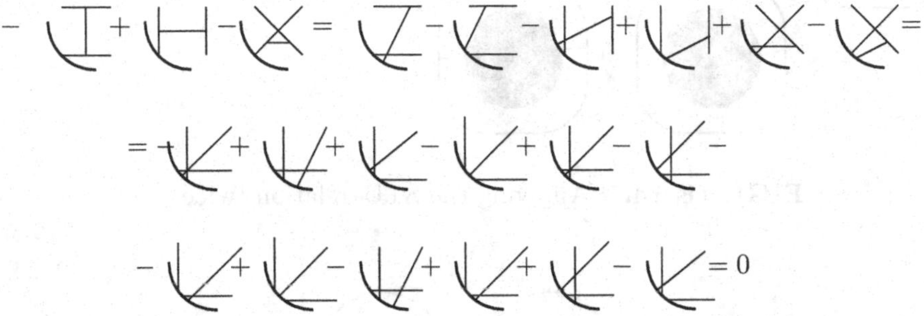

FIGURE 14.10: The STU–relation implies the IHX relation

for all obtained diagrams. Finally, we get many triples of diagrams for each of which we have to check the IHX relation. For each of them, we have to consider only the partial case. The last step is shown in Fig. 14.10.

$\square$

Remark 14.2. *Note that the 1–term relation does not spoil the bialgebra structure; the corresponding bialgebra is obtained by a simple factorisation.*

14.3   Coproduct for Feynman diagrams

Now, let us define the coproduct in the algebras $\mathcal{A}^c$ and $\mathcal{A}^t$ of chord and Feynman diagrams.

Remark 14.3. *Within this section, we consider the algebras not factorised by the 1T–relation.*

As shown above, these algebras are isomorphic. Thus, $\mathcal{A}^t$ has a Hopf algebra structure as well. Let us describe this structure explicitly.

Let D be a Feynman diagram and let $V(D)$ be *connected components* of the diagram; i.e., connected components of the graph obtained from D by deleting the circle. Let $J \subset V(D)$ be a subset of $V(D)$. This subset defines

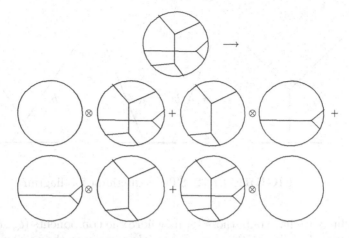

FIGURE 14.11: Coproduct of a Feynman diagram

a Feynman diagram C_J, whose connected components lie in $V(D)$; i.e., the Feynman diagram consisting of the circle of the diagram D and those connected components of the graph $V(D)$ belonging to J.

Let us define the *coproduct* $\tilde{\mu}(D)$ as

$$\tilde{\mu}(D) = \sum_{J \subset V(D)} C_J \otimes C_{V(D) \backslash J}.$$

Example 14.1. *In Fig. 14.11 we illustrate the coproduct operation for a Feynman diagram.*

Theorem 14.4. *The coproduct defined below coincides with that for $\mathcal{A}^c$; i.e. $\mu \equiv \tilde{\mu}$.*

Proof. We have to show that for each Feynman diagram D, its coproduct coincides with the linear combination of coproducts of chord diagrams that D can be decomposed into.

We shall use induction on the number k of interior vertices of the diagram. For $k = 0$, there is nothing to prove.

Suppose that the statement is true for all Feynman diagrams with n interior vertices. Consider a Feynman diagram D with $(n+1)$ interior vertices. We have to show that $\mu(D) = \tilde{\mu}(D)$. According to the STU–relation, the diagram D can be represented as a difference $D_+ - D_-$, as shown in Fig. 14.12; in this case the connected component of D corresponds to a pair of connected components for each of D_+, D_-.

Let us choose the components (a_+, b_+) and (a_-, b_-) for the diagrams D_+ and D_-. These components are obtained by resolving a point of D. Each of $D_\pm$ has n interior vertices. Thus, the claim of the theorem is true for them: $\mu(D_\pm) = \tilde{\mu}(D_\pm)$. Let us now write $\mu(D) = \mu(D_+) - \mu(D_-)$. In each of these

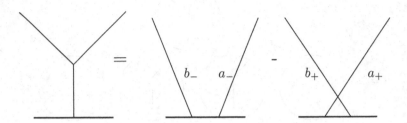

FIGURE 14.12: STU-reduction of a diagram

coproducts we have only those terms where the components (a_+, b_+) (respectively, (a_-, b_-)) lie on the same side with respect to the $\otimes$ sign. Obviously, these terms collected together give $\tilde{\mu}(D)$ (in the previous sense of the coproduct). It is easy to see that the remaining terms give us zero. In fact, suppose we have a splitting of the Feynman diagram D_+ into two diagrams, where a_+ belongs to one of them and b_+ belongs to the other. Then if we divide D_- in just the same way as we did with D_+ with respect to all other connected components and take a_- to be the first multiplicator of the tensor product and b_- to be the second one, we obtain two coinciding tensor products.

Collecting all previous statements together, we obtain the statement of the theorem.

$\square$

14.4   Lie algebra representations, chord diagrams, and the four colour theorem

There is a beautiful idea connecting the representation theory of Lie algebras, and knot theory. It was popularised in [BN1]; for further developments see e.g. [BN4, BLT, CV, Vog2, Vog3].

The motto is: contract trivalent tensors along graphs.

Remark 14.4. *Within this section, we do not take into account the 1T–relation. We work only with the 4T–relation (or STU–relation for Feynman diagrams).*

In more detail, having a trivalent graph and a trivalent tensor, we can set this tensor to each vertex of the graph, and then contract the tensors along the edges of the graph. Clearly, we need some metric; besides we must be able to switch indices. All these conditions obviously hold for the case of semisimple Lie algebras: we can take the structure constant tensor C_{ijk} (with all lower indices) and the metric g_{ij} that is not degenerate. As a trivalent graph, one can take a Feynman diagram together with its rotation structure at vertices.

The beautiful observation is that the STU–relation (as well as the IHX–relation) for Feynman diagrams represents the Jacobi identity for Lie algebras. Thus, the constructed numbers (obtained after all contractions) are indeed invariant under the STU–relation.

This construction is the simplest case of the general construction; it deals only with adjoint representations of Lie algebras. In the general case, one should fix the circle of the Feynman diagram and the representation R of the Lie algebra G. Then, along the circle, we put elements of the representation space such that any two adjacent elements are obtained from each other by the action of the Lie algebra element associated with the edge outgoing from the point connecting these two arcs.

Consider the adjoint representation of $SO(3)$. In this case, it is very easy to calculate the contractions. Namely, we have three elements a, b, c and the following contraction law: $[a, b] = c; [b, c] = a; [c, a] = b$.

Definition 14.7. By a *planar map* we mean a cubic graph embedded in $\mathbb{R}^2$ (this graph divides the plane into cells which are called *regions*).

Now, suppose we wish to colour the map with four colours. Let us take them from the palette $\mathbb{Z}_2 \oplus \mathbb{Z}_2$: they are $(0,0),(0,1),(1,0)$, and $(1,1)$.

Definition 14.8. The map is *four colourable* if one can associate one of the four colours to each region in such a way that no two adjacent regions have the same colour

The four colour theorem claims that every planar map (without loops) is four colourable.

It remained unsolved for a long time. Its first solution [AH] is very technically complicated and contains numerous combinatorial constructions to work with. Below, we give some sufficient condition for a map to be four colourable [BN4].

Suppose we have some colouring of some map. Then, we can colour each edge by an element from $\mathbb{Z}_2 \oplus \mathbb{Z}_2$ which is a sum of the elements associated with the adjacent regions.

Now, it is obvious that the map is four colourable if the edges of it can be coloured only with the colours $(0,1),(1,0)$, and $(1,1)$ in such a way that no two adjacent edges have the same colour. So, the edges should be *three colourable*. The inverse statement is also true: if edges are three–colourable then the map is four–colourable.

Now, we can think of each map M as a Feynmann diagram and associate a number to it with respect to the adjoint representation of $SO(3)$. Denote this number by $I_3(N)$.

Theorem 14.5 (BN5). *If* $I_3(M) \neq 0$ *then* M *is four colourable.*

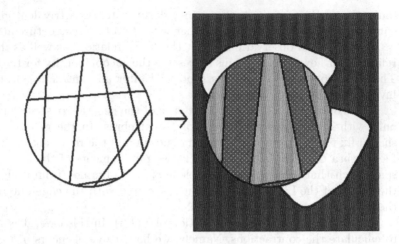

FIGURE 14.13: Each map coming from d–diagram is four colourable

Proof. Indeed, we can use the basis a, b, c for calculating $I_3(M)$. Since $I_3(M) \neq 0$, there exists at least one contraction that gives a non–zero element. Taking into account the law for $SO(n)$, we see that the edges are three colourable (triples with at least two equal elements give zero!).Thus, M is four colourable. □

Suppose we have a d–diagram embedded in $\mathbb{R}^2$. Then, the corresponding map is four colourable. Actually, our plane is divided into two parts by the circle. A simple observation shows that each of these parts (interior and exterior) is two colourable; see Fig. 14.13. Thus, the whole picture is four colourable.

Thus, one can ask the following question. Consider a map (or planar Feynman diagram). Can one recognise a d–diagram in it? In other words, can one select some subset edges that compose a cycle in such a way that all other edges connect points of this cycle?

This problem is very famous and is still unsolved. It was first stated by W.R.Hamilton, and the cycle we are looking for is called a *Hamiltonian cycle* . To date, only some (positive) solutions for some classes of maps are known.

One can easily see that the unsolved problem on Hamiltonian curves leads to the positive solution of the four colour problem via d–diagrams.

However, the way proposed by Bar–Natan is not a criterion. Actually, having a Feynmann diagram D, for which $I_3(D) = 0$, the corresponding map can have a Hamiltonian curve and thus be four colourable. The point is that while calculating $I_3(D)$, some terms (corresponding to proper four colourings) give a positive contribution and the others give a negative contribution and the total number can be equal to zero.

14.5 Dimension estimates for $\mathcal{A}_d$. A table of known dimensions.

We are going to talk about the lower and upper bounds for the dimensions of spaces Δ_n. Later, we shall prove that Δ_n is precisely $\mathcal{V}_n/\mathcal{V}_{n-1}$.

14.5.1 Historical development

A priori it is obvious that the cardinality of the set of all chord diagrams on d chords does not exceed $(2d - 1)!! = 1 \cdot 3 \cdots (2d - 1)$.

This gives the first evident upper bound. After this, the following results appeared (results listed according to [CDM]).

1. (1993) Chmutov and Duzhin in [CD1] proved that $dim\mathcal{A}_d < (d - 1)!$.

2. (1995) K. Ng in [NgK] replaced $(d - 1)!$ by $\frac{(d-2)!}{2}$

3. (1996) A. Stoimenov [Sto] proved that $dim\mathcal{A}_d$ grows slower than $\frac{d!}{a^d}$, where $a = 1.1$.

4. (2000) B. Bollobás and O. Riordan [BR] obtained the asymptotical bound $\frac{d!}{(2\ln(2)+o(1))^d}$ (approximately $\frac{d!}{1.38^d}$).

5. (2001) D. Zagier in [Zag] improved the result to $\frac{6^d\sqrt{d}\cdot d!}{\pi^{2d}}$, which is asymptotically smaller than $\frac{d!}{a^d}$ for any constant $a < \frac{\pi^2}{6} = 1.644...$

The history of lower bounds was developing as follows.

1. (1994) Chmutov, Duzhin and Lando [CDL] gave a lower approximation for the number of primitive elements $\mathcal{P}_n$ ("forest elements"): $dim\mathcal{P}_d \geq 1$ for $d > 1$.

2. (1995) $dim\mathcal{P}_d \geq [\frac{d}{2}]$ (see Melvin–Morton [MeMo] and Chmutov–Varchenko [CV]).

3. (1996) $dim\mathcal{P}_d \gtrsim \frac{d^2}{96}$, see Duzhin [Duz].

4. (1997) $dim\mathcal{P}_d \gtrsim d^{logd}$, see Chmutov–Duzhin, [CD2].

5. (1997) $dim\mathcal{P}_n > e^{\pi\sqrt{n/3}}$, see Kontsevich [Kon1].

6. (1997) $dim\mathcal{P}_n > e^{C\sqrt{n}}$ for any constant $C < \pi\sqrt{2/3}$ (Dasbach, [Das]).

Below, we are going to prove the simplest upper bound from [CD1] and give an idea that leads to the lower bound estimates from [CD2]. The ideas of [Das] generalise the techniques from [CD2] by adding some more low–dimensional topology. For more details, see [CDM] or the original works (for all the other estimates).

14.5.2   An upper bound

First, let us discuss the upper bound [CD1]. They state that $dim\Delta_n \leq (n-1)!$. Namely, they present a set of $(n-1)!$ generating elements for Δ_n.

Definition 14.9. A chord diagram is called a *spine* if it contains a chord of it intersecting all other chords.

Theorem 14.6. *The set of all spine chord diagrams generates* Δ_n.

Rather than proving this theorem, we shall divide it into small steps (according to [CD2]); each of these steps can be easily proven by the reader as a simple exercise. While performing these steps, we shall use induction on some parameters.

Definition 14.10. Given a chord diagram D, let d be a chord of diagram D. The *degree* of d is the number of chords intersecting d. The degree of the chord diagram D is the maximal degree of its chords.
  Notation: $deg(d), deg(D)$.

Fix a diagram D. By definition, if D has n chords and $deg(D) = n - 1$ then D is a spine diagram.
  We shall use induction on the degree of D and prove that if $deg(D) < n-1$ then D can be represented as a linear combination of diagrams of greater degrees.
  Suppose $deg(D) = f$. Choose a chord d such that $deg(d) = f$. After a rotation, we can assume that the chord f is vertical and the right part of D (with respect to d) contains more chord ends than the left one.
  Then there exists a chord d' with both ends lying in the right part. The top end of it lies at some distance from the top end of f. Denote the number of points between them by k.

Exercise 14.3. *If $k > 0$ then the diagram D can be represented as a linear combination of chord diagrams of greater degree and diagrams of the same degree with smaller k.*

Thus we can assume that $k = 0$.
  Now let us consider the lower end of d'.

Definition 14.11. The lower end of any chord which intersects d' and does not intersect d is *bound*. A point that lies on the lower arc between d and d' and is not bound will be referred as *loose*.

Let l be the number of loose points in the chord diagram, and b be the number of bound points between the lower end of d' and the first loose point. *The index of the chord diagram is the pair (l, b).* The index will be used as the induction parameter with respect to the following lexicographical ordering: $(l_1, b_1) > (l_2, b_2)$ if and only if either $l_1 > l_2$ or $l_1 = l_2, b_1 > b_2$.
  Now we are going to show that each non–zero degree diagram can be

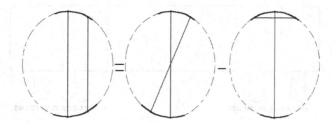

FIGURE 14.14: A linear combination of diagrams of greater degree

represented as a linear combination of diagrams of the same degree and lower index and some diagrams of greater degree.

This consists of two induction steps, both of which are left to the reader. Each of them follows straightforwardly from applying the $4T$–relation.

The first step.

Exercise 14.4. *If $l > 0, b = 0$, then the diagram can be represented as a linear combination of diagrams of greater degrees and diagrams of the same degree and smaller l.*

The second step.

Exercise 14.5. *If $l > 0, b > 0$ then the diagram can be represented as a linear combination of chord diagrams of greater degree or the same degree and smaller index.*

Now, if $l = 0, b \neq 0$ then each chord intersecting d intersects d' as well and there are chords intersecting d' but not d. Thus, the degree of the diagram is indeed greater than f, so this is not the case.

If $l = 0, b = 0$, then we have two "parallel chords" and the following $4T$–relation (together with a $1T$–relation that we do not illustrate on the picture) completes the proof of the theorem; see Fig. 14.14.

14.5.3 A lower bound

We are going to present a lower bound for the dimension of $\mathcal{A}_n$ according to [CD2]. As the authors say, "the story of lower bounds for the Vassiliev invariants is more enigmatic" [than that of upper bounds]. The first estimate was proposed by Bar–Natan in [BN1] as follows: $dim \mathcal{V}_n > e^{c\sqrt{n}}, n \to \infty$. This estimate comes from the connection between the Vassiliev knot invariants and Lie algebras that was discussed in the Section 14.4 of this chapter.

Definition 14.12. By a *Jacobi diagram* we mean the same as a Feynman diagram but without the oriented circle; namely, a graph whose vertices have valency one or three; each vertex of valency three should be endowed with a

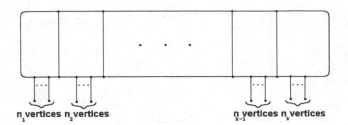

FIGURE 14.15: The diagram $B_{n_1,\ldots,n_k}$

cyclic order of outgoing edges. The *degree* of a Jacobi diagram is half of the total number of its vertices.

For Jacobi diagrams, we can consider the IHX relation and the antisymmetry relations. Obviously, they are both homogeneous with respect to the graduation described above. Like Feynman diagrams, Jacobi diagrams have multiplication and comultiplication, which are defined even more simply than that for Feynman diagrams: we have no oriented circle, so the multiplication is just the disconnected sum, and the comultiplication is defined by splittings into connected components.

Thus, we know which Jacobi diagrams are primitive: they are just connected uni–trivalent graphs for which each trivalent vertex is endowed with a cyclic order.

To go on, we shall need to introduce some notions. Consider the space of all primitive Jacobi diagrams. Each primitive Feynman diagram has an even number of vertices. Denote half of this number by d. Both IHX and antisymmetry relations are homogeneous with respect to d. The space $\mathcal{C}$ is thus bigraded: $\mathcal{C} = \oplus \mathcal{C}_{d,n}$, where $\mathcal{C}_{d,n}$ is a subspace of the space $\mathcal{C}$ generated by primitive Jacobi diagrams with a total of $2d$ vertices, precisely n of which are univalent.

Let us define a family of *Baguette diagrams*.

Definition 14.13. A *Baguette diagram* $B_{n_1,\ldots,n_k}$ is a Jacobi diagram shown in Fig. 14.15.

The baguette diagram $B_{n_1,\ldots,n_k}$ has $2(n_1 + \cdots + n_k - k - 1)$ vertices, out of which $n_1 + \cdots + n_k$ are univalent.

Theorem 14.7 (Main theorem,[CD2]). *Let $n = n_1 + \cdots + n_k$ and $d = n + k - 1$. The elements $B_{n_1 \ldots n_k}$ defined as above are linearly independent in $\mathcal{C}_{d,n}$ if $n_1, \ldots, n_k$ are all even and satisfy the following conditions:*

$$n_1 < n_2$$

$$n_1 + n_2 < n_3$$

$$n_1 + n_2 + n_3 < n_4$$

$$\cdots$$

$$n_1 + n_2 + \cdots + n_{k-2} < n_{k-1}$$

$$n_1 + n_2 + \cdots + n_{k-2} + n_{k-1} < \frac{n}{3}.$$

The proof of this theorem involves the techniques of Bar–Natan. Namely, instead of chord diagrams (Feynman diagrams) we consider the corresponding polynomials coming from the natural representation of $SL(n)$. Thus, we obtain a polynomial in N. The polynomials corresponding to baguette diagrams are linearly independent; thus, so are the diagrams themselves. In addition, one can consider uni–trivalent diagrams, which are in natural correspondence with linear combinations of Feynman diagrams, see [CDM].

Theorem 14.8. *([CD2]) For any fixed value of $k = d - n + 1$ we have the following asymptotic inequality as d tends to ∞:*

$$dim\mathcal{C}_{d,n} \gtrsim \frac{1}{2^{\frac{k(k-1)}{2}} 3^{k-1}(k-1)!}(d - k + 1)^{k-1}.$$

Proof. This theorem follows from Theorem 14.7. Actually, we have to count the number of integer points with even coordinates belonging to the body in $\mathbb{R}^{k-1}$ described by the set of inequalities above. Asymptotically, the number of such points is equal to the volume of the body divided by 2^{k-1}.

To find this volume, let us note that the condition $n_1 + \cdots + n_{k-1} < \frac{n}{3}$ specifies the interior part of a $(k-1)$–simplex in $\mathbb{R}^{k-1}$ that has $(k-1)$ sides of length $\frac{n}{3}$ and all right angles between sides. Obviously, its volume is equal to $\frac{(n/3)^{k-1}}{(k-1)!}$. The inequality $n_1 < n_2$ cuts exactly one half of this body, the next equality cuts a quarter of the obtained half, and so on, and the last one cuts $\frac{1}{2^{k-2}}$–th part of the result obtained at the previous step.

Summarising the results above, we obtain the statement of the theorem. $\qquad\square$

14.5.4 A table of dimensions

The first precise calculation for the dimensions $dim\ \mathcal{A}_n$ were made by Bar–Natan. To implement his algorithm and to calculate the dimensions of $\mathcal{A}_n$ up to $n = 9$ he borrowed extra RAM for his computer. The program had been working for several days.

Below, we give the table of dimensions up to $n = 12$; see [Kne].

n	0	1	2	3	4	5	6	7	8	9	10	11	12
$dim\ \mathcal{P}_n$	0	0	1	1	2	3	5	8	12	18	27	39	55
$dim\ \mathcal{A}_n$	1	0	1	1	3	4	9	14	27	44	80	132	232
$dim\ \mathcal{V}_n$	1	1	2	3	6	10	19	33	60	104	184	316	548

The answer for $n = 10, 11, 12$ was obtained by using a thin technique using special structures on the set of primitive diagrams.

Chapter 15

The Kontsevich integral and formulae for the Vassiliev invariants

The Kontsevich integral was first invented by M.L. Kontsevich [Kon1] in 1992. It was based on a remarkable construction of the product integral, better known as Chen construction or iterated integration formula. Kontsevich used the integration in the way proposed by Knizhnik and Zamolodchikov [KZ].

After Kontsevich's original proof, some other sympathetic (mostly combinatorial) constructions describing the same knot invariant arose; see the works of Cartier and Piunikhin [Car93, Piu]. The work by Le and Murakami [LM] proposes a concrete method of calculation of the Kontsevich integral. See also [Lan, CD3].

A very fundamental approach to Kontsevich's integral is presented in the book by Chmutov and Duzhin [CDM].

First, recall some definitions. Given a Vassiliev invariant V of degree n, then its $(n+1)$-th derivative equals zero. The value of the n-th derivative of the invariant V (the symbol of V) depends only on the passing order of singular points; thus it can be considered as a function on chord diagrams. It was shown that each such function satisfies the one–term and the four–term relations (such functions are called weight systems).

Theorem 15.1. *(1) (V.A. Vassiliev) Each symbol of an invariant of degree n comes from some element of graduation n of the chord diagram algebra (with $1T$ and $4T$–relations).*

(2) (M.L. Kontsevich) All elements of Δ_n are symbols of the Vassiliev knot invariants of degree n.

The first part of the theorem follows from Vassiliev's works [Vas1, Vas2]. We have already proved it previously.

The main goal of this chapter is to prove the second part of this theorem.

15.1 Preliminary Kontsevich integral

Definition 15.1. The completion $\overline{\Delta}$ of $\Delta = \oplus_{m=0}^{\infty}\Delta_m$ is the set of all formal series $\sum_m c_m a_m$, where $c_m \in \mathbb{C}$ are numeric coefficients, and $a_m \in \Delta_m$ are elements of the space of degree m chord diagrams.

Let us think of the space $\mathbb{R}^3$ as a Cartesian product of $\mathbb{C}^1$ with the coordinate z and $\mathbb{R}^1$ with the coordinate t.

Given an oriented knot K in $\mathbb{R}^3 = \mathbb{C}_z \times \mathbb{R}_t$, by a small motion in $\mathbb{R}^3$ (without changing the knot isotopy type), we can make the coordinate t a *simple Morse function* on the knot K. This means that all critical points of t on the knot K are regular and all critical points have different critical values.

Remark 15.1. *Later on, such embeddings will be called* Morse knots.

Definition 15.2. The *preliminary Kontsevich integral* of a knot K is the following element of $\overline{\Delta}$:

$$Z(K) = \sum_{m=0}^{\infty} \frac{1}{(2\pi i)^m} \int_{\substack{c_{min}<t_1<\cdots<t_m<c_{max} \\ t_j \text{ non-critical}}} \sum_{P=\{(z_j,z_j')\}} (-1)^{\downarrow} D_P \bigwedge_{j=1}^{m} \frac{dz_j - dz_j'}{z_j - z_j'}$$

(15.1)

We decree the coefficient of the "empty" chord diagram to be equal to one.

Let us discuss the formula (15.1) in more detail.

The real numbers c_{min} and c_{max} are maximal and minimal values of the function t on the knot K.

The integration domain is an n–simplex $c_{min} < t_1 < \cdots < t_m < c_{max}$. This domain is divided into connected components. Herewith, z_i and dz_i should be understood as functions of the corresponding t_i. For instance, for the unknot shown in Fig. 15.1 and $m = 2$ the integration domain consists of 6 components and looks as shown in Fig. 15.1.

The number of summands is constant for each connected component, but it can vary when passing from one component to another. The part of the knot lying inside the margin between two adjacent critical levels is a set of curves; each of these curves is uniquely parametrised by t.

Let us fix m and choose m horizontal planes $\{t = t_i\}, i = 1, \ldots, m$, each of which does not contain critical points and lies between the minimal and the maximal levels. Later, we shall take the sum over all natural m. At each plane $\{t = t_i\} \subset \mathbb{R}^3$, let us choose an unordered pair (z_i, t_i), (z_i', t_i) of different points lying on K. Denote by $P = \{(z_i, z_i')\}$ the system of m such pairings. Fix a pairing P. If we think of a knot as a circle and then connect the points of the circle corresponding to z_i, z_i' of the same pair (according to P) we obtain a chord diagram. Denote this diagram by D_P.

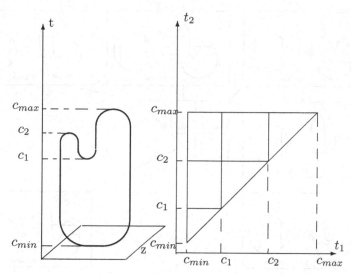

FIGURE 15.1: Integration domains for the Kontsevich integral

Now, under the integral we have the sum of such diagrams corresponding to different pairings P. The coefficients are obtained in the following way. Choosing any arbitrary connected component, the choice of P means that for each t_i, some pair of knot branches is taken. Thus, choosing m planes, we get m pairs of points.

As a matter of fact, after we have chosen all pairings, the diagram D_P is defined; thus we should integrate not chord diagrams, but only the form $(-1)^{\downarrow} \wedge_{j=1}^{m} \frac{dz_j - dz_j'}{z_j - z_j'}$. The obtained integral will give us the coefficient of our chord diagram D_P. Later, we shall collect similar terms.

In the example shown above, the connected component $\{c_{min} < t_1 < c_1, c_2 < t_2 < c_{max}\}$ corresponds to a unique pair of points at the levels $\{t = t_1\}$ and $\{t = t_2\}$. In this case, the desired sum consists of a unique summand. For the component $\{c_{min} < t_1 < c_1, c_1 < t_2 < c_2\}$, we have a unique choice at the level $\{t = t_1\}$, but the plane $\{t = t_2\}$ intersects the knot at four points; thus we have $C_4^2 = 6$ possible pairings (z_2, z_2'), and the total number of summands equals six. For the component $\{c_1 < t_1, t_2 < c_2\}$ we have 36 summands, among them the most interesting case of $\bigotimes$ appears. In each part of the figure, we choose exactly one pairing and show the corresponding chord diagram.

It is easy to see that in all cases except $\{c_1 < t_1 < t_2 < c_2\}$ we obtain the chord diagram $\bigodot\!\!\bigodot$ with two non–intersecting chords. These diagrams are equal to zero modulo one–term relation. Thus, the integration can be reduced to the small simplex $\{c_1 < t_1 < t_2 < c_2\}$.

The symbol $\downarrow$ for a given set choice of P denotes the number of points

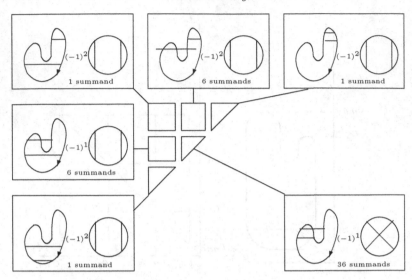

FIGURE 15.2: Integration domain and chord diagrams

(z, t_i) or (z', t_i) of P, where the coordinate t is decreasing while moving along the knot according to its orientation. In Fig. 15.2, the diagrams corresponding to different integration domains are shown.

Now we have the following questions to answer.

1. Do the coefficients of $\bar{\Delta}$ in the formula (15.1) converge?

2. Is the obtained element a knot invariant?

3. How is it related to the Vassiliev invariants?

4. How do we calculate this integral?

Theorem 15.2 ([Kon1], see also [BN1]). *All coefficients of* (15.1) *are finite.*

Definition 15.3. A *horizontal deformation* is an isotopy of a Morse embedding of a curve in $\mathbb{R}^3$ that does not change the setup of singular points.

The horizontal deformation can be expressed as a composition of moves shown in Fig. 15.3.

Theorem 15.3 ([BN1]). *The function* $Z(K)$ *is invariant under horizontal deformations of a knot and under the transformation shown in Fig. 15.4, but not invariant under the transformation* (*)*, shown in Fig. 15.5.*

Denote the knot representing the closure of the arc shown in Fig. 15.5 by A.

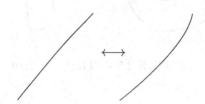

FIGURE 15.3: Horizontal deformation

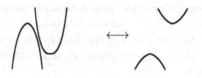

FIGURE 15.4: Moving critical values

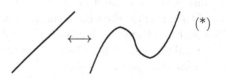

FIGURE 15.5: Forbidden transformation

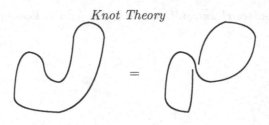

FIGURE 15.6: The "∞" knot

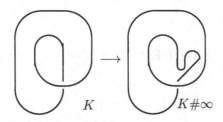

FIGURE 15.7: Transformation (*)

We can consider the simplest realisation of the unknot (with one minimum and one maximum) and the realisation given by ∞; see Fig. 15.6. It is easy to see that $Z(K)$ for the simplest realisation is equal to one (i.e., the series consisting of the only diagram without chords with coefficient one). Moreover, $Z(\infty)$ is not equal to $1 = \bigcirc$.

Thus we see that Z is not a knot invariant.

On the other hand, one can prove the following theorem.

Theorem 15.4. *If the knot K' is obtained from K by using (*) then $Z(K') = Z(K) \cdot Z(\infty)$.*

Proof. First, let us note that ∞ is obtained from the knot A by using allowed moves, thus $Z(\infty) = Z(A)$.

Let us consider now the connected sum of K with a "small" knot ∞ in such a way that the interval of the coordinate t, corresponding to the knot A, has no critical points of the knot K. In this case, just two new critical points — one maximum and one minimum are added to this knot; see Fig. 15.7.

By virtue of the previous theorems, the Kontsevich integral of the obtained knot coincides with the Kontsevich integral for K' that is obtained from $K \# \infty$ by using horizontal deformations. Comparing the Kontsevich integral for the initial knot and for the knot K, we see that each term for the integral of the knot K corresponds to the same term multiplied by the Kontsevich integral for A (in the integral of K'). Consequently, $Z(K') = Z(K) \cdot Z(\infty)$.

□

15.2 $Z(\infty)$ and the normalisation

Thus, the change of the preliminary integral $Z(\cdot)$ under $(*)$ is not difficult: the value is just multiplied by $Z(\infty)$. Now let K be a Morse embedding of S^1 in $\mathbb{R}^3$, and c be the number of critical points of t on K.

Let us consider now the preliminary Kontsevich integral as a formal series. Hence this series consists of elements of a graded algebra and its initial element is the unit element of this algebra. Then one can inverse such rows by

$$(1+a)^{-1} = 1 - a + a^2 - a^3 + \dots,$$

where a^i is the formal series for the i-th power of the series a. Furthermore, one can formally multiply such series.

Definition 15.4. The *universal Vassiliev–Kontsevich invariant* of a knot K is the following element of the completion of the chord diagram algebra:

$$I(K) = \frac{Z(K)}{Z(\infty)^{\frac{c}{2}-1}}. \tag{15.2}$$

Remark 15.2. *Here the degree* $(\frac{c}{2}-1)$ *is taken for the following majors. In the case of the simplest embedding representing the unknot we wish to have* $I(\bigcirc) = 1$. *For one maximum and one minimum we have* $\frac{c}{2} - 1 = 0$.

Remark 15.3. *Obviously, if* (15.1) *converges, then* (15.2) *makes sense: it is just the fraction of two series.*

Thus we obtain the following theorem.

Theorem 15.5. *The Kontsevich integral* $I(\cdot)$ *is a knot invariant.*

Proof. Indeed, by virtue of Theorem 15.4 we see that $Z(K)$ depends not on the configuration of critical points but only on their quantity.

It is easy to check that two Morse embeddings represent the same knot if and only if one can be transformed to the other by means of moves not changing the setting of critical points and moves shown in Figs. 15.4 and 15.5.

Taking into account the invariance of Z under all moves but the last one, we obtain the statement of the theorem. $\qquad\square$

The invariant $I(\cdot)$ is called *the universal Vassiliev–Kontsevich invariant.*

Now it remains to formulate and to prove the most important theorem.

Let W be a weight system of degree m. Decree that $W(d) = 0$ for all diagrams d with the number of chords not equal to m.

Theorem 15.6 (Kontsevich, see also [BN1]). *The invariant* $W(I(\cdot))$ *is a Vassiliev invariant with symbol* W; *i.e.,*

$$V(W)(K) = W(I(K))$$

for each knot K.

This theorem implies the second (difficult) part of the Vassiliev–Kontsevich theorem about the existence of Vassiliev invariants corresponding to any given weight system.

We shall prove Theorems 15.2, 15.3 and 15.6 later.

15.3   Invariance of the Kontsevich integral

We are now going to prove theorems 15.2, 15.3 and 15.6.

Remark 15.4. *By $Z_m(K)$ and $I_m(K)$ we mean the m–th graded summand of $Z(K)$ and $I(K)$, respectively.*

First, let us prove Theorem 15.2 which states that the series for each coefficient at each term of (15.1) converges.

Proof. Consider a Morse knot K in $\mathbb{R}^3$. Let us fix $m \in \mathbb{N}$ and choose some m planes not intersecting K at critical points.

Choose some chord diagram D and consider the coefficient at this diagram. It is obtained by integrating the form

$$\bigwedge_{j=1}^{m} \frac{dz_j'(t_j) - dz_j(t_j)}{z_j'(t_j) - z_j(t_j)}$$

over the part of the simplex $\{c_{min} < t_1 < \cdots < t_m < c_{max}\}$ corresponding to the chord diagram D.

Let us consider the singular points of the form, namely, those where the condition $z_j = z_j'$ holds for some j. The integral of the form might diverge only in the neighbourhood of these points. Consider such pairs of points z_j, z_j' closed to the singular position.

Then we have the two possibilities:

1. The arc between z_j and z_j' contains other ends of chords (as shown in Fig. 15.8). Then the integration domain (where we integrate $z_j - z_j'$) has smallness of higher order than $z_j - z_j'$ because the singular point is not degenerate. Consequently, this part of (15.2) gives no divergence.

2. The arc between z_j and z_j' has no other chord ends; see Fig. 15.9. Then the chord $z_j z_j'$ of the diagram D is isolated; thus, the diagram D equals zero modulo 1T–relation.

This completes the proof of the theorem.   □

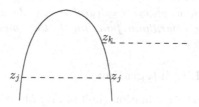

FIGURE 15.8:

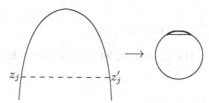

FIGURE 15.9:

15.3.1 Integrating holonomies

In order to prove the remaining two theorems, we shall have to integrate holonomies and introduce the so called *Knizhnik–Zamolodchikov connection*.

First, let us recall some constructions.

Definition 15.5. Let X be a smooth manifold and let $\mathcal{U}$ be an associative topological algebra with the unit element (considered over $\mathbb{R}$ or $\mathbb{C}$).

Then a $\mathcal{U}$*–connection* Ω on the manifold X is a 1–form Ω on X with coefficients from $\mathcal{U}$.

The *curvature* of the connection Ω is the 2–form

$$F_\Omega = d\Omega + \Omega \wedge \Omega.$$

The connection is *flat* if its curvature equals zero.

Definition 15.6. Let $B : I \to X$ be a smooth mapping of the interval $[a, b]$ to the space X. Let Ω be a $\mathcal{U}$–connection on X.

Let us define the *holonomy* $h_{B,\Omega}$ of the form Ω along the path B as the solution of the differential equation $\frac{\partial}{\partial t} h_{B,\Omega}(t) = \Omega(B'(t)) \cdot h_{B,\Omega}(t), t \in I$, with the initial condition $h_{B,\Omega}(a) = 1$.

Remark 15.5. *It is easy to show that if the connection Ω is flat, then the holonomy is defined only by the ends of a path and the homotopy type of this path.*

This is quite analogous to the Gauss–Ostrogradsky formula in the commu-tative case. Then the multiplicative integral is just the exponent of the Riem-manian integral for the logarithmic function. In the non–commutative case the extra term $\Omega \wedge \Omega$ arises.

15.3.1.1   The product integration

The solution to such an equation (holonomy) often exists. It is called the *product* or *multiplicative integral* of the form Ω. In many cases, the holonomy can be calculated according to the following *iterated formula*:

$$h_{B,\Omega}(t) = 1 + \sum_{m=1}^{\infty} \int_{a \le t_1 \le t_2 \le \cdots \le t_m \le t} (B^*\Omega)(t_m) \ldots (B^*\Omega)(t_1). \qquad (15.3)$$

In order to clarify the situation, let us consider the following simple con-struction.

Example 15.1. *Let*
$$Y' = AY$$
be a differential equation with the initial condition $Y(0) = 1$, say, in $n \times n$ matrices.[1]. Obviously, its solution $Y(t)$ is the product of "infinitely many" elements "infinitely close to the unit element". This is naturally called the product integral *of A and denoted by*

$$Y(x) = \int_0^{x \cap} (E + A(t)dt).$$

Note that in order to calculate $Y(x)$, one can use the following formula

$$Y(x) = E + \int_0^x A(t_1)dt_1 + \int_0^x A(t_1) \int_0^{t_1} A(t_2)dt_2 dt_1 + \cdots \qquad (15.4)$$

if the series (15.4) converges.

Remark 15.6. *In all "normal" cases this series actually converges.*

Actually, while integrating the series, each next term becomes equal to the previous one multiplied by A.

Each term of the iterated integral (15.4) can be considered as an integral over some simplex.

The formula (15.3) is completely analogous to the formula (15.4).
The theory of product integration is well described in [DF, ManO2, MaMa].

---

[1]or in any other topological algebra: all we need here is linear operations over the main field, multiplication, and an intrinsic topology (in order to consider limits of sequences).

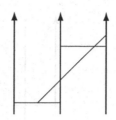

FIGURE 15.10: An element $\mathcal{A}^{KZ}$

Remark 15.7. *In the normal (convergent) case it is obvious that the formula actually gives a solution to the differential equation. The initial condition evidently holds. The derivative of the m–th integral gives the $(m-1)$–th integral with coefficient $\Omega(\dot{B})$.*

15.3.1.2 The Knizhnik–Zamolodchikov connection

Denote by $\mathcal{D}_n^{KZ}$ the set of all diagrams consisting of n ascending infinite arrows (in Fig. 15.10 they are shown by thick lines) and a finite number of edges such that:

1. each end point of each edge either lies on the arrow or is a trivalent vertex (with two other ends of edges);

2. one point on the arrow is incident to no more than one interval (only one end of this edge can coincide with this point).

Such diagrams are considered up to combinatorial equivalence.

Let $\mathbb{C}$ be the main field. Consider the set $\mathcal{A}_n^{KZ} = span(\mathcal{D}_n^{KZ})/\{STU-$ relations$\}$. The STU–relation means the same as for the Feynman diagrams (by "multiplication" of all "partial" integrals), where we consider a part of an arrow instead of part of an oriented circle. Note that the STU–relation is local. When we finally close the "arrow" diagrams in order to obtain the Feynman diagram, we get the STU–relation as well.

For a fixed n, the set $\mathcal{A}_n^{KZ}$ admits an algebraic structure: the product means the juxtaposition of one diagram over the other.

Example 15.2. *For $n = 3$ such a multiplication for $\mathcal{A}_n^{KZ}$ is shown in Fig. 15.11.*

For a fixed n, the algebra $\mathcal{A}_n^{KZ}$ is graded: the order of an element is equal to half of the total number of vertices.

For $1 \leq i, j \leq n$, let us define $\Omega_{ij} \in \mathcal{A}_n^{KZ}$ as the element with only one edge connecting the arrows i and j.

Remark 15.8. *It is easy to see that if $\{i,j\} \bigcap \{k,l\} = \emptyset$ then Ω_{ij} and Ω_{kl} commute.*

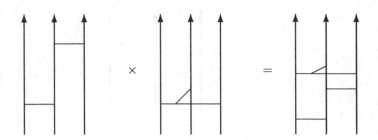

FIGURE 15.11: Multiplication in $\mathcal{A}^{KZ}$

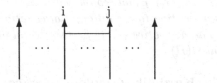

FIGURE 15.12: The element Ω_{ij}

Let X_n be the configuration space of n pairwise different points on $\mathbb{C}^1$. Let ω_{ij} be the following 1–form on X_n:

$$\omega_{ij} = d(\ln(z_i - z_j)) = \frac{dz_i - dz_j}{z_i - z_j}.$$

Let us define the formal Knizhnik–Zamolodchikov connection Ω_n with coefficients in $\mathcal{A}_n^{KZ}$ as $\Omega_n = \sum_{1 \le i < j \le n} \Omega_{ij}\omega_{ij}$ on X_n.

Theorem 15.7. *This connection is flat. More precisely $\Omega_n \wedge \Omega_n = 0$ and $d\Omega_n = 0$.*

Proof. The last statement is evident. Indeed, $d\omega_{ij} = d^2(\ln(z_i - z_j))$ and this vanishes by definition of d.

Let us prove the first statement. Consider the element

$$\Omega_n \wedge \Omega_n = \sum_{i<j;k<l} \Omega_{ij}\Omega_{kl}\omega_{ij}\omega_{kl} \qquad (15.5)$$

and the set $\{i,j,k,l\}$. If this set consists of two or four elements then the corresponding term of the sum equals zero (this case is commutative). Consequently, the desired sum equals the sum along all i,j,k,l, where the set $\{i,j,k,l\}$ consists of three elements. Consider, e.g., the set $\{i,j,k,l\} = \{1,2,3\}$ and all corresponding terms in the sum (15.5). In this case we get:

$$\sum_{\{i,j,k,l\}=\{1,2,3\}} \Omega_{ij}\Omega_{kl}\omega_{ij}\omega_{kl} = (\Omega_{12}\Omega_{23}-\Omega_{23}\Omega_{12})\omega_{12}\wedge\omega_{23}+\langle\text{cyclic permutations}\rangle.$$

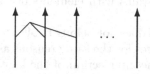

FIGURE 15.13: The element Ω_{123}

By using the STU–relation, we see that the desired sum equals

$$-\Omega_{123}(\omega_{12} \wedge \omega_{23} + \omega_{23} \wedge \omega_{31} + \omega_{31} \wedge \omega_{12}),$$

where Ω_{123} is the element shown in Fig. 15.13.

Exercise 15.1. *(V.I. Arnold's identity.)*
   Show that $\omega_{12} \wedge \omega_{23} + \omega_{23} \wedge \omega_{31} + \omega_{31} \wedge \omega_{12} = 0$.

Remark 15.9. *This identity appeared in [Arn1] when Arnold studied the cohomologies of the pure braid group.*

Thus if some set $\{i, j, k, l\}$ consists of precisely three different members, it gives no contribution. We have considered all possible cases. Thus, $\Omega_n \wedge \Omega_n = 0$. $\qquad\square$

Remark 15.10. *The connection Ω_n can be slightly modified for the case of the algebra $\mathcal{A}_{nn}^{KZ}$. This algebra is generated by arrow diagrams with $2n$ arrows (the first n arrows oriented upwards and the last n arrows oriented downwards). The STU–relation for such diagrams depends on the direction of the chord that the relation has to be applied to.*

Let $\Omega_{nn} = \sum_{1 \le i < j \le n} s_i s_j \Omega_{ij} \omega_{ij}$, where s_i equals 1 for $i \le n$ and -1 for $i > n$.

Exercise 15.2. *Show that the connection Ω_{nn} is flat.*

Now, let us prove the invariance theorem (Theorem 15.3).

Proof. First, let us prove that the preliminary Kontsevich integral $Z(K)$ is invariant under the transformation preserving the critical points.

The point is that the Kontsevich integral for the whole knot can be decomposed into a product of similar integrals for *parts* of this knot; each of these parts represents an element of some Knizhnik–Zamolodchikov algebra with ascending and descending arrows; being connected together, they constitute a normal chord diagram. Thus, the product of elements in some A^{KZ} is thought to be an element of $\bar{\Delta}$.

Let $c_{min} \le a < b \le c_{max}$. Let us define $Z(K, [a, b])$ just as was done in (15.1), but taking the integration domain to be $\{a < t_1 < \cdots < t_n < b\}$, and replacing the chord diagrams with elements of the Knizhnik–Zamolodchikov algebra.

Although $Z(K, [a, b])$ does not belong to $\overline{\mathcal{A}^c}$, the corresponding series (evaluated at a knot) converges for the same reasons as Z. Since the interval (a, b) has no critical points, the intersection of the knot with the margin $\mathbb{C} \times (a, b)$ is a set of oriented curves without horizontal tangent lines. Suppose that the number of such curves equals $2n$. Obviously, n of them are ascending and the other n are descending. Let us fix the lower points $a_1, \ldots, a_{2n}$ and the corresponding upper points $b_1, \ldots, b_{2n}$, where the first n coordinates correspond to ascending curves and the other ones correspond to descending curves. The convergence of the integral can be proved in the same manner as before. One should, however, introduce an analogue of the one–term relation taking all diagrams with a "solitary" chord (with one end on an ascending chord and one end on a descending arc) to zero.

Now, the integral $Z(K[a, b])$ can be represented as the holonomy of the connection Ω_{nn} along the path from $(a_1, \ldots, a_{2n})$ to $(b_1, \ldots, b_{2n})$ by virtue of the iteration formula (15.3). Actually, the m–th term of the iteration formula for Ω_{nn} corresponds to the m–th term of the Kontsevich integral because in both cases we integrate the form

$$\sum_{P=\{(z_j, z'_j)\}} (-1)^{\downarrow} \Omega_{jj'} \bigwedge_{j=1}^{m} \frac{dz_j - dz'_j}{z_j - z'_j}, \tag{15.6}$$

where $(-1)^{\downarrow}$ corresponds to the sign of the product $s_j s'_j$.

Recall that the STU–relation for Feymann diagrams is "the same" as the $4T$–relation for chord diagrams. This is just the place when we use the $4T$–relation (in its STU–form).

Since the curvature of the connection Ω_{nn} is zero, the integral (15.6) is invariant under homotopies of the integration path with fixed endpoints; i.e., under horizontal isotopies of the part of the knot lying inside $t \in (a, b)$.

It is not difficult to show that for arbitrary $a < b < c$ (possibly, critical), we have $Z(K, [a, c]) = Z(K, [a, b]) \cdot Z(K, [b, c])$. Thus we conclude that the integral $Z(K)$ which is a product $Z(K, [c_i, c_{i+1}])$, where c_i, c_{i+1} are all pairs of "adjacent" critical points, is invariant under horizontal deformation in the intervals not containing critical points.

Now, let us consider the cases when critical points are moving during the knot isotopy.

1. The critical point is moving, but the disposition of all critical points stays the same; see Fig. 15.14.

2. The order of applicates of two critical points changes; see Fig. 15.15.

As shown in Figs. 15.14 and 15.15, one can first perform the transformation that does not change $Z(K)$ to obtain a knot with a thin "needle". Let us show that the removal of this needle changes the m–th graduation term of the

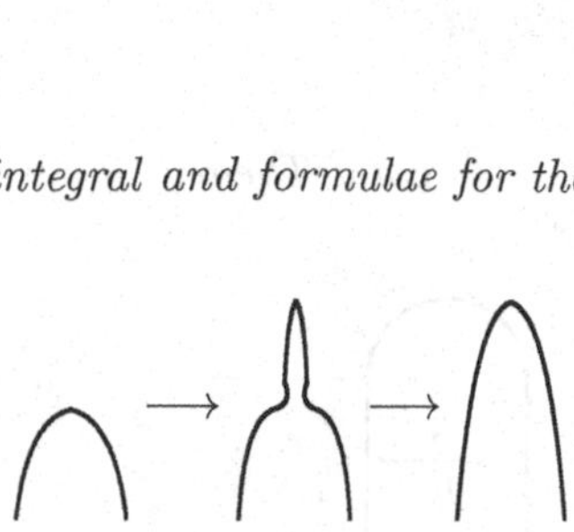

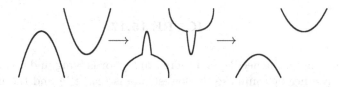

FIGURE 15.14:

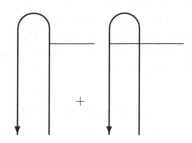

FIGURE 15.15:

Kontsevich integral by some infinitely small ε depending on the diameter of the needle.

Actually, let K be a knot and let K' be the knot obtained from the knot K by means of adding a vertical needle somewhere.

Obviously, the difference $Z(K) - Z(K')$ contains only the terms corresponding to the diagrams with ends lying inside the needle. Suppose that the width of the needle equals ε. Let us show that $Z_m(K) - Z_m(K') = O(\varepsilon)$.

Actually, consider all the chords incident to the needle. If the upper chord has both ends on the needle then the chord diagram equals zero modulo $1T-$relation. If there are no chords with all ends lying on the margin, the situation is quite simple as well: the term shown in Fig. 15.16 should have smallness of order ε: while integrating the left and the right part, the numbers $\uparrow$ have difference 1; thus we obtain a contraction because for each term there exists a "mirror" term; see Fig. 15.16.

Thus, we only have to consider the case when the upper chord (z_i, z_i') has

FIGURE 15.16:

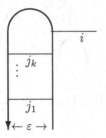

FIGURE 15.17:

one end lying on the needle, and there are k chords lying under this with both ends on the needle. Suppose the lowest one is (z_{j_1}, z'_{j_1}) and the upper one is (z_{j_k}, z'_{j_k}); see Fig. 15.17.

We may assume that (z_i, z'_i) is the only chord such that one end of it lies on the needle. If we delete such chords, we multiply the final integral by some number bounded from zero and the infinity.

Let $\delta_\alpha = |z_{j_\alpha} - z_{j'_\alpha}|$. Then the difference $Z(K') - Z(K)$ is bounded by some constant multiplied by

$$\int_0^\varepsilon \frac{d\delta_1}{\delta_1} \int_0^{\delta_1} \frac{d\delta_2}{\delta_2} \cdots \int_0^{\delta_{k-1}} \frac{d\delta_k}{\delta_k} \int_{z_{j_k}}^{z_{j'_k}} \frac{dz_i - dz'_i}{z_i - z'_i}.$$

The integral has smallness of the order $\tilde{\varepsilon}$. Actually, the last integral has smallness of the order of δ_k. Consequently, the term δ_k is reduced in the penultimate integral, so this integral has smallness of δ_{k-1}, and so on. Finally, the total integral has smallness of $\delta_1 \sim \varepsilon$.

Since ε is arbitrarily small, we conclude the desired invariance.   □

Thus, we have proved that $I(\cdot)$ is a knot invariant.

Now, let us prove Theorem 15.6 that for each weight system W, the function $W(I(\cdot))$ generates a Vassiliev invariant with symbol W.

Proof. Without loss of generality, we might assume that our knots are not only Morse embedded in $\mathbb{R}^3$ but their projections on some vertical plane (say, Oxz) represent planar knot diagrams (in the ordinary sense). Let W be a weight system of order m. In order to prove the theorem, we have to show that if D is a chord diagram of degree m and K_D is a Morse embedding of the singular curve (curve with intersection) in $\mathbb{C}_z \times \mathbb{R}_t$ (the singular knot corresponds to D) then we have

$$I(K_D) = \bar{D} + \langle \text{terms of order} \geq m \rangle,$$

where $\bar{D}$ is the equivalence class of the chord diagram D and $I(K_D)$ is defined to be the alternating sum of I evaluated at 2^m knots generating the singular knot K_D.

If two Morse knots K_1 and K_2 in $\mathbb{C}_z \times \mathbb{R}_t$ coincide everywhere except for a small part, where the branches of K_1 form an overcrossing (with respect to the projection on a vertical plane) and those of K_2 form an undercrossing, then the values $Z(K_2)$ and $Z(K_1)$ differ only in those chord diagrams, for which some point(s) on this branches is (are) paired with other point(s).

By virtue of Vassiliev's relation, the singular knot K_D is an alternating sum of 2^m knots that differ in small neighbourhoods of m points. Note that the sign of this alternating sum is regulated by the multiplicator $(-1)^{\downarrow}$ in (15.1).

Arguing as above, we conclude that $Z(K_D)$ has non-zero coefficients only at those chord diagrams obtained by pairing points for each neighbourhood. Thus, chord diagrams with non-zero coefficients must have at least m chords.

For chord diagrams of degree m this coefficient is not equal to zero only for the diagram K_D.

Let us calculate this coefficient.

At each of m vertices we obtain the difference of the integrals of the differential form $\frac{dz_i - dz_i'}{z_i - z_i'}$.

This difference equals the integral of $\frac{dz}{z}$ along the circuit passing once around zero. According to Cauchy's theorem, this integral equals $2\pi i$. Because the number of such contours equals m, the coefficients should be multiplied. Thus we obtain the multiplication factor $(2\pi i)^m$ that is cancelled by the denominator of (15.1). This means that

$$Z(K_D) = \bar{D} + \langle \text{terms of order} \geq m \rangle.$$

Taking into account $I(K) = \frac{Z(K)}{Z(\infty)^{\frac{c}{2}-1}}$, we have

$$I(K_D) = \bar{D} + \langle \text{terms of order} \geq m \rangle.$$

Consequently, $W(I(K_D)) = W(D)$, and the Vassiliev invariant $W(I(\cdot))$ of order m has the symbol W. This completes the proof.   □

The calculation of the Kontsevich integral is, however, very difficult. For instance, it was quite a complicated problem to calculate the integral (preliminary) of ∞. The form (in the Feynman diagram) of the integral was conjectured by Bar–Natan, Garoufalidis, Rozansky, Thurston [BGRT] and finally proved in [BN6].

The formula is represented in terms of Feynman diagrams. It looks like

$$I(\infty) = exp \sum_{n=0}^{\infty} b_{2n} w_{2n} = 1 + \left(\sum_{n=0}^{\infty} b_{2n} w_{2n} \right) + \frac{1}{2} \left(\sum_{n=0}^{\infty} b_{2n} w_{2n} \right)^2 + \ldots.$$

Here b_{2n} are modified Bernoulli numbers; i.e., the coefficients of the Taylor series:

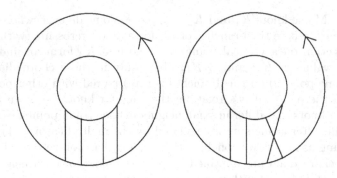

FIGURE 15.18:

$$\sum_{n=0}^{\infty} b_{2n} x^{2n} = \frac{1}{2} \ln \frac{e^{x/2} - e^{-x/2}}{x/2},$$

and w_{2n} are wheels.

Each wheel w_{2n} is $\frac{1}{(2n)!}$ multiplied by the sum of $(2n!)$ Feynman diagrams. Each of these diagrams consists of one exterior circle, one interior circle (treated just as a circular set of interior edges), and $2n$ chords connecting fixed $2n$ points on the first one with fixed $2n$ points on the second one. These points can be connected according to arbitrary permutation from S_{2n}. Thus, we have $(2n)!$ summands and take their average.

For instance, if we consider w_4, we see that eight summands represent the diagram D_x shown in the left part Fig. 15.18 and another sixteen summands represent the diagram D_y, see the right part of Fig. 15.18.

In the terms of chord diagrams w_4 can be represented as follows:

$$w_4 = \bigotimes - \frac{10}{3}\bigoplus + \frac{4}{3}\bigcirc.$$

Analogously (in fact, even more easily) one can find the expression for w_2 and w_2^2.

Exercise 15.3. *Prove the formulae above.*

The first terms of the final result look like:

$$I(\infty) = 1 + \frac{1}{48} w_2 + \frac{1}{4608} w_2^2 - \frac{1}{5760} w_4 + \ldots$$

or, in terms of chord diagrams,

$$I(\infty) = 1 - \frac{1}{24}\bigotimes - \frac{1}{5760}\bigotimes + \frac{1}{1152}\bigoplus + \frac{1}{2880}\bigcirc + \ldots$$

Besides this, Le and Murakami [LM] constructed a generalisation of the Kontsevich integral for the case of so–called *tangles* — one–dimensional manifolds lying between two horizontal planes and incident to these planes only at a finite number of points. A tangle is a common generalisation of both knots and braids, and the computation of the Kontsevich integral for the case of braids is much easier. In fact, tangles appeared indirectly in the text while calculating $Z[a, b]$ for some interval $[a, b]$. By using their own techniques, they calculated $Z(\infty)$. Later, S.D. Tyurina calculated such integrals for various knots, see [Tyu1, Tyu2].

15.4   Vassiliev's module

It is not known whether the Vassiliev knot invariants distinguish all isotopy classes of knots and link. This is conjectured and known as *Vassiliev's conjecture*.

Let us introduce the *Vassiliev module* where two knots are taken to be different if they are distinguished by some Vassiliev invariant having order not higher than some fixed order. Besides, each knot can be decomposed into a finite sum of generators of the module.

Let us give now the precise definition.

Definition 15.7. The *Vassiliev module* of order n is the module over $\mathbb{Z}$ (or $\mathbb{Q}$) generated by isotopy classes of oriented knots and singular knots modulo the following relations:

1. $\bigcirc = 0$, where $\bigcirc$ is the unknot.

2. The Vassiliev relation.

3. $K_m = 0$ for $m > n$, where K_m is an arbitrary singular knot of order m.

The following theorem holds.

Theorem 15.8 (Decomposition theorem). *In the Vassiliev module of order n, each knot K has the following decomposition:*

$$K = \sum_{i=1}^{r+s} v_i(K)K_i,$$

where r is the dimension of the set of Vassiliev invariants having order less than or equal to $(n-1)$, and s is the dimension of weight systems of order n; all v_i's are Vassiliev's invariants of order less than or equal to n, and K_i are some fixed basic knots independent of the knot K.

This theorem follows straightforwardly from the definitions.

Definition 15.8. For any n, the *actuality table* of a Vassiliev invariant of type n is the set of its values on all basic knots.

The simplest decomposition in Vassiliev's module [Lan] of order two is the following:

$$K = V_2(K) \,\vcenter{\hbox{}},$$

where V_2 is the second coefficient of the Conway polynomial.

For more details see, e.g. [CDM, Tyu1, Tyu2].

Chapter 16

Atoms, height atoms and knots

In the present chapter, we shall talk about an alternative way for encoding knots and links (different from planar diagrams and closures of braids). Namely, all knots can be encoded by so-called "atoms" and d–diagrams.

Atoms are combinatorial objects that arose several years ago in [Fom] for purposes of classification of integrable Hamiltonian systems of low complexity. d–diagrams are special chord diagrams closely connected with atoms.

Atoms play a crucial role for the construction of Khovanov homology.

By using this approach, we are going to prove several theorems on knots and curves: Kauffman–Murasugi's theorem on alternating links, the criterion for embeddability of special graphs, etc. We shall also describe a way of encoding knots by words in a finite alphabet via d–diagrams ("bracket calculus"). For a review on the bracket calculus see [Man5].

16.1   Atoms and height atoms

Let us start with definitions and introduce the notation.

Definition 16.1. An *atom* is a pair: a connected 2–manifold M^2 without boundary and a graph $\Gamma \subset M^2$ such that $M^2 \backslash \Gamma$ is a disconnected union of cells that admit a chessboard colouring (with black and white colours).

The graph Γ is said to be the *frame* of the atom. The *genus* (respectively, *Euler characteristic*) of the atom is that of its first component.

The *complexity* of the atom is the number of vertices of its frame.

Atoms are considered up to natural isomorphism: two atoms are called *isomorphic* if there exists a one–to–one map of their first components taking frame to frame and black cells to black cells.

Atoms can be generated by Morse functions on 2–surfaces: an atom's frame is just the critical level with several critical points on it.

Definition 16.2. An atom is called a *height* (or a *vertical*) atom if it is isomorphic to an atom obtained by the third projection function on some closed 2–manifold embedded in $\mathbb{R}^3$.

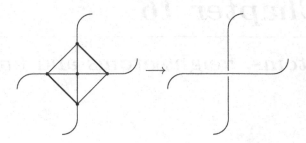

FIGURE 16.1: A part of a knot diagram constructed by a frame embedding

Each atom (more precisely, its equivalence class) can be completely restored from the following combinatorial structure:

1. the frame (four–valent graph);

2. the A–structure (dividing the outgoing half–edges into two pairs according to their disposition on the surface); and the

3. B–structure (for each vertex, we indicate some two pairs of adjacent half–edges (also: two angles) that constitute a part of the boundary of black cells).

In [Man2, Man'1] the following criterion is proved.

Theorem 16.1. *An atom V is a height atom if and only if its frame Γ is embeddable in $\mathbb{R}^2$ with respect to the A–structure (i.e., the intrinsic A–structure on the surface coincides with that induced from the plane).*

It turns out that height atoms are closely connected with knots unlike the non-height ones. Having a height atom V, one can construct a knot diagram as follows. Consider the frame Γ of V and let us embed Γ in the plane with respect to the A–structure on V. Then, the B–structure of this atom can be illustrated on the plane: if a pair of edges outgoing from a vertex is adjacent in V, it remains so on the plane. Thus, one can locally indicate the structure of supercritical levels on the plane.

Thus, we have a four–valent graph on the plane with endowed B–structure. This B–structure allows us to construct a link diagram as shown in Fig. 16.1.

Remark 16.1. *In Fig. 16.1, the angles of the supercritical level (in the left part) are marked by additional thick lines.*

For each vertex A we set the crossing type in such a way that while turning inside the supercritical angle clockwise, one passes from the undercrossing to the overcrossing.

Thus, having an embedding of the frame of an atom with respect to the A–structure, one can construct a knot (link) diagram.

Let an atom be given. Assume that for the A-structure of the atom there exists an orientation of all edges of the atom such that at each vertex two opposite edges are emanating and two other opposite edges are coming.

Definition 16.3. We call this structure the *source–sink* structure.

Remark 16.2. *The same structure was investigated in the theory of virtual knots by Kamada, see, e.g. [KamN1, KamN2]. This structure was called an* alternating orientation *for a graph (in the present work we call this graph a frame of an atom).*

Exercise 16.1. *The frame of an atom admits a source–sink structure if and only if the atom is orientable.*

Remark 16.3. *It follows from Exercise 16.1 that if an atom (M, Γ) with a frame Γ is orientable, then each atom (M', Γ) with the same frame and A-structure is orientable, too.*

Remark 16.4. *The source–sink structure given on the whole atom defines an orientation for circles at all states of the Kauffman bracket polynomial of the corresponding link. Thus, if one constructs a diagram obtained by smoothing of some crossings and deleting unlinked circles not being incident to chosen crossings, then the frame of the atom corresponding to the new diagram will inherit the source–sink structure from the initial one. Therefore, the obtained atom will be orientable.*

16.2 Theorem on atoms and knots

It turns out that all knots can be encoded (not uniquely) by height atoms. In fact, consider a height atom V. Let us embed its frame in $\mathbb{R}^2$ while preserving the A–structure of the atom.

Furthermore, the following theorem holds.

Theorem 16.2 (Theorem on atoms and knots, [Man2]). *Let V be an atom. Then the planar link diagrams obtained from V by using the algorithm above generate diagrams representing the same link isotopy type.*

Thus, we can say that an atom *generates* a knot (link).

The proof of the theorem on atoms and knots follows from a well–known theorem:

Theorem 16.3. *If two planar graphs are isomorphic, then their embeddings in one and the same plane $\mathbb{R}^2$ are homeomorphic in $\mathbb{R}^3$.*

This theorem allows one to restore the homeomorphism of knots in the ambient space from embeddings of the atom's frame. For more details see e.g. [Man2].

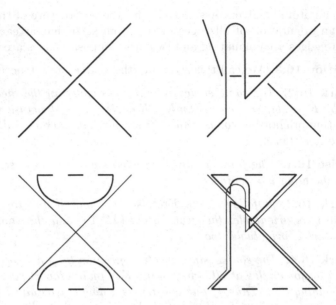

FIGURE 16.2: Link diagram above and supercritical circles below

---

16.3 Encoding of knots by d–diagrams

We begin with a theorem from [Man3].

Theorem 16.4. *For each link isotopy class L, there exists a height atom V that encodes a link from this class and has only one supercritical circle.*

Proof. Consider an arbitrary diagram of link L and the corresponding height atom V_1. Suppose that the atom V_1 has k supercritical circles. If $k = 1$ then there is nothing to prove.

If $k > 1$ then there exists a vertex A of V_1 such that the two supercritical angles of this vertex correspond to two arcs of different supercritical circles.

Let us apply the move Ω_2 to the initial diagram, as shown in Fig. 16.2.

It is easy to see that after such transformation, the two circles shown in Fig. 16.2 are transformed into one circle. Thus, we decrease the number of supercritical circles by one without changing the link isotopy class. Reiterating this operation many times, we obtain a diagram with precisely one supercritical circle. □

Consequently, d–diagrams encode all knot and link isotopy classes.

Let us give some examples. Consider the simplest planar diagram of the left trefoil knot. The corresponding height atom has two supercritical circles.

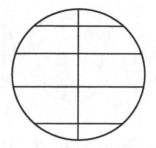

FIGURE 16.3: d-diagram

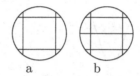

a   b

FIGURE 16.4: d–Diagrams of trefoil and figure eight knot

Thus, by applying one move Ω_2, we can obtain a diagram of the same trefoil for which the corresponding atom has one supercritical circle. The corresponding d–diagram is shown in Fig. 16.3.

Exercise 16.2. *Show that d–diagrams shown in Figs. 16.4.a, 16.4.b encode the right trefoil and the figure eight knot.*

From the arguments described above, we conclude that if a link L has a planar diagram with n crossings such that the corresponding height atom $V(L)$ has k supercritical circles, then L can be encoded by a d–diagram having $n + 2(k - 1)$ chords.

It is easy to see that the total number of sub– and supercritical circles of $V(L)$ does not exceed $\chi(V) - n + (2n) \leq n + 2$, where $\chi(V)$ is the Euler characteristic of $V(L)$. Since V has at least one subcritical circle then k does not exceed $(n + 1)$. So, we obtain an upper bound for the minimal number of chords of the d–diagram corresponding to our knot: it does not exceed $3n$.

Exercise 16.3. *Show that a chord diagram is a d–diagram if and only if it does not contain the subdiagrams shown in Fig. 16.5 ($2n + 1$–gons).*

We have constructed the map from the set of all d–diagrams to the set of all link isotopy classes. This map is not injective (for instance, by adding a solitary chord to a d–diagram we do not change the link isotopy type). In fact, d–diagrams do not encode all the planar diagrams of links but only those corresponding to atoms with a unique supercritical circle. To simplify the situation, let us generalise the notion of d–diagram as follows.

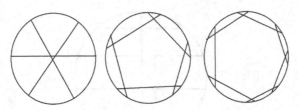

FIGURE 16.5: Chord diagrams which are not d–diagrams

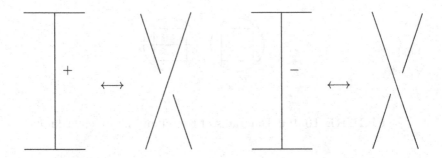

FIGURE 16.6: Crossings corresponding to chords

Definition 16.4. A *marked* or *labelled d–diagram* is a d–diagram where each chord is endowed with a label "+" or "−". Unlabelled d–diagrams are taken to have all labels positive.

Having a d–diagram C, one can construct a link diagram as follows. Let us split chords of the diagram into two families of non-intersecting chords. Then, let us embed C into the plane: chords of the first family are embedded inside the circle; chords of the second family are embedded outside the circle. Then we replace all chords together with small pieces of arcs by crossings, as shown in Fig. 16.6.

It is easy to see that this definition coincides with the old one in the case of an unlabelled d–diagram.

Exercise 16.4. *Show that the link isotopy class constructed in this way (for a labelled d–diagram) does not depend on the splitting of chords into two families.*

Theorem 16.5 ([Man'2]). *Each planar link diagram can be obtained from some labelled d–diagram in the way described above.*

Proof. Let L be a planar diagram of some link. Let us construct a circuit of this diagram as follows. Let us choose a vertex V and an edge e outgoing from

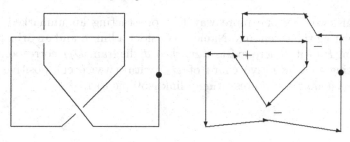

FIGURE 16.7: A circuit for the trefoil

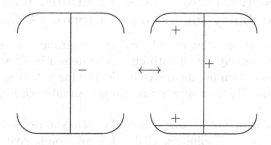

FIGURE 16.8: Twice cut positive chord and negative chord

this vertex. Then, let us move along e. When we meet a vertex, we turn to one of the two possible directions (not opposite to the direction where we have come from).

Exercise 16.5. *Show that one can choose the directions of our turns in such a way that we return to V after passing each edge once and each vertex twice.*

Such a circuit generates some chord diagram. Actually, it represents a circle together with a rule for identifying points on it: we identify pre-images of vertices. Denote this diagram by C. Obviously, C is a d–diagram (there is a natural splitting of chords into interior and exterior ones). To each chord of C, there corresponds a crossing of L.

Here, the circuit at vertices looks as shown in the left part of Fig. 16.6. We set the positive label in this case, and the negative label otherwise.

By construction, the obtained labelled d–diagram encodes the link diagram L.   □

Definition 16.5. A chord a is said to be *cut* if one "small" positive chord intersecting only a is added to it.

A chord is said to be *cut twice* if it is cut from both ends.

Exercise 16.6. *Show that if we replace a negative chord of a marked d–diagram with a positive chord cut twice (see Fig. 16.8), we obtain a d–diagram representing the same link isotopy type.*

Thus, we give one more way for constructing an unmarked d–diagram representing the given link. Namely, consider a link L and an arbitrary planar diagram P of it. There exists a marked d–diagram D_M corresponding to P. Replacing each negative chord of D_M with a twice cut positive chord, we obtain a d–diagram representing a link isotopic to L.

16.4 d–Diagrams and chord diagrams. Criterion of embeddability for a curve in terms of chord diagrams

The method of encoding links by using d–diagrams can be considered for a simpler object, namely, on smooth curves immersed in $\mathbb{R}^2$ where only double transverse intersection points are available. Having a d–diagram, we construct such a curve just like a knot diagram: we put an intersection instead of crossing.

In terms of d–diagrams, one can easily solve the realisability problem for a *Gauss* diagram (see Definition 17.6). One solution is given in [Burm].

The main features of our d–diagram method of recognition realisability are the following. We look at a diagram of a curve (disposition of its crossings) from two points of view: Gaussian (when we go along the curve transversely) or d–diagram (when we should always turn left or right). In the second point of view, only d–diagrams represent realisable curves. So, we just have to translate Gauss diagrams into the language of d–diagrams and see what happens there.

First, let us solve this problem for the case of regularly immersed curves. To do this, we shall not pay attention to crossing types. In this way, each d–diagram encodes not a link but one or several immersed curves (immersion is thought to be regular if it has no tangencies and no intersections of multiplicity more than two). Moreover, instead of Gauss diagrams, we deal just with chord diagrams corresponding to such immersions.

Definition 16.6. Such immersion are called *proper*; the corresponding chord diagrams are called *realisable*.

Suppose G is a chord diagram of an immersed curve. Then we can construct a circuit of this diagram according to "d–diagram rules". Namely, let us choose an arc a of G and let us go along this arc in an arbitrary direction. When we get to some vertex V of the diagram G, we have to turn right or left (in terms of planar diagrams). In the language of Gauss diagrams, this means the following. Suppose the vertex V is incident to the chord X of G; the two arcs incident to V are a and b; the arcs incident to the other end of the chord X are c and d. Thus, we must choose one of the two arcs: c or d. Then, we move along the chosen arc and come to some vertex. We must repeat the same situation. It is obvious that we can arrange our circuit in such a way that all arcs of the diagram G are passed once and finally we come to the point we started from.

In this way, we can construct a chord diagram (that should be in fact a d–diagram). Namely, we just give a new enumeration for the edges according to our circuit and compose a circle of them. Then, if a quadruple of edges (arcs of the diagram to be constructed) p, q, r, s corresponds to a chord (say p, q are incident to one end of it, and r, s are incident to the other end), then the same quadruple will give us a couple of points for the new diagram (say, p, r and q, s). These points must be connected by a chord.

This is the algorithm of translating a chord diagram to the "d–diagram language". By construction, we have the following statement.

Statement 16.1. *If G is a realisable chord diagram then the corresponding diagram is a d–diagram.*

The condition above is, however, not sufficient. In fact, for the diagram ⊕ it does not hold.

If we perform the algorithm above, we obtain the same diagram. However, this is not a realisable Gauss diagram. The reason is obvious. Let us call this algorithm the *A–algorithm*.

Remark 16.5. *Note that this algorithm is not uniquely defined.*

Definition 16.7. A chord d of a chord diagram D is called *even* if the number of chord ends lying in an arc between the ends of this chord is even. Otherwise, the chord is called *odd*.

Obviously, this is well defined (does not depend on the choice of one of the two arcs between two points).

Thus, for each realisable chord diagram each chord is even.

Exercise 16.7. *Prove this fact.*

It turns out that the two conditions described above are sufficient. In fact, the following theorem is true.

Theorem 16.6. *Let G be a chord diagram. Let D be a diagram obtained from G by applying the A–algorithm. Then G is realisable if and only if each chord of G is even and D is a d–diagram.*

Before proving this theorem, we formulate a corollary from it.

Corollary 16.1. *If G is a chord diagram with all chords even, then diagrams that can be obtained from G by applying the A–algorithm are either all d–diagrams, or not d–diagrams.*

In order to prove Theorem 16.6, we shall construct the reversed algorithm (how to obtain the chord diagram of a curve from its d–diagram).

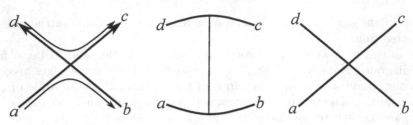

FIGURE 16.9: The "bad" case

Let D be a d–diagram. Then it generates some curve K, which is standardly immersed in $\mathbb{R}^2$. In order to construct the chord diagram corresponding to K, we must find some *unicursal* circuit for D (as in the case of the A–algorithm). More precisely, let us choose some vertex V_1 of D and some edge a outgoing from it (say, in the clockwise direction). Then we come to some other vertex V_2. After it, our algorithm is uniquely defined: we have only the possibility to go forward (not right or left). This means that for the chord c incident to V_2 we go to the opposite side, and then proceed moving in the opposite direction of our circuit (i.e., counterclockwise, if the initial moving was clockwise). After moving to each next vertex, we jump to the other side of the chord and change the direction. We can choose our circuit in such a way that finally we successfully return to the vertex V_1 with the initial direction after we shall have passed all arcs precisely once.

Let us call this algorithm the B–*algorithm*.

If we have a realisable chord diagram C then, for each diagram D obtained from C by applying the A–algorithm, the diagram obtained from D by the B–algorithm will be just the diagram C. However, for some diagrams this is not so. For instance, if we take the diagram then the A–algorithm will give us ⊕, and the B–algorithm applied to the "new ⊕" will give ⊗.

The only thing that remains to prove is that it is not the case for diagrams all chords of which are even.

The reason is the following. Consider a chord diagram C, and a vertex V of it. There are four arcs incident to V and the opposite (by edge) vertex. Denote them by a, b, c, d. Suppose that a is opposite to c (i.e., next on the diagram C). Then, for the d–diagram $D(C)$ obtained from C by the A–algorithm a is adjacent either to b or to d. Without loss of generality, suppose a is adjacent to b. Then there are two hypothetical possibilities for the B–algorithm to restore the opposite chord for a: it will be either c (as it must be) or d; see Fig. 16.9.

The B–algorithm is uniquely defined. Thus, we need to find a condition for the initial diagram C such that this algorithm always restores the opposite edge correctly (in our case c for a).

Now, suppose that C is a chord diagram, all edges of which are even. Then, for any circuit of C constructed according to the A–algorithm, orient all edges of the diagram according to this circuit. A vertex is said to be *good*

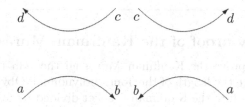

FIGURE 16.10: Local structure of edges and arcs.

if it is either a source (both incident edges are outgoing) or a strain (both are incoming). It is easy to see that for C having all even edges, all vertices are good.

The case of good edges is just what we wanted: in this case, the algorithm B will restore the initial diagram. Let us consider this fact in detail.

Let C, V, a, b, c, d be defined as above. Consider the d–diagram $D(C)$ obtained from the diagram C by applying the A–algorithm. Without loss of generality, assume that in the diagram C, the arc c goes after a, and the arc d goes after b. Suppose we have chosen a way of applying the A–algorithm such that in the diagram $D(C)$, the arc b is adjacent to a and d is adjacent to c. Moreover, suppose that b follows a in our circuit (in the d–diagram D). Thus, the vertex V is a strain. The vertex V' connected with V by a chord (in C) is thus a source. Thus we conclude that the chord d follows c. So, when we draw our d–diagram D on the plane, we see the picture shown in Fig. 16.10 (in the right part or in the left part).

This means that the arc c is opposite to the arc a (and the analogous situation is true at all vertices), and after applying the B–algorithm, we obtain the initial diagram C. This completes the proof of the theorem.

One can easily modify this algorithm for the case of knots and Gaussian curves. Obviously, it is sufficient to consider only the case of connected Gauss diagrams.

First, having a Gauss diagram G one should forget about its labels and arrows and consider the chord diagram C. If it is not realisable then G is not realisable either.

In the case when C is realisable, one should apply the first algorithm to it and obtain the d–diagram D. It is not difficult to consider all embeddings of D in the plane. Then one should try to set all crossings according to the labels on the arcs and check carefully whether there is no contradiction with the directions of these arcs (the arcs can be set with respect to the initial point, thus one should consider several cases). If there are no contradictions then the diagram G is realisable.

16.5 A new proof of the Kauffman–Murasugi theorem

In order to prove the Kauffman–Murasugi theorem (Theorem 7.5), one should appreciate the length of the Jones polynomial. Obviously, it is just the same as the length of the Kauffman bracket divided by four.

Consider formulae (6.4) and (6.5) for the definition of the Jones–Kauffman polynomial. We are interested in the states that give the maximal and the minimal possible degree of monomials in the sum (6.4).

Definition 16.8. Now, let the *minimal state* of L be the state where all crossings are resolved positively, and the *maximal state* be the state of L where all crossings are resolved negatively.

In order to estimate these degrees, we shall use atoms. Namely, consider a diagram L of a link (all link diagrams are thought to be connected). It has an intrinsic A–structure as any four–valent graph embedded in the plane. It also has a B–structure of the atom that can be restored from it.

Suppose the diagram L is prime. The remaining case will be considered later.

It is easy to see that the maximal possible monomial degree in formula (6.4) corresponds to the maximal state and the minimal possible degree corresponds to the minimal state. These easy facts are left for the reader as exercises.

Denote these states by s_{max} and s_{min} and the corresponding numbers of circles by γ_{max} and γ_{min}, respectively.

Let us calculate the desired maximal and minimal monomial degrees. We have: $n + 2(\gamma_{max} - 1)$ and $-n - 2(\gamma_{min} - 1)$. Here "2" and "minus 2" come from the exponent $a^{\pm 2(\gamma(s)-1)}$.

The difference (the upper bound) equals

$$2n + 2(\gamma_{min} + \gamma_{max}) - 4.$$

Now, let us return to the atom V corresponding to L. Its Euler characteristic obviously equals $-n + \gamma_{min} + \gamma_{max}$. Taking into account that this is less than or equal to two, we conclude that our upper bound does not exceed $4n$.

Thus, we have proved the first part of the Kauffman–Murasugi Theorem.

Now, the question is: when can we get this upper bound? First, the atom should be a spherical one.

So, we must present a B–structure of a spherical atom with respect to the A–structure of the shadow of L in order to obtain a spherical atom. There are two such structures corresponding to the two alternating diagrams with the same shadow.

But, since L is prime, **there are no other B–structures** creating spherical atoms. Thus, the diagram L is alternating.

Now, let us check that the only obstruction for $X(L)$ to have length $4n$ is the existence of splitting points. Obviously, if a diagram has a splitting point, it can be represented as a connected sum of two diagrams having a smaller number of crossings. Thus, its length cannot be equal to $4n$.

If the length of the Kauffman bracket for an alternating link diagram with n crossings is less than $4n$, then either the leading or the lowest coefficient coming from γ_{max} is cancelled by some other term. Without loss of generality, assume that the first one is the case. Then, there exists a state different from s_{max} that gives the same exponent of a: $n + 2\gamma_{max} - 1$. Suppose it differs from s_{max} at some crossings. Let us choose one of them and denote it by X. It is obvious that the state s'_{max} that differs from s_{max} at the only crossing X also gives the power $n + 2\gamma_{max} - 1$. So, $\gamma(s'_{max}) = \gamma(s_{max}) + 1$. This means that when we change the state s_{max} at the vertex X, one circle is divided into two circles. Taking into account that the initial diagram is alternating, we conclude that X is a splitting point.

Finally, if L is not a prime diagram, we can decompose it into prime components. Taking into account the multiplicativity of the Jones (or Kauffman) polynomial, we see that they all are alternating diagrams without splitting points. This completes the proof of the Kauffman–Murasugi theorem.

16.6 Representation of long links by words in a finite alphabet

As shown above, all links can be represented by d–diagrams. One can view a d–diagram as follows. First, fix the way of splitting chords of d–diagrams into two families. Choose a point of a d–diagram different from any chord end.

d–diagrams with a marked point admit a simple combinatorial representation by words in the four-bracket alphabet. Indeed, while "reading" the chord diagram starting from the given point, we can write down a round bracket when encountering an end of chord belonging to the first family and a square bracket when we meet an end of chord from the second family. Thus we get what is called a "balanced bibracket structure".

Definition 16.9. A *balanced bibracket structure* is a word in the alphabet $(,) , [,]$ such that:

1. in each initial subword a' of the word a the number of ")" does not exceed that of "(", and the number of "]" does not exceed that of "[";

2. in the word a the number of "(" equals the number of ")", and that of "[" equals that of "]".

It is obvious that the d–diagram with initial point can be uniquely restored from the corresponding balanced bibracket structure.

d–diagrams encode links. Thus, d–diagrams with a fixed point encode links

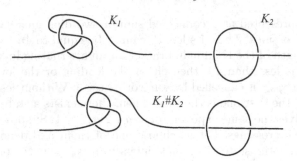

FIGURE 16.11: Product of two long links

with a fixed point on the oriented component (or, what is just the same) *long links*.

Definition 16.10. By a *long link* is meant a smooth compact 1–manifold with boundary, embedded in $\mathbb{R}^3$ coinciding with Ox outside some ball centred at O and isotopic to the disconnected sum of one line and several (possibly, no) circles.

A long link consisting of one component is called a *long knot*.

Long links are considered up to natural isotopy.

Now, let is define *the semigroup K of long links* as follows.

1. Elements of K are the isotopy classes of long links.

2. The unit of K is the equivalence class of the manifold, given by $\{y = 0, z = 0\}$. This equivalence class is called the *long unknot* or the *long trivial knot*.

3. The product of two elements $L_1, L_2 \in K$ is defined as follows. First, we choose representatives of these classes: some links K_1 and K_2. They coincide with Ox outside the balls centred at zero; the radii of these balls are some R_1 and R_2, respectively. Then we construct a long link consisting of the three following parts. One part of it lies in $\{x > 2R_2\}$ and $\{x < -2R_1\}$ and coincides with Ox there. Another part lies inside $B_{R_2}(R_2, 0, 0)$ and coincides there with the shift of $K_2 \cap B_{R_2}(0, 0, 0)$ along the vector $(R_2, 0, 0)$. The third part lies inside $B_{R_1}(-R_1, 0, 0)$ and coincides there with the shift of $K_1 \cap B_{R_1}(0, 0, 0)$ along the vector $(-R_1, 0, 0)$. See Fig. 16.11. The isotopy class of the constructed long link is decreed to be the product of $L_1 \cdot L_2$. Obviously, this isotopy class depends neither on the randomness of K_1 and K_2 nor on the randomness of the radii R_1 and R_2. Thus, the product in K is well defined.

It is obvious that the concatenation of balanced bibracket structures corresponds to the connected sum operation for long links.

Thus, one can say that the long link semigroup can be encoded in terms of balanced bibracket structures. This is called the *bracket calculus*.

All balanced bibracket structures themselves form a semigroup, where the empty word plays the role of the unit element, and the multiplication is expressed by concatenation. Denote this group by G.

The main problem of the bracket calculus is to describe the equivalence of long links in terms of bibracket structures. This was done in [Man3].

The main idea is to describe the "elementary" isotopy moves in terms of bracket structures.

These isotopies originate from Reidemeister moves and one special move that corresponds to the "circuit change". They are completely enumerated in [Man5].

The semigroup of balanced bibracket structures factorised by these relations is thus isomorphic to the semigroup K.

Each of these relations is an identity $A = B$, where A, B are some (possibly, non–balanced) words in the bracket alphabet. Such a relation means that for any balanced bibracket structure xAy, the structure xBy is balanced as well, and represents the same long link isotopy class (the same statement is true for xBy, xAy). We factorise the group G by all relations $xAy = xBy$ for all x, y such that xAy is balanced which are taken with respect to the relation $A = B$ taken from the given list. Some parts of this list are given below.

First Reidemeister move:

$$(\,) = \text{empty word}$$

$$[\,] = \text{empty word}$$

$$[\, (\,] \, [\,) \,] = \text{empty word}$$

$$(\, [\,) \, (\,] \,) = \text{empty word}.$$

Second Reidemeister move:

$$(\, [\, (\,] \, A \, [\,) \,] \,) = A$$

$$[\, (\,] \, (\, A \,) \, [\,) \,] = A,$$

In each of the two relations above, A has balanced round–bracket structure.

$$[\, (\, [\,) \, A \, (\,] \,) \,] = A$$

$$(\, [\,) \, [\, A \,] \, (\,] \,) = A.$$

Here A has balanced square–bracket structure.

Third Reidemeister move:

$$(\, (\, [\,) \, [\, (\,] \, (\, [\,) \, (\, [\,) \, [\,) \,] \,) \, (\,] \,) = (\, [\, (\,] \, (\, [\,) \, [\,) \,] \, [\, (\,] \, (\, [\,) \, [\,) \,] \,)$$

$$(\, [\,) \, (\, (\, [\,) \, (\, [\, (\,] \, (\,] \,) \, (\,] \,) \, [\,) \,] = [\, (\, [\,) \, (\, (\,] \,) \, [\, (\,] \,].$$

For the change of circuits we use the following four moves:

$$[A(]C) = [(]C([)[)]A(]),$$
$$([]A(()C) = (C([)]A(]),$$
$$[A[(]]C[)] = [(]C[[)]A],$$
$$([]A[(](]C[)] = (C[]A].$$

In each of these relations, A should have balanced square bracket structure and C has balanced round bracket structure.

For more details and all proofs, see [Man5].

16.7   Representation of links by quasitoric braids

In the present section, we are going to describe how knots can be represented by closures of a small class of braids, and the class of d–diagrams, generating these braids, see [Man6].

16.7.1   Definition of quasitoric braids

We recall that toric braids (depending on the two parameters p and q, where p is the number of strands) are given by the following formula:

$$T(p,q) = (\sigma_1 \ldots \sigma_{p-1})^q$$

and have an intuitive interpretation; see Fig. 16.12.a.

Definition 16.11. A braid β is said to be *quasitoric* of type (p,q) if it can be expressed as $\beta_1 \ldots, \beta_q$, where for each $\beta_j = \sigma_1^{e_{j1}} \ldots \sigma_{p-1}^{e_{j,p-1}}$, each e_{jk} is either 1 or -1. In other words, a quasitoric braid of type (p,q) is a braid obtained from the standard diagram of the toric (p,q) braid by switching some crossing types; see Fig. 16.12.b.

It is easy to see that the product of quasitoric n-strand braids is a quasitoric n-strand braid.

In fact, a more precise statement can be made.

Proposition 16.1. *For every $p \in \mathbb{N}$, p-strand quasitoric braids make a subgroup in B_p.*

To prove this result, we only have to prove the following lemma.

Lemma 16.1. *For every p, the inverse of a p-strand quasitoric braid is quasitoric.*

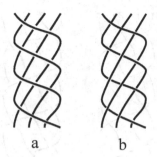

a   b

FIGURE 16.12: Toric and quasitoric braids

Proof. We have to prove that for the braid $\delta = \sigma_1^{e_1} \ldots \sigma_{p-1}^{e_{p-1}}$, where each e_i equals 1 or -1, the braid δ^{-1} is a quasitoric braid. In this case, the proof of the lemma follows straightforwardly, since each quasitoric braid is just a product of positive powers of such braids.

In other words, we have to prove that there exists a quasitoric braid η, such that $\eta \cdot \delta$ is the trivial braid.

We are going to prove that there exist braids $\delta_1, \ldots \delta_{p-1}$, where $\delta_i = \sigma_1^{e_{i1}} \ldots \sigma_{p-1}^{e_{i,p-1}}$, such that for $\eta = \delta_1, \ldots, \delta_{p-1}$ we have $\eta\delta = e$ is the trivial braid.

Let us consider the shadow S of the standard toric braid diagram of type (p,p) that will be the shadow of our diagram $\delta_1 \ldots \delta_{p-1}\delta$. The lower part of this diagram has crossing types coming from δ. So we only have to set the rest of the crossings (i.e. to define $\delta_1 \ldots \delta_{p-1}$ in order to get the trivial braid $\delta_1 \ldots \delta_{p-1}\delta$). The lower part of the shadow S consists of p strands; one of them (denote it by x) intersects all other strands once; other strands do not intersect each other.

If we set all crossing types for the lower part of S as in δ then we see that some strands in the lower part come over x, and the others come under x; see Fig. 16.13.a.

Let us denote strands from the first set of strands by $y_1, \ldots y_k$, and strands from the second set by $z_1, \ldots z_l$.

Let us say that for a pure r-strand braid β_1 with strands $a_i, i = 1 \ldots r$, the order of strands is $a_1 > a_2 > \cdots > a_r$ if at each crossing X involving $a_i, a_j, i < j$, the strand a_i comes over a_j. It is obvious that in this case the braid β_1 is trivial.

Now we can easily set all crossing types for the upper part of S in such a way that for the braid $\delta_1\delta_2 \ldots \delta_{p-1}\delta$ the order of strands is $y_1 > y_2 > y_3 > \cdots > y_k > x > z_1 > \cdots > z_l$; see Fig. 16.13.b. So the braid $\eta\delta$ is trivial, which completes the proof of the lemma. $\square$

Thus, we have the *group of quasitoric braids*. Now, let us state the main theorem of this section.

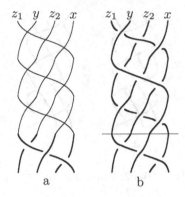

z_1 y z_2 x z_1 y z_2 x

a   b

FIGURE 16.13: Inverting a quasitoric braid

Theorem 16.7. *Each knot isotopy class can be obtained as a closure of some quasitoric braid.*

16.7.2   Pure braids are quasitoric

First, note that, by Alexander's theorem, for a given knot K there exists a braid β, whose closure is isotopic to K. Our goal is to transform β in a proper way in order to obtain a quasitoric braid.

Let us prove the following lemma.

Lemma 16.2. *Every braid β is Markov–equivalent to an r-strand braid whose permutation is a power of the cyclic permutation $(1, 2, \ldots, r)$ for some r.*

Proof. Suppose β has n strands. Consider the permutation α corresponding to it, and orbits of the action of α on the set $(1, \ldots, n)$.

These orbits might contain different numbers of elements. Now, let us apply Markov's move for transforming these orbits. The first Markov move conjugates the braid, thus, it conjugates the corresponding permutation. So, the number of elements in orbits does not change, but elements in orbits permute. The second Markov move increases the number of strands by one, adds the element $(n + 1)$ to the orbit, containing the element n and does not change other orbits. Thus, by using Markov's moves, one can re–enumerate elements in such a way that the smallest orbit contains n, and then increase the number of elements in this orbit by one. Reiterating this operation many times, we finally obtain the same number of elements for all orbits. Suppose the permutation corresponding to the obtained km-strand braid β acts on km elements in such a way that each of k orbits of the permutation contains m elements. By conjugating β, we can get the corresponding permutation equal to the m-th power of the cyclic permutation $(1\ 2\ \ldots\ km)$, $k, m \in \mathbb{N}$. $\square$

Denote the obtained braid by γ. As shown in Lemma 16.2, γ is Markov–equivalent to β.

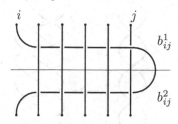

FIGURE 16.14: Generator b_{ij} of the pure braid group

The next step is to prove that γ is a quasitoric braid.

So, let us change the braid diagram of γ, without changing its isotopy type. To complete the proof of the theorem, we have to prove the following lemma.

Lemma 16.3. *An r–strand braid whose permutation is a power of the cyclic permutation $(1\,2\,\ldots\,r)$ is quasitoric.*

Proof. Let γ be an r-strand braid, having the permutation $(1\,2\,\ldots\,r)^s$. Consider the braid $\gamma' = \gamma \cdot T(r,1)^{-s}$.

Then, γ' is a pure braid. Besides, the braid γ' is quasitoric if and only if γ is quasitoric (since $T(r,1)^s$ is a quasitoric braid).

Thus, it remains to prove the following

Lemma 16.4. *Every pure braid is quasitoric.*

Recall that one can choose generators $b_{i,j}, 1 \leq i < j \leq r$, of the pure r-strand braid group, as shown in Fig. 16.14.

Now, we only have to show that all generators b_{ij} are quasitoric braids.

Actually, for all i and j between 1 and n, $i < j$, the braid b_{ij} is a product of the two braids $b_{ij}^1 \cdot b_{ij}^2$ (they are shown in Fig. 16.14 above and below the horizontal line), where the first braid $b_{ij}^1 = \sigma_i^{-1}\ldots\sigma_{j-2}^{-1}\sigma_{j-1}$ has ascending order of generators, and $b_{ij}^2 = \sigma_{j-1}\sigma_{j-2}\ldots\sigma_i$ has descending order of generators; see Fig. 16.14. Now, consider only strands numbered from i-th to j-th. Then, we can introduce the analogous definition of quasitoric braids on strands from the i-th to j-th (i.e. with other strands vertical). It is evident that both b_{ij}^1 and b_{ij}^2 are (i,j)-quasitoric braids on the strands from i-th to j-th (for b_{ij}^1 it is clear by definition, and b_{ij}^2 is the inverse to a quasitoric braid).

Definition 16.12. For $1 \leq i < j \leq n$ an n-strand braid ζ is said to be (i,j)–*quasitoric* if it has a diagram with strands of it except those numbered from i-th to j-th going vertically and unlinked with the other strands, and strands from i to j forming a quasitoric braid (in the standard sense). Such a diagram is called a *standard* diagram of an (i,j)-quasitoric braid.

To complete the proof of Lemma 16.4 and the main theorem, it suffices to prove the following lemma.

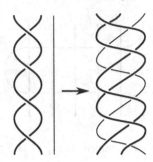

FIGURE 16.15: Adding a "thin" strand to a quasitoric braid

Lemma 16.5. *Assume* $1 \leq i < j \leq r$. *Then every* (i,j)-*quasitoric pure* r-*strand braid is quasitoric.*

Proof. We use induction on $r - (j - i + 1)$. We have to show that by adding a separate vertical strand to the standard diagram of a quasitoric braid, we obtain a diagram of a quasitoric braid.

Consider a standard quasitoric q-strand braid diagram ρ and add a separate strand on the right hand (the case of a left–handed strand can be considered analogously). Let the initial braid diagram be obtained from the toric braid (q, ql) by switching some crossings.

Consider the standard diagram of the pure toric braid $(q + 1, (q + 1)l)$ and the first q strands of it. Obviously, they form a toric braid diagram of type (q, ql). Let us set the crossing types of these strands as in the case of the diagram ρ, and let us arrange the additional strand under all the others. Obviously, we get a diagram, isotopic to that obtained from ρ by adding a separate strand on the right hand; see Fig. 16.15. $\qquad\square$

Thus, the standard generators of the pure braid group P_n for arbitrary n are quasitoric, hence, by Proposition 16.1, so is every pure braid. $\qquad\square$

So, by using Markov's moves and braid diagram isotopies to the initial braid diagram, we obtain a quasitoric braid ζ, whose closure is isotopic to K and this completes the proof of the main theorem.

16.7.3 d–diagrams of quasitoric braids

Toric (and quasitoric) braids in their natural representation allow us to consider two ways of encoding links: by braids and by d–diagrams together.

Namely, the following statement is true.

Statement 16.2. *Standard diagrams of closures of quasitoric braids (with odd* p) *are the only link diagrams that can be obtained from labelled* d–*diagrams and braided around the centre of the corresponding circle.*

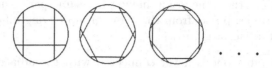

FIGURE 16.16: Quasitoric d–diagrams for $q = 3$

FIGURE 16.17: Quasitoric d–diagrams for $q = 5$

More precisely, we require that the circle is standardly embedded in $\mathbb{R}^2$: $x^2 + y^2 = 1$, and that ends of chords are uniformly distributed along the circle; chords of one (interior) family of a d–diagram are taken to be straight lines (thus we require the absence of diametral chords), and chords of the other family are the images of straight lines inverted in the circle. One should also make one more correction. Namely, let D be a d–diagram, and let a and b be some two intersecting chords of D belonging to different families, such that one end a_1 of a and one end b of b_1 are adjacent vertices. Let us choose points P, Q on the chords a, b and a point R on the arc a_1b_1. Denote the corresponding unit tangent vectors (at these points) by $t(P), t(Q), t(R)$. Then the vectors $OP \times t(P)$ and $OQ \times t(Q)$ (where O is the centre of the circle) are collinear, and the vector $OR \times t(R)$ has opposite direction.

So, we shall delete such an arc; i.e., construct the link diagram by a d–diagram in such a way that one half of a_1b_1 is deleted together with the chord a, and the other is deleted together with the chord b. Within this chapter, we accept these corrections.

To check the Statement 16.2, one should only study the property of shadows of such link diagrams. The proof of the statement is left for the reader.

The d–diagram corresponding to the toric braid $T(p, q)$ is constructed as follows. Let $p = 2m + 1$. Let us mark the $4mq$ points on the sphere, split into $2q$ groups of $2m$ adjacent points in each group. Enumerate the points in each group by numbers from 1 to $2m$; in "even groups" we enumerate clockwise; in odd groups we enumerate counterclockwise.

Each marked point is connected with a point from an adjacent group having the same number. The adjacent group is chosen according to the following rules:

1. Points from the same group having the same parity have the same adjacent groups; points from the same group having different parity have different adjacent groups.

2. Point number one is never connected with the adjacent point (on the circle).

Examples of these d–diagrams are shown in Fig. 16.16 for $p = 3$ and Fig. 16.17 for $p = 5$.

Obviously, "quasitoric" d–diagrams with odd q are obtained from these "toric" ones by marking some chords as "negative".

Part IV

Virtual knots

Chapter 17

Virtual knots. Basic definitions and motivation

Virtual knot theory was proposed by Kauffman [Kau3]. This theory arises from the theory of knots in thickened surfaces $S_g \times I$, first studied by Kauffman, Jaeger, and Saleur, see [JKS]. Virtual knots (and links) appear by projecting knots and links in S_g to $\mathbb{R}^2$ and hence, $S_g \times \mathbb{R}$ onto $\mathbb{R}^3$. By projecting link diagrams (i.e., graphs of valency four with over– and undercrossing structures at vertices) in S_g onto $\mathbb{R}^2$, one obtains diagrams on the plane. Virtual crossings arise as artefacts of such a projection; i.e., intersection points of images of arcs, non-intersecting in S_g and classical crossings appear just as projections of crossings.

In the very beginning of this theory, the creators have proposed generalisations of some basic knot invariants: the knot quandle, the fundamental group, the Jones polynomial [Kau3]. For further developments see [MI, Man4, Man9, Man10, Man13, Man'3, FJK, HK, Hre, Kau7, Kau10, KK, KNS, Saw, SW, Kup1] On the other hand, see [GPV], virtual knots arise from non-realisable Gauss diagrams: having a non-realisable (by embedding) diagram, one can "realise it" by means of immersion; the "new" intersection points are marked by virtual crossings. We recall that realisability of Gauss diagrams was described in Chapter 16.

17.1 Combinatorial definition

Let us start with the definitions and introduce the notation.

Definition 17.1. A *virtual link diagram* is a planar graph of valency four endowed with the following structure: each vertex either has an over– and undercrossing or is marked by a virtual crossing, (such a crossing is shown in Fig. 17.1).

All crossings except virtual ones are said to be *classical*.

Two diagrams of virtual links (or, simply, *virtual diagrams*) are said to be *equivalent* if there exists a sequence of *generalised Reidemeister moves*, transforming one diagram to the other one.

FIGURE 17.1: Virtual crossing

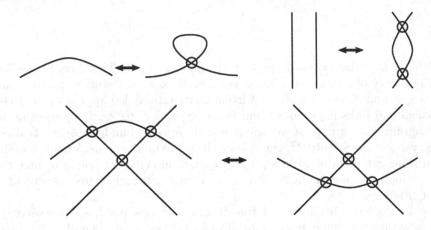

FIGURE 17.2: Moves $\Omega_1', \Omega_2', \Omega_3'$

As in the classical case, all moves are thought to be performed inside a small domain; outside this domain the diagram does not change.

Definition 17.2. Here we give the list of *generalised Reidemeister moves*:

1. Classical Reidemeister moves related to classical vertices.

2. Virtual versions $\Omega_1', \Omega_2', \Omega_3'$ of Reidemeister moves; see Fig. 17.2.

3. The "semivirtual" version of the third Reidemeister move; see Fig. 17.3,

Remark 17.1. *The two similar versions of the third move shown in Fig. 17.4 are* forbidden, *i.e., they are not in the list of generalised moves and cannot be expressed via these moves.*

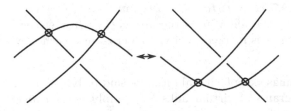

FIGURE 17.3: The semivirtual move Ω_3''

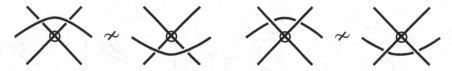

FIGURE 17.4: The forbidden move

Definition 17.3. A *virtual link* is an equivalence class of virtual diagrams modulo generalised Reidemeister moves.

One can easily calculate the number of components of a virtual link. A *virtual knot* is a one–component virtual link.

Exercise 17.1. *Show that any virtual link having a diagram without classical crossings is equivalent to a classical unlink.*

Remark 17.2. *Formally, virtual Reidemeister moves give a new equivalence relation for classical links: there exist two isotopies for classical links, the classical one that we are used to working with, and the virtual one. Later we shall show that this is not the case, see also [Man27].*

Remark 17.3. *Actually, the forbidden move is a very strong one. Each virtual knot can be transformed to another one by using all generalised Reidemeister moves and all versions of the forbidden moves. This was proved by Sam Nelson in [Nel] by using Gauss diagrams of virtual links.*

The idea was that one can make each pair intersecting chords of Gauss diagrams of a virtual knot having adjacent ends non-intersecting. See also [Man27].

If we allow only the forbidden move shown in the left part of Fig. 17.4, we obtain what is called welded *knots. Some initial information on this theory can be found in [KamS1]. Welded knots can be interpreted as isotopy classes of toral surfaces in $\mathbb{R}^4$, see [Sat, Rou].*

Definition 17.4. By a *mirror image* of the virtual link diagram we mean the diagram obtained from the initial one by switching all types of **classical** crossings (all virtual crossings stay on the same positions).

17.2   Projections from handlebodies

The choice of generalised Reidemeister moves is very natural. Namely, it is the complete list of moves that may occur while considering the projection of $S_g \times I$ to $\mathbb{R} \times I$ (or, equivalently, $\mathbb{R}^3$). Obviously, all classical Reidemeister moves can be realised on a small part of any S_g that is homeomorphic to a 3-ball B^3. The other moves, namely, the semivirtual move and purely

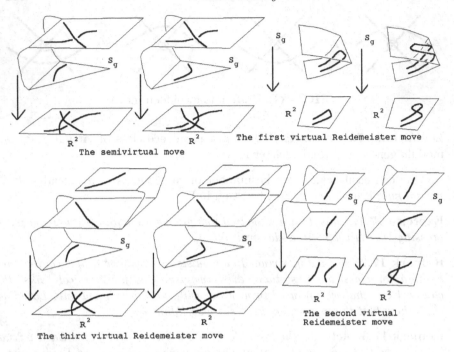

The semivirtual move

The first virtual Reidemeister move

The second virtual Reidemeister move

The third virtual Reidemeister move

FIGURE 17.5: Generalised Reidemeister moves and thickened surfaces

virtual moves, are shown in Fig. 17.5 together with the corresponding moves in handlebodies.

There exists a more intuitive topological interpretation for virtual knot theory in terms of embeddings of links in thickened surfaces [Kau5, Kau7]. For more details see [MI]. Regard each virtual crossing as a shorthand for a detour of one of the arcs in the crossing through a 1–handle that has been attached to the 2–sphere of the original diagram. The two choices for the 1–handle detour are homeomorphic to each other (as abstract manifolds with boundary). By interpreting each virtual crossing in such a way, we obtain an embedding of a collection of circles into a thickened surface $S_g \times \mathbb{R}$, where g is the number of virtual crossings in the original diagram L and S_g is the orientable 2–manifold homeomorphic to the sphere with g handles. Thus, to each virtual diagram L we obtain an embedding $s(L) \rightarrow S_{g(L)} \times \mathbb{R}$, where $g(L)$ is the number of virtual crossings of L and $s(L)$ is a disjoint union of circles. We say that two such stable thickened surface embeddings are stably equivalent if one can be obtained from the other by isotopy in the thickened surface, homeomorphisms of surfaces, and the addition of substraction or handles not incident to images of curves.

Theorem 17.1. *Two virtual link diagrams generate equivalent (isotopic) vir-*

tual links if and only if their corresponding surface embeddings are stably equivalent.

This result was sketched by Kauffman in [Kau6], see also [Kau10, GPV, Kup1].

A hint to this proof is demonstrated in Fig. 17.5.

Here we wish to emphasise the following important circumstance.

Definition 17.5. A virtual link diagram is *minimal* if no handles can be removed after a sequence of Reidemeister moves.

An important Theorem by Kuperberg [Kup1] says the following.

Theorem 17.2. *For a virtual knot diagram K there exists a unique minimal surface in which an I–neighbourhood of an equivalent diagram embeds and the embedding type of the surface is unique.*

17.3 Gauss diagram approach

Definition 17.6. A *Gauss diagram* of a (virtual) knot diagram K is an oriented circle (with a fixed point) where pre-images of over– and undercrossing of each crossing are connected by a chord. Pre-images of each **classical** crossing are connected by an arrow, directed from the pre-image of the overcrossing to the pre-image of the undercrossing. The sign of each arrow equals the local writhe number of the vertex. The signs of chords are defined as in the classical case. Note that arrows (chords) correspond to classical crossings only. This means that virtual knot theory is essentially defined by means of classical crossings.

Remark 17.4. *For classical knots this definition is just the same as before.*

Given a Gauss diagram with labelled arrows, if this diagram is realisable then it (uniquely) represents some classical knot diagram. Otherwise one cannot get any classical knot diagram.

Herewith, the four–valent graph represented by this Gauss diagram and not embeddable in $\mathbb{R}^2$ can be *immersed* to $\mathbb{R}^2$. Certainly, we shall consider only "good" immersions without triple points and tangencies.

Having such an immersion, let us associate virtual crossings with intersections of edge images, and classical crossings at images of crossing; see Fig. 17.6.

Thus, by a given Gauss diagram we have constructed (not uniquely) a virtual knot diagram.

Theorem 17.3 ([GPV]). *The virtual knot isotopy class is uniquely defined by this Gauss diagram.*

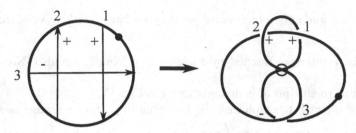

FIGURE 17.6: A virtual knot and its Gauss diagram

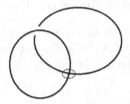

FIGURE 17.7: A virtual link that is not isotopic to any classical link

Exercise 17.2. *Prove this fact.*

Hint 17.1. *Show that purely virtual moves and the semivirtual move are just the moves that do not change the Gauss diagram at all.*

High-dimensional virtual knots were considered by Kamada [KamS2].

17.4 Virtual knots and links and their simplest invariants

There exist a lot of simple combinatorial ways for constructing virtual knot and link invariants.

Consider the virtual link shown in Fig. 17.7.

It is intuitively clear that this link cannot be isotopic to any classical one because of "the linking number".

Exercise 17.3. *Define accurately the linking number for virtual links and prove that the link shown in Fig. 17.7 is not isotopic to any classical link.*

The next simplest invariant is the colouring invariant: we take the three colours (as before) and associate a colour with each (long) arc (i.e., a part of the diagram going from one undercrossing to the next undercrossing; this part might contain virtual crossings). Then we calculate the number of proper colourings. This invariant will be considered in more detail together with the quandle and fundamental group.

Exercise 17.4. *Prove the invariance of the colouring invariant.*

17.5 Invariants coming from the virtual quandle

Later, we shall define the quandle for virtual knots in many ways. In the present sections, we are going to construct invariants "coming from this quandle"; however, we are going to describe them independently.

17.5.1 Fundamental groups

Though virtual knots are not embeddings in $\mathbb{R}^3$, one can easily construct a generalisation of the knot complement fundamental group (or, simply, the *knot group*) for virtual knots. Namely, one can modify the Wirtinger presentation for virtual diagrams. Consider a diagram $\bar{L}$ of a virtual link L. Instead of arcs we shall consider long arcs of $\bar{L}$. We take these arcs as generators of the group to be constructed. After this, we shall write down the relations at classical crossings just as in the classical case: if two long arcs a and c are divided by a long arc b, whence a lies on the right hand with respect to the orientation of b, then we write down the relation $c = bab^{-1}$.

The invariance of this group under classical Reidemeister moves can be checked straightforwardly: the combinatorial proof of this fact works both for virtual and classical knots (see Exercise 4.7). For the semivirtual move and purely virtual moves there is nothing to prove: we shall get the same presentation.

However, this invariance results from a stronger result: invariance of the virtual knot quandle, which will be discussed later.

Definition 17.7. The group defined as above is called the *group of the link* L.

Obviously, the analogue of the colouring invariant Lemma 4.6 is true for virtual knots. Its formulation and proof literally coincide with the formulation and proof of Lemma 4.6.

17.5.2 Strange properties of virtual knots

Some virtual links may have properties that do not occur in the classical case. For instance, both in the classical and the virtual case one can define "upper" and "lower" presentations of the knot group (the first is as above,

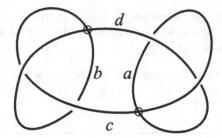

FIGURE 17.8: A virtual knot with different upper and lower groups

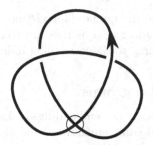

FIGURE 17.9: The virtual trefoil

the second is just the same for the knot (or link) where all classical types are switched). In the classical case, these two presentations give the same group (for geometric reasons). In the virtual case it is however not so. The example first given in [GPV] is as follows.

In fact, taking the arcs a, b, c, d shown in Fig. 17.8 as generators, we obtain the following relations:

$$b = dad^{-1}, \quad a = bdb^{-1}, \quad d = bcb^{-1}, \quad c = dbd^{-1}.$$

Thus, a and c can be expressed in the terms of b and d. So, we obtain the presentation $\langle b, d | bdb = dbd \rangle$. So, this group is isomorphic to the trefoil group.

Exercise 17.5. *Show that the group of the mirror virtual knot is isomorphic to* $\mathbb{Z}$.

This example shows us that the knot shown above is not a classical knot. Moreover, it is a good example of the existence of a non-trivial virtual knot with group $\mathbb{Z}$ (the same as that for the unknot). The latter cannot happen in the classical case.

The simplest example of the virtual knot with group $\mathbb{Z}$ is the virtual trefoil; see Fig. 17.9.

The fact that the virtual trefoil is not the unknot will be proved later.

Besides this example, one encounters the following strange example (*Kishino knot*): the connected sum of two (virtual) unknots is not trivial;

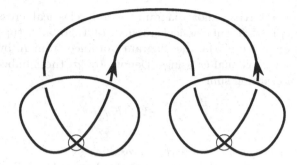

FIGURE 17.10: Non-trivial connected sum of two unknots

see Fig. 17.10. This example was considered in [Kis], see also [Kim]. We shall discuss this problem later, while speaking about long virtual knots.

It is well known that the complement of each classical knot is an Eilenberg–McLane space $K(\pi, 1)$, for which all cohomology groups starting from the second group, are trivial. However, this is not the case for virtual knots: if we calculate the second cohomology the $K(\pi, 1)$ space where π is some virtual knot group, we might have some torsion. In [Kim] one can find a detailed description of such torsions.

17.6 Vassiliev's invariants for virtual links

There are two approaches to the finite type invariants of virtual knots: the one proposed by Goussarov, Polyak and Viro [GPV] and the one proposed by Kauffman [Kau6]. They both seem to be natural because they originate from the formal Vassiliev relation but the invariants proposed in [GPV] are not so strong.

Below, we shall give the basic definitions and some examples.

17.6.1 The Goussarov–Viro–Polyak approach

First, we shall give the definitions we are going to work with, see also [GPV]. We introduce the *semivirtual crossing*. This crossing still has an overpass and an underpass. In a diagram, a semivirtual crossing is shown as a classical one but encircled. Semivirtual crossings are related to the crossings of other types by the following formal relation:

$$\otimes = \diagup\!\!\!\!\diagdown - \bigotimes \qquad (17.1)$$

Let K be a virtual knot diagram with n classical crossings, and let $\{v_1, v_2, \ldots, v_n\}$ be different classical crossings of it. For an n-tuple $\{\sigma_1, \ldots, \sigma_n\}$ of zeros and ones, let K_σ be the diagram obtained from K by switching all v_i's with $\sigma_i = 1$ to virtual crossings. Denote by $|\sigma|$ the number of ones in σ. The formal alternating sum

$$\sum_\sigma (-1)^{|\sigma|} K_\sigma$$

is called a *diagram with n semivirtual crossings.*

Denote by $\mathcal{K}$ the set of all virtual knots. Let $\nu: \mathcal{K} \to G$ be an invariant of virtual knots with values in an abelian group G. Extend this invariant to $\mathbb{Z}[\mathcal{K}]$ linearly.

The next definition is due to Goussarov, Polyak and Viro.

Definition 17.8. We say that ν is an *invariant of finite type* or a *finite-type invariant* if for some $n \in \mathbb{N}$, it vanishes on any virtual knot with more than n semivirtual crossings, see [GPV]. The minimal such n is called *the degree* of the invariant ν.

The formal Vassiliev relation in the form $\bigotimes = \bigotimes - \bigotimes$ together with the relation defining a virtual crossing implies the relation

$$\times = \bigotimes - \bigotimes \tag{17.2}$$

Remark 17.5. *Note that singular knots are not considered here as independent objects having a geometrical sense of knots with some singularities, but just as linear combinations of simpler objects.*

It is obvious that for any finite-type invariant ν of the virtual theory in the sense described above, its restriction for the case of classical knots is a finite-type invariant in the ordinary sense.

However, not every classical finite-type invariant can be extended to a finite-type invariant in this virtual sense. For instance, there are no invariants of order two for (compact) virtual knots; see, for more details, [GPV, Oht, PV].

Starting from the formal relation defining a virtual crossing and the Vassiliev relation, Polyak constructed the Polyak algebra [GPV] that gave formulae for all finite-type invariants of virtual knots. Besides this, they give explicit diagrammatic formulae for some of them and also construct some finite-type invariants *for long virtual knots.*

Following Goussarov, Polyak and Viro [GPV], let us describe diagrammatic formulae for classical long knots. We need some definitions and constructions.

For a classical (virtual) long knot diagram we can construct the Gauss diagram; i.e. the line parametrizing the knot together with signed arrows connecting the preimages of each classical crossing.

Definition 17.9. An *arrow diagram* (on a circle) is an abstract diagram, which consists of an oriented circle with pairs of distinct points connected by dashed arrows. Each arrow is equipped with a sign. The *group of arrow diagrams* $\mathcal{A}$ is the free abelian group generated by all arrow diagrams.

Denote the set of all Gauss diagrams (non-realizable diagrams are allowed) by $\mathcal{D}$ (here all diagrams have thick arrows). Starting from any Gauss diagram we get an arrow diagram just by making all its arrows dashed. The extension of this map to $\mathbb{Z}[\mathcal{D}]$ defines a natural isomorphism $i\colon \mathbb{Z}[\mathcal{D}] \to \mathcal{A}$.

There is another important map $I\colon \mathcal{D} \to \mathcal{A}$, assigning to a Gauss diagram D the sum of all its subdiagrams and then making each of them dashed:

$$I(D) = \sum_{D' \subset D} i(D')$$

(here D' is a subdiagram of D if all the arrows of D' belong to D, and we write $D' \subset D$). Extend I to $\mathbb{Z}[\mathcal{D}]$ by linearity.

The following proposition is left to the reader as an exercise.

Proposition 17.1 ([GPV]). *There exists the inverse map* $I^{-1}\colon \mathcal{A} \to \mathbb{Z}[\mathcal{D}]$ *which is defined on the generators of $\mathcal{A}$ by the formula:*

$$I^{-1}(A) = \sum_{A' \subset A} (-1)^{|A-A'|} i^{-1}(A'),$$

where $|A - A'|$ is the number of arrows of A which do not belong to A'. Therefore, $I\colon \mathbb{Z}[\mathcal{D}] \to \mathcal{A}$ is an isomorphism.

Since the group $\mathcal{A}$ has a distinguished basis, consisting of arrow diagrams, there is a natural orthonormal scalar product $(\cdot, \cdot)$ on $\mathcal{A}$. Namely, on the generators of $\mathcal{A}$ we put (A_1, A_2) to be 1, if $A_1 = A_2$, and 0 otherwise, and then extend $(\cdot, \cdot)$ bilinearly. This allows us to define the pairing $\langle \cdot, \cdot \rangle \colon \mathcal{A} \times \mathcal{D} \to \mathbb{Z}$ by putting

$$\langle A, D \rangle = (A, I(D))$$

for any $D \in \mathcal{D}$ and $A \in \mathcal{A}$. Informally speaking, we count subdiagrams of D with weights, where the weight of a diagram D' is the coefficient of $i(D')$ in A.

Let us consider the case of (classical) long knots. The following theorem shows that any Vassiliev invariant can be calculated as a function of arrow polynomials evaluated on the knot diagram.

Theorem 17.4 (Goussarov et al. [GPV]). *Let G be an abelian group, and let ν be a G-valued invariant of degree n of classical long knots. Then there exists a function $\pi\colon \mathcal{A} \to G$ such that $\nu = \pi \circ I$ and π vanishes on any arrow diagram with more than n arrows.*

We immediately get the following corollary.

FIGURE 17.11: Forbidden situations.

Corollary 17.1 ([GPV]). *Any integer-valued finite-type invariant of degree n of classical long knots can be presented as $\langle A, \cdot \rangle$, where A is a linear combination of arrow diagrams on a line with at most n arrows.*

Let us prove Theorem 17.4 by following along the lines of [GPV, Theorem 3.A]. Consider *long virtual singular knot diagrams*; i.e. we have three types of crossings. We equip each double point with a sign as follows. The branches at a double point are ordered and the sign is the intersection number of the branches (taken in this order).

On the Gauss diagram of a long singular knot, each double point is shown by a dashed chord equipped with the above sign.

Definition 17.10. A diagram D' is called a *subdiagram* of a diagram D if D' consists of all the chords and some arrows of D.

Definition 17.11. A diagram of a classical long knot is *descending* if when going along the knot in the positive direction we pass first along overcrossing and then undercrossing. In terms of Gauss diagrams it means that all the arrows are directed to the right.

Let us now extend this notion to virtual long knots with double points. We still require that all the arrows are directed to the right. There is also an additional condition: There is no chord whose left endpoint neighbors with an endpoint of an arrow from left. In Fig. 17.11 forbidden situations are shown.

It is not difficult to see that a classical long knot with a Gauss diagram of this type can be presented by a diagram such that

1. all the double points are in the left half-plane,

2. all the crossings are in the right half-plane,

3. the intersection of the diagram with the left half-plane is an embedded tree,

4. the intersection with the right half-plane is an ordered collection of arcs; each of them is descending and lies below all the previous ones.

An example of such a diagram is given in Fig. 17.12.

It is easy to see that the chord part of the Gauss diagram of a descending classical long knot diagram with singular crossings determines the isotopy class of the classical long knot completely. As a result, we get the following lemma.

FIGURE 17.12: A descending classical long knot diagram and its Gauss diagram.

Lemma 17.1 ([GPV]). *Let D_1 and D_2 be Gauss diagrams of descending classical long knots (with singular crossings), and let ν be an invariant of long knots. If the chord parts of D_1 and D_2 coincide, then $\nu(D_1) = \nu(D_2)$.*

Remark 17.6. *Lemma 17.1 is not true for descending virtual long knots with singular crossings. Namely, we cannot determine a virtual long knot with singular crossings by just knowing the chord part of the Gauss diagram of its descending diagram. Therefore, the proof of the Goussarov theorem given above cannot be straightforwardly generalized for the virtual case.*

The next step of the proof is to show that the Gauss diagram of a classical long knot with double points can be represented as a linear combination of descending diagrams. There is an algorithm allowing us to do this. This algorithm consists of steps of two types. At each step, one inspects the Gauss diagram from the left to the right looking for the first fragment where the diagram fails to be descending. Such a fragment may either be a bad arrow or a bad chord.

Definition 17.12. An arrow is *bad* if it is directed to the left and a bad chord is depicted in Fig. 17.11.

In the case of a bad arrow the step of the algorithm is the replacement of the diagram with the sum of two diagrams according to the formula

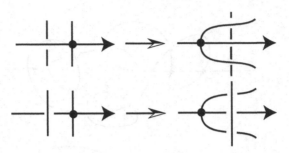

FIGURE 17.13: The case of a bad chord.

In terms of Gauss diagrams this replacement is as follows:

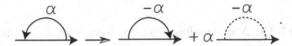

In the case of a bad chord the step of the algorithm is the pulling of the crossing over or under the appropriate branch by isotopy; see Fig. 17.13 (knot diagrams) and Fig. 17.14 (Gauss diagrams).

Denote by $\mathcal{D}_n$ the free abelian group generated by Gauss diagrams of virtual long singular knots with at most n chords (note that $\mathbb{Z}[\mathcal{D}] = \mathcal{D}_0 \subset \mathcal{D}_n$). We shall think of a step of the algorithm as an operator acting on $\mathcal{D}_n$. Denote this operator by P. By the definition of P, for any descending Gauss diagram D we have $P(D) = D$.

Lemma 17.2 ([GPV]). *For any diagram $D \in \mathcal{D}_n$ there exists m such that $P^m(D)$ is a sum of descending diagrams.*

This lemma can be proved by considering the number $l(D)$ of chords of D which have one of the endpoints to the left of the first bad fragment. It is not easy to see that this number does not decrease after applying the operator P, and the number of such chords in a non-descending diagram is at most n. Recall that we deal with an invariant of degree n, the diagrams with more than n chords are disregarded. Thus, when one applies a step of the algorithm to a bad arrow in a diagram with n chords, the summand with $n + 1$ chords disappears. To complete the proof of the lemma one can show that the diagram cannot change infinitely many times in subsequent iterations of P without changing l.

Let us extend an invariant ν of degree at most n to all virtual knot diagrams.

Denote by $\mathcal{D}_n^{re}$ the subgroup of $\mathcal{D}_n$ generated by Gauss diagrams of *classical* long singular knots. Any finite-type invariant of classical knots of degree at most n extends to $\mathcal{D}_n^{re}$ by linearity. The next lemma is obvious.

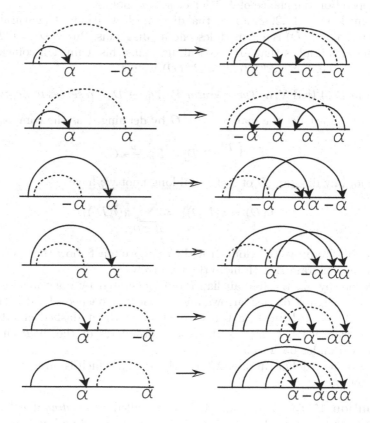

FIGURE 17.14: The case of a bad chord in terms of Gauss diagrams.

Lemma 17.3 ([GPV]). *The operator $P: \mathcal{D}_n \to \mathcal{D}_n$ preserves $\mathcal{D}_n^{re}$. The restriction of P to $\mathcal{D}_n^{re}$ preserves any invariant of degree at most n.*

Let us first consider virtual descending long knot diagrams. By turning all the virtual crossings of such a diagram D into appropriate classical ones, we get a descending long classical knot diagram D^{re} with the same double points. Put $\nu(D) = \nu(D^{re})$ (note that this operation is not well defined on virtual knots, it is defined just on diagrams). By Lemma 17.1, $\nu(D^{re})$ does not depend on the choice of D^{re} for classical knots.

Using Lemma 17.2, for any virtual diagram D we can find a natural number m such that $P^m(D)$ is a sum of descending diagrams. Put $\nu(D) = \nu(P^m(D))$. Lemma 17.3 implies that for classical diagrams this definition coincides with the initial one. Since $P^{m+1}(D) = P^m(D)$ we get

Lemma 17.4 ([GPV]). *The operator $P: \mathcal{D}_n \to \mathcal{D}_n$ preserves ν, i.e. $\nu \circ P = \nu$.*

Let us construct the map $\pi: \mathcal{A} \to G$ by defining it as the composition

$$\mathcal{A} \xrightarrow{I^{-1}} \mathbb{Z}[\mathcal{D}] \subset \mathcal{D}_n \xrightarrow{\nu} G.$$

Then for any diagram D of a classical long knot we have

$$\nu(D) = \pi(I(D)) = \sum_{D' \subset D} \pi(i(D')).$$

In order to prove the Goussarov theorem, we must show that $\pi(A) = 0$ for any arrow diagram A with more than n arrows.

Denote by $\mathcal{A}_n$ the free abelian group generated by diagrams on the line containing signed dashed arrows and at most n dashed chords. The maps $i, I: \mathbb{Z}[\mathcal{D}] \to \mathcal{A}$ defined on Gauss diagrams without dashed chords extend to isomorphisms $i, I: \mathcal{D}_n \to \mathcal{A}_n$ (the chord parts of the diagrams remain untouched under both i and I).

Let us now define an operator $Q: \mathcal{A}_n \to \mathcal{A}_n$, which is an analogue of the operator P.

Definition 17.13. A diagram $A \in \mathcal{A}_n$ is called *descending* if $i^{-1}(A)$ is descending. A fragment of A is called *bad* if the corresponding fragment of $i^{-1}(A)$ is bad.

Put $Q(A) = A$ if A is descending. Otherwise, find the leftmost bad fragment of A. If it is a bad arrow, we define $Q(A) = iPi^{-1}(A)$. If it is a bad chord, put $Q(A) = \sum A'$ where the sum runs over all the subdiagrams of $iPi^{-1}(A)$, each of which contains all the arrows not shown in Fig. 17.14, all the chords and at least one more arrow. In other words, we sum up all seven subdiagrams of $iPi^{-1}(A)$ which contain all the arrows and chords also belonging to A plus at least one more arrow. Here we need that the number of arrows is not decreased by Q, though this map is not invariant.

The next lemma is obvious.

Lemma 17.5 ([GPV]). *For any diagram $A \in \mathcal{A}_n$, the total number of arrows and chords in each diagram appearing in $Q(A)$ is at least the total number of arrows and chords in A.*

The following lemma is analogous to Lemma 17.2.

Lemma 17.6 ([GPV]). *For any diagram $A \in \mathcal{A}_n$, there exists m such that $Q^m(A)$ is a sum of descending diagrams.*

Lemma 17.7 ([GPV]). *For any non-descending diagram $D \in \mathcal{D}_n$, there is a splitting $I(D) = U + V$ with $U, V \in \mathcal{A}_n$ such that*

$$I(P(D)) = Q(U) + V \qquad (17.3)$$

and $U = i(D) + U'$, where U' is a sum of diagrams each of which has fewer arrows than D.

Proof. Let U be the sum of all the subdiagrams of $i(D)$ which include the first bad fragment of $i(D)$. These subdiagrams contain the same bad fragment as the whole diagram $i(D)$. Here $Q(U)$ is the sum of all subdiagrams of diagrams in $iP(D)$ which are not subdiagrams of $i(D)$. Then V is the sum of the subdiagrams of $i(D)$ which do not contain the arrow from the bad fragment and these subdiagrams of $i(D)$ remain unchanged, when one applies P to D. Thus $I(P(D)) = Q(U) + V$. $\qquad \square$

Though the operator Q is not invariant under the Reidemeister moves (sometimes we remove one term from the summation), the following lemma holds.

Lemma 17.8 ([GPV]). *The operator $Q\colon \mathcal{A}_n \to \mathcal{A}_n$ preserves π, i.e. $\pi \circ Q = \pi$.*

Proof. Let $A \in \mathcal{A}_n$ be a diagram and $D = i^{-1}(A)$. Let us prove that $\pi(Q(A)) = \pi(A)$ by induction on the number of arrows in A.

The induction base. If this number equals 0, then A is descending and $Q(A) = A$ by definition of Q.

The induction step. Suppose inductively that the statement is correct for any diagram whose number of arrows is less than the number of arrows in A, and let us prove the statement for A. Apply π to the equality (17.3):

$$\pi \circ Q(U) + \pi(V) = \pi \circ I \circ P(D) = \nu \circ P(D).$$

Since the operator P preserves ν, we get

$$\nu \circ P(D) = \nu(D) = \pi \circ I(D) = \pi(U) + \pi(V).$$

Thus $\pi \circ Q(U) = \pi(U)$. By the induction assumption, $\pi \circ Q(U') = \pi(U')$, where $U' = U - A$, and we obtain the desired equality $\pi(Q(A)) = \pi(A)$. This completes the induction step. $\qquad \square$

Lemma 17.9 ([GPV]). *Let $A \in \mathcal{A}_n$ be a descending diagram such that the total number of arrows and chords in A is greater than n. Then $\pi(A) = 0$.*

Proof. Let $D = i^{-1}(A)$. We have

$$\pi(A) = \nu \circ R(A) = \sum_{D' \subset D} (-1)^{|D-D'|} \nu(D').$$

Since any subdiagram D' of D is descending and has the same chord part, $\nu(D') = \nu(D)$ by the construction of ν. Therefore,

$$\pi(A) = \left(\sum_{D' \subset D} (-1)^{|D-D'|} \right) \nu(D).$$

As one can easily check by induction on the number of arrows in A, the sum in parentheses is equal to 1 if A has no arrows and is 0 otherwise. Since all the diagrams in $\mathcal{A}_n$ have at most n chords and the total number of arrows and chords in A is greater than n, it has at least one arrow. Hence $\pi(A) = 0$. □

Lemma 17.10 ([GPV]). *Let $A \in \mathcal{A}_n$ be a diagram such that the total number of arrows and chords in A is greater than n. Then $\pi(A) = 0$.*

Proof. Let m be the number which exists for A by Lemma 17.6. By Lemma 17.8, $\pi(A) = \pi(Q^m(A))$. By Lemma 17.5, the expansion of $Q^m(A)$ contains only descending diagrams with the total number of chords and arrows greater than n. Then by Lemma 17.9, $\pi(A) = 0$. □

The last lemma completes the proof of the Goussarov theorem.

17.6.2   The Kauffman approach

Kauffman starts from the formal definition of a singular virtual knot (link).

Definition 17.14. A *singular virtual link diagram* is a four–valent graph in the plane endowed with orientations of unicursal curves and crossing structure: each crossing should be either classical, or virtual, or singular.

Definition 17.15. A *singular virtual knot* is an equivalence class of virtual knot diagrams by generalised Reidemeister moves and *rigid vertex isotopy*, shown in Fig. 17.15

Now, the definition of the Vassiliev knot invariants is literally the same as in the classical case. For each invariant f of virtual links, one defines its formal derivative f' by Vassiliev's rule $f'(\times) = f(\times) - f(\times)$ and says that the invariant f has order less than or equal to n if $f^{(n+1)} \equiv 0$.

The space of finite type invariants of virtual knots (by Kauffman) is much more complicated than that of classical knots. For instance, the space of invariants of order zero is infinite–dimensional because there are infinitely many

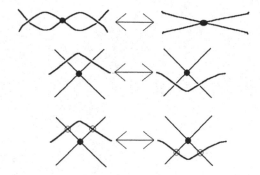

FIGURE 17.15: Rigid vertex isotopy moves for virtual knots

classes of virtual knots that can not be obtained from each other by using isotopy and classical crossing switches (one can separately define the value of an invariant on each of these equivalence classes). The structure of higher order invariants is even more complicated.

The Jones–Kauffman invariant in the form [Kau6] is weaker than the finite type invariants. Namely, one can transform it (by the exponential variable change) into a series with finite type invariant coefficient just as was done in Chapter 13.

It is easy to see that many such invariants are not of finite type in the sense of [GPV]. This observation is due to Kauffman [Kau5].

FIGURE 15.

Chapter 18

Invariant polynomials of virtual links

We have already considered some generalisations of basic knot invariants: the colouring invariant and the fundamental group. In Section 18.2 below we shall consider a generalisation of the Jones–Kauffman polynomial [Kau5].

All the generalisations described above were constructed by using the following idea: one thinks of a virtual link diagram as a set of classical crossings provided with the information about how they are connected on the plane and one does not pay attention to virtual crossings. Thus, for instance, the generators of the fundamental group that correspond to arcs of the diagram may pass through virtual crossings, and all relations are taken only at classical crossings.

In the present chapter, we shall describe the invariants proposed in the author's papers [Man4, Man9, Man10, Man13, MI] (for short versions see in [Man7, Man11]). Interested readers may also read the excellent review of Kauffman [Kau7] and his works with Radford [KR] and the author [KM1] about so-called "biquandles". Polynomial invariants of virtual knots and links were also constructed in [BF, Saw, SW].

In the present chapter, we are going to modify these invariants in the following manner: we find a way that a virtual crossing can have impact on the constructed object (e.g., for the case of the fundamental group and what it does with the generator) and then prove its invariance.

The main results present here can be found in [Man4].

The main idea of this construction is the following: while constructing the invariant of the virtual link (or braid), we have taken into consideration that virtual crossings can change the corresponding element, say, by multiplying one of them by q, and the other one will be denoted by q^{-1} (if we deal with some group structures). This idea of adding a new "variable" will be the main one in this chapter.

It turns out that in some cases (e.g. the Alexander module) this new variable plays a significant role and allows us to construct a virtual knot invariant that *is not* a generalisation of any classical knot invariant.

Throughout the present chapter, all knots and links (virtual or classical) are thought to be oriented, unless otherwise specified.

By a *homomorphism* of two objects $\mathcal{O}_1$ and $\mathcal{O}_2$ both endowed with a set of operations $(o_1, \ldots, o_m)$ we mean a map from $\mathcal{O}_1$ to $\mathcal{O}_2$ with respect to all these operations.

18.1   The virtual groupoid (quandle)

We recall that the notion of quandle (also known as a distributive groupoid) first appeared in the pioneering works of Matveev [Mat1] and Joyce [Joy]. They have proved that it is a complete invariant of knots.

We recall that a *quandle* is a set M together with the operation $\circ$ that is

1. idempotent: $\forall a \in M : a \circ a = a$,

2. right self–distributive:

$$\forall a, b, c \in M : (a \circ b) \circ c = (a \circ c) \circ (b \circ c), \tag{18.1}$$

   and

3. left–invertible: $\forall a, b \in M : \exists! x \in M : x \circ a = b$. This element is denoted by b/a.

All these conditions are necessary and sufficient for the constructed quandle to be invariant under the Reidemeister moves. More precisely, for each knot (link) one can construct the knot (link) quandle.

Let L be an oriented virtual link diagram.

Definition 18.1. A *Kauffman arc* or *long arc* of this diagram is an oriented interval (piece of a curve) between two adjacent undercrossings (i.e., while walking along this arc, we make only overcrossings or virtual crossings).

With each arc $a_i, i = 1, \ldots, n$, we associate an element x_i of the quandle to be constructed. First, we take the free quandle generated by $a_1, \ldots, a_n$. Then, if three arcs a_1, a_2, a_3 meet each other at a classical crossing as shown in Fig. 18.1, we write down the relation

$$a_{i_1} \circ a_{i_2} = a_{i_3}. \tag{18.2}$$

Definition 18.2. The *Kauffman quandle* of L is the formal quandle, generated by $a_i, i = 1, \ldots, n$, and all relations (18.2) for all classical vertices.

More precisely, elements of such a quandle are equivalence classes of words obtained from a_i by means of $\circ$ and $/$, where equivalence is defined by crossing relations.

The invariance of this quandle under purely virtual moves and the semivirtual move comes straightforwardly: the representation stays the same.

Remark 18.1. *The invariance of this quandle under the classical Reidemeister moves can be checked straightforwardly just as in the classical case.*

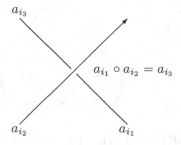

FIGURE 18.1: Crossing relation for the quandle

Denote the obtained quandle by $Q_K(L)$.

Obviously, for the unknot U we have $Q_K(L) = \{a\}$.

Thus, we have constructed a link invariant. By definition, it coincides with the classical quandle [Mat1, Joy] on the classical links.

It might seem that for the classical links there are two equivalences: the classical one and the virtual one. However, this is not the case.

Theorem 18.1. *[GPV] Let L and L' be two (oriented) classical link diagrams such that L and L' are equivalent under generalised Reidemeister moves. Then L and L' are equivalent under classical Reidemeister moves.*

Proof. Note that longitudes (see definition on page 29) are preserved under virtual moves (adding a virtual crossing to the diagram does not change the expression for a longitude). Thus, an isomorphism for $Q_K(L)$ and $Q_K(L')$ induced by generalised Reidemeister moves preserves longitudes. Since the isomorphism class of the quandle plus longitudes classifies classical knots, we conclude that L and L' are classically equivalent. $\qquad\square$

However, unlike the classical case, this quandle is rather weak in the virtual sense: there are different simple knots that cannot be recognised by it.

Indeed, consider the knot diagram K shown in the left part Fig. 18.2. The quandle Q_K corresponding to this diagram has two generators a, b and two relations $a \circ b = b$ and $b \circ b = a$. Obviously, they imply $a = b$. Thus this quandle is the same as that for the unknot.

Later, we shall see that this non-trivial knot is recognised by the "virtual quandle" to be constructed and thus, it is not trivial.

Besides the virtual quandle to be constructed, there is another generalisation of the quandle: the so-called *biquandle* (due to Kauffman and Radford, [KR], and to Kauffman and the author [KM1]). Unlike our "virtual quandle" approach, where we add a new operation at a virtual crossing, Kauffman and Radford extend the algebraic structure at a classical crossing: the algebraic element associated to a part of an arc "before" the (under)crossing differs from that associated to the part "after" the undercrossing. The comparison of these two approaches seems to be quite interesting for further investigations.

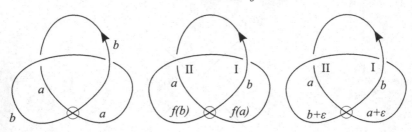

FIGURE 18.2: The virtual trefoil and its labellings

Now, let us construct the modified (virtual) quandle (we shall denote it just by Q unlike Kauffman's quandle Q_K).

Definition 18.3. A *virtual quandle* is a quandle $(M, \circ)$ endowed with a unary operation f such that:

1. f is invertible; the inverse operation is denoted by f^{-1};

2. $\circ$ is distributive with respect to f:

$$\forall a, b \in M : f(a) \circ f(b) = f(a \circ b). \qquad (18.3)$$

Remark 18.2. *The equation* (18.3) *easily implies for all* $a, b \in M$:

$$f^{-1}(a) \circ f^{-1}(b) = f^{-1}(a \circ b),$$

$$f(a)/f(b) = f(a/b),$$

and

$$f^{-1}(a)/f^{-1}(b) = f^{-1}(a/b).$$

For a given virtual link diagram L, let us construct its virtual quandle $Q(L)$ as follows.

First, let us choose a diagram L' in such a way that it can be divided into long arcs in a proper way. Such diagrams are called *proper*. Obviously, a proper diagram with m crossings has m long arcs.

More precisely, we need the result that each long arc has two different final crossing points. For some diagrams this is not true. However, this can easily be done by slight deformations of the diagram; see Fig. 18.3.

Exercise 18.1. *Show that equivalent proper diagrams of virtual links can be transformed to each other by generalised Reidemeister moves in the class of proper diagrams.*

Remark 18.3. *Later in this chapter, all diagrams are taken to be proper, unless otherwise specified.*

circular long arc   proper diagram

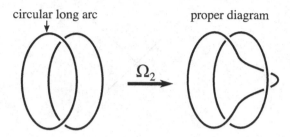

FIGURE 18.3: Reconstructing a link diagram in a proper way

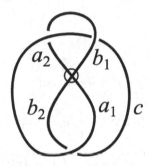

FIGURE 18.4: A knot diagram and its arcs

Let L' be a virtual diagram; let us think of its undercrossings as broken (disconnected) lines as they are drawn on the plane. Let $\hat{L}'$ be the set obtained from L' by removing all virtual crossings (vertices).

Definition 18.4. An *arc* of L' is a connected component of $\hat{L}'$.

Exercise 18.2. *The knot shown in Fig. 18.4 has three classical crossings, three arcs (a_1 and a_2; b_1 and b_2; c), and five virtual arcs (a_1, a_2, b_1, b_2, c).*

Remark 18.4. *The knot in Fig. 18.2 has thus four arcs (see the middle picture): each of the two "former" arcs a and b is now divided into two arcs by the virtual crossing.*

The invariant $Q(L)$ is now constructed as follows. Consider all arcs $a_i, i = 1, \ldots, n$, of the diagram L'. Consider the set of formal words $X(L')$ obtained inductively from a_i by using $\circ, /, f, f^{-1}$. In order to construct $Q(L')$ we shall factorise $X(L')$ by some equivalence relations.

First of all, for each $a, b, c \in X(L')$ we identify:

$$f^{-1}(f(a)) \sim f(f^{-1}(a)) \sim a;$$

$$(a \circ b)/b \sim a;$$

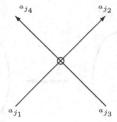

FIGURE 18.5: Relation for a virtual crossing

$$(a/b) \circ b \sim a;$$

$$a \sim a \circ a;$$

$$(a \circ b) \circ c \sim (a \circ c) \circ (b \circ c);$$

$$f(a \circ b) \sim f(a) \circ f(b).$$

Thus, we get the "free" quandle with generators $a_1, \ldots, a_n$. The following factorisation will be done with respect to the structure of the diagram L'.

For each classical crossing we write down the relation (18.2) just as in the classical case. For each virtual crossing V we also write relations. Let $a_{j_1}, a_{j_2}, a_{j_3}, a_{j_4}$ be the four arcs incident to V as it is shown in Fig.18.5.

Then, let us write the relations:

$$a_{j_2} = f(a_{j_1}) \tag{18.4}$$

and

$$a_{j_3} = f(a_{j_4}). \tag{18.5}$$

So, the virtual quandle $Q(L)$ is the quandle generated by all arcs $a_i, i = 1, \ldots, n$, all relations (18.2) at classical vertices and all relations (18.4), (18.5) at virtual crossings.

Theorem 18.2. *The virtual quandle $Q(L)$ is a virtual link invariant.*

Proof. The invariance under classical Reidemeister moves is just the same as in the classical case (cf. Remark 18.1).

First, let us note that two proper diagrams generate isotopic virtual links if and only if one of them can be deformed to the other by using a sequence of virtual Reidemeister moves. Indeed, if a circular long link occurs during the isotopy, then we can modify the isotopy by applying the first classical Reidemeister move to this long arc and subdividing it into two parts.

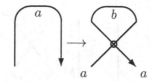

FIGURE 18.6: Invariance of Q under the first virtual move

Now, we have to show that by applying a Reidemeister move to a link diagram, we transform our quandle to an isomorphic one.

Consider some virtual Reidemeister move. Let $\bar{L}$ and $\bar{L}'$ be two virtual link diagrams obtained from each other by applying this move. Since this move is performed inside a small circle C, all arcs of L and L' can be split into three sets: the common set E of exterior arcs belonging to both L and L', the common set S of arcs intersecting the circle C, and the sets I, I' of interior arcs belonging to L and L', respectively. Hence, the quandle $\mathcal{M}(L)$ has the following generators and relations.

First, we have the relations to be denoted by ε (distributivity and idempotence) and E, S, I, and $\Gamma(L')$ is generated by ε and E, S, I'.

Relations (crossings) for diagrams L, L' are also divided into two sets: exterior ones R_E which are common for L and L' and interior R_I, R'_I related to L and L', respectively. Besides them, each quandle has general quandle relations (left–invertibility and right self–distributivity). Later in the proof, by *relation* we shall mean only those relations that come from crossings (not idempotence or distributivity).

Now, it is easy to see that for each concrete generalised Reidemeister move, by using R_I one can remove the generators I by expressing them in terms of S. Actually, this will add some "interior" relations R_S for S. The same can be done for I'. Denote these relations by R'_S. So, we transform both quandles $\Gamma(L)$ and $\Gamma(L')$ into isomorphic quandles $\tilde{\Gamma}(L)$ and $\tilde{\Gamma}(L')$. The latter ones are generated only by E, S (and ε). They have a common set of exterior relations R_E.

The only thing to show is that relations R_S and R'_S determine the same equivalence on S (by means of ε).

Let us perform it for concrete versions of Reidemeister moves; the other cases are completely analogous to those to be described.

We have to show that $Q(L)$ is invariant under virtual Reidemeister moves.

The invariance of Q under all classical moves is checked in the same way as that of Q.

Let us now check the invariance of Q under purely virtual Reidemeister moves.

The first virtual Reidemeister move is shown in Fig. 18.6. In the initial local picture we have one local generator a. Here we just add a new generator b and two coinciding relations: $b = f^{-1}(a)$. Thus, it does not change the virtual quandle at all.

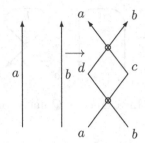

FIGURE 18.7: Invariance of Q under the second virtual move

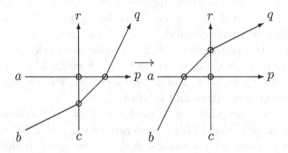

FIGURE 18.8: Invariance of Q under the third virtual move

The case of inverse orientation at the crossings gives us $b = f(a)$ which does not change the situation.

For each next relation, we shall check only one case of arc orientation.

The second Reidemeister move (see Fig. 18.7) adds two generators c and d and two pairs of coinciding relations: $c = f(a), d = f^{-1}(b)$. Thus, the quandle Q stays the same.

In the case of the third Reidemeister move we have six "exterior arcs": three incoming (a, b, c) and three outgoing (p, q, r), see Fig 18.8. In both cases we have $p = f^2(a), q = b, r = f^{-2}(c)$. The three interior arcs are expressed in a, b, c, and give no other relations.

Finally, let us check the mixed move. We are going to check the only version of it; see Fig. 18.9

In both pictures we have three incoming edges a, b, c and three outgoing edges p, q, r. In the first case we have relations: $p = f(a), q = b, r = f^{-1}(c) \circ a$. In the second case we have: $p = f(a), q = b, r = f^{-1}(c \circ f(a))$.

The distributivity relation $f(x \circ y) = f(x) \circ f(y)$ implies the relation $f^{-1}(c) \circ a = f^{-1}(c \circ f(a))$. Hence, two virtual quandles before the mixed move and after the mixed move coincide.

The other cases of the mixed move lead to other relations all equivalent to $f(x \circ y) = f(x) \circ f(y)$.

This completes the proof of the theorem.

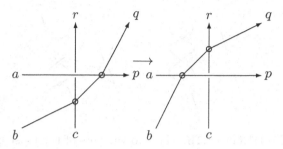

FIGURE 18.9: Invariance of Q under the mixed move

☐

Remark 18.5. *Note that for the classical links, Q_K can be easily restored from Q. In the case of virtual links Q is, indeed, stronger: having Q, one easily obtains Q_K by putting $\forall x \in Q f(x) \equiv x$.*

Example 18.1. *Consider the virtual knot K (middle part) shown in Fig. 18.2. It has four arcs: $a_1 = a, b_1 = f(b)$ (before the virtual crossing) and a_2, b_2 (after the virtual crossing).*

The relations in this quandle are: $a_2 = f(a_1), b_1 = f(b_2), a_2 \circ b_2 = b_1$ and $b_2 \circ b_1 = a_1$. Rewriting the last two relations for the two generators a_1, b_2, we get: $f(a_1) \circ b_2 = f(b_2)$ and $b_2 \circ f(b_2) = a_1$.

Obviously, two quandles are not easy to compare. Of course, neither are virtual quandles. And we cannot show just now why $Q(K) \neq \{a\}$. However, we shall soon describe some simplifications of Q which are weaker, but easier to compare. They will show us that $Q(K) \neq \{a\}$, and consequently K is indeed knotted.

The idea of applying some relation to virtual crossings can be transformed to the idea of endowing classical crossings with additional information (which is somehow related to the virtual crossing count).

Here we briefly mention *parity* (see [MM1, MM2, Man22, Man23, Man24, Man25, Man28, CM, MI, IMN1, IMN2, CFM, IM1, Man31, IMN3, FM]).

In fact, classical crossings can be endowed with some much more powerful information that just numbers, namely, we can endow classical crossings with pictures [KM2, KM3, Man30, MN].

18.2 The Jones–Kauffman polynomial

The Kauffman construction for the Jones polynomial for virtual knots [Kau5] works just as well as in the case of classical knots. Namely, we first

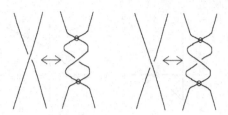

FIGURE 18.10: The two variants of the twist move

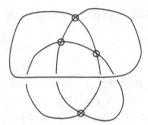

FIGURE 18.11: A virtual knot reduced to the unknot by the virtualisation and the generalised Reidemeister moves.

consider an oriented link L and the corresponding unoriented link $|L|$. After this, we smooth all classical crossings of $|L|$ just as before (obtaining *states* of the diagram). In this way, we obtain a diagram without classical crossings, which is an unlink diagram. The number of components of this diagram (for a state s) is denoted by $\gamma(s)$. Then we define the Kauffman bracket by the same formula

$$X(L) = \sum_s (-a)^{3w(L)} a^{\alpha(s)-\beta(s)} (-a^2 - a^{-2})^{\gamma(s)-1}, \qquad (18.6)$$

where $w(L)$ is the writhe number taken over all classical crossings of L, and $\alpha(s), \beta(s)$ are defined as in the classical case.

The invariance proof for this polynomial under classical Reidemeister moves is just the same as in the classical case; under purely virtual and semivirtual moves it is clearly invariant term–by–term.

However, this invariant has a disadvantage [Kau5]: invariance under a move that might not be an equivalence.

Exercise 18.3. *Prove that the polynomial X is invariant under the following local moves — twist move or virtual switch move or virtualisation; see Fig. 18.10.*

Example 18.2. *Let us consider the virtual knot diagram shown in Fig. 18.11. This knot was first considered by Kauffman.*

This knot can be reduced to the unknot with virtualisations and generalised Reidemeister moves; see Fig. 18.12.

In Fig. 18.12 by the transformation B' we mean a move applied to one classical and one virtual crossing; it represents a composition of the virtualisation and the second Reidemeister move; see Fig. 18.13. For each of the transformations shown in Fig. 18.12, we pick out a domain which this transformation is applied to.

Thus, the Jones–Kauffman polynomial of the knot depicted in Fig. 18.11 coincides with the polynomial of the unknot. One can show though (e.g. using the techniques of virtual quandles) that this virtual knot is not trivial.

18.3 Presentations of the quandle

The quandle admits some presentations such as the fundamental group, the Alexander polynomial, and the colouring invariant. Here we shall show how to construct analogous presentations for the quandle Q.

18.3.1 The fundamental group

We recall that the fundamental group G of the complement to an oriented link L is obtained from its quandle $Q_K(L)$ as follows. Instead of elements a_i of the quandle Q we write elements of the formal group G (to be constructed), and instead of the operation $\circ$ we write the conjugation operation: $x \circ y$ becomes yxy^{-1}. It is easy to check that this presentation of the operation $\circ$ preserves the idempotence property and the relation (18.2). Besides, it has an evident inverse operation, namely: in the group a/b is going to be $b^{-1}ab$.

Thus, having written all relations for all classical vertices of the link l, we get a group $G(L)$ that is called the *fundamental group of the complement to* L. In the case of classical knots, this group has a real geometric sense.

Obviously, we can do just the same for the case of virtual knots. In this case we also get a virtual link invariant, called the *Kauffman fundamental group of a virtual link*.

Actually, each such presentation makes the initial invariant weaker. Thus, Kauffman's virtual fundamental group does not distinguish the "virtual trefoil" knot K and the unknot.

Now, we construct the presentation of the invariant Q in the category of groups.

We have already seen that the conjugation plays the role of the operation $\circ$. So, we only have to find an appropriate operation to present $f(\cdot)$.

This operation can be taken as follows: we just add a new generator q and say that $f(a) = qaq^{-1}$.

These two operations together make a groupoid from each group.

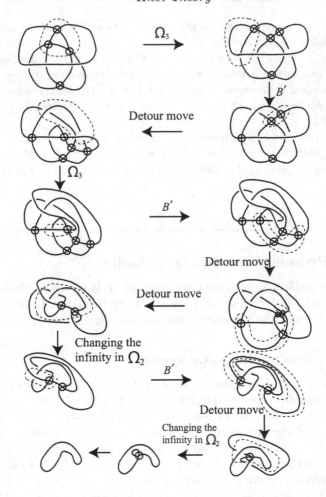

FIGURE 18.12: Reducing to the unknot by virtualisations and generalised
Reidemeister moves.

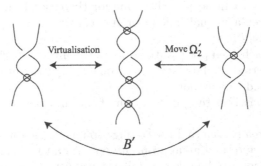

FIGURE 18.13: The move B' is expressed in terms of the virtualisation.

In fact, the following lemma holds.

Lemma 18.1. *For each group G, the group $G * \{q\}$ (free product) with the two operations $\circ, f(\cdot)$ defined as $a \circ b = bab^{-1}, f(c) = qcq^{-1}$ for all $a, b, c \in G * \{q\}$, is a virtual groupoid.*

Proof. Indeed, we just have to show that $f(a \circ b) = f(a) \circ f(b)$. Actually, $f(a \circ b) = f(bab^{-1}) = qbab^{-1}q^{-1} = qbq^{-1}(qaq^{-1})qb^{-1}q^{-1} = f(b)f(a)f(b)^{-1} = f(a) \circ f(b)$. $\qquad\square$

Now, let us construct the *fundamental group $G(L)$* of a virtual link diagram L. Let us enumerate all arcs of L by $a_i, i = 1, \ldots, n$. So, G is the group generated by $a_1, \ldots, a_n, q$ with the relations obtained from (18.2), (18.4), (18.5) by putting $f(x) = qxq^{-1}, y \circ z = zyz^{-1}$.

Thus, we obtain the following important theorem.

Theorem 18.3. *The group $G(L)$ is an invariant of virtual links.*

Proof. The proof follows immediately from Theorem 18.2 and Lemma 18.1. $\qquad\square$

Obviously, for the unknot U we have $G(U) = \langle a, q | \rangle$ is a free group with two generators.

Exercise 18.4. *Calculate the fundamental group $G(K)$ of the virtual trefoil (together with the element q in it) and prove that this group together with q distinguishes K from the unknot.*

Thus, the knot K is not trivial.

Besides the fundamental group, for each knot one can construct another invariant group by using the following relations:

$$x \circ y = y^p x y^{-p}, \quad f(x) = qxq^{-1}$$

where p is a fixed integer, and q is the fixed group element. The proof is quite analogous to the previous one.

Another way to construct an invariant group is the following:

$$x \circ y = yx^{-1}y, f(x) = qxq^{-1}$$

or

$$x \circ y = yx^{-1}y, f(x) = qx^{-1}q.$$

18.3.2   The colouring invariant

The idea of colouring invariant is very simple. We take a presentation of some finite quandle Q' (say, obtained from a finite group G') by generators $a_1, \ldots, a_k$ and relations.

More precisely, the following lemma holds.

Lemma 18.2. *Let Q' be a virtual quandle. Then the set of homomorphisms $Q(L) \to Q'$ is an invariant of link L.*

This claim is obvious.

Now, let us prove the following lemma.

Lemma 18.3. *For each finite virtual quandle Q' the number of homomorphisms $Q(L) \to Q'$ is finite for each link L.*

Proof. Indeed, consider a link L and a proper diagram $\bar{L}$ of it. In order to construct a homomorphism $h : Q(L) \to Q'$ we only have to define the images $h(a_i)$ of those elements of $Q(L)$ that correspond to arcs. Since the number of arcs is finite, the desired number of homomorphisms is finite.   □

The two lemmas proved above imply the following theorem.

Theorem 18.4. *Let Q' be a finite virtual quandle. Then the number of homomorphisms $Q(L) \to Q'$ is an integer–valued invariant of L.*

The sense of this invariant is pretty simple: it is just the proper colouring number of arcs of L by elements of Q'; the colouring is proper if and only if it satisfies the virtual quandle condition.

How do we construct finite virtual quandles? Let us generalise the ideas for ordinary quandle construction from [Mat1] for the virtual case. Here are some examples.

Let G be a finite group, $g \in G$ be a fixed element of it, and n be an integer number. Then the set of elements $x \in G$ equipped with the operation $x \circ y = y^n x y^{-n}, f(x) = gxg^{-1}$ is a virtual groupoid.

Another way for constructing virtual groupoids by using groups is as follows: for a group G with a fixed element $g \in G$, we set $x \circ y = yx^{-1}y, f(x) = gxg^{-1}$.

These examples give two series of integer–valued virtual link invariants.

For further reading see [KM1, Car12].

18.4   The VA-polynomial

In the present section we give a generalisation of the Alexander module. This module leads to the construction of the so–called VA–*polynomial* that has no analogue in the classical case.

Consider a module R over the ring of Laurent polynomials in the variable t over $\mathbb{Q}$.

Remark 18.6. *Here we take the field $\mathbb{Q}$ (instead of the ring $\mathbb{Z}$ as in the classical case) in order to get a graded Euclidean ring of polynomials. In our case $\deg P = length(P)$ where length means the difference between the leading degree and the lowest degree.*

For any two elements of this module we can define the operation $\circ$ as follows:

$$a \circ b = ta + (1-t)\,b. \tag{18.7}$$

Obviously, this operation is invertible; the inverse operation (denoted by "/") is given by the formula

$$a/b = \frac{1}{t}a + \left(1 - \frac{1}{t}\right)b. \tag{18.8}$$

Clearly, $a \circ a = a$. The self–distributivity of (18.7) can be easily checked.

Thus, having a classical link diagram we can define a module over the ring of Laurent polynomials by the following rule: the generator system of this module consists of elements a_i, corresponding to arcs of the diagram; at each classical crossing we write down a relation (18.2), where the operation $\circ$ is taken from (18.7).

Remark 18.7. *In the sequel, all modules are right–left modules; i.e., one can multiply elements of the module by an element of the ring on the right or left hand; thus one obtains the same result.*

Thus, the defined module is a link invariant. However, this module allows us to extract a more visible invariant, called *the Alexander polynomial*. This can be done as follows.

For a diagram of a classical link, the system of relations defining this module is a linear systems of n equations on n variables $a_i, i = 1, \ldots, n$. Thus, we get an $n \times n$ matrix of relations. It is also called *the Alexander matrix* $M(L)$. Hence for each equation the sum of the coefficients equals zero, the rows of this matrix are linearly–dependent, and thus the determinant of this matrix equals zero.

It is not difficult to prove that all minors of order $n - 1$ of this matrix are the same up to multiplying by $\pm t^k$. The *Alexander polynomial* is just this minor (defined up to $\pm t^k$). The complete proof of invariance for the classical Alexander polynomial can be read, in e.g., [Man'2].

Remark 18.8. *Note that $\pm t^k$ are precisely all invertible elements in the ring of Laurent polynomials.*

Now, let us generalise the Alexander approach for the case of virtual knots. We shall use the same ring of Laurent polynomials over t. Fortunately, this is quite easy. Indeed, to define the *virtual Alexander module*, we need to find a "good" presentation for the function f, such that

$$f(a \circ b) = f(ta + (1-t)b) = tf(a) + (1-t)f(b). \qquad (18.9)$$

Here we can just set $f(a) \equiv a + \varepsilon$, where ε is a new vector (it is the fixed vector in the new module). More precisely, consider a module R over the ring of polynomials of t, t^{-1} (say, with rational coefficients) and set

$$\forall a, b, c \in R: \quad a \circ b = ta + (1-t)b, \quad f(c) = c + \varepsilon. \qquad (18.10)$$

In this case, the formula (18.10) follows straightforwardly.

Thus, having a virtual link L diagram, we can define the *virtual diagram Alexander module* $M(L)$ over the Laurent polynomial ring taking arcs of the diagrams as generators and (18.2), (18.4), (18.5) as relations (in the form (18.10)). In this module there exists a fixed element denoted by ε.

This definition together with Theorem 18.2 implies the following Theorem.

Theorem 18.5. *The pair consisting of the virtual Alexander module together with the fixed element $(M(L), \varepsilon)$ is a virtual link invariant.*

Now, let us see what happens with the Alexander polynomial.

Remark 18.9. *For the sake of simplicity, we shall think of arcs of the diagram as elements of the Alexander module (i.e. we shall not introduce any other letters).*

Let L be a proper virtual link diagram with m crossings. Thus, it has precisely m long arcs; each of the long arcs is divided into several arcs. According to the operation $f(\cdot)$, we see that if two arcs p and q belong to the same long arc, they satisfy the relation

$$p = q + r\varepsilon.$$

For each long arc, choose an arc of it. Since we have m long arcs, we can denote chosen arcs by $b_1, \ldots, b_m$. All other arcs are, hence, equal to $b_i + p_{ij}\varepsilon$, where $p_{ij} \in \mathbb{Z}$.

Now, we can write down the relations of the Alexander polynomial. For each classical crossing v we just write the relation (18.2) in the form (18.7). Thus, we obtain an $m \times m$ matrix, where rows correspond to classical crossings, and columns correspond to long arcs. Thus, the element of the j-th column and i-th string is something like $P(t) \cdot (b_j + p_{ji}\varepsilon)$, where $P(t)$ is a polynomial in t.

Now, we can take all terms containing ε and move them to the right part. Thus we obtain an equation

$$(A) b = c\varepsilon, \qquad (18.11)$$

where b is the column $(b_1, \ldots, b_m)^*$, c is the column $(c_1, \ldots c_m)^*$ of coefficients for ε, and A is a matrix.

The system (18.11) completely defines the virtual Alexander module M together with the element $\varepsilon \in M$.

As in the classical case, A is called *the Alexander matrix* of the diagram L. It is easy to see that in both the classical and in the virtual case A is degenerate: for the vector $x = (1, \ldots, 1)^*$ we have

$$Ax = 0.$$

However, unlike the classical case, here we have a non–homogeneous system (18.11) of equations.

Since A is degenerated, it has rank at most $m - 1$. Thus, the equation (18.11) implies some condition on ε: since strings of A are linearly–dependent, so are $c_1\varepsilon, \ldots, c_m\varepsilon$. This means that our Alexander module might have a relation $V(t)\varepsilon = 0$, where $V(t)\varepsilon = 0$. Here we write "might have" because it can happen that $V(t)\varepsilon = 0$ implies $V(t) = 0$.

The set of all $i \in R$ such that $i\varepsilon = 0$ forms an ideal $I \subset R$. The ring R is Euclidean, thus the ideal I is a principal ideal. It is characterised by its minimal polynomial $VA \in I$ that is defined up to invertible elements of R.

Definition 18.5. The minimal polynomial $VA(t) \in R$ of the ideal I is called the VA–*polynomial* of the virtual link diagram L.

Remark 18.10. *Obviously, VA is defined up to invertible elements of R; i.e., up to $\pm t^k$.*

The polynomial $VA(L)$ depends only on the module $M(L)$ with the selected element ε.

Thus, we obtain the following theorem.

Theorem 18.6. *The polynomial VA (defined up to invertible elements of R) is a virtual link invariant.*

Now, let us calculate the value of the VA–polynomial on some virtual knots.

Exercise 18.5. *Consider the "virtual trefoil knot". Let us calculate the virtual quandle of it. In Fig. 18.2, rightmost picture, we have two classical vertices: I and II. They give us a system of two relations:*

$$I : (a + \varepsilon)t + b(1 - t) = b + \varepsilon$$

$$II : bt + (b + \varepsilon)(1 - t) = a,$$

or:

$$at - bt = \varepsilon(1 - t)$$
$$b - a = \varepsilon(t - 1).$$

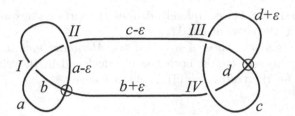

FIGURE 18.14: One way to construct $K \# K$

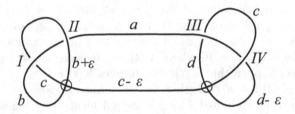

FIGURE 18.15: Another way to construct $K \# K$

Multiplying the second equation by t and adding it to the first one, we get: $0 = \varepsilon(t-1)(1-t)$. *Thus,* $VA(K) = (1-t)^2$.

Having two oriented virtual knots K_1 and K_2, one can define the *connected sum* of these knots as follows. We just take their diagrams, break each of them at two points close to each other and connect them together according to the orientation. This construction is well known in the classical case. In the classical case, the product is well defined (i.e., it does not depend on the broken point).

Let us see what happens in the virtual case. For example, let us take the two copies of one and the same knot K (shown in Fig. 18.2) and attach them together in two different ways as shown in Figs. 18.14 and 18.15.

We have two systems of equations. The first knot gives:

$$I : a - b = t\varepsilon$$

$$II : -ta + tc = \varepsilon$$

$$III : -tc + td = (1-2t)\varepsilon$$

$$IV : tb + (1-t)c - d = -t\varepsilon.$$

The zero linear combination is $t^2(I) + t(II) + (III) + t(IV) = 0$. Thus, the VA–polynomial equals $t^2(t) + t(1) + (1-2t) + t(-t) = (t-1)^2(t+1)$. The second knot gives:

$$I : a - b = t\varepsilon$$

$$II : -ta + tc = (1 - t)\varepsilon$$

$$III : tb + (1 - t)c - d = 0$$

$$IV : (1 - t)b - c + dt = -t\varepsilon.$$

The linear combination is $(t^3 - t^2 + t)(I) + (t^2 - t + 1)(II) + t^2(III) + t(IV) = 0$. Thus, the VA–polynomial equals $(t - 1)^2(t^2 + 1)$.

These two polynomials are not proportional with invertible coefficient λt^k. Thus, the two connected sums are not equivalent.

Definition 18.6. By an oriented *long virtual knot diagram* we mean an immersion of the oriented line $\mathbb{R}^1$ into $\mathbb{R}^2$ with double crossing points, endowed with crossing structure at each intersection point (classical or virtual). We also require that outside an interval the image coincides with the line Ox where the abscissa increases while walking along the line according to its orientation.

Definition 18.7. By an *oriented long virtual knot* we mean an equivalence class of oriented long virtual knot diagrams modulo Reidemeister moves.

A long virtual knot can be obtained from an ordinary virtual knot by breaking it at a point and taking the free ends to infinity (say, $+\infty$ and $-\infty$ along Ox). It is well known that the theory of classical long knots is isomorphic to that of classical ordinary knots; i.e. the long knot isotopy class does not depend on the choice of the break point.

The two examples shown above demonstrate that there exist two long virtual knots K_1, K_2 shown in Fig. 18.16 and obtained from the same virtual knot K_1 which are not isotopic. This fact was first mentioned in [GPV].

Indeed, if K_1 and K_2 were isotopic, the virtual knot shown in Fig. 18.14 would be isotopic to that shown in Fig. 18.15. The latter claim is however not true. The long knots K_1 and K_2 can be distinguished also by the long quandle invariant, see Section 20.2.

Thus, long virtual knot theory differs from ordinary virtual knot theory.

18.4.1　Properties of the VA–polynomial

Theorem 18.7. *For each virtual knot K the polynomial $VA(K)$ is divisible by $(t - 1)^2$.*

Proof. Let $\bar{K}$ be a virtual knot diagram. Choose a classical crossing V_1 of it. Let X be a long arc outgoing from V_1, and let x be the first arc of X incident to V_1. Denote this arc by a_1. By construction, all arcs belonging to X are

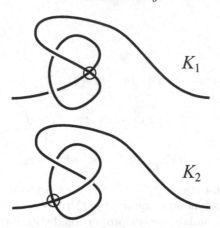

FIGURE 18.16: Two different long virtual knots coming from the same knot

associated with $a_1 + k\varepsilon, k \in \mathbb{N}$. Let the last arc of X be marked by $a_1 + k_1\varepsilon$. Denote the final point of it by V_2. Now, let us take the first arc outgoing from V_2 and associate $a_2 + k_1\varepsilon$ to it. Then, we set the labels $a_2 + k\varepsilon$ for all arcs belonging to the same long arc. Let the last arc have the label $a_2 + k_2\varepsilon$ and have the final point at V_3. Then, we associate $a_3 + k_2\varepsilon$ with the first arc outgoing from V_3, and so on.

Finally, we shall come to V_1. Let us show that the process converges; i.e. the label of the arc coming in V_1 has the label $a_{j+1} + 0 \cdot \varepsilon$, where j is the total number of long arcs.

Indeed, let us see the ε–part of labels while walking along the diagram from V_1 to V_1. In the very beginning, it is equal to zero by construction. Then, while passing through each virtual crossing, it is increased (or decreased) by one. But each virtual crossing is passed twice; thus each $+\varepsilon$ is compensated by $-\varepsilon$ and vice versa. Thus, finally we come to V_1 with $0 \cdot \varepsilon$.

Note that the process converges if we do the same, starting from any arc with an arbitrary integer number as a label.

In this case, each relation of the virtual Alexander module has the right part divisible by $\varepsilon(t-1)$: the relation

$$(a_i + p\varepsilon)t + (a_j + q\varepsilon)(1-t) = (a_k + p\varepsilon)$$

is equivalent to

$$a_i t + a_j(1-t) - a_k = (t-1)(q-p)\varepsilon. \tag{18.12}$$

This proves that $VA(K)$ is divisible by $(t-1)$.

Denote the summands for the i–th vertex in the right part of (18.12) by q_i and p_i, respectively.

Let us seek relations on rows of the virtual Alexander matrix. Each relation holds for arbitrary t, hence for $t = 1$.

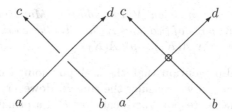

FIGURE 18.17: Labels of $\bar{K}$ and $\bar{K}'$

Denote rows of M by M_i, $i = 1, \ldots, n$. So, if for the matrix M we have $\sum_{i=1}^{n} c_i M_i = 0$ then $\sum_{i=1}^{n} c_i|_{t=1} M_i|_{t=1} = 0$.

The matrix $M(\bar{K})|_{t=1}$ is very simple. Each row of it (as well as each column of it) consists of 1 and -1 and zeros. The relation for rows of this matrix is obvious: one should just take the sum of these rows that is equal to zero. So, $\forall i, j = 1 \ldots, n : c_i|_{t=1} = c_j|_{t=1}$.

Each relation for ε looks like $\left(\sum_{i=1}^{n} c_i(q_i - p_i)\right)(t - 1) = 0$.

Since we are interested in whether this expression is divisible by $(t-1)^2$, we can easily replace c_i with $c_i|_{t=1}$. Thus, it remains to prove that $\sum_{i=1}^{n}(q_i - p_i) = 0$ for the given diagram $\bar{K}$.

Let us prove it by induction on the number n of classical crossings.

For $n = 0$, there is nothing to prove.

Now, let $\bar{K}$ be a diagram with n classical crossings, and $\bar{K}'$ be a diagram obtained from $\bar{K}$ by replacing a classical crossing by a virtual one.

Consider the case of the positive classical crossing X (the "negative" case is completely analogous to this one); see Fig. 18.17.

Denote the lower–left arc of both diagrams by a, and other arcs by b and c, d (for $\bar{K}$ we have $a = d$); see Fig. 18.17. Assign the label 0 to the arc a of both diagrams. Let us calculate $\sum(q_i - p_i)$ for $\bar{K}'$ and $\bar{K}$. By the induction hypothesis, for $\bar{K}'$ this sum equals 0.

Denote the label of b for the first diagram by l_{b1} and that for the second diagram by l_{b2}.

The crossing X of $\bar{K}$ has $q = l_{b1}, p = 0$, thus, its impact is equal to l_{b1}.

The other crossings of $\bar{K}$ (classical or virtual) are in one–to–one correspondence with those of $\bar{K}'$. Let us calculate what the difference is between the p's and q's for these two diagrams. The difference comes from classical crossings. Their labels differ only in the part of the diagram from d to b. While walking from d to b, we encounter classical and virtual crossing. The total algebraic number is equal to zero. The algebraic number of virtual crossings equals $-q$. Thus the algebraic number of classical crossings equals q. Each of them impacts -1 to the difference between $\bar{K}$ and $\bar{K}'$. Thus, we have $q - q = 0$ which completes the induction step and hence, the theorem. $\qquad\square$

The following theorem can be proven straightforwardly.

Theorem 18.8. *The invariant VA is additive. More precisely, for any connected sum $K = K_1 \# K_2$ of two links K_1 and K_2 there exist invertible elements $\lambda, \mu \in R : VA(K) = \lambda VA(K_1) + \mu VA(K_1)$.*

One should also mention that the VA–polynomial can be defined more precisely (up to $\pm t^k$) if we consider the ring R. However, this approach works for knots when the corresponding ideal over R is a principal ideal.

18.5   Multiplicative approach

18.5.1   Introduction

Here we are going to use a construction quite analogous to the previous ones; however, instead of an extra "additive" element ε added to the module we shall add some "multiplicative" elements to the basic ring.

Throughout the section, we deal only with oriented links. In the sequel, we deal only with proper diagrams.

18.6   The two–variable polynomial

Below, we construct two invariant polynomials of virtual links ([Man9], for a short version see [Man11]). In this section, the first one (in two variables) will be constructed. The second one, which will be described in Section 18.7, deals with coloured links: with each n–component coloured link we associate an invariant polynomial in $n + 1$ variables. The first invariant can be easily obtained from the second one by a simple variable change.

Let $\bar{L}$ be a proper diagram of a virtual link L with n classical crossings.

Let us construct an $n \times n$–matrix $M(\bar{L})$ with elements from $\mathbb{Z}[t, t^{-1}, s, s^{-1}]$ as follows.

First, let us enumerate all classical crossings of L by integer numbers from one to n and associate with each crossing the outgoing long arc. Each long arc starts with a (short) arc. Let us associate the label one with this arc. All other arcs of the long arc will be marked by exponents $s^k, k \in \mathbb{Z}$, as follows; see Fig. 18.18. While passing through the virtual crossing, we multiply the label by s if we pass from the left to the right (assuming that the arc we pass by is oriented upwards) or by s^{-1} otherwise.

Since the diagram is proper, our labelling is well defined. Consider a classical crossing V_i with number i. It is incident to some three arcs p, q, r, belonging to long arcs with numbers i, j, k; whence the number j belongs to the arc pass-

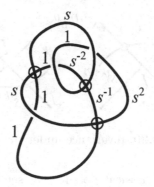

FIGURE 18.18: Symbols s^k on arcs

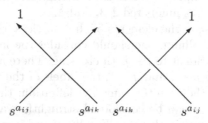

FIGURE 18.19: Arcs incident to classical crossings

ing through V_i. Denote the exponent of s of the label corresponding to q by a_{ij}, and that of the label corresponding to r by a_{ik}; see Fig. 18.19.

Let us define the i–th row of the matrix $M(\bar{L})$ as the sum of the following three rows y_1, y_2, y_3 of length n. Each of these rows has only one non-zero element. The i–th element of the row y_1 is equal to one. If the crossing is positive, we set

$$y_{2k} = -s^{a_{ik}}t, y_{3j} = (t-1)s^{a_{ij}};\qquad(18.13)$$

otherwise we set

$$y_{2k} = -s^{a_{ik}}t^{-1}, y_{3j} = (t^{-1}-1)s^{a_{ij}}.\qquad(18.14)$$

Let $\zeta(\bar{L})$ be equal to $detM(\bar{L})$. Obviously, $\zeta(\bar{L})$ does not depend on the enumeration of rows of the matrix.

Theorem 18.9. *For each two diagrams $\bar{L}$ and $\bar{L}'$ of the same virtual link L we have $\zeta(\bar{L}) = t^l\zeta(\bar{L}')$ for some $l \in \mathbb{Z}$.*

Proof. Note that purely virtual moves do not change the matrix $M(\bar{L})$ and hence ζ do not change.

While applying the semivirtual third Reidemeister move, we multiply one row of the matrix by $s^{\pm1}$ and one column of the matrix by $s^{\mp1}$. Indeed, let

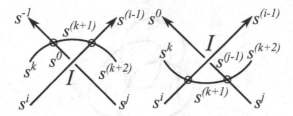

FIGURE 18.20: Invariance under the semivirtual move

I be the number of the classical crossing, the semivirtual move is applied to. Then we have the following two diagrams, L and L'; see Fig. 18.20.

For the sake of simplicity, let us assume that arcs labelled by s^i, s^j and s^k correspond to long arcs numbered 2, 3, and 4.

In order to compare the elements of the matrices $M(L)$ and $M(L')$ in the p–th row and q–th column, one should consider the labels of (short) arcs of the q–th long arc incident to the p–th crossing. There are four different cases. In the simplest case $p \neq 1$ and $q \neq 1$, the labels of the two arcs are the same. If $p = 1$ and $q = 1$ they are the same by definition: they both equal s^0. The first row $p = 1, q \neq 1$ can be considered straightforwardly: we have at most three non-zero elements in this row. Finally, if $q = 1, p \neq 1$, then we deal with the long arc outgoing from the first crossing. Consider the domain D of the semivirtual Reidemeister move, shown in Fig. 18.20. In the case of L, the first long arc leaves this domain with label s^{-1} on some (short) arc, and in the case of L' it leaves this domain with label s^0. So, all further labels of this long arc (e.g. all those containing crossings except the first one) will be different. Namely, the label for L' will be equal to that for L multiplied by s.

This allows us to conclude the following.

The two matrices $M(L)$ and $M(L')$ will both have 1 on the place $(1,1)$ and the same elements except for $(1,p)$ or $(q,1)$ for $p \neq 1$ and $q \neq 1$.

In the case $p \neq 1$ we have $M(L')_{p1} = s \cdot M(L)_{p1}$ and for $q \neq 1$ we have $M(L')_{1q} = s^{-1}M(L)_{1q}$.

Thus, the semivirtual move does not change the determinant either.

Now let us consider classical Reidemeister moves. We begin with the first move Ω_1. Suppose we add a loop dividing some arc labelled by s^i of the first long arc into two long arcs. One of them lies "before" the loop of the move, the other one lies "after" the loop. These parts correspond to some columns in the matrix M. Denote these columns by A and B, respectively.

After performing this move, we obtain one more row and one more column in the matrix. The row will correspond to the added vertex. Let us renumber vertices by $j \to j + 1$, and associate the number one with the added vertex. This row will contain only two non-zero elements on places 1 and 2. Besides this, instead of columns A and B we have columns A and $s^{-i} \cdot B$ because, according to our rules, the part corresponding to B will start not from s^i but from $s^0 = 1$.

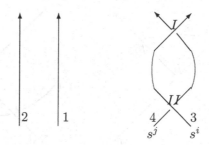

2   1

4   3

s^j s^i

FIGURE 18.21: The second move

So, in each of the four cases of the first Reidemeister move, our transformation will look like this:

$$\left(A + B \quad * \right) \rightarrow \begin{pmatrix} x & s^i y & 0 \\ Bs^{-i} & A & * \end{pmatrix}.$$

Here x and y are some functions depending only on t. Indeed, consider the two arcs of the diagram with curl. If the first arc is incident to the first crossing once, then the element x equals 1, and y equals -1: it will be the sum of two elements $-t$ and $(t-1)$ or $-\frac{1}{t}$ and $(\frac{1}{t}-1)$. If the first arc is incident twice, we may have two different possibilities: $x = t, y = -t$, or $x = \frac{1}{t}, y = -\frac{1}{t}$.

Now, it can be checked straightforwardly that in each of the four cases the determinant will either stay the same or be multiplied by $t^{\pm 1}$.

Let us now consider the move Ω_2. We shall perform all calculations just for the one case shown in Fig. 18.21.

This move adds two new crossings (they are numbered by one and two, and all other numbers are increased by two). Let us look at what happens with the matrix.

Assume that the initial diagram has n crossings. Denote the first two columns of the first matrix by A and $B+C$, where B and C have the following geometric meaning. The column $B + C$ corresponds to the long arc; i.e., to all crossings of the arc. After performing the second Reidemeister move, this long arc breaks at some interval, thus, all its incidences with crossings can be divided into two parts: those before and those after. Accordingly, the column will be decomposed into the sum of two columns which are denoted by B and C.

Thus, the first matrix looks like:

$$(A \quad B + C \quad *).$$

In right part of Fig. 18.21, we have a $(n+2) \times (n+2)$–matrix. The matrix will look like:

$$\begin{pmatrix} 1 & -t & (t-1)s^i & 0 & 0\ldots0 \\ 0 & 1 & (\frac{1}{t}-1)\,s^i & -\frac{s^j}{t} & 0\ldots0 \\ Cs^{-j} & 0 & A & B & * \end{pmatrix}.$$

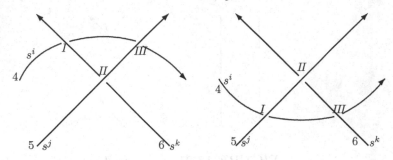

FIGURE 18.22: The third Reidemeister move

We only have to show that the initial and the transformed matrices have equal determinants.

We shall do this in the following way, transforming the second matrix. First, we add the second column multiplied by $s^i \left(\frac{t-1}{t} \right)$ to the third one. Thus, the elements $(1,3)$ and $(2,3)$ vanish. We get:

$$\begin{pmatrix} 1 & -t & 0 & 0 & 0 \ldots 0 \\ 0 & 1 & 0 & -\frac{s^j}{t} & 0 \ldots 0 \\ Cs^{-j} & 0 & A & B & * \end{pmatrix}.$$

Now, let us add the first column multiplied by s^j to the fourth one. We get:

$$\begin{pmatrix} 1 & -t & 0 & s^j & 0 \ldots 0 \\ 0 & 1 & 0 & -\frac{s^j}{t} & 0 \ldots 0 \\ Cs^{-j} & 0 & A & B+C & * \end{pmatrix}.$$

Finally, we add the second column multiplied by $\frac{s^j}{t}$ to the fourth one. We obtain the matrix

$$\begin{pmatrix} 1 & -t & 0 & 0 & 0 \ldots 0 \\ 0 & 1 & 0 & 0 & 0 \ldots 0 \\ Cs^{-j} & 0 & A & B+C & * \end{pmatrix}.$$

The determinant of this matrix obviously coincides with that of the first matrix.

Now, it remains to prove the invariance of ζ under Ω_3. As before, we are going to consider only one case. We shall perform explicit calculation for the case, shown in Fig. 18.22.

The three crossings are marked by Roman numbers I, II, III, so one can uniquely restore the numbers of outgoing arcs and their labels. All the other long arcs are numbered 4, 5, 6 with labels s^i, s^j, s^k at their last arcs.

We have two matrices M_1 and M_2:

$$M_1 = \begin{pmatrix} 1 & \left(\frac{1}{t}-1\right) & 0 & -\frac{s^i}{t} & 0 & 0 & 0\ldots0 \\ 0 & 1 & 0 & 0 & s^j(t-1) & -ts^k & 0\ldots0 \\ -\frac{1}{t} & 0 & 1 & 0 & s^j\left(\frac{1}{t}-1\right) & 0 & 0\ldots0 \\ 0 & 0 & 0 & & & & \\ \vdots & \vdots & \vdots & & * & & \\ 0 & 0 & 0 & & & & \end{pmatrix}$$

$$M_2 = \begin{pmatrix} 1 & 0 & 0 & -\frac{s^i}{t} & s^j\left(\frac{1}{t}-1\right) & 0 & 0\ldots0 \\ 0 & 1 & 0 & 0 & s^j(t-1) & -ts^k & 0\ldots0 \\ -\frac{1}{t} & 0 & 1 & 0 & 0 & s^k\left(\frac{1}{t}-1\right) & 0\ldots0 \\ 0 & 0 & 0 & & & & \\ \vdots & \vdots & \vdots & & * & & \\ 0 & 0 & 0 & & & & \end{pmatrix}.$$

To show that these two matrices have equal determinants, we shall perform the following operations with rows and columns (by using only the first three rows having zeros at positions ≥ 7 and only the six columns (the first of them has zeros at positions ≥ 4).

First, let us transform the first matrix as follows. Add the first column multiplied by $s^j(1-t)$ to the fifth column, and the first column multiplied by $s^k(t-1)$ to the sixth column. We get the matrix

$$M_1' = \begin{pmatrix} 1 & \left(\frac{1}{t}-1\right) & 0 & -\frac{s^i}{t} & s^j(1-t) & s^k(t-1) & 0\ldots0 \\ 0 & 1 & 0 & 0 & s^j(t-1) & -ts^k & 0\ldots0 \\ -\frac{1}{t} & 0 & 1 & 0 & 0 & s^k\left(\frac{1}{t}-1\right) & 0\ldots0 \\ 0 & 0 & 0 & & & & \\ \vdots & \vdots & \vdots & & * & & \\ 0 & 0 & 0 & & & & \end{pmatrix}.$$

We see that all elements of M_1' and M_2, except for those lying in the first row, coincide. Now, it can be easily checked, that if we add the second row of M_2 multiplied by $\left(\frac{1}{t}-1\right)$ to the first row of M_2, we obtain just the first row of M_1'. Thus, $\det M_1 = \det M_1' = \det M_2$, which completes the proof. $\square$

Let m, M be the leading and the lowest exponents of t in monomials of $\zeta(L)$. Define $\xi(L) = t^{-\left(\frac{m+M}{2}\right)}\zeta(L)$. By construction, ξ is a virtual link invariant.

The following properties of ξ hold.

Theorem 18.10. *For a virtual link L isotopic to a classical one, we have* $\xi(L) = 0$.

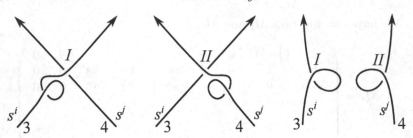

FIGURE 18.23: A modified Conway triple

Proof. For a diagram $\bar{L}$ of L having no virtual crossings, arcs coincide with long arcs; hence all labels equal $s^0 = 1$. Thus, the matrix M has the eigenvector $\underbrace{(1, \ldots, 1)}_{n}$ with zero eigenvalue. Thus $det M(\bar{L}) = 0$. $\square$

Theorem 18.11. *For any Conway triple L_+, L_-, L_0 there exist $p, q \in \mathbb{Z}$ such that $t^p \xi(L_+) - t^q \xi(L_-) = (1 - t)\xi(L_0)$.*

Proof. In view of Theorem 18.9, we can slightly modify the Conway triple by performing the first Reidemeister move, and then check the conditions of the theorem for triples of diagrams locally looking as shown in Fig. 18.23.

Here we enumerate crossings by Roman letters. We also mark by labels and numbers the arcs not starting from selected crossings. Such a "coordinated" enumeration is possible always when all arcs represented in the picture are different. This is the main case when each diagonal element of the matrix is equal to one. In the other case, the labelling shown in Fig.18.23 may not apply; the statement, however remains true.

Denote the matrices corresponding to the three diagrams in Fig. 18.23 by M_+, M_-, M_0. As before, we shall restrict our calculations to small parts of the matrices. In this case, these matrices differ only in rows 1 and 2. In these rows, all elements but those numbered $1, 2, 3$, and 4 are equal to zero. So, we shall perform calculations concerning 2×4 parts of M_+, M_-, M_0.

Initially, these parts are as follows:

$$M_+ \to \begin{pmatrix} 1 & t-1 & 0 & -ts^j \\ 0 & 1 & -s^i & 0 \end{pmatrix};$$

$$M_- \to \begin{pmatrix} 1 & 0 & 0 & -s^j \\ \frac{1}{t}-1 & 1 & -\frac{s^i}{t} & 0 \end{pmatrix};$$

$$M_0 \to \begin{pmatrix} 1 & 0 & -s^i & 0 \\ 0 & 1 & 0 & -s^j \end{pmatrix}.$$

Let us add the second row of the first matrix to the first row of it. For the second matrix, let us multiply the first row by t and then add the second row multiplied by t to the (modified) first row. For the third matrix, let us add the second row multiplied by t to the first row.

After performing these operations, we obtain three matrices M'_+, M'_-, M'_0 with common first rows. The submatrices 2×4 we work with are as follows:

$$M'_+ \to \begin{pmatrix} 1 & t & -s^i & -ts^j \\ 0 & 1 & -s^i & 0 \end{pmatrix};$$

$$M'_- \to \begin{pmatrix} 1 & t & -s^i & -ts^j \\ \frac{1}{t} - 1 & 1 & -\frac{s^i}{t} & 0 \end{pmatrix};$$

$$M'_0 \to \begin{pmatrix} 1 & t & -s^i & -ts^j \\ 0 & 1 & 0 & -s^j \end{pmatrix}.$$

Obviously, $det(M'_+) = det(M_+), det(M'_0) = det(M_0), det(M'_-) = t \cdot det(M_-)$.

Now, for the matrix M'_-, let us add the first row multiplied by $(1 - \frac{1}{t})$, to the second row. We obtain the matrix M'' such that $det M'' = det M'_-$. The matrices M'_+, M'', M'_0 differ only in the second row. Their second rows look like:

$$p = (0, 1, -s^i, 0, 0, \ldots, 0),$$

$$q = (0, t, -s^i, (1 - t)s^j, 0, \ldots, 0),$$

$$r = (0, 1, 0, -s^j, 0, \ldots, 0).$$

Taking into account that $p - q = (1 - t)r$, one obtains the claim of the theorem.

$\square$

The polynomial ξ allows us to distinguish some virtual links that cannot be recognised by the Jones polynomial V introduced in [Kau3] (e.g. the trivial two–component link and the closure of the two–strand virtual braid $\sigma_1 \zeta_1 \sigma_1^{-1} \zeta_1$) and the VA–polynomial (the disconnected sum of the "virtual trefoil" with itself and with the unknot).

18.7 The multivariable polynomial

The multivariable polynomial is constructed quite analogously to the previous one, see [Man9, Man10]. Let L be a k–component link. Let $\bar{L}$ be a proper link diagram with n classical crossings representing L. Let us associate with each component $K_i, i = 1, \ldots, k$, of L the letter s_i. Consider a component $\bar{K}_i$ of $\bar{L}$ and let us mark its arcs by monomials which are products of $s_1, \ldots, s_k, s_1^{-1}, \ldots, s_k^{-1}$. As above, each long arc of K_i starts with a (short) arc. Let us associate the label 1 with the latter. All other arcs of the long arc will be marked by monomials as follows. While passing through the virtual

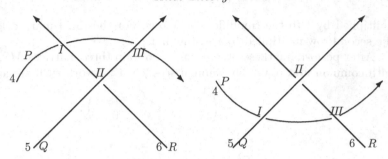

FIGURE 18.24: The third move in the multivariable case

crossing with j–th component we multiply the label by s_j if we pass from the left to the right or by s_j^{-1} otherwise.

As before, we construct the $n \times n$ matrix according to the same rule. In this case, elements of the matrix belong to $\mathbb{Z}[t, t^{-1}, s_1, \ldots, s_k, s_1^{-1}, \ldots, s_k^{-1}]$.

The matrix M will be constructed just as in the case of the 2–variable polynomial. To define it, we just modify formulae (18.13) and (18.14) by replacing exponents of s by monomials in s_i. Let us define $\chi(\bar{L}) = det M(\bar{L})$. Obviously, $\chi(\bar{L})$ does not depend on the enumeration of rows of the matrix.

Theorem 18.12. *The polynomial χ is invariant under all generalised Reidemeister moves but the first classical one. The first classical Reidemeister move either does not change the value of χ or multiplies it by $t^{\pm 1}$.*

Proof. First, it is evident that purely virtual moves do not change the matrix at all.

Let us consider the case of the semivirtual move shown in Fig. 18.20. Let m be the number of the component that takes part in the semivirtual move and has two virtual crossings with the other components.

Thus, after applying the semivirtual move, the first row of the matrix is multiplied by $s_m^{\pm 1}$, and the first column is multiplied by $s_m^{\mp 1}$. The proof of this fact is quite analogous to that in the case of Theorem 18.9. One should just look at Fig. 18.20 and consider the labels shown in it. In the multivariable case, the arbitrary arcs will have labels P, Q, R instead of s^i, s^j, s^k, where all P, Q, R are some monomials in $s_1, s_2 \ldots, s_k$.

The remaining part of the proof (of the invariance under classical Reidemeister moves) repeats that of Theorem 18.9. One should just replace arbitrary powers of s by some monomials T_p in many variables s_i.

Here we consider only the most interesting case; i.e., the third Reidemeister move. Let us consider the move shown in Fig. 18.24.

In this case, we have the following two matrices:

$$\begin{pmatrix} 1 & \left(\frac{1}{t}-1\right) & 0 & -\frac{P}{t} & 0 & 0 & 0\ldots0 \\ 0 & 1 & 0 & 0 & Q(t-1) & -Rt & 0\ldots0 \\ -\frac{1}{t} & 0 & 1 & 0 & Q\left(\frac{1}{t}-1\right) & 0 & 0\ldots0 \\ 0 & 0 & 0 & & & & \\ \vdots & \vdots & \vdots & & * & & \\ 0 & 0 & 0 & & & & \end{pmatrix}$$

and

$$\begin{pmatrix} 1 & 0 & 0 & -\frac{P}{t} & Q\left(\frac{1}{t}-1\right) & 0 & 0\ldots0 \\ 0 & 1 & 0 & 0 & Q(t-1) & -Rt & 0\ldots0 \\ -\frac{1}{t} & 0 & 1 & 0 & 0 & R\left(\frac{1}{t}-1\right) & 0\ldots0 \\ 0 & 0 & 0 & & & & \\ \vdots & \vdots & \vdots & & * & & \\ 0 & 0 & 0 & & & & \end{pmatrix}.$$

Consider the first matrix. Adding the first column multiplied by $Q(1-t)$ to the fifth one, and the first column multiplied by $R(t-1)$ to the sixth one, we get

$$\begin{pmatrix} 1 & \left(\frac{1}{t}-1\right) & 0 & -\frac{P}{t} & Q(1-t) & R(t-1) & 0\ldots0 \\ 0 & 1 & 0 & 0 & Q(t-1) & -Rt & 0\ldots0 \\ -\frac{1}{t} & 0 & 1 & 0 & 0 & R\left(\frac{1}{t}-1\right) & 0\ldots0 \\ 0 & 0 & 0 & & & & \\ \vdots & \vdots & \vdots & & * & & \\ 0 & 0 & 0 & & & & \end{pmatrix}.$$

The same matrix can be obtained if we replace the first row of the second matrix with the sum of the first row and the second row of the second matrix by $\left(\frac{1}{t}-1\right)$ to the first row of it. □

One can also prove the analogue of Theorem 18.10 for the polynomial χ. Also, the normalization for χ can be done in the same manner as that for ζ. Namely, let m, M be the leading and the lowest powers of t in monomials of $\chi(L)$. Define $\eta(L) = t^{-\left(\frac{m+M}{2}\right)}\chi(L)$. By construction, η is a virtual link invariant.

Theorem 18.13. *For a virtual link L isotopic to a classical one, we have $\eta(L) = 0$.*

□

The following statement follows from the construction.

Statement 18.1. *For any k-component link L, we have*

$$\eta(L)|_{s=s_1=\cdots=s_k} = \xi(L).$$

Knot Theory

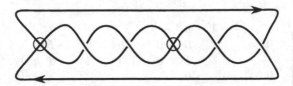

FIGURE 18.25: A link for which $\xi(L) = 0$.

This shows that $\eta(L)$ is at least as strong as ξ. In fact, it is even stronger. Consider the link L shown in Fig. 18.25.

It is not difficult to calculate that for this link L, the polynomial $\eta(L)$ is divisible by $(s_2 - s_1)$ and is not equal to zero. Thus, $\xi(L) = 0$, so η is strictly stronger than the invariant ξ.

Chapter 19

Generalised Jones–Kauffman polynomial

In the present chapter, we are going to give a generalisation of the Jones–Kauffman polynomial for virtual knots by adding some "extra information" to it, namely, some objects connected with curves in 2–surfaces (for a short version see [Man8]). In the second part of the chapter, we are going to consider the minimality aspects in virtual knot theory and give a proof of the generalised Murasugi theorem (short version in [Man12]).

Note that unlike all invariants constructed before and valued in rings of (Laurent) polynomials, the invariant constructed here is valued in *pictures*, thus containing a substantial amount of information about the diagram of the initial knot.

This picture-valued invariant due to the author [Man8] was one of the first incidences of picture-valued invariants of virtual knots.

This theme was later developed by using parities [Man22, Man30, MN, IMN3], Kuperberg brackets [KM2, KM3].

By just looking at a polynomial invariant we can make some numerical estimates of the complexity of diagrams: crossing number, virtual crossing number etc. As for the *shape* of the diagram, we can judge about it quite implicitly.

However, when dealing with picture-valued invariants, we can make conjectures of the form:whatever the diagram D of a given knot K is, it it will contain a subdiagram of a certain shape.

This immediately results in various consequences about many previously known invariants. In particular, by using picture-valued invariants the author has proved that virtual crossing number of virtual knots may have quadratic growth with respect to classical crossing number.

19.1   Introduction. Basic definitions

Virtual equivalence and classical equivalence for classical knots coincide [GPV] and the set of all classical knots is a subset of the set of all virtual knots. Thus, each invariant of virtual links generates some invariant of classical links.

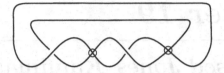

FIGURE 19.1: A virtual link with trivial Jones–Kauffman polynomial

In the previous chapter, we described some virtual link polynomials vanishing on classical links.

Below, we shall construct an invariant polynomial of virtual links that equals the classical Jones–Kauffman polynomial on classical links.

Let us first recall how one defines the classical Jones–Kauffman polynomial [Kau3] for the case of virtual links. Let L be an oriented virtual link diagram with n classical crossings. Denote by $|L|$ the diagram obtained from L by "forgetting" the orientation.

Just as in the classical case, for the non–oriented virtual link diagram $|L|$, one can "smooth" each classical crossing of $|L|$ in two possible ways, called $A : \times \to)($ and $B : \times \to \asymp$.

After such a smoothing of all classical crossings, one obtains a non–oriented diagram that does not contain classical crossings. Hence, this diagram generates the trivial virtual link.

Recall that a *state* of $|L|$ is a choice of smoothing type for each classical crossing of $|L|$. Thus, $|L|$ has 2^n states. Each state s has the following three important characteristics: the number $\alpha(s)$ of smoothings of type A, the number $\beta(s) = n - \alpha(s)$ of smoothings of type B, and the number $\gamma(s)$ of link components of the smoothed diagram.

The Jones–Kauffman polynomial for virtual links [Kau3] is given by

$$X(L) = (-a)^{-3w(L)} \sum_{s} a^{\alpha(s)-\beta(s)} (-a^2 - a^{-2})^{\gamma(s)-1}. \qquad (19.1)$$

Here the sum is taken over all states of $|L|$; $w(L)$ is the writhe number of L.

In [Kau3], Kauffman shows the invariance of X under generalised Reidemeister moves. However, he indicates a significant disadvantage of X: this polynomial is invariant under the virtualisation move shown in Fig. 18.10 which is not an isotopy.

Thus, the Jones–Kauffman polynomial does not distinguish the trivial two–component virtual link and the virtual link Λ shown in Fig. 19.1.

Let us construct now a modification of the Jones–Kauffman invariant. Let S be the set of all pairs (M, γ) where M is a smooth orientable surface without boundary (possibly, not connected) and γ is an unordered finite system of unoriented closed curves immersed in M.

Let us define the equivalence on S by means of the following elementary equivalences:

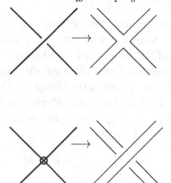

FIGURE 19.2: Local structure of the surface M'

1. Two pairs $(M, \gamma), (M', \gamma')$ are equivalent if there exists a homeomorphism $M \to M'$ identifying γ with γ'.

2. For a fixed manifold M, if the set γ is homotopic to the set γ' in M then the pairs (M, γ) and (M, γ') are equivalent.

3. Two pairs (M, γ) and (M, γ') are said to be equivalent if γ' is obtained from γ by adding a curve bounding a disk and not intersecting all other curves from γ'.

4. Pairs (M, γ) and (N, γ) should be equivalent, if N is a manifold obtained from M by cutting two disks not intersecting the curves from γ, and attaching a handle to boundaries of these disks.

5. Finally, for any closed compact orientable 2–manifold N, pairs (M, γ) and $(M \sqcup N, \gamma)$ are equivalent.

Here $\sqcup$ means the disjoint sum of M (with all curves of γ lying in it) and N without curves.

Denote the set of equivalence classes on S by $\mathfrak{S}$. There are several algorithms to distinguish elements of this set; the first follows from B. L Reinhart's work [Rein].

The basic idea of this invariant is the construction of a $\mathfrak{S}\mathbb{Z}[a, a^{-1}]$–valued invariant function on the set of virtual links; values of this function should be linear combinations of elements from $\mathfrak{S}$ with coefficients from $\mathbb{Z}[a, a^{-1}]$.

Let L be a virtual link diagram. Let us construct a 2–manifold M' as follows. At each classical crossing of the diagram we draw a cross (the upper picture of Fig. 19.2), and at each virtual crossing we set two non-intersecting bands (the lower picture). Connecting these crosses and bands by bands going along link arcs, we obtain a 2–manifold with boundary. This manifold is obviously orientable.

One can naturally project the diagram of L to M' in such a way that arcs of the diagram are projected to middle lines of bands; herewith classical crossings generate crossings in "crosses". Thus, we obtain a set of curves $\gamma \subset M'$.

Attaching discs to boundary components of M', one obtains an orientable manifold $M = M(L)$ together with the set γ of circles immersed in it.

Now, each state of the diagram L can be considered directly on M because to each local neighbourhood of a classical crossing of L, there corresponds an intersection point of one or two curves from γ. Thus, to each state s of L there corresponds the set $\Gamma(s)$ of "smoothed" curves in M. The manifold M with all curves belonging to $\gamma \sqcup \Gamma(s)$ generates some element of $p(s) \in \mathfrak{S}$.

Now, let us define $\Xi(L)$ as follows.

$$\Xi(L) = (-a)^{-3w(L)} \sum_s p(s) a^{\alpha(s) - \beta(s)} (-a^2 - a^{-2})^{\gamma(s) - 1}. \qquad (19.2)$$

Theorem 19.1. *The function $\Xi(L)$ is invariant under generalised Reidemeister moves; hence, it is a virtual link invariant.*

Proof. It is obvious that purely virtual Reidemeister moves and the semivirtual move applied to L do not change $\Xi(L)$ at all: by construction, all terms of (19.2) stay the same.

The proof of the invariance of $\Xi(L)$ under the first and the third classical Reidemeister moves is quite analogous to the same procedure for the classical Jones–Kauffman polynomial; one should accurately check that the corresponding elements of $\mathfrak{S}$ coincide.

In fact, if L and L' are two diagrams obtained one from the other by some first or third Reidemeister move, then for the diagrams $|L|$ and $|L'|$, the corresponding surfaces M' are homeomorphic, and the behaviour of the system of curves γ for $M(L)$ and $M(L')$ differs only inside the small domain where the Reidemeister moves take place.

For the first move, the two situations (corresponding to the twisted curls with local writhe number $+1$ or -1) are considered quite analogously. Let L be a diagram and L' be the diagram obtained from L by adding such a curl. To each state s of $|L|$ there naturally corresponds two states of $|L'|$. Fix one of them and denote it by s'. Let $L \sqcup \bigcirc$ be the disconnected sum of L and a small circle. Then we have:

$$p(s) = p(s').$$

Indeed, both surfaces for $|L|$ and $|L'|$ are the same and the only possible difference between corresponding curve systems is one added circle (elementary equivalence No. 3, see page 357). So, we have to compare terms with the same coefficients from $\mathfrak{S}$. The comparison procedure coincides with that for the classical Jones–Kauffman polynomial.

Now, if we consider two diagrams L and L' obtained one from the other by using the third Reidemeister move, we see again that their surfaces M coincide. Let us select the three vertices P, Q, R of the diagram L and the corresponding vertices P', Q', R' of the diagram L', as shown in the upper part of Fig. 19.3.

So, the diagrams L, L' differ only inside a small disc D in the plane. The

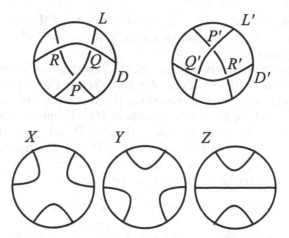

FIGURE 19.3: Diagrams and lines after smoothings

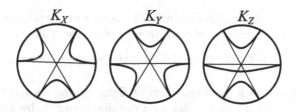

FIGURE 19.4: Parts of diagrams K_X, K_Y, K_Z

same can be said about the system of curves corresponding to some states of them: they differ only inside a small disc D_M in M. Thus, one can indicate six points on the boundary ∂D such that all diagrams of smoothings (in M) of L, L' pass through these and only these points of ∂D.

Consider the three possibilities X, Y, Z of connecting these points shown in the lower part of Fig. 19.3. In fact, there are other possibilities to do it but only these will play a significant role in the future calculations.

Let $|L_X|, |L_Y|, |L_Z|$ be the three planar diagrams of unoriented links coinciding with $|L|$ outside D and coinciding with X, Y, and Z inside D, respectively.

We shall need the following three elements from $\mathfrak{S}$ represented by K_X, K_Y, and K_Z; see Fig. 19.4. The element K_X contains the three lines of the third Reidemeister move (with fixed six endpoints) inside D_M. It also contains X. Analogously, K_Y contains the three lines and Y, and K_Z contains the three lines and Z. The only thing we need to know about the behaviours of K_X, K_Y, and K_Z outside D_M is that they coincide.

We have to prove that $\Xi(L) = \Xi(L')$. Obviously, we have $w(L) = w(L')$. So, we have to compare the terms of (19.2) for $|L|$ and $|L'|$. With each state of L, one can naturally associate a state for L'. For each state of $|L|$ having the crossing P in position A, the corresponding state of $|L'|$ gives just the same contribution to (19.2) as $|L|$ since diagrams $|L|$ and $|L'|$ after smoothing P in position A coincide.

So, we have to compare all terms of (19.2) corresponding to the smoothing of P in position B. We shall combine these terms (for $|L|$ and $|L'|$) in fours that differ only in the way of smoothing the vertices R and S. Now, let us fix the way of smoothing for A and A' outside D and compare the corresponding four terms. If we delete the interior of the disc D and insert there X, Y, or Z, we obtain a system of curves in the plane. Denote the numbers of curves in these three systems by ν_X, ν_Y, and ν_Z, respectively.

Now, the four terms for $|L|$ give us the following:

$$aK_X(-a^2-a^{-2})^{(\nu_X-1)} + a^{-1}(K_Z(-a^2-a^{-2})^{(\nu_Z-1)}$$

$$+K_X(-a^2-a^{-2})^{\nu_X}) + a^{-3}K_X(-a^2-a^{-2})^{(\nu_X-1)} =$$

$$= a^{-1}(-a^2-a^{-2})^{(\nu_Z-1)}K_Z.$$

Analogously, for $|L'|$ we have a similar formula with terms containing K_Z and K_Y. The latter terms are reduced, so we obtain the same expression:

$$a^{-1}(-a^2-a^{-2})^{(\nu_Z-1)}K_Z.$$

Let us now check the invariance of Ξ under the second classical Reidemeister move. Let L' be the diagram obtained from L by applying the second classical Reidemeister move adding two classical crossings. Obviously, $w(L) = w(L')$.

Consider the manifold $M(L)$. The image of L divides it into connected components. We have two possibilities. In one of them, the Reidemeister move is applied to one and the same connected component. Then $M(L')$ is homeomorphic to $M(L)$, and curves from the set γ get two more crossings. In this case the proof of the equality $\Xi(L) = \Xi(L')$ is just the same as in the classical case (the reduction here treats not polynomials but elements from $\mathfrak{G}$ with polynomial coefficients). Moreover, the proof is even simpler than that for the third move: we have to consider the sum of four summands for $|L|$ and $|L'|$. In each case, three of them vanish, and the remaining ones (one for $|L|$ and one for $|L'|$) coincide. Taking into consideration that $w(L) = w(L')$, we get the desired result.

Finally, let us consider the case of the second Reidemeister move, where $M(L')$ is obtained from $M(L)$ by adding a handle. On this handle, two extra points P and Q appear; see Fig. 19.5.

Consider all states of the diagram $|L'|$. They can be split into four types depending on smoothing types of the crossings P and Q. Thus, each state s of $|L|$ generates four states s_{++}, s_{--}, s_{-+}, and s_{+-} of $|L'|$. Note that $p(s) = p(s_{+-})$ (this follows from handle removal; see Fig. 19.5), and $p(s_{++}) = p(s_{--}) = p(s_{-+})$.

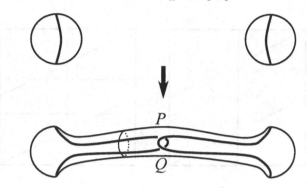

FIGURE 19.5: Adding a handle while performing Ω_2

Besides, for each s, we have the following equalities:

$$\alpha(s) - \beta(s) = \alpha(s_{+-}) - \beta(s_{+-}), \gamma(s) = \gamma(s_{+-}),$$

$$\gamma(s_{++}) = \gamma(s_{--}) = \gamma(s_{-+}) - 1.$$

Thus, all terms of (19.2) for L' corresponding to s_{--}, s_{++}, and s_{-+} will be reduced because of the identity $a^2 + a^{-2} + (-a^2 - a^{-2}) = 0$. The terms corresponding to s_{+-} give just the same as (19.2) for L. $\square$

We can mention many other "chnancements" of Jones–Kauffman polynomial: Miyazawa [Miy], magnetic [IKK], arrow polynomial [DK2, DK3], etc.

19.2 An example

Let $P \in \mathfrak{S}$ be the element represented by the sphere without curves. It is obvious that for each classical link L, $\Xi(L) = P \cdot V(L)$. So, for the two–component unlink L we have $\Xi(L) = P \cdot (-a^2 - a^{-2})$.

It is known that the two–component unlink L and the closure Λ shown in Fig. 19.1 have the same Jones polynomial.

Consider the following two elements from $\mathfrak{S}$ (for the sake of simplicity, we shall draw the elements of S); see Fig. 19.6. Here we consider the torus as the square with identified opposite sides.

The element $Q \in \mathfrak{S}$ is initially represented by the same diagram shown in Fig. 19.6 with two additional circles that can be removed by equivalence No. 3, see page 357.

Let us show that $Q \neq P, R \neq P$ and $Q \neq R$ in $\mathfrak{S}$. Actually, $Q \neq P$ because Q has two curves with non-zero intersection ($+2$ or -2 according to

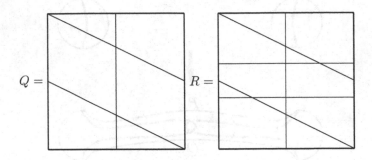

FIGURE 19.6: Two elements from S

the orientation); thus, none of these curves can be removed by the equivalences described above. So, $R \neq P$ either. Besides, $R \neq Q$ because R contains
three different curves on the torus (in coordinates from Fig. 19.6 they are
$(0,1)$, $(1,0)$, and $(2,1)$); each two of them have a non-zero intersection. Thus,
none of them can be removed. So, the simplest diagram of $[R]$ in S cannot
have less than three curves.

Now, for the link Λ, we have

$$\Xi(\Lambda) = Qa^2 + 2R(-a^2 - a^{-2}) + Qa^{-2} = (2R - Q)(-a^2 - a^{-2}).$$

Thus, $\Xi(\Lambda) \neq \Xi(L)$.

19.3   Atoms and virtual knots. Minimality problems

In this section, we shall not distinguish virtual diagrams that can be obtained from each other by using only purely virtual and semivirtual moves.
Such diagrams are called *strongly equivalent*; they have the same Gauss diagram. Furthermore, equivalent diagrams are thought to be different if in order
to show their equivalence one needs some classical Reidemeister move. In this
sense, a virtual knot can be completely generated by the setup of its classical
crossings and lines connecting them (in the one-component case this means
that the Gauss diagrams coincide).

Like classical links, virtual links may or may not be oriented. The *complexity* of a virtual diagram is the number of its classical crossings. Obviously,
diagrams of complexity zero generate unlinks. We are interested in the minimality (in the sense of the absence of diagrams with smaller complexity) of
diagrams realizing the given link.

In Chapter 7 we formulated the Kauffman-Murasugi theorem that was proved in Chapter 16. We are going to generalise this result for the case of virtual links. The notions of primitivity and splitting point are well known in the classical case (see Definition 2.10 in page 22 and page 89). Their virtual analogues will be defined later.

In order to prove Theorem 7.5, K. Murasugi used some properties of the Jones polynomial. Let K be a virtual diagram and let X be a crossing of K. Now, $\rho_X(K)$ is the diagram where the small neighbourhood of the crossing X is transformed as shown in Fig. 18.10 (this transformation is called the *twist move*). It is easy to see that if we apply the twist move twice to the same crossing, we get the initial diagram (in the sense that the obtained diagram is strongly equivalent to the initial one). Denote the set of all diagrams obtained from K by arbitrary twists ρ, by $[K]$.

Definition 19.1. By a *splitting point* of a virtual diagram L we mean a classical crossing X of it such that for any diagram L' strongly equivalent to L, the removal of the small neighbourhood of the corresponding crossing X' divides the diagram.

Definition 19.2. A virtual diagram L is said to be *non-primitive* if for some diagram L' strongly equivalent to it, there exists a closed simple curve separating some non-empty set of classical crossings of L' from the remaining non-empty set of classical crossings, and intersecting the diagram L' precisely in two points. In this case the virtual diagram L' can be represented as a *connected sum of two* virtual diagrams.

Analogously, a virtual diagram L is called *disconnected* if there exists a diagram L' that is strongly equivalent to L such that L' can be divided into two parts $L'_1 \sqcup L'_2$ such that L'_1 and L'_2 lie inside two open non-intersecting sets on the plane. All diagrams we shall deal with, are thought to be connected.

There are exactly two (up to combinatorial equivalence) ways for embedding a primitive diagram of a classical link into the sphere with respect to the opposite outgoing edge structure. These two embeddings coincide up to the orientation of the sphere.

A virtual link L' is called *quasi-alternating* if there exists a classical alternating diagram L such that $L' \in [L]$.

One of the main results of the present chapter is the following

Theorem 19.2. *Any quasi-alternating diagram without splitting points is minimal.*

An analogue of the minimality theorem was proved for the case of knots in $\mathbb{R}P^3$, see [Dro].

Let us first prove the analogue of the Murasugi theorem (on the Jones polynomial) for the case of virtual links. In the proof, we use the techniques proposed in [Man'2] (which differ from the original Murasugi techniques, [Mur1]).

Theorem 19.3. *Let L be a connected virtual diagram of complexity n. Then*

$span(V(L)) \leq n$. Moreover, the equality $span(V(L)) = n$ holds only for virtual diagrams representing a connected sum of some quasi-alternating diagrams without splitting points.

Consider formula (19.1). We are interested in the states that give the maximal and the minimal possible degree of monomials in the sum (19.1). It is easy to check the fact that the maximal state gives the maximal possible degree, and the minimal state gives the minimal possible degree.

In order to estimate these degrees, we shall need the notions of atom and d–diagram. Recall that an atom is a two–dimensional connected closed manifold without boundary together with an embedded graph of valency four (frame) that divides the manifold into cells that admit a chessboard colouring. Atoms are considered up to the natural equivalence. An atom (more precisely, its equivalence class) can be completely restored from the following combinatorial structure: the frame (four–valent graph), the A–structure (dividing the outgoing half–edges into two pairs according to their disposition on the surface), and the B–structure (for each vertex, we indicate two pairs of adjacent half–edges that constitute a part of the boundary of black cells). A height (vertical) atom (see Definition 16.2) is an atom whose frame is embeddable in $\mathbb{R}^2$ with respect to the A–structure. Note that the Turaev genus (i.e. atom genus) of a height atom can be greater than 0.

In the case of an arbitrary atom one should replace embeddings by regular immersions. There might be many immersions for a given frame. The point is that having some A–structure $(1, 3), (2, 4)$ at some vertex, there can be two different dispositions of this order on the place: 1, 2, 3, 4 or 1, 4, 3, 2 (counterclockwise).

Obviously, the obtained knot diagrams are defined up to strong equivalence and (possibly) twist moves (see Fig. 18.10) at some classical crossings.

First, let us consider the case of primitive diagrams. Consider the maximal and the minimal states of the diagram $\bar{L}$. Let us define the numbers of link components corresponding to them by γ_{max} and γ_{min}, respectively. Thus, the length of the Kauffman bracket $\langle \bar{L} \rangle$ is going to be $l = 2n + 2(\gamma_{max} + \gamma_{min} - 2)$. The diagram $\bar{L}$ has intrinsic A–structure of some atoms: having it, one can construct an atom with γ_{max} black cells, γ_{min} white cells thinking of the circles of minimal and maximal states to be boundaries of the 2–cells to be attached. The Euler characteristic of the constructed manifold equals $n - 2n + \gamma_{min} + \gamma_{max}$. Since it does not exceed two, we see that $l \leq 4n$.

The equality can take place only in the case when the obtained surface is a sphere. In this case the B–structure of $\bar{L}$ corresponds to some planar atom.

These structures correspond to an embedding of the frame. These structures correspond to embeddings of the frame in S^2 with respect to the A–structure. There are only two such embeddings; they correspond to alternating diagrams. Thus, the diagram $\bar{L}$ has one of these two B–structures. So, it can differ from an alternating diagram only by virtualisations and hence, it is quasi-alternating.

If the diagram is not primitive, it is sufficient to decompose it into a connected sum and apply the multiplicativity of the Jones polynomial.

Thus we have proved Theorem 19.3. Theorem 19.2 is just a simple corollary of it.

Chapter 20

Long virtual knots and their invariants

20.1   Introduction

Long virtual knots and their invariants first appeared in [GPV]. The present chapter consists of the author's results. The two main arguments that can be taken into account in the theory of "long" virtual knots and could not be used before, are the following:

1. One can indicate the initial and the final *arcs* (which are not compact) of the quandle; the elements corresponding to them are invariant under generalised Reidemeister moves.

2. One can take two different quandle–like structures of the same type at vertices depending on which arc is "before" and which is "after" according to the orientation of a long knot.

As shown in the previous chapter, the procedure of breaking a virtual knot is not well defined: breaking the same knot diagram at different points, we obtain different long knots. Moreover, a "virtual" unknot diagram broken at some point can generate a non-trivial long knot diagram. The aim of this chapter is to construct invariants of long virtual knots that feel "the breaking point".

Remark 20.1. *Throughout the chapter, we deal only with long virtual knots, not links.*

Remark 20.2. *We shall never indicate the orientation of the long knot, assuming it to be oriented from the left to the right.*

Throughout this section, R will denote the field of rational functions in one (real) variable t: $R = \mathbb{Q}(t)$.

Let us recall the definitions of virtual long link.

Definition 20.1. By a *long virtual knot diagram* we mean a smooth immersion f of the oriented line $L_x, x \in (-\infty, +\infty)$ in $\mathbb{R}^2$, such that:

1. outside some big circle, we have $f(t) = (t, 0)$;

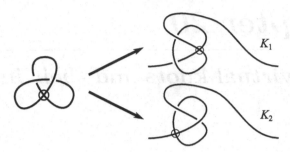

FIGURE 20.1: Transforming a diagram into a long diagram

2. each intersection point is double and transverse;

3. each intersection point is endowed with a classical or virtual crossing structure.

Definition 20.2. A *long virtual knot* is an equivalence class of long virtual knot diagrams modulo generalised Reidemeister moves.

Long virtual knots admit a well–defined concatenation operation: for $K_1 \# K_2$ we just put a diagram of K_2 after a diagram of K_1.

Thus, we can define the *semigroup* $\mathfrak{W}$ of virtual knots where the long unknot plays the role of the unit element.

An *arc* and a *long arc* of a long virtual knot diagram are just the same as in the ordinary case.

Obviously, having a virtual knot diagram, we can break it at some "interior" point in order to get a long virtual knot diagram; see Fig. 20.1.

It is known that in the classical case the result (i.e. the isotopy class of the obtained long knot) does not depend on the choice of the break point.

We shall give one more proof that in the virtual case this is not so. We are going to present an invariant of long virtual knots by using the ideas of the previous paragraph.

20.2 The long quandle

Definition 20.3. A *long quandle* is a set Q equipped with two binary operations $\circ$ and $*$ and one unary operation $f(\cdot)$ such that $(Q, \circ, f)$ is a virtual quandle and $(Q, *, f)$ is a virtual quandle and the following two relations hold:

$$\forall a, b, c \in Q : (a \circ b) * c = (a * c) \circ (b * c),$$
$$\forall a, bc \in Q : (a * b) \circ c = (a \circ c) * (b \circ c)$$

(new distributivity relations) and

$$\forall x, a, b \in Q : x\alpha(a \circ b) = x\alpha(a * b)$$

$$\forall x, a, b \in Q : x\beta(a/b) = x\beta(a//b),$$

where α and β are some operations from the list $\circ, *, /, //$.

The inverse operation for $\circ$ is $/$ and the inverse operation for $*$ is $//$.

Remark 20.3. *It might seem that the last two relations hold only in the case when $\circ$ coincides with $*$. However, the equation $(a \circ b) = c$ has the only solution in a, not in b!*

Consider a diagram $\bar{K}$ of a virtual long knot and arcs of it. Let us fix the initial arc a and the final arc b.

Now, we construct the long quandle of it by the following rule. First, we take all arcs of it including a and b and consider the *free long quandle*, just by using formal operations $\circ, *, /, //, f$ factorised only by the quandle relations (together with the new relations).

After this, we factorise by relations at crossings. At each virtual crossing, we do just the same as in the case of a virtual quandle. At each classical crossing we write the relation either with $\circ$ or with $*$, namely, if the overcrossing is passed **before** the undercrossing (with respect to the orientation of the knot) then we use the operation $\circ$ (respectively, $/$); otherwise we use $*$ (respectively, $//$).

After this factorisation, we obtain an algebraic object M equipped with the five operations $\circ, /, *, //$, and f and two selected elements a and b.

Definition 20.4. Denote the obtained object by $Q_L(\bar{K})$.

Call $Q_L(\bar{K})$ *the long quandle of K*.

Obviously, for the *long unknot U* (represented by a line without crossings) we have for $a, b \in Q_L(U) : a = b$.

Theorem 20.1. *The quandle Q_L together with selected elements a, b is invariant with respect to generalised Reidemeister moves.*

Proof. The proof is quite analogous to the invariance proof of the virtual quandle. Thus, the details will be sketched. The invariance under purely virtual moves and the semivirtual move goes as in the classical case: we deal only with f and one of the operations $*$ or $\circ$. Only one of $*, \circ$ appears when applying the first or the second classical Reidemeister move.

So, the most interesting case is the third classical Reidemeister move. In fact, it is sufficient to consider the following four cases shown in Fig. 20.2 (a,b,c,d).

In each of the four cases everything the relations hold for p and q (p does not change and q is operated on by p in the same manner on the right hand and on the left hand). So, one should only check the transformation for r.

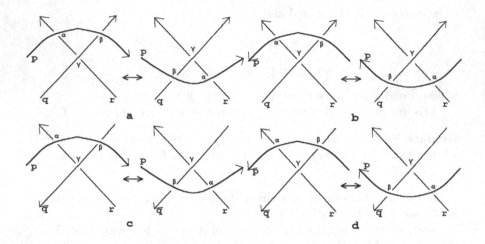

FIGURE 20.2: Checking the move Ω_3

In each picture, at each crossing we put some operation α, β or γ. This means one of the operations $\circ, *, /, //$ (that will be applied to the arc below to obtain the corresponding arc above).

Consider the case a. We have: each α, β, γ is a multiplication $\circ$ or $*$.

Thus, at the upper left corner we shall have: $(r\gamma q)\alpha p$ in the left picture and $(r\alpha p)\gamma(q\beta p)$. But, by definition, $(r\gamma q)\alpha p = (r\alpha p)\gamma(q\alpha p)$. The latter expression equals $(r\alpha p)\gamma(q\beta p)$ according to the "new relation" (because both β and α are multiplications).

Now, let us turn to the case b. Here γ is multiplications and α, β are divisions. Thus, the same equality holds: $(r\gamma q)\alpha p = (r\alpha p)\gamma(q\alpha p) = (r\alpha p)\gamma(q\beta p)$.

The same equation is true for the cases shown in pictures c and d: the only important thing is that α and β are either both multiplications (as in the case c) or both divisions (as in the case d). The remaining part of the statement follows straightforwardly.

$\square$

As an example of a long quandle (see [Man15]) one can consider the ring $\mathbb{Z}_m$ and operations in it:

$$\begin{cases} a \circ b = pa + (1-p)b, \\ a * b = qa + (1-q)b, \\ f(a) = ka, \end{cases} \tag{20.1}$$

where k, p, q are invertible elements in the ring, and $(1-p)(p-q) = (1-q)(p-q) = 0$. The axioms of a quandle are checked straightforwardly.

Let us call long quandles of such type *linear long quandles*. As it turns out one can recognize non-triviality of some long knots having the trivial closure with the help of linear long quandles.

Let $\mathcal{R}$ be a ring with a unit, and p and q be two fixed invertible elements satisfying the equation $(p-1)(q-1) = (q-1)(p-q) = 0$. Let k be an invertible element also. For a long virtual knot K, denote by $\widetilde{M}(K)$ the module over $\mathcal{R}$ generated in the way described above (the generators are arcs, the relations at crossings are (20.1)) with two distinguished elements corresponding to the initial and final arcs.

20.3 Colouring invariant

Let us consider one example: the colouring function. Namely, let $Q_L(K, a_1, a_2)$ be the long virtual quandle of the long K, with operations $f, \circ, *, /, //$, where a_1 and a_2 are the elements of Q corresponding to the initial and the final arc, respectively.

Let G be a finite virtual quandle. Let g_1, g_2 be two elements of G. Then the following theorem holds.

Theorem 20.2. *The number of homomorphisms from $Q(K)$ to G such that $Q(g_1) = a_1$ and $Q(g_2) = a_2$ is finite; besides, it is an invariant of the long knot K.*

The proof of this theorem is obvious. However, it allows us to emphasise the following effect: for long links, each finite virtual quandle G generates not only one colouring function, but a *matrix of colouring functions*. Namely, we enumerate elements of G by integers $1, \ldots, n$ and set M_{ij} to be the total number of proper colourings such that the initial arc has colour i and the final arc has colour j. Denote the obtained matrix for a long virtual knot K by $M(K)$.

The following theorem is obvious by construction.

Theorem 20.3. *For any two long virtual knots K_1, K_2 we have $M(K_1 \# K_2) = M(K_1) \cdot M(K_2)$.*

This means that each finite virtual quandle defines a *representation* of the semigroup $\mathfrak{W}$.

20.4 The $\mathfrak{V}$–rational function

Two arcs of each long knot diagram are special: those containing the two infinite points.

Consider a diagram $\bar{K}$ of a long knot K. Let us construct the virtual

Alexander module of it. For the sake of simplicity, we shall preserve the previous notation. This module (which is now a linear space over R) will be denoted by $\mathcal{M}$.

Suppose we have $n+1$ long arcs (this case corresponds to n long arcs in the classical case). Each of the two infinite long arcs has one infinite arc. Denote the arc containing $-\infty$ by a_1, and that containing $+\infty$ by a_{n+1}.

For each of the remaining $n-1$ long arcs, choose an arc of it. Denote these chosen arcs by $a_2, \ldots, a_n$.

Now, let us construct the linear space over R. First, consider the $(n+2)$–dimensional space S generated by $a_1, \ldots, a_{n+1}, \varepsilon$.

Now, let us define $\mathcal{M}$ as the factor space obtained from S by factorizing it by relations just as in the classical case. We get the invariant triple $\mathcal{M}, a_1, a_{n+1}$. Obviously, $a_{n+1} - a_1 = k \cdot \varepsilon$. By definition, this k is a long knot invariant. Denote it by $\mathfrak{V}(K)$.

Obviously, the $\mathfrak{V}$ polynomial has the following property.

Theorem 20.4. *For any long classical knot L we have $\mathfrak{V}(L) = 0$.*

Now, let us consider the following example. In [Man4] it was shown that if we break the virtual trefoil in two different ways (see Fig 20.1), we obtain two different long virtual knots. This fact was proved by using the VA polynomial of connected sums of virtual knots. Let us prove this fact now by using $\mathfrak{V}$.

For the knot K_1 we have:

$$a_1(1 - t) + a_2 t - a_3 = \varepsilon(1 - 2t)$$

$$a_1 t + a_3(1 - t) - a_2 = \varepsilon t.$$

Multiplying the second equation by t and adding it to the first one, we get

$$\mathfrak{V}(K_1) = \frac{a_3 - a_1}{\varepsilon} = -\frac{(t - 1)^2}{t^2 - t + 1}.$$

For K_2 we have:

$$t a_1 + (1 - t)(a_3 + \varepsilon) - a_2 = 0$$

$$t(a_2 + \varepsilon) + (1 - t)a_3 - (a_3 + \varepsilon) = 0.$$

Multiplying the first equation by t and adding it to the second equation, we get

$$\mathfrak{V}(K_2) = -\frac{(t - 1)^2}{t^2}.$$

Thus, we see that $\mathfrak{V}$ distinguishes long virtual knots, corresponding to one and the same (ordinary) virtual knot.

The connected sum # of long virtual knots is well defined. Obviously, the following theorem is true.

a   b

FIGURE 20.3: Two long virtual knots obtained by breaking the unknot

Theorem 20.5. *For any two long virtual knots K_1 and K_2 we have*
$\mathfrak{V}(K_1 \# K_2) = \mathfrak{V}(K_1) + \mathfrak{V}(K_2)$.

20.5   Virtual knots versus long virtual knots

We have already given some examples that show that when breaking the same virtual knot diagram at different points, we obtain different (non-equivalent) long virtual knots. The simplest and, probably, most interesting example is the Kishino knot (Fig. 17.10). The (non–trivial) virtual knot represented there is the connected sum of two unknots. In particular, this means that the corresponding long virtual knots are not trivial.

Consider the unknots shown in Fig. 20.3, a and b. Let us show that they are not isotopic to the trivial knot. To do this, we shall use the presentation of the long virtual quandle to the module over $\mathbb{Z}_{16}$ by:

$$a \circ b = 5a - 4b, \ a * b = 9a - 8b$$

$$f(x) = 3 \cdot x.$$

It can be readily checked that these relations satisfy all axioms of the long quandle.

Let us show that for none of these two knots $a = b$. Indeed, for the first knot (Fig. 20.3.a), denote by c the next arc after a. Then we have:

$$9a - 8 \cdot (3c) = c, 5b - 4 \cdot (3c) = c \Longrightarrow b = 9a.$$

For the second knot (Fig. 20.3.b), denote by c the upper (shortest) arc. We have:

$$5 \cdot (3b) - 4a = c, 9 \cdot (3a) - 8b = c \Longrightarrow b = 9a.$$

As we can see, in none of these cases does $a = b$. Besides, the expressions of b via a are different. Thus, none of the two long knots shown in Fig. 20.3.a and Fig. 20.3.b are trivial.

20.6   The question about commutative of long knots

It is known (see, e.g. [CF]) that classical long knots commute. Moreover, the following theorem is true.

Theorem 20.6. *Let a long knot K have no virtual crossings. Then for any long virtual knot K' the commutative property $K \# K' = K' \# K$ holds.*

Proof. Indeed, let us make a diagram of K very small and start pulling it through a diagram of K'. When pulling it through the virtual crossings we shall use the detour move; see Fig. 20.4, and when we have a pulling through classical crossings, we shall use the classical Reidemeister moves.

As a result, we have the desirable equivalence.   □

Remark 20.4. *This proof does not work in the case when the knot K has virtual crossings, since in this case we cannot draw K through arcs of K' consisting of classical crossings. We should have used the forbidden move; see Fig. 17.4.*

Let us show that there are long virtual knots which do not commute with each other. This fact was first discovered in [Man15].

Example 20.1. *Let us consider the long virtual knots K_1 and K_2 depicted in Fig. 20.5.*

Consider the linear long quandle with the ring $\mathcal{R} = \mathbb{Z}_{11^2 \cdot 19^2}$ and the parameters $p = 20 + 121 \cdot 19$, $q = 20$, $k = 70$. Thus, in $\widetilde{M}(K)$ the operations look like $a * b = 20a - 19b$, $a \circ b = (20 + 121 \cdot 19)a - (122 \cdot 19)b$ and $f(a) = 70a$. One note that in the module $\widetilde{M}$ for the knot $K_1 \# K_2$ the initial arc must be divided by 121, while for the module corresponding to the knot $K_2 \# K_1$ there exists a homomorphism to the ring $\mathbb{Z}_{11^2 \cdot 19^2}$ under which the initial arc is sent to $11 \cdot 19^2$.

Namely, for the knot K_1 (see upper part of Fig. 20.5) we have the following relations in the linear quandle $\widetilde{M}(K_1)$: $a * (70c) = c = b \circ 70c$; the first relation means that $20a = (19 \cdot 70 + 1)c = 1331c$; since the element 20 is invertible in the ring $\mathbb{Z}_{11^2 \cdot 19^2}$, the elements a is divided by 121.

For the knot K_2 (see the lower part in Fig. 20.5) we have: $(70x) * y = z = (70y) \circ x$. It means that we can set $x = 11 \cdot 19^2$, $y = 0$ (the coefficient in x in the expression $(70x) * y - (70y) \circ x$ is divided by 11); i.e. we can map $\widetilde{M}(K_2)$ to the linear quandle $\mathbb{Z}_{11^2 \cdot 19^2}$ with the same operations. Further, in the knot $K_2 \# K_1$ we can set all remaining arcs (belonging to the long knot K_1) equal to zero; this can be done since $y = 0$ (the value z is calculated).

The other way to formulate the arguments above is: The set of homomorphisms $\mathcal{H}$ from $Q_L(K_1 \# K_2)$ to the linear quandle which is the ring $\mathbb{Z}_{11^2 \cdot 19^2}$ with the operations (20.1) and $p = 20 + 121 \cdot 19$, $q = 20$, $k = 70$ such that

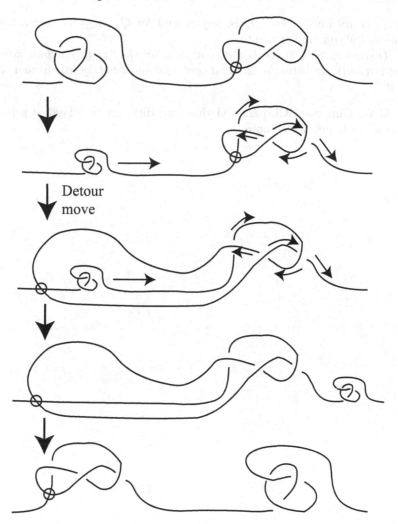

Detour move

FIGURE 20.4: A classical long knot commutes with any long knot.

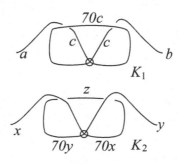

FIGURE 20.5: Labeling the knots K_1 and K_2.

$\mathcal{H}(a_1)$ *is not divided by 121, is empty, and for* $Q_L(K_2 \# K_1)$ *the set of such homomorphisms is not empty.*

Therefore, these knots do not commute, this fact confirms their difference, non-triviality, and also the fact that each of them is not equivalent to a classical knot.

M.W. Chrisman [Chr] showed that two different nonclassical prime long virtual knots never commute.

Chapter 21

Virtual braids

Just as classical knots can be obtained as closures of classical braids, virtual knots can be similarly obtained by closing *virtual braids*. Virtual braids were suggested by Vladimir V. Vershinin, [Ver].

21.1 Definitions of virtual braids

As well as virtual knots, virtual braids have a purely combinatorial definition. Namely, one takes virtual braid diagrams and factorises them by virtual Reidemeister moves (all moves with the exception of the first classical and the virtual moves; the latter moves do not occur).

Definition 21.1. A *virtual braid diagram* on n strands is a graph lying in $[1, n] \times [0, 1] \subset \mathbb{R}^2$ with vertices of valency one (there should be exactly $2n$ such vertices with coordinates $(i, 0)$ and $(i, 1)$ for $i = 1, \ldots, n$) and a finite number of vertices of valency four. The graph is a union of n smooth curves without vertical tangent lines connecting a point on the line $\{y = 1\}$ with those on the line $\{y = 0\}$; their intersection makes crossings (four–valent vertices). Each crossing should be either endowed with a structure of over– or undercrossing (as in the case of classical braids) or marked as a virtual one (by encircling it).

Definition 21.2. A *virtual braid* is an equivalence class of virtual braid diagrams by planar isotopies and all virtual Reidemeister moves except the first classical move and the first virtual move.

Like classical braids, virtual braids form a group (with respect to juxtaposition and rescaling the vertical coordinate). The generators of this group are:

$\sigma_1, \ldots, \sigma_{n-1}$ (for classical crossings) and $\zeta_1, \ldots, \zeta_{n-1}$ (for virtual crossings).

The inverse elements for the σ's are defined as in the classical case. Obviously, for each $i = 1, \ldots, n - 1$ we have $\zeta_i^2 = e$ (this follows from the second virtual Reidemeister move).

One can show that the following set of relations [Ver] is sufficient to generate this group:

1. (Braid group relations):

$$\sigma_i\sigma_j = \sigma_j\sigma_i$$

for $|i - j| \geq 2$;

$$\sigma_i\sigma_{i+1}\sigma_i = \sigma_{i+1}\sigma_i\sigma_{i+1};$$

2. (Permutation group relations):

$$\zeta_i\zeta_j = \zeta_j\zeta_i$$

for $|i - j| \geq 2$;

$$\zeta_i\zeta_{i+1}\zeta_i = \zeta_{i+1}\zeta_i\zeta_{i+1};$$

$$\zeta_i^2 = e;$$

3. (Mixed relations):

$$\sigma_i\zeta_{i+1}\zeta_i = \zeta_{i+1}\zeta_i\sigma_{i+1};$$

$$\sigma_i\zeta_j = \zeta_j\sigma_i$$

for $|i - j| \geq 2$.

The proof of this fact is left to the reader.

21.2   Burau representation and its generalisations

In [Ver], the following generalisation of the Burau representation is given. The virtual braid group $VB(n)$ is represented by $n \times n$ matrices where the generators σ_i, ζ_i are represented by block–diagonal matrices with the only nontrivial block on lines and columns $(i, i - 1)$. The block for σ_i's is just as in Chapter 9. For ζ_i we use simply permutations, namely, the matrix

$$\begin{pmatrix} 0 & 1 \\ 1 & 0 \end{pmatrix}.$$

The proof that it really gives a representation is left to the reader as an exercise.

However, this representation is rather weak. It is easy to check that for the non–trivial virtual two–strand braid represented by the word $b = (\sigma_1^2\zeta_1\sigma_1^{-1}\zeta_1\sigma_1^{-1}\zeta_1)^2$ we have $f(b) = f(e)$.

The generalisation of this Burau representation is the following: we take polynomial matrices in two variables, t and q, and construct the following 2×2 blocks: the same for σ and

$$\begin{pmatrix} 0 & q \\ q^{-1} & 0 \end{pmatrix}$$

for ζ.

Denote the map, defined above on generators of the braid group, by R.

Theorem 21.1. *The map R can be generated as a representation of the braid group.*

Proof. Obviously, the matrix $R(\sigma_i)$ is invertible, and for the matrix $R(\zeta_i)$ we have $(R(\zeta_i))^2 = e$.

Furthermore, the relations of the braid group for the σ's can be easily checked as in the case of the "weaker" Burau representation.

So, we only have to check the relations $R(\zeta_i \zeta_{i+1} \zeta_i) = R(\zeta_{i+1} \zeta_i \zeta_{i+1})$ and $R(\zeta_i \zeta_{i+1} \sigma_i)$
$= R(\sigma_{i+1} \zeta_i \zeta_{i+1})$.

They can be checked straightforwardly by direct calculation with 3×3 matrices.

$\square$

Now, we can prove the following theorem.

Theorem 21.2. *The group $Br(3)$ is naturally embedded in the virtual braid group $VB(3)$.*

Proof. Actually, let β_1, β_2 be some braid–words written in $\sigma_1, \sigma_2, \sigma_1^{-1}, \sigma_2^{-1}$. Suppose they represent the same braid in $VB(n)$. Then their Burau matrices coincide. Hence the Burau representation of the classical braid group is faithful for the case of three strands, and we conclude that β_1 and β_2 represent the same word in $Br(n)$.

$\square$

21.3   Invariants of virtual braids

In this section, we are going to present an invariant of virtual braids proposed by the author in [Man'3] and show that the classical braid group is a subgroup of the virtual one. For an elementary proof of this fact see [Man'5]. More precisely, we give a generalisation of the complete braid invariant described before for the case of virtual braids. The new "virtual invariant" is very strong: it is stronger than the Burau representation, the Jones–Kauffman polynomial. A simple computer program written by the author recognises all virtual braids on three and four crossings, given by the author. The question of whether the invariant is complete was answered negatively by O.Chterental [Cht1]. The completeness of the multi-variable extension of the invariant (see [Man9]) is unknown.

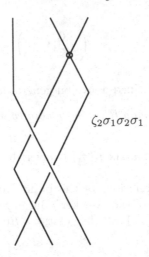

$$\zeta_2\sigma_1\sigma_2\sigma_1$$

FIGURE 21.1: A virtual braid diagram and the corresponding braid word

A virtual braid diagram is called *regular* if any two different crossings have different ordinates.

Let us start with basic definitions and introduce the notation.

Remark 21.1. *In the sequel, the number of strands for a virtual braid diagram is denoted by* n, *unless otherwise specified.*

Remark 21.2. *In the sequel, regular (virtual) braid diagrams and corresponding braid words (see definition below) will be denoted by Greek letters (possibly, with indices). Virtual braids will be denoted by Latin letters (with indices, maybe).*

Remark 21.3. *We shall also treat braid words and braids familiarly, saying, e.g. "a strand of a braid word" and meaning "a strand of the corresponding braid".*

Let us describe the construction of the word by a given regular virtual braid diagram as follows. Let us walk along the axis Oy from the point $(0,1)$ to the point $(0,0)$ and watch all those levels $y = t \in [0,1]$ having crossings. Each such crossing permutes strands $\#i$ and $\#(i+1)$ for some $i = 1,\ldots,n-1$. If the crossing is virtual, we write the letter ζ_i, if not, we write σ_i if overcrossing is the "northwest–southeast" strand, and σ_i^{-1} otherwise.

Thus, we have got a braid word by a given regular virtual braid diagram; see Fig. 21.1.

Thus the main question is the word problem for the virtual braid group: **How to recognise whether two different (regular) virtual braid diagrams** β_1 **and** β_2 **represent the same braid** b[1]. One can apply the virtual

---

[1]The recognition problem for virtual braids was solved by O.Chterental [Cht2].

braid group relations to one diagram without getting the other and one does not know whether he has to stop and say that they are not isomorphic or he has to continue.

A partial answer to this question is the construction of a virtual braid group invariant; i.e., a function on virtual braid diagrams (or braid words) that is invariant under all virtual braid group relations. In this case, if for an invariant f we have $f(\beta_1) \neq f(\beta_2)$ then β_1 and β_2 represent two different braids.

Here we give the generalisation of the complete classical braid group invariant, described in Chapter 9 for the case of virtual braids.

Let G be the free group in generators $a_1 \ldots, a_n, t$. Let E_i be the quotient set of right residue classes $\{a_i\}\backslash G$ for $i = 1, \ldots, n$.

Definition 21.3. A *virtual n–system* is a set of elements $\{e_1 \in E_1, e_2 \in E_2, \ldots, e_n \in E_n\}$.

The aim of this subsection is to construct an invariant map (non–homomorphic) from the set of all virtual n–strand braids to the set of virtual n–systems.

Let β be a braid word. Let us construct the corresponding virtual n–system $f(\beta)$ step–by–step. Namely, we shall reconstruct the function $f(\beta\psi)$ from the function $f(\beta)$, where ψ is σ_i or σ_i^{-1} or ζ_i .

First, let us take n residue classes of the unit element of G: $\langle e, e, \ldots, e \rangle$. This means that we have defined

$$f(e) = \langle e, e, \ldots, e \rangle.$$

Now, let us read the word β. If the first letter is ζ_i then all words but e_i, e_{i+1} in the n–systems stay the same, e_i becomes equal to t and e_{i+1} becomes t^{-1} (here and in the sequel, we mean, of course, residue classes, e.g. $[t]$ and $[t^{-1}]$. But we write just t and t^{-1} for the sake of simplicity).

Now, if the first letter of our braid word is σ_i, then all classes but e_{i+1} stay the same, and e_{i+1} becomes a_i^{-1}. Finally, if the first letter is σ_i^{-1} then the only changing element is e_i: it becomes a_{i+1}.

The procedure for each next letter (generator) is the following. Denote the index of this letter (the generator or its inverse) by i. Assume that the left strand of this crossing originates from the point $(p, 1)$, and the right one originates from the point $(q, 1)$. Let $e_p = P, e_q = Q$, where P, Q are some words representing the corresponding residue classes. After the crossing of all residue classes but e_p, e_q should stay the same.

Then if the letter is ζ_i then e_p becomes $P \cdot t$, and e_q becomes $Q \cdot t^{-1}$. If the letter is σ_i then e_p stays the same, and e_q becomes $QP^{-1}a_p^{-1}P$. Finally, if the letter is σ_i^{-1} then e_q stays the same, e_p becomes $PQ^{-1}a_qQ$. Note that this operation is well defined.

Actually, if we take the words $a_p^l P, a_q^m Q$ instead of the words P, Q, we get: in the first case $a_p^l Pt \sim Pt, a_q^m Qt^{-1} \sim Qt^{-1}$, and in the second case we obtain

$a_p^l P \sim P, a_q^m Q P^{-1} a_p^{-l} a_p^{-1} a_p^l P = a_q^m Q P^{-1} a_p^{-1} P \sim Q P^{-1} a_p^{-1} P$. In the third case we obtain $a_p^l P Q^{-1} a_q^{-m} a_q^{-1} a_q^m P = a_p^l P Q^{-1} a_q^{-1} Q \sim P Q^{-1} a_q^{-1} Q, a_q^m Q \sim Q$.

Thus, we have defined the map f from the set of all virtual braid diagrams to the set of virtual n–systems.

Theorem 21.3. *The function f, defined above, is a braid invariant. Namely, if β_1 and β_2 represent the same braid β then $f(\beta_1) = f(\beta_2)$.*

Proof. We have to demonstrate that the function f defined on virtual braid diagrams is invariant under all virtual braid group relations. It suffices to prove that, for the words $\beta_1 = \beta\gamma_1$ and $\beta_2 = \beta\gamma_2$ where $\gamma_1 = \gamma_2$ is a relation we have proved, we can also prove $f(\beta_1) = f(\beta_2)$. During the proof of the theorem, we shall call it the A–statement.

Indeed, having proved this claim, we also have $f(\beta_1\delta) = f(\beta_2\delta)$ for arbitrary δ because the invariant $f(\beta_1\delta)$ (as well as $f(\beta_2\delta)$) is constructed step–by–step; i.e., knowing the value $f(\beta_1)$ and the braid word δ, we easily obtain the value of $f(\beta_1\delta)$. Hence, for braid words β, δ and for each braid group relation $\gamma_1 = \gamma_2$ we prove that $f(\beta\gamma_1\delta) = f(\beta\gamma_2\delta)$. This completes the proof of the theorem.

Now, let us return to the A–statement.

To prove the A–statement, we must consider all virtual braid group relations. The commutation relation $\sigma_i\sigma_j = \sigma_j\sigma_i$ for "far" i, j is obvious: all four strands involved in this relation are different, so the order of applying the operation does not affect on the final result. The same can be said about the other commutation relations, involving one σ and one ζ or two ζ's.

Now let us consider the relation $\zeta_i^2 = e$ which is pretty simple too.

Actually, let us consider a braid word β, and let the word β_1 be defined as $\beta\zeta_i^2$ for some i. Let $f(\beta) = (P_1, \ldots, P_n), f(\beta_1) = (P_1', \ldots, P_n')$. Let p and q be the numbers of strands coming to the crossing from the left side and from the right side. Obviously, for $j \neq p, q$ we have $P_j = P_j'$. Besides, $P_p' = (P_p \cdot t) \cdot t^{-1} = P_p, P_q' = (P_q \cdot t^{-1}) \cdot t = P_q$.

Now let us consider the case $\beta_1 = \beta \cdot \sigma_i \cdot \sigma_i^{-1}$ (obviously, the case $b_1 = \beta\sigma_i^{-1}\sigma_i$ is quite analogous to this one).

As before, denote $f(\beta)$ by $(\ldots P_i \ldots)$, and $f(\beta_1)$ by $(\ldots P_i' \ldots)$, and the corresponding strand numbers by p and q. Again, we have: for $j \neq p, q : P_j' = P_j$. Moreover, $P_p = P_p'$ by definition of f (since the p-th strand makes an overcrossing twice), and $P_q' = (P_q P_p^{-1} a_p^{-1} P_p) P_p^{-1} a_p P_p = P_q$.

Now let us check the invariance under the third Reidemeister move. Let β be a braid word, $\beta_1 = \beta\zeta_i\zeta_{i+1}\zeta_i$, and $\beta_2 = \beta\zeta_{i+1}\zeta_i\zeta_{i+1}$. Let p, q, r be the global numbers of strands occupying positions $n, n+1, n+2$ at the bottom of b.

Denote $f(\beta)$ by $(P_1, \ldots, P_n)$, $f(\beta_1)$ by $(P_1^1, \ldots, P_n^1)$, and $f(\beta_2)$ by $P_1^2, \ldots, P_n^2$. Obviously, $\forall i \neq p, q, r$ we have $P_i = P_i^1 = P_i^2$. Direct calculations show that $P_p^1 = P_p^2 = P_p \cdot t^2, P_q^1 = P_q^2 = P_q$ and $P_r^1 = P_r^2 = P_r \cdot t^{-2}$.

Now, let us consider the mixed move by using the same notation:

$\beta_1 = \beta\zeta_i\zeta_{i+1}\sigma_i$, $\beta_2 = \sigma_{i+1}\zeta_i\zeta_{i+1}$. As before, $P_j^1 = P_j^2 = P_j$ for all $j \neq p, q, r$. Now, direct calculation shows that $P_p^1 = P_p t^2, P_q^1 = P_q t^{-1}, P_r^1 = P_r t^{-1}(P_q t^{-1})^{-1} a_q^{-1}(P_q t^{-1}) = P_r P_q a_q^{-1}\hat{P}_q t^{-1}$ and $P_p^2 = P_p t^2, P_q^2 = P_q t^{-1}, P_r^2 = P_r P_q^{-1} a_q^{-1} P_q$.

Finally, consider the "classical" case $\beta_1 = \beta\sigma_i\sigma_{i+1}\sigma_i, \beta_2 = \beta\sigma_{i+1}\sigma_i\sigma_{i+1}$; the notation is the same. Again $\forall j \neq p, q, r : P_j^1 = P_j^2 = P_j$. Besides this, since the p-th strand forms two overcrossings in both cases then $P_p^1 = P_p^2 = P_p$. Then, $P_q^1 = P_q P_p^{-1} a_p^{-1} P_p, P_r^1 = (P_r P_p^{-1} a_p^{-1} P_p) \cdot (P_q P_p^{-1} a_p^{-1} P_p)^{-1} a_q^{-1}$ $(P_q P_p^{-1} a_p^{-1} P_p) = P_r P_p^{-1} a_p^{-1} P_q P_p^{-1} a_p^{-1} P_p$ and $P_q^2 = P_q P_p^{-1} a_p^{-1} P_p, P_r^2 = P_r P_q^{-1} a_q^{-1} P_q P_p^{-1} a_p^{-1} P_p$.

As we see, the final results coincide and this completes the proof of the theorem. □

Thus, we have proved that f is a virtual braid invariant; i.e., for a given braid the value of f does not depend on the diagram representing b. So, we can write simply $f(b)$.

Remark 21.4. *In fact, we can think of f as a function valued not in $(E_1, \ldots, E_n)$, but in n copies of G: all these invariances were proved for the general case of $(G, \ldots, G)$. The present construction of $(E_1, \ldots, E_n)$ is considered for the sake of simplicity.*

As well as classical knots, classical braids (i.e., braids without virtual crossings) can be considered up to two equivalences: classical (modulo only classical moves) and virtual (modulo all moves). Now, we prove that they are the same (as in the case of classical knots). This fact is not new. It follows from [FRR]. An elementary proof was given in [Man'5].

Theorem 21.4. *Two virtually equal classical braids b_1 and b_2 are classically equal.*

Proof. Since b_1 is virtually equal to b_2, we have $f(b_1) = f(b_2)$. Now, taking into account that f is a complete invariant on the set of classical braids, we have $b_1 = b_2$ (in the classical sense). □

As in the case of virtual knots, in the case of virtual braids there exists a forbidden move, namely, $X = \sigma_i\sigma_{i+1}\zeta_i = \zeta_{i+1}\sigma_i\sigma_{i+1} = Y$. Now, we are going to show that it cannot be represented by a finite sequence of the virtual braid group relations.

Theorem 21.5. *A forbidden move (relation) cannot be represented by a finite sequence of legal moves (relations).*

Proof. Actually, let us calculate the values $f(\sigma_1\sigma_2\zeta_1)$ and $f(\zeta_2\sigma_1\sigma_2)$. In the first case we have:

$$(e, e, e) \to (e, a_1^{-1}, e) \to (e, a_1^{-1}, a_1^{-1}) \to (e, a_1^{-1}t, a_1^{-1}t^{-1}).$$

In the second case we have:

$$(e, e, e) \rightarrow (e, t, t^{-1}) \rightarrow (e, t, t^{-1}a_1^{-1}) \rightarrow (e, ta_1^{-1}, t^{-1}a_1^{-1}).$$

As we see, the final results are not the same (i.e., they represent different virtual n–systems); thus, the forbidden move changes the virtual braid. □

Remark 21.5. *If we put $t = 1$, the results $f(X)$ and $f(Y)$ become the same. Thus that is the variable t that "feels" the forbidden move.*

Definition 21.4. For a given braid diagram β and two numbers $1 \le i < j \le 2$ let us define *the linking coefficient* (see, e.g. [GPV]) $l_{i,j}(\beta)$ as follows. Let us watch all those crossings where the i–th strand is the undercrossing, and the j–th strands is the overcrossing, and take the algebraic sum of all these crossings (-1 if the crossing is negative, and 1 if it is positive).

We shall show now that this function can be calculated by using only $f(\beta)$, thus, it is a braid invariant.

Now consider a braid b and the value $f(b)$. It consists of n terms $(e_1, \ldots, e_n)$. For $1 \le p \ne q \le n$ let f_{pq} be the algebraic number of entrances of a_q in e_p. All numbers f_{pq} are well–defined from f: each element e_i is defined up to multiplication by a_i from the left side; such a multiplication does not change f_{ij} for $j \ne i$.

Thus, $\forall 1 \le p \ne q \le n$ the function f_{pq} is a virtual braid invariant.

Theorem 21.6. *The invariant f_{pq} coincides with the linking coefficient $-l_{pq}$ of strands p and q.*

Proof. Actually, let us consider a braid word β and let us construct $f(\beta)$ step–by–step. Note that f_{pq} demonstrates the "abelinisation" of the invariant f: instead of multiplication of generators, we just add them and watch the corresponding coefficients. Let us study the subject more precisely. Now we begin to prove the statement of the theorem using induction on the length of β. For the case of zero crossings everything is obvious. Now let us add a new crossing and see what happens.

The case of a virtual crossing does not change the linking coefficient (which is constructed by taking the algebraic sum for classical crossings). This letter (crossing) does not change f_{ij} either: two elements e_i, e_j are multiplied by $t^{\pm 1}$, but the number of entrances of $a_k, k = 1 \ldots, n$, stays the same. By adding some letter σ_i, we change $f_{kl}, l = 1, \ldots, n$, as follows. Let p be the number of strands coming to this crossing from the right side, and q be the number of the strand coming from the left side. Then the only thing changing here is e_q. It is to be multiplied by $e_p^{-1}p^{-1}e_p$. While calculating the algebraic number of entrances of some letters, e_p on the right hand cancels the effect of e_p^{-1} on the left hand. Thus, f_{qp} becomes $f_{qp} - 1$, the coefficient l_{qp} becomes $l_{qp} + 1$, and all other coefficients l_{xy}, f_{xy} stay the same. The same thing happens with the linking coefficient l_{pq}.

The case of σ_i^{-1} can be considered analogously.

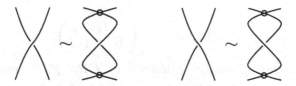

FIGURE 21.2: Pairs of diagrams not distinguished by the Kauffman polynomial

We have proved the induction step and, thus, the theorem is proved.

□

So, we have shown that f is stronger than the well–known invariant, called the linking coefficient.

Here we give two more examples showing the advantages of the invariant f.

Consider the 3–strand braid $b = \zeta_2\sigma_2^{-1}\zeta_2\sigma_1\sigma_2\zeta_1\sigma_1\zeta_1\sigma_2^{-1}\sigma_1^{-1}$. A computer shows that for this braid $f(b) \neq f(e)$. However, this braid is not distinguished by the virtual Jones–Kauffman polynomial proposed in [Kau5]. More precisely, consider the link $L(b)$ obtained as a closure of b and the Kauffman polynomial $K(L(b))$ of this link. It is well known that the Kauffman polynomial does not distinguish links that differ as shown in Fig. 21.2.

Thus, it is easy to see that if for some braid we substitute $\sigma_i^{\pm 1}$ for $\zeta_i\sigma_i^{\pm 1}\zeta_i$ then the closures of both braids will have the same Kauffman polynomial.

So,

$$K(L(b)) = K(L(\sigma_2^{-1}\sigma_1\sigma_2\sigma_1\sigma_2^{-1}\sigma_1^{-1})).$$

The transformed braid is trivial, so $K(L(b)) = K(L(e))$.

Now, we give another example of the strength of the invariant f. Consider the Burau representation of the virtual braid group $VB(n)$, see, e.g. [Ver], more precisely, the representation of $VB(2)$, generated by two matrices:

$$R(\sigma_1) = \begin{pmatrix} 1-t & t \\ 0 & 1 \end{pmatrix}, \quad R(\zeta_1) = \begin{pmatrix} 0 & 1 \\ 1 & 0 \end{pmatrix}.$$

It is easy to see that the matrix $R(\sigma_1)$ has the following eigenvalues: 1 and $-t$. More precisely,

$$CR(\sigma_1)C^{-1} = \begin{pmatrix} 1 & 0 \\ 0 & 1-t \end{pmatrix}$$

for

$$C = \begin{pmatrix} 0 & 1 \\ 1 & -1 \end{pmatrix}.$$

In this case,

$$CR(\zeta_1)C^{-1} = \begin{pmatrix} 1 & t-1 \\ 0 & 1 \end{pmatrix}.$$

Now, let us write simply: ζ instead of $CR(\zeta_1)C^{-1}$ and σ instead of $CR(\sigma_1)C^{-1}$.

Thus we have: $F(k,l,m) = \sigma^k \zeta \sigma^l \zeta \sigma^m \zeta$ is an upper–triangular matrix with 1 and -1 on the main diagonal if $k+l+m = 0$. Assume $k = 2, l = -1, m = -1$. Then $F(2,-1,-1)^2 = e$.

It is easy to check that for the non–trivial virtual braid $b = (\sigma_1^2 \zeta_1 \sigma_1^{-1} \zeta_1 \sigma_1^{-1} \zeta_1)^2$ we have $f(b) \neq f(e)$.

In this sense, the invariant f is stronger than the Burau representation even for the case of two strands.

21.3.1   How strong is the invariant f?

As we have shown above, the new invariant is stronger than link coefficients, sometimes it recognises virtual braids, which cannot be recognised by the Jones–Kauffman polynomial or by the Burau representation.

Besides this, the restriction of the invariant f for the case of classical braids (also denoted by f) coincides with the complete classical braid group invariant, described in Chapter 9.

The invariant f gives us an example of a map from one algebraic object (braid group) to another algebraic object (n copies of a free group or n residue classes in free groups). However, this map is not homomorphic.

Thus, in order to understand the strength of the invariant f, we are going to establish some properties of this map.

Fortunately, there are some properties that make f similar to a homomorphic map. Namely, the following lemma holds.

Lemma 21.1. *If $f(b_1) = f(b_2)$ for some braids b_1, b_2 then for any two braids a and c we have $f(ab_1c) = f(ab_2c)$ (all braids are taken to have the same number of strands).*

Proof. We shall prove the lemma in two steps. The first step is to prove that $f(ab_1) = f(ab_2)$. The second step is to prove that if $f(a_1) = f(a_2)$ then $f(a_1c) = f(a_2c)$. If we prove both statements, then, substituting ab_1 for a_1 and ab_2 for a_2, we obtain the statement of the theorem.

The second step is obvious, and it was already proved while proving Theorem 21.3 as the A–statement.

So, let us prove that if $f(b_1) = f(b_2)$ then $f(ab_1) = f(ab_2)$. Let us consider some words α, β_1, and β_2 representing the braids a, b_1, and b_2, respectively. We are going to apply the induction method on the length of α. If α has length zero, then $a = e$, and then $f(ab_1) = f(ab_2)$ by the main assumption.

Now, let us consider the case when α has length one; i.e., it is just a letter. Suppose $\alpha = \zeta_i$. Then instead of the system of generators $(a_1, \ldots, a_n, t)$

of the group G we can consider the system $a_1, \ldots, a_{i-1}, ta_{i+1}t^{-1}, t^{-1}a_i t,$
$a_{i+2}, \ldots, a_n, t$. Obviously, these $n + 1$ generators are independent and they
generate the same group G. Denote $f(\beta_j)$ by $(P_1^j, \ldots, P_n^j)$ for $j = 1, 2$ (P
without upper index concerns β without lower index).

Obviously, each P_j^i depends on "the old generators": $P_j^i = P_j^i(a_1, \ldots, a_n, t)$.
Now let us, for a given function $X(a_1, \ldots, a_n)$, define X' as the value

$$X(a_1, \ldots, a_{i-1}, ta_{i+1}t^{-1}, t^{-1}a_i t, a_{i+2}, \ldots, a_n),$$

i.e., just by substituting the new generators for the old ones. Now, we state
that $\forall \beta \ f(\alpha\beta) = (P_1', \ldots, P_{i-1}', tP_i', t^{-1}P_{i+1}', \ldots, P_n')$.

This can be easily checked by using the induction method on the length of
β. But here the set of P_j' depends only on the set of P_j and can be uniquely
restored from it (strictly speaking, one should also check that multiplying
some P_j by a_j^l on the left side, all residue classes of P_k stay the same, but this
can be checked straightforwardly). This shows that the map $f(\beta) \to f(\alpha\beta)$ is
well–defined and injective.

Thus, if $f(\beta_1) = f(\beta_2)$ then $f(\alpha\beta_1) = f(\alpha\beta_2)$.

The same reasons are true in the cases when $\alpha = \sigma_i^{\pm 1}$. Here we just indicate
the way of transforming the P_i's.

In the case of $\alpha = \sigma_i$ the generators are: $a_1, \ldots, a_{i-1}, a_i a_{i+1} a_i^{-1},$
$a_i, a_{i+2}, \ldots, a_n, \ f(\alpha\beta) = (P_1', \ldots, P_{i-1}', P_i', a_i^{-1}P_{i+1}', \ldots, P_n')$ (here and later
by P_i' are meant the result of substituting the new generators for the old
ones).

In the case of $\alpha = \sigma_i^{-1}$ the generators are: $a_1, \ldots, a_{i-1}, a_{i+1}, a_{i+1}^{-1}a_i a_{i+1},$
$a_{i+2}, \ldots, a_n, \ f(\alpha\beta) = (P_1', \ldots, P_{i-1}', a_{i+1}P_i', P_{i+1}', \ldots, P_n')$

Thus, in the three cases described above the word $f(\alpha\beta)$ can be uniquely
restored from α and $f(\beta)$. Therefore $f(\alpha\beta_1) = f(\alpha\beta_2)$.

So, we have established the induction basis. Suppose the statement is
true for any word with length less than k for some given $k \geq 1$. Let α
be a word of length k. Then $\alpha = \alpha'\psi$, where ψ is the last letter of α
and α' has length $k - 1$. Let $\beta_1' = \psi\beta_1, \beta_2' = \psi\beta_2$. By the induction hypothesis, $f(\beta_1') = f(\beta_2')$. Applying again the induction hypothesis, we get
$f(\alpha\beta_1) = f(\alpha'\beta_1') = f(\alpha'\beta_2') = f(\alpha\beta_2)$.

This completes the proof of the first step and the lemma. Combining it
with the second step (already proved), we obtain the desired result. $\qquad\square$

Corollary 21.1. *If for some braid a we have $f(a) = f(e) = (e, \ldots, e)$ then
for any braid b: $f(b^{-1}ab) = f(e)$.*

The next step is now to describe all possible values of the invariant f. In
the general case this problem is very difficult; we restrict ourselves only to the
case of $n = 2$ strands. We shall consider an even simpler problem, concerning
a simpler invariant.

Notation change: instead of generators a_1, a_2 we shall write a, b; instead
of σ_1, ζ_1 we write simply σ, ζ.

Let $\mathcal{F}$ be an invariant, obtained from f by putting $t = 1$. Denote the free group with generators a, b by G', and let $E'_1 = \{a\}\backslash G, E'_2 = \{b\}\backslash G$.

In the case of two strands, $\mathcal{F}$ is a map from $VB(2)$ to (E'_1, E'_2) or, simply, to (G', G').

For a braid α, denote $\mathcal{F}(\alpha)$ by $(P(\alpha), Q(\alpha))$.

First, let us consider some examples of virtual two–strand braid words and values of $\mathcal{F}$ on them:

1. for the trivial word we have (e, e);

2. for σ we have $(1, a^{-1})$;

3. for σ^{-1} we have $(b, 1)$;

4. for ζ we have (t, t^{-1}).

It is not difficult to prove the following.

Theorem 21.7. *Let β be a braid word. Then $P(\beta)Q(\beta)^{-1} = a^k b^l$ for some k, l.*

Proof. We shall use the induction method on the number of crossings. For zero crossings there is nothing to prove. Now, let β be a braid with n crossings, $\beta' = \beta\alpha$, whence $\alpha = \zeta, \sigma$ or σ^{-1}. Let $P(\beta)Q(\beta)^{-1} = a^n b^m$.

For $\alpha = \zeta$ we have $P(\beta') = P(\beta)t, Q(\beta') = Q(\beta)t^{-1}$, thus $P(\beta')Q(\beta')^{-1} = a^n b^m$.

For $\alpha = \sigma$, the word β is even, and we have: $P \mapsto P, Q \mapsto QP^{-1}a^{-1}P, PQ^{-1} \mapsto a^{-1}PQ^{-1}$; for odd β: $Q \mapsto Q, P \mapsto PQ^{-1}b^{-1}Q, PQ^{-1} \mapsto PQ^{-1}b^{-1}$.

For $\alpha = \sigma^{-1}$, β is even: $Q \mapsto Q, P \mapsto PQ^{-1}bQ, PQ^{-1} \mapsto PQ^{-1}b$, for odd β: $P \mapsto P, Q \mapsto QP^{-1}aP, PQ^{-1} \mapsto aPQ^{-1}$.

Thus, we have made the induction step that completes the proof of the theorem. $\qquad\square$

The condition on PQ^{-1} is, indeed, quite natural. It means that $\exists g \in G' :$ $g \in [P] \in E'_1$ and $g \in [Q] \in E'_2$. Obviously, this element g is unique. Thus, g can be considered as an invariant of the group $VB(2)$.

Indeed, the situation in the group $VB(2)$ is quite simple.

Obviously, for any braid b we have $\mathcal{F}(b) = \mathcal{F}(b\zeta)$. Besides, for each even virtual braid b in $VB(2)$ there exist the unique braid $b\zeta$, corresponding to it. Thus, it is actual to consider only the even subgroup $EVB(2)$ of the group $VB(2)$.

Theorem 21.8. *The invariant g (as well as the invariant $\mathcal{F}$) of the virtual braid group $EVB(2)$ is complete.*

It suffices to prove that g is complete. To prove this theorem, we shall need an auxiliary lemma.

Lemma 21.2. *For any even two–strand braid words* π, ρ *we have* $g(\pi\rho) = g(\rho)g(\pi)$ *and* $g(\pi)^{-1} = g(\pi^{-1})$, *thus* g *is an antihomomorphism.*

Proof. First, let us note that the group $EVB(2)$ is a free group with two generators $\alpha = \zeta\sigma$ and $\beta = \zeta\sigma^{-1}$.

It can easily be checked that $g(e) = e, g(\alpha) = a, g(\beta) = b^{-1}, g(\alpha^{-1}) = a^{-1}, g(\beta^{-1}) = b$.

It can also be checked straightforwardly that $\forall\rho \; : \; g(\rho\alpha) = g(\alpha)g(\rho), g(\rho\beta) = g(\beta)g(\rho), g(\rho\alpha^{-1}) = g(\alpha^{-1})g(\rho), g(\rho\beta^{-1}) = g(\beta^{-1})g(\rho)$.

Let us first prove that $g(\pi\rho) = g(\rho)g(\pi)$. We shall do it by using the induction method on the length of ρ (by "length" we mean here the minimal number of $\alpha, \beta, \alpha^{-1}$ and β^{-1} in the decomposition of ρ). For $\rho = e$ there is nothing to prove.

For the word ρ having length one, it can be checked straightforwardly that:

$$\forall\alpha : g(\rho\alpha) = g(\alpha)g(\rho), g(\rho\beta) = g(\beta)g(\rho),$$

$$g(\rho\alpha^{-1}) = g(\alpha^{-1})g(\rho), g(\rho\beta^{-1}) = g(\beta^{-1})g(\rho).$$

Now, assume that the word ρ has length $k+1 > 1$, and for each word ρ' having length $\leq k$ we have $g(\pi\rho') = g(\rho')g(\pi)$. So, let $\rho = \rho_1\rho_2$, where ρ_1 has length 1 and ρ_2 has length k.

Then,

$$g(\pi\rho) = g(\pi\rho_1\rho_2) = g((\pi\rho_1)\rho_2)$$

by the induction hypothesis for ρ_2

$$= g(\rho_2)g(\pi\rho_1) =$$

again by the induction hypothesis for ρ_1

$$= g(\rho_2)g(\rho_1)g(\pi) =$$

and again by induction hypothesis for ρ_1

$$= g(\rho_1\rho_2)g(\pi) = g(\rho)g(p),$$

Q.E.D.

Now, since $g(e) = e$, we have:

$$e = g(e) = g(\rho\rho^{-1}) = g(\rho)^{-1}g(\rho),$$

and thus we obtain the second statement of the lemma.

$\square$

Proof of the Theorem. The lemma shows that g is an antihomomorphic map mapping the free group $EVB(2)$ to the free group with generators a, b. This map maps generators α, β to generators a, b^{-1}. Thus, it has no kernel. So, g is a complete invariant of $EVB(2)$, and so is $\mathcal{F}$. $\square$

Certainly, f is a complete invariant of the group $EVB(2)$ too. Besides, this invariant "feels" multiplication by ζ on the right side, thus f recognises all elements of $VB(2)$ as well. In order to recognise whether a pair of elements $(e_1 \in E_1, e_1 \in E_1)$ is a value of the invariant f on some braid, we just factorise them by t, take the pre-image b of the obtained couple (e'_1, e'_2) under $\mathcal{F}$, and see whether $f(b) = (e_1, e_2)$ or $f(b\zeta) = (e_1, e_2)$.

Certainly, the group $VB(2)$ is simple to recognise: it is just a free product of $\mathbb{Z}$ (generator σ) and $\mathbb{Z}_2$ (generator ζ).

So, the simplest example of $(e_1 \in E_1, e_2 \in E_2)$ that is not a value of f on a virtual braid is (b, a). In this case $PQ^{-1} = ba^{-1}$ which is not equal to $a^k b^l$ for any integer numbers k, l.

21.3.2   A $2n$-variable generalization

Problem. Understand the geometrical meaning of this invariant similar to $< \cdots >$.

Here $< \cdots >$ is the usual Hurwitz action — Artin invariant on the free group

This invariant was invented by the author soon after the paper [Cht1] was published; however, the definition remained unpublished since the invariant of $(n + 1)$ variables itself was conjecturally complete.

Once Oleg Chterental in the first arxiv version of $< \ldots >$ showed the incompleteness, the author pointed out to him the existence of this invariant.

So, the first instance of $< \ldots >$ appeared in Chterental's paper.

Now, we formulate the following.

Conjecture 21.1. $<>$ *is complete.*

21.4   Virtual links as closures of virtual braids

Analogously to classical braids, virtual braids admit *closures* as well; see Fig. 21.3. The obtained virtual link diagram will be *braided* with respect to some point A.

The closure of a virtual braid is a virtual link diagram. Obviously, isotopic virtual braids generate isotopic virtual links. Furthermore, all virtual link isotopy classes can be represented by closures of virtual braids.

Exercise 21.1. *Construct the analogues of the Alexander and Vogel algorithms.*

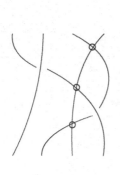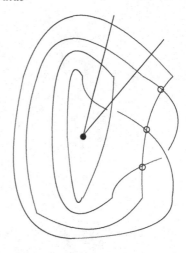

FIGURE 21.3: Closure of a virtual braid

21.5 An analogue of Markov's theorem

In [KamS1], Seiichi Kamada proved an analogue of Markov's theorem for the case of virtual braids. Namely, he proved the following.

Theorem 21.9. *Two virtual braid diagrams have equivalent (isotopic) closures as virtual links if and only if they are related by a finite sequence of the following moves (VM0)–(VM3).*

(VM0) braid equivalence;

(VM1) a conjugation (in the virtual braid group);

(VM2) a right stabilisation (adding a strand with additional positive, negative or **virtual** *crossing) and its inverse operation;*

(VM3) a right/left virtual exchange move; see Fig. 21.4.

The moves (VM0)–(VM2) are analogous to those in the classical case. The "new" move has two variants: the right one and the left one.

The necessity of the moves listed above is obvious; it is left for the reader as a simple exercise, see also [Man29]. For sufficiency, we refer the reader to the original work [KamS1].

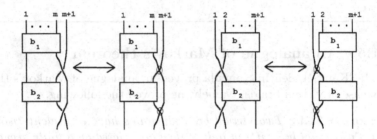

FIGURE 21.4: The virtual exchange move

Chapter 22

Khovanov homology of virtual knots

22.1 Introduction

In the present chapter, we construct the Khovanov homology for virtual knots. The main difficulty is algebraic: for virtual knots which do not admit source-sink structure ("orientable atoms"), the differential complex defined "in a natural way", does not satisfy $\partial^2 = 0$. To overcome this difficulty, we introduce twisted coefficients (see Section 22.7.2).The results of this chapter are due to the author [Man18, Man19]. We follow closely Chapter 5 of the book [MI].

Since the Khovanov homology theory for virtual knots appeared, it was natural to look for Lee–Rasmussen's theory.

Note that when we restrict ourselves to virtual knots with oriented atoms and a special sort of cobordism where all sections are virtual knots with oriented atoms, the results of Chapter 7 can be extended verbatim. For general virtual knots, there are two generalisations of Lee–Rasmussen theory which give bounds for slice genus estimates. The theory due to Dye, Kaestner and Kauffman [DKK] is based on the result of the present section.

The theory due to William Rushworth relies on another complex called *double Khovanov complex*, see [Rush].

Recall that the Khovanov chain complex (for classical knots) is defined by the axioms:

$$[[\emptyset]] = (0 \to \mathbb{Z} \to 0), \qquad [[\bigcirc K]] = V \otimes [[K]],$$

$$[[\overset{\curlyvee}{\underset{\curlywedge}{\times}}]] = \mathcal{F}\left(0 \to [[\,)(\,]] \overset{d}{\to} [[\asymp]]\{1\} \to 0\right).$$

Here V is a vector space of graded dimension $q + q^{-1}$, the operator $\{1\}$ is the operation of grading shift by 1, $\mathcal{F}$ is the flatten operation which sets a double complex to a single complex by taking direct sums along diagonals, and d is a differential. The Khovanov invariant is the homology of a renormalization of the Khovanov complex. The Khovanov invariant is indeed a link invariant and its graded Euler characteristic is the unnormalised Jones polynomial.

This passage from polynomials to (bi)graded complexes is also called *categorification*: Complexes form a category in which there are natural morphisms generated, for example, by cobordisms.

This theory has many generalisations and led to solutions of many problems in classical knot theory (for example, a simple proof of Milnor's conjecture about the Seifert genus of torus links, Chapter 8).

An important generalisation in the theory of extraordinary homology of links was the construction of categorification for a set of polynomials of type HOMFLY, made by Khovanov and Rozansky [KhR1, KhR2]. Polynomials of type HOMFLY have more complicated relations and the problem of categorification for them was elegantly solved by means of instruments of *matrix factorisations* and *Koszul complex*. Khovanov and Rozansky [KhR3] devoted their paper to the categorification of the $so(N)$-type Kauffman polynomial in which virtual knots are also used besides matrix factorisations.

The Khovanov homology possesses important properties coming from algebraic topology: the (projective) functoriality. In the given case, the morphisms are cobordisms of knots. Thus, the Khovanov homology is extended to invariants of knot cobordisms representing two-dimensional surfaces with boundary in $\mathbb{R}^3 \times I$. The projective functoriality (i.e. functoriality up to the overall minus sign) was first established by Jacobsson [Jac], see also [BN6, CMW, McD].

The functoriality allows one to construct invariants of cobordisms of two-dimensional surfaces in $\mathbb{R}^4$ from the Khovanov complex; a particular case of cobordisms is the cobordism between two links consisting of an empty set of components. In the case of projective functoriality, a cobordism invariant is defined up to an inverse element of the main ring. In this case the Khovanov construction gives an invariant of two-dimensional knots, and two-dimensional surfaces embedded in $\mathbb{R}^3 \times I \subset \mathbb{R}^4$. The accurate functoriality was established in [CMW], see also [Bla]. For Lee theory, such a functoriality is described explicitly in Chapter 8.

One of the most natural problems in the theory of virtual knots is the problem of generalisation of the Khovanov complex for virtual knots. An immediate attempt to generalise the theory leads to an algebraic difficulty: By writing down all necessary equations for the Khovanov complex to be invariant, we conclude that the main ring of coefficients should be the two-element ring. The indicated generalisation was done in [Man14]. Some difficulties of the immediate approach can be avoided by using geometrical constructions related to atoms (Chapter 16).

The main goal of this chapter is the construction of a chain complex for a virtual diagram with the homology being invariant under the generalised Reidemeister moves.

Note that the Khovanov homology for knots in thickened surfaces and in bundles over surfaces S_g whose fiber is an interval (by using some additional gradings for curves in a *given* surface) was also constructed by Asaeda, Przytycki and Sikora [APS], see also [MN]. This homology does not lead to the Khovanov homology for virtual knots, since it depends on a concrete surface S_g and is not invariant under destabilisations and homeomorphisms of the surfaces onto itself.

A further development of the Khovanov homology theory for virtual knots representing a generalisation of the paper [APS], and the results of this chapter, are given in [Man20, Man21], see also [DKM, MN]. In these papers topological and combinatorial coefficients at terms in the Kauffman bracket polynomial are "lifted" to new gradings in the Khovanov homology.

Below, we shall first describe four ways of constructing the Khovanov complex for virtual knots with some restrictions. First, we construct the Khovanov complex for arbitrary virtual knots with coefficients in $\mathbb{Z}_2$; in the second case, we show how one can construct the Khovanov complex for framed virtual links (by means of double diagrams) with coefficients in an arbitrary ring; in the third case, we construct the Khovanov complex of two-sheeted coverings over virtual knots (in the sense of atoms) with coefficients in an arbitrary field. The fourth way arises from the projective map, which sends all virtual knots to knots with orientable atoms and does not change knots with orientable atoms. This projective map allows one to "lift" all invariants defined for virtual knots with orientable atoms to all virtual knots.

In the second part of the chapter with each diagram of a virtual link we associate a complex with the homology being invariant under the generalised Reidemeister moves (this construction first appeared in [Man18, Man19]). Moreover, in the classical case, the complex has the same homology as the ordinary Khovanov complex, and the particular cases constructed in the present chapter give the complexes with the homology being isomorphic to the homology constructed for all virtual knots. The graded Euler characteristic of this complex coincides with the Jones polynomial $\hat{J}$ of the virtual link. Proceeding with this construction and using the parity arguments, we get the invariance of the Khovanov homology of two-sheeted coverings over virtual knots with coefficients from an arbitrary ring.

The main difficulty in constructing a Khovanov homology for virtual knots is how to define the differential for complexes corresponding to arbitrary virtual knots. Here one must consider many more cases than for classical knots (the corresponding atoms are considered in Sec. 7.6.3). This difficulty is overcome by means of a construction of a new complex having the same homology as the usual Khovanov complex. The first key idea is to change the basis of the Frobenius algebra representing the Khovanov homology of the unknot (it is connected with a choice of a local orientation of the corresponding circle originating from the crossing) as we pass from one crossing of the knot diagram to another. The second key idea is to replace the usual tensor product (corresponding to several circles in a given state) by the *exterior product* of the corresponding graded spaces. This enables us to avoid the "artificial" procedure of transforming the commutative cube into an anticommutative one, as was done in [BN1, Kho1] and in Chapter 7.

We mention some important properties of this construction.

1. The construction of the complex uses *atoms*. The complex is invariant under virtualisation. This is proved in Lemma 22.5.

2. There is a natural map from the set of "twisted virtual knots" in the

sense of Bourgoin and Viro [Bour, Vir2] (see below) to the set of virtual knots modulo virtualisation. Therefore, our approach yields invariants of twisted virtual knots. The set of twisted virtual knots (knots in oriented thickenings of non-orientable two-dimensional surfaces up to stabilisation) contains all knots in the punctured three-dimensional projective space. A particular case of this theory is the theory of knots in the three-dimensional projective space $\mathbb{R}P^3$. Note that the Kauffman bracket polynomial for knots in $\mathbb{R}P^3$ was constructed by Drobotukhina in [Dro]. Moreover, this theory admits different generalisations constructed for the ordinary Khovanov homology: Lee's theory [Lee1, Lee2], Wehrli's and Champanerkar–Kofman spanning tree expansion [ChKo, Weh], etc.

3. For the coefficient field $\mathbb{Z}_2$, our complex coincides with the complex constructed in Sec. 22.3.

4. For orientable atoms (in particular, for classical knots), the homology of our complex is the same as the homology of the complex constructed in Sec. 22.4.

5. The proof of invariance of the homology is *local*. It repeats the proof of the invariance in the classical case. The main difficulty is in defining the differential: How can one choose signs that make the cube anticommutative? We overcome this difficulty by constructing a new complex which is homotopy equivalent to Khovanov's original complex.

Theorems 22.10 and 22.10 are the main results of this chapter.

In Section 22.9 we show that the approach using atoms can be applied for the general Khovanov homology theory (Frobenius extensions) [Man16], and we describe algebraic equations and structures which appear under the attempt of generalising the universal theory of the Khovanov homology directly (as opposed to the simplest (initial) Khovanov complex it turns out that this theory is richer).

An important question in the theory of classical and virtual knots is the problem of defining minimal diagrams of links, diagrams with minimal number of classical crossings in a given class.

At the end of the chapter, we consider the construction of a spanning tree for the Khovanov complex by the author [Man17], this construction literally the same as one introduced for classical knots in Section 7.6.2. We show how one can establish the minimality of link diagrams by using the Khovanov complex. Different minimality theorem [Man'4, Man17] will be formulated which are based on the Jones polynomial as well as the Khovanov complex.

22.2   Basic constructions: The Jones polynomial $\hat{J}$

In the sequel, we shall deal with bigraded complexes $\mathcal{C} = \bigoplus_{i,j} C^{i,j}$, where i is called the *height*, and j is called the *(quantum) grading*. The differential in the complex does not change the grading and increases the height by one. As before the height is also called the *homological grading*.

As usual, we make the substitution $a = \sqrt{(-q^{-1})}$ in the Kauffman bracket. Then, instead of the Jones polynomial we shall get its modified version J. Let us consider the polynomial $\hat{J} = J \cdot (q + q^{-1})$. More precisely, $\hat{J}$ is defined as follows. Let K be an oriented virtual diagram, and let $|K|$ be the corresponding unoriented virtual diagram obtained from K by forgetting the orientation, let n_+ and n_- be the numbers of positive and negative classical crossings of K, and $n = n_+ + n_-$ be the total number of crossings. We set:

$$\hat{J}(K) = (-1)^{n_-} q^{n_+ - 2n_-} [K],$$

where $[K]$ is the modified Kauffman bracket defined according to the rule $[\bigcirc] = (q + q^{-1})$, $[K \sqcup \bigcirc] = (q + q^{-1}) \cdot [K]$, $[\overset{\frown}{\smile}] = [\asymp] - q[\,)(\,]$.

The polynomial $\hat{J}$ has the following conceptually important description in terms of the state cube. Taking away the normalizing factor $(-1)^{n_-} q^{n_+ - 2n_-}$, we get a (slightly modified) Kauffman bracket $\sum_s (-q)^{\beta(s)} (q + q^{-1})^{\gamma(s)}$. This means that we take the sum over all vertices of the cube, of the following products $(-q)^h \times (q + q^{-1})^{\#\bigcirc}$, where h is the height of the vertex, and $\#\bigcirc$ is the number of circles in the state corresponding to the given vertex of the cube.

Thus, in order to compute the polynomial, one has to associate with every circle the Laurent polynomial $(q + q^{-1})$, and then multiply these polynomials taken with some coefficients of the form $\pm q^k$, and take the sum of the obtained polynomials over all vertices of the cube.

Consequently, the Jones polynomial can be restored from the information about the *number* of circles in each of the Kauffman states. If we also take into account *how these circles interfere* when passing from one state to another, we would be able to construct the Khovanov complex.

22.3   Khovanov homology with $\mathbb{Z}_2$-coefficients

Let K be an oriented diagram of a virtual link with n classical crossings.

Consider the bifurcation cube (see Definition 7.3) of K. As usual (see page 94), with each circle in each state of the cube we associate the linear space V over the field $\mathbb{Z}_2$ generated by two vectors v_+ and v_- having grading

± 1, resp. Thus, $\operatorname{qdim} V = (q + q^{-1})$. For each vertex $s = \{a_1, \ldots, a_n\}$ of the cube, we have a certain number of circles to be denoted by $\gamma(s)$. With such a vertex, we associate the vector space $V^{\otimes \gamma(s)}\{\sum_{i=1}^n a_i\}$ obtained from the tensor power of the space V by a grading shift.

Remark 22.1. *In the sequel, we shall use the same notation V for the two-dimensional free module generated by the elements v_+, v_- of grading ± 1 considered over an arbitrary ring of coefficients.*

Remark 22.2. *In this section of the chapter, we consider the symmetric tensor product for which for elements $x_i \in V_i$, $i = 1, \ldots, n$, the following equality $x_{\sigma(1)} \otimes \cdots \otimes x_{\sigma(n)} = x_1 \otimes \cdots \otimes x_n$ holds for any arbitrary permutation σ. We shall also call this product* unordered. *In Section 22.7, we shall consider the tensor product where the sign is the sign of the permutation when identifying products in different orders (this is also called* signed tensor product*).*

We have defined the chain groups of our graded complex. This yields that whatever differentials we take for this complex (provided that $\partial^2 = 0$), the Euler characteristic of this complex will not depend on them. Namely, $\chi(\operatorname{Kh}(K)) = \hat{J}(K)$, where $\operatorname{Kh}(K)$ denote the bigraded homology of the complex we are going to construct.

As in Chapter 7, define the *partial differentials* between the chain groups, acting along the edges of the cube according to the edge directions; i.e. from a smoothing of type A to a smoothing of type B, in the following way. Let an edge of the bifurcation cube correspond to a passage from a state s to a state s' in such a way that l circles are not incident to the crossing in question. These circles do not change when passing from s to s'. At the crossing of $|K|$, corresponding to the edge either one circle splits into two circles or two circles merge into one. In the first two cases, we shall define the partial differential as it was defined in the case of classical knots [BN5], namely, on an edge increasing the number of circles we set $\Delta \otimes \operatorname{Id}^{\otimes l}\{1\}$ and on an edge decreasing the number of circles we set $m \otimes \operatorname{Id}^{\otimes l}\{1\}$. Here the identical mapping Id is referred to the circles which are not incident to the crossing in question, and the maps $m \colon V \otimes V \to V$ and $\Delta \colon V \to V \otimes V$ are defined by formulas (7.1) and (7.2).

For those chains corresponding to the fixed vertex of the cube, the differential ∂ is a sum of all partial differentials (each to be denoted by ∂', possibly, with an index indicating to the edge along which the partial differential acts) along all edges emanating from the given vertex of the cube (oriented in a way increasing the sum of coordinates).

In the general case, the main problem is to define the differential of type $(1 \to 1)$ in a way compatible with differentials of types $(1 \to 2)$ and $(2 \to 1)$ to make the cube anticommutative. For coefficients from $\mathbb{Z}_2$ this difficulty is easy to overcome.

Namely, in the case of bifurcation of type $(1 \to 1)$ we define the partial differential on the edge as the map taking the whole space to zero. Thus, we get

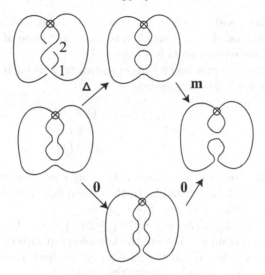

FIGURE 22.1: The commutativity check for a 2-face of the cube.

the *bifurcation cube*, where in comparison with the state cube we additionally indicate how the partial differentials ∂' act. Denote the obtained set of the bigraded groups (the cube) by $[[K]]$. In order for the differential to be well defined, the cube has to be *anticommutative*, i.e. for every two-dimensional face of the cube, the composition of the maps corresponding to one pair of consecutive edges is equal to minus the composition of the maps corresponding to the other pair of consecutive edges connecting the same pair of points. Note that in this case (for the field $\mathbb{Z}_2$) the anticommutativity and commutativity are the same.

Let us define the differential ∂ as the sum of all differentials ∂'.

Lemma 22.1. *The cube $[[K]]$ defined above is commutative.*

This statement is verified by a routine check analogous to that from Chapter 7. It is left to the reader as an exercise. Namely, we check the anticommutativity for every face of the cube.

Here we give an example of such a check (the most interesting one); see Fig. 22.1.

Later on, we shall see (Secs. 22.4 and 22.7.2) that every 2-face of the cube generates a certain atom.

In the present case (Fig. 22.1), it is necessary to check that the map $m \circ \Delta\colon V \to V$ takes the whole space V to zero. Indeed, for such a map we have: $v_- \mapsto v_- \otimes v_- \mapsto 0$, $v_+ \mapsto v_+ \otimes v_- + v_- \otimes v_+ \mapsto 2v_- = 0$ over $\mathbb{Z}_2$.

Note that this case is the only essential "non-classical" case where a bifurcation of type $1 \to 1$ takes place. Indeed, from the parity arguments it follows that on every 2-face of the cube the number of $1 \to 1$-bifurcations is either

equal to zero or it is at least two. For more details see Sec. 22.7.2. If there are
no such bifurcations, then the problem is reduced to one of the classical cases
(all such cases were considered in Chapter 7).

If we consider the case when there are two or four such bifurcations, then
in the 2-face of the cube in question,

$$
\begin{array}{ccc}
V^{\otimes a}\{1\} & \xrightarrow{\ s\ } & V^{\otimes b}\{2\} \\
r\uparrow & & \uparrow t \\
V^{\otimes c} & \xrightarrow{\ p\ } & V^{\otimes d}\{1\},
\end{array}
$$

either each of the compositions $t \circ p$ and $s \circ r$ contains a zero map corresponding
to $1 \to 1$-bifurcation (for example, in the case $a = b$, $c = d$ the maps p and s
are both zero) or the above case takes place.

We set (cf. [Kho1]), $\mathcal{C}(K) = [[K]]\{n_+ - 2n_-\}[-n_-]$. In this case $\mathcal{C}(K)$ is a
well-defined chain complex. Denote the homology groups of the complex $\mathcal{C}(K)$
by $\mathrm{Kh}(K)$ (or by $\mathrm{Kh}_{\mathbb{Z}_2}(K)$ in the case when we have to emphasise that the
Khovanov complex is considered over the field $\mathbb{Z}_2$).

Theorem 22.1 ([Man14, Man'4]). *The graded homology* $\mathrm{Kh}(K)$ *is an invariant of the link* K; *the graded Euler characteristic* $\chi(\mathrm{Kh}(K))$ *is equal to the
Jones polynomial.*

The second statement of the theorem follows from the fact that the Euler
characteristic defined as the alternating sum of (graded) dimensions of homology groups is equal to the alternating sum of the graded dimensions of chain
spaces.

The proof for the homology to be invariant under the Reidemeister moves
just repeats the proof for the case of classical links (see Theorem 7.8).

Definition 22.1. Recall that the *height* $h(\mathrm{Kh}(K))$ of the Khovanov homology
of a virtual link K is the difference between the leading and lowest non-zero
quantum gradings of non-zero Khovanov homology of K (cf. Definition 7.6).

By construction it is clear that

$$
h(\mathrm{Kh}(K)) - 2 \geqslant \frac{\mathrm{span}\langle K \rangle}{2}.
$$

Note that the complex $\mathcal{C}(K)$ splits into the direct sum of two complexes:
the complex with an even grading and the complex with an odd grading (recall
that the differential preserves the grading).

We get two types of the Khovanov homology: the *even* one Kh^e and the
odd one Kh^o.

They correspond to monomials of the Jones polynomial, having degrees
congruent to two modulo four (Kh^o), and monomials the degrees of which are
divisible (congr$\equiv 0 \mod 4$) by four (Kh^e). A classical (or *even virtual*; i.e.
virtual link having a diagram with orientable atom) link has only one of these
two types, more precisely, the following theorem holds.

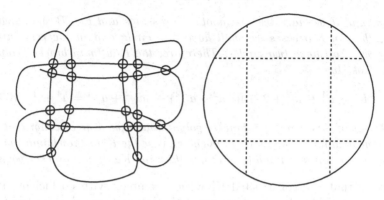

FIGURE 22.2: A virtual knot with orientable atom with genus 2.

Theorem 22.2. *For a classical (and even virtual) link with even number of components the isomorphism* $Kh^o \cong 0$ *holds. For a classical link with odd number of components the isomorphism* $Kh^e \cong 0$ *holds.*

This theorem is completely analogous to Theorem 7.4 about degrees of monomials occurring in the Jones polynomial.

Moreover, it is easy to check that this theorem is true not only for classical links but also for virtual links having a diagram with orientable atom.

Example 22.1. *Let us consider the diagram K depicted in Fig. 22.2 (left).*

The chord diagram corresponding to the leading state of the Kauffman bracket polynomial is depicted in the picture on the right. In this state there exists one circle, and in any of four crossings this circle can be transformed into one circle by using the corresponding dashed chord (with framing 1).

We assert that this link has no diagrams with orientable atoms. Indeed, for the given diagram both complexes Kh^o and Kh^e (with coefficients in $\mathbb{Z}_2$) have non-trivial homology. Actually, the A-state of the diagram with one circle with a label 1 gives a non-trivial cycle (since all differentials coming from the A-state to neighboring states are zero). Further, in states where one crossing is B-smoothed and the other three crossings are A-smoothed there exists exactly one circle. Let us consider the chain equal to the sum of chains having label 1 at each of these four states. It is easy to check that this chain is a cycle. Further, it cannot be a boundary, since all chains in the A-state are cycles.

Thus, there are two homology groups, whose quantum gradings differ by 1, therefore, the link has no diagram with orientable atoms.

In particular, we have shown that the atom genus (the Turaev genus) of the link (see Chapter 16) is equal to one.

Note that this fact cannot be revealed by using the Kauffman bracket polynomial. Indeed, in the A-state (as well as in the B-state) there exists exactly one circle, at each state with one (or three) crossing A-smoothed we have one

circle, and if we have two A-smoothed crossings and two B-smoothed cross-
ings, then in two cases we shall have one circle and in the remaining four
cases we shall have two circles. Therefore, the Kauffman bracket polynomial
of K looks like:

$$\langle K \rangle = a^4 + 4a^2 + 2 + 4(-a^2 - a^{-2}) + 4a^{-2} + a^{-4} = a^4 + 2 + a^{-4}.$$

All terms of this Kauffman bracket polynomial have degrees congruent to each
other modulo four. Therefore, in the given case the Khovanov homology is more
sensitive to non-orientability of atoms than the Kauffman bracket polynomial.

Note that the constructed Khovanov complex with coefficients in $\mathbb{Z}_2$ is
completely defined by the structure of the bifurcation cube and the numbers
n_+, n_-. Therefore, the Khovanov $\mathbb{Z}_2$-homology does not change under the
virtualisation of the given link.

In the next section, we shall give another approach to the construction of
the Khovanov complex (for framed links) which is sensitive to the virtualisation. The Khovanov complex given here coincides with the general Khovanov
complex with coefficients in $\mathbb{Z}_2$ in the classical case; in this case it is easy to
overcome the difficulty with bifurcations of type $1 \to 1$. Later on, we shall
construct the Khovanov complex for not all diagrams of virtual links but
only for "right" virtual diagrams, which have no partial differentials of type
$1 \to 1$ on the cube. As we shall see later, "right" virtual diagrams are those
diagrams which orientable atoms correspond to. Then we shall construct a
"right" virtual diagram for each virtual diagram by some invariant way and
see how the Khovanov homology of the corresponding "right" virtual diagram
changes under the generalised Reidemeister moves applied to the initial diagram (not necessarily "right"). In the next section we shall construct the
Khovanov complex for framed links where the double diagram plays the role
of a "right" diagram.

Example 22.2. *Let us take the virtual knot diagram considered in Exam-*
ple 18.2 (see Fig. 18.11). This knot can be reduced to the unknot with virtuali-
sations and generalised Reidemeister moves. Thus, the Khovanov $\mathbb{Z}_2$-homology
of the knot depicted in Fig. 18.11 coincides with the Khovanov $\mathbb{Z}_2$-homology
of the unknot.

22.4 Khovanov homology of double knots

In the next three sections, we shall use the construction connecting atoms
with virtual knots. Recall this construction given in Chapter 16 which assigns
to a height atom a classical link. This construction is as follows. We embed
the frame of an atom in the plane with its A-structure preserved, and each

crossing is equipped with the over/undercrossing structure according to the B-structure of the atom.

Let an arbitrary atom be given. Let us immerse its frame in the plane with the A-structure preserved, construct a virtual diagram K from the atom in the way given in Chapter 16.

The equivalence class of K is well defined up to virtualisations.

Let K be a virtual diagram with an orientable atom.

Define the complex $\mathcal{C}(K)$ as follows. Fix a ring R of coefficients and two-dimensional free module V over this ring such that $\operatorname{qdim} V = q + q^{-1}$.

The chain space of our complex is the same as in the case of coefficients from $\mathbb{Z}_2$. After that a differential is defined as the sum of partial differentials with signs, and partial differentials are defined with the maps m and Δ.

In the case of coefficients from the field $\mathbb{Z}_2$ the commutativity of each face is equivalent to its anticommutativity. In the case of coefficients from $\mathbb{Z}$ one can make an anticommutative cube from a commutative cube in the following way.

As in Chapter 7, assign to all edges of the cube $\{0, 1\}^n$ sequences consisting of elements from $\{0, 1, *\}$ and having length n and one element $*$. Each such edge connects two vertices obtained by replacing $*$ by one and zero.

Thus, if we denote the map corresponding to an edge ξ by ∂'_ξ, then the differential looks like:

$$\partial^r = \sum_{\{|\xi|=r\}} (-1)^\xi \partial'_\xi.$$

Now we have to explain what the sign $(-1)^\xi$ means and define the map ∂_ξ. To well define the operator ∂ such that the property $\partial \circ \partial = 0$ holds, it is sufficient to show that partial differentials ∂'_ξ on two-dimensional faces of the cube are anticommutative diagrams.

Now, we slightly reformulate the conventions from Chapter 7. A commutative cube can be transformed to an anticommutative cube as follows. First, we have to construct maps on edges such that each two-dimensional face is a commutative diagram, and then we shall equip partial differentials ∂'_ξ with signs. A sign is defined by the following rule. Vertices of the cube are ordered (the homology will not depend on an order). To each vertex of the cube we assign the numbers of all its unit coordinates in the increasing order: $i_1, i_2, \ldots, i_k$ and the formal exterior product $x_{i_1} \wedge x_{i_2} \wedge \cdots \wedge x_{i_k}$. For example, for $n = 3$ we assign to the vertex $\{1, 0, 1\}$ the exterior product $x_1 \wedge x_3$.

Each edge of the cube, increasing some jth coordinate, can be treated as the exterior multiplication on the right by x_j. If as a result of application of this exterior multiplication to a "lower" vertex we get an exterior product assigned to an "upper" vertex, we put the sign "plus" on the edge, and the sign "minus" otherwise. For example, for the edge $\{1, *, 1\}$ we have the sign minus since $(x_1 \wedge x_3) \wedge x_2 = -x_1 \wedge x_2 \wedge x_3$.

Thus, we got a collection of chain groups $[[K]]$ with the differential ∂.

From Exercise 16.1, it follows that if the atom corresponding to a virtual diagram is orientable, then there is no bifurcation of type $1 \to 1$ in the bifurcation cube corresponding to the diagram. Indeed, let us consider the frame Γ of the corresponding atom. Each state of the diagram is an atom having the frame Γ. Circles of the state serve for pasting black cells to the frame Γ. According to Exercise 16.1, the new atom is also orientable. Therefore, this atom cannot have a black cell approaching to itself in the non-orientable way (the way the smoothing at the crossing where this cell touches itself, does not change the number of circles).

Thus, bifurcation cubes are well defined for virtual diagrams with orientable atoms, namely, all bifurcations have the following types $1 \to 2$ and $2 \to 1$; partial differentials are defined by the maps m and Δ; the differential is defined as the sum of partial differentials with signs, and the statement that $\partial^2 = 0$ is checked analogously to the classical case.

Note the following two important lemmas.

Lemma 22.2. *Let K be a virtual diagram with an orientable atom. Then the collection of the groups $[[K]]$ together with the differential ∂ gives a complex; i.e. $\partial^2 = 0$.*

Proof. We have to check that each two-dimensional face of the cube $[[K]]$ is anticommutative. This is equivalent to the verification of the commutativity of two-dimensional faces before putting the signs ± 1.

Each two-dimensional face of the cube $[[K]]$ represents the atom with two vertices. Each two-dimensional face of the cube corresponds to a smoothing of some $(n - 2)$ classical crossings of the diagram K; see Fig. 22.11. The remaining two crossings can be smoothed arbitrarily; four possibilities of such a smoothing correspond to vertices of the two-dimensional face.

In these four states there are some number of common circles not being incident to the two crossings under consideration. After deleting these circles, we get an atom with two vertices.

Thus, we have to check that each two-dimensional face which can correspond to some atom with two vertices represents an anticommutative diagram.

Since the atom corresponding to K is orientable, then the atom corresponding to any two-dimensional face of the corresponding complex is also orientable (according to Remark 16.4).

Let us now use the theorem from [Man2] which tells us that all orientable atoms with two vertices are height atoms.

This means that each atom corresponding to a two-dimensional face of the bifurcation cube corresponding to an orientable atom occurs in the classical case. All such two-dimensional faces are sorted out in [BN1] and for them the commutativity of the corresponding diagrams is proved (before placing signs in differentials).

After that the proof follows line-by-line the proof in the classical case (see, e.g. [BN1]) and from the verification of properties of the maps m and Δ.

Thus, we have shown that the collection of chains $[[K]]$ with the differential

∂ represents an anticommutative cube. Therefore, the complex $\mathcal{C}(K)$ is well defined.

Denote the homology of the complex by $\mathrm{Kh}(K)$. □

Lemma 22.3. *Let K, K' be two virtual diagrams with orientable atoms, herewith K' differs from K by applying a detour move or one of the three classical Reidemeister moves. Then there exists an isomorphism of the Khovanov homology $\mathrm{Kh}(K) \cong \mathrm{Kh}(K')$.*

Proof. By applying the detour move, the structure of classical crossings does not change. Thus, the state cube does not change either, and, therefore, the complex does not change.

In the case of the *classical Reidemeister moves* we use the same proof based on the cancellation principle which was earlier used for the Khovanov homology of classical links. It is local; i.e. it uses only the local structure of Reidemeister moves (not depending on the fixed part of the link under the move). Therefore, the proof passes verbatim for virtual knots under the condition that all complexes are well defined. □

Exercise 22.1. *Let K be a diagram of a virtual link. Then the atom corresponding to the double diagram $D_2(K)$ is orientable.*

Taking into account Exercise 22.1 and Lemma 22.2 we conclude that the Khovanov complex for cables $D_{2n}(K)$ is well defined for any ring of coefficients. The map $K \mapsto D_{2n}(K)$ is almost invariant (it is invariant under all combinations of Reidemeister moves which do not change the writhe number). Therefore, it is natural to expect that the homology of the Khovanov complex for double diagrams of a knot is an invariant of framed links. Namely, the following statement is true.

Exercise 22.2. *Let K, K' be two diagrams of equivalent framed virtual links. Then there exists a collection of diagrams $D_2(K) = K_0, K_1, \ldots, K_n = D_2(K')$ such that:*

1 all atoms corresponding to the diagrams K_i are orientable;

2 for each $i = 0, \ldots, n-1$ the diagram K_{i+1} is obtained from the diagram K_i by applying one of the generalised Reidemeister moves.

Note that it suffices to consider only classical Reidemeister moves, since the detour move does not change an atom.

So, let us consider all classical Reidemeister moves.

If diagrams K and K' differ by applying the first or third Reidemeister move, then the local source–sink structure for the diagram K is in one-to-one correspondence with the local source–sink structure for K' such that outside the domain of the application of the move these diagrams coincide. Here the source–sink structure of lines depicted by dashed lines is defined as opposite to "thick" lines joining to them; see Fig. 22.3.

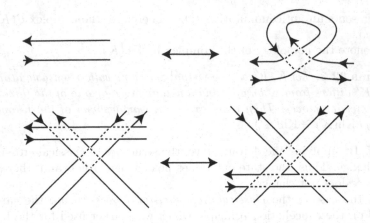

FIGURE 22.3: Labeling for the doubling moves Ω_1 and Ω_3.

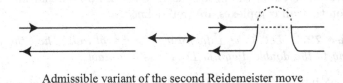

Admissible variant of the second Reidemeister move

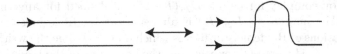

Inadmissible variant of the second Reidemeister move

FIGURE 22.4: Labeling for the doubling move Ω_2.

The second Reidemeister move has two principal different cases, depicted in Fig. 22.4. In the first case (the upper picture), we have two opposite directed arcs (according to the orientation of the source–sink structure), and in the second case we have two arcs going in the same direction.

In the first case, it is mentioned how the local labeling and the source–sink structure change.

The second case is not possible; i.e. it can lead to the fact that after applying the second Reidemeister move the atom becomes non-orientable.

Thus, the (increasing) second Reidemeister move is the only move from the classical Reidemeister moves which can violate the orientability of the atom. All moves from Exercise 22.2 do not violate the orientability.

Remark 22.3. *One can consider the set of diagrams of virtual knots with orientable atoms and the set of moves on it consisting of all Reidemeister* moves not violating the property of orientability (i.e. the detour move, the first and third classical Reidemeister moves and the "orientable" version of the second classical Reidemeister move).

This set was investigated by Kamada under the name alternating virtual links.

In particular, from the arguments given above (Lemma 22.3), it follows that the Khovanov complex is well-defined over any ring of coefficients and invariant in the category of orientable virtual links.

Theorem 22.3. Let n be a natural number. Then $\mathrm{Kh}(D_{2n}(K))$ is an invariant of framed virtual links.

Proof. According to Exercise 22.1, $\mathcal{C}(D_{2n}(K))$ is a well-defined complex. Let K, K' be two diagrams of equivalent framed virtual links. Then by virtue of Exercise 22.2, there exists a collection of virtual diagrams $D_{2n}(K) = K_0, \ldots, K_m = D_{2n}(K')$ corresponding to orientable atoms such that the diagram K_{i+1} is obtained from the diagram K_i by applying generalised Reidemeister moves. By Lemma 22.2 for each of the diagrams K_j the homology $\mathrm{Kh}(D_{2n}(K_j))$ is well defined. The invariance of the homology Kh under the detour move is obvious by construction. Thus, by virtue of Lemma 22.3 (which asserts the invariance under the classical Reidemeister moves), we get $\mathrm{Kh}(D_{2n}(K)) = \mathrm{Kh}(K_1) = \cdots = \mathrm{Kh}(D_{2n}(K'))$. $\square$

Note that the double diagram of K and the double diagram of K' obtained from K by virtualizing one crossing, have different state cubes. Thus the complex constructed in the section can *a priori* distinguish framed virtual diagrams obtained from each other by virtualisation.

However, the "double" Khovanov complex constructed in this section essentially differs from the "general" Khovanov complex for classical knots. In the classical case as well as in the virtual case we have to double and after that we have to calculate the Khovanov homology.

It is natural to raise the question whether the "general" Khovanov homology $\mathrm{Kh}(K)$ is invariant in the case of diagrams with orientable atoms. The positive answer to this question will be given (with some restrictions) in the next section and (completely) in the sections devoted to the Khovanov homology for virtual links.

22.5 Khovanov homology of two-sheeted coverings and atoms

The main goal of this section is the construction of the Khovanov homology by means of two-sheeted coverings. This will lead us to the following

Statement 22.1. *Let* F *be a field, and let* K, K' *be two equivalent virtual diagrams with orientable atoms. Then there exists an isomorphism of graded homology* $\mathrm{Kh_F}(K) \cong \mathrm{Kh_F}(K')$.

Note that the general assertion about the invariance of this homology with arbitrary coefficients follows from the parity arguments (see below) and from the explicit construction of the Khovanov homology for virtual knots.

The main construction is as follows. For each virtual diagram K one can consider the atom $\mathrm{At}(K)$ corresponding to it. Later on, we shall use the techniques of *orientable covering*. Namely, if the atom $\mathrm{At}(K)$ is orientable, we consider two copies of $\mathrm{At}(K)$; if it is not, then we consider the atom $\widetilde{\mathrm{At}}(K)$ which is the orientable two-sheeted covering over the atom $\mathrm{At}(K)$. It is defined as the two-sheeted covering over the corresponding surface; here, the preimage of the frame is a graph which we consider as the frame, the preimage of a black cell is a pair of black cells, and the preimage of a white cell is a pair of white cells. The atom obtained in such a way can be either two-component or one-component, and it depends on the orientation of the initial atom.

Denote the virtual diagram corresponding to the atom $\widetilde{\mathrm{At}}(K)$ by $\widetilde{K}$.

If we apply a classical Reidemeister move Ω_i to the initial diagram K, then the move Ω_i will be applied to the diagram $\widetilde{K}$ in two places; here in the case of the move Ω_2, the *admissible* variant of the second Reidemeister move will be applied to $\widetilde{K}$ twice.

This construction can be treated as follows: We consider two sets of vertices of the atom with the A-structure at them and connect vertices by edges.

Thus, for each virtual knot we can consider its "covered version":

$$K \to \mathrm{At}(K) \mapsto \widetilde{\mathrm{At}}(K) \mapsto \mathrm{Kh_F}(\widetilde{K}).$$

In terms of a knot diagram, this construction is described as follows. Let a virtual diagram K be given; this diagram has n classical crossings $v_1, \ldots, v_n$. These crossings are connected with each other in some way. Thus, we have a graph Γ immersed in the plane. Each crossing v_i has four (adjacent) ends v_{i1}, v_{i2}, v_{i3}, v_{i4} enumerated, for example, in clockwise manner, with crossings connected by branches of the diagram which edges of the atom correspond to. Let an edge e_j connect the ends $v_{j_1 j_2}$ and $v_{j_3 j_4}$, where j_2, $j_4 \in \{1, 2, 3, 4\}$.

The diagram $\widetilde{K}$ is constructed as follows. It contains $2n$ crossings $v_1', \ldots, v_n', v_1'', \ldots, v_n''$, which are connected by edges. Each edge e_j of the initial diagram has two preimages: e_j^1 and e_j^2. Each of two edges e_j^i connects an end $v_{j_1 j_2}'$ or $v_{j_1 j_2}''$ with an end $v_{j_3 j_4}'$ or $v_{j_3 j_4}''$. For each edge e_j^1 we have to choose which ends are connected (v' or v''). Here we have an ambiguity. The matter is that before describing edges we have not had a natural ordering of vertices: Which of the vertices v_i' or v_i'' is the "first" and which one is the "second"? To overcome this difficulty let us choose some spanning tree T for Γ and say that all edges e_j^1 corresponding to edges of this graph connect ends $v_{j_1 j_2}'$ with $v_{j_3 j_4}'$ (thereby edges e_j^2 connect ends $v_{j_1 j_2}''$ and $v_{j_3 j_4}''$).

Another choice of the tree will correspond to some change of notation:

v'_j and v''_j swap places in some pairs. After that the rule for connecting the remaining ends by edges e^1_i and e^2_i follows. In order to indicate which pairs of ends are connected by an edge e^α_i, we shall either connect them by the edge e^1_i or e^2_i: the "symmetric" pair of ends corresponding to it obtained by swapping $v' \longleftrightarrow v''$ will also be connected by an edge. Henceforth, for constructing a virtual diagram it is not important for us to remember the notation for these edges.

We shall not pay attention to how we place edges e^α_i on the plane. The resulting class of virtual link will not depend on it (by construction, diagrams will differ from each other by applying a finite sequence of detour moves).

So, we have fixed a maximal tree $T \subset \Gamma$. Each edge e_j not belonging to this tree represents the minimal cycle on the subgraph $T \cup e_j \subset \Gamma$. In the case when this cycle is *good* (see below), we connect the ends $v'_{j_1 j_2}$ and $v'_{j_3 j_4}$ by the edge e^1_j, and the ends $v''_{j_1 j_2}$ and $v''_{j_3 j_4}$ by the edge e^2_j. In the case of a *bad* cycle we connect the ends $v'_{j_1 j_2}$ and $v''_{j_3 j_4}$ by the edge e^1_j, and the ends $v''_{j_1 j_2}$ and $v'_{j_3 j_4}$ by the edge e^2_j. The notion of *good and bad* edges goes back to orientable and non-orientable cycles on the corresponding atom. An edge is called *good* if the corresponding cycle is orientable. Under the covering of the atom, orientable cycles are taken to cycles, and non-orientable cycles are sent to paths (with some ends in v'_k, v''_k). Let us define the notion of a *good edge* (for edges not belonging to T), and the notion of a *good (orientable)* cycle in terms of a diagram of the virtual link. For this we consider all edges of the given cycle $e_{j_1}, e_{j_2}, \ldots, e_{j_k}, e_{j_{k+1}} = e_{j_1}$, where edges $e_{j_i}, e_{j_{i+1}}$ meet at a vertex (indices i are taken modulo k) and let us try to define locally the source–sink structure along them. Let us orient the edge e_{j_1} in some way. Further, if the edge e_{j_2} is opposite to the edge e_{j_1} at a vertex, then we orient e_{j_2} such that either both edges e_{j_1} and e_{j_2} come into the vertex, or both edges emanate from it; in the case when the edges are not opposite, we shall make one of them come into the vertex and the other emanate from it. Further, we do the same for the orientation of $e_{j_3}, e_{j_4}, \ldots$. If the process converges; i.e. we have the orientation of $e_{j_{i+1}} = e_{j_1}$ coincides with the initial one, we call the cycle *good*, and *bad* otherwise. Namely, a cycle is called *good (orientable)* if the number of its transversal passages through classical crossings, vertices of the atom, is even.

Remark 22.4. *For a plane diagram the parity of the number of transverse passages through classical crossings coincides with the parity of passages through virtual crossings (all these passages are transverse).*

It is easy to check that this definition of a good cycle coincides with the definition of an orientable cycle on the atom defined by the A-structure. Setting successively orientations of edges according to the source–sink structure, we define orientations of black cells approaching (locally) to these edges. The first vector of the basis is directed along the orientation of the edge, and the second one is directed inward the black cell. If we return to the initial edge with the same orientation, then this means that we have traveled along an orientable cycle, and a non-orientable cycle otherwise. Indeed, if we pass through a classical crossing, then orientations of neighboring cells defined in such a way, are opposite to each other. Thus, getting a compatible orientation means precisely that our path goes transversely evenly many times.

So, we have defined the notion of a *good* (*orientable*) cycle and a *good* edge (for edges not belonging to the tree T). Therefore, we have completely constructed the virtual diagram $\widetilde{K}$. Note that the definition of a good cycle does not depend (up to detour moves) on the choice of the tree T.

Moreover, from the atom $\widetilde{At}(K)$ the knot corresponding to the two-sheeted covering is restored up to virtualisations; we have already mentioned the explicit way of constructing the diagram $\widetilde{K}$ with the diagram K; it corresponds to some immersion of the frame of the atom $\widetilde{At}(K)$ (with preserving the A-structure).

It is easy to see that the detour move in the initial diagram K of the link induces some combinations of the detour moves on the diagram $\widetilde{K}$. Moreover, the following lemma takes place.

Lemma 22.4 ([MI]). *By applying one of the classical Reidemeister moves to a diagram K the diagram $\widetilde{K}$ will change in the following way: The same Reidemeister move is applied to it in two places. Herewith, the atom corresponding to the "middle" diagram obtained from $\widetilde{K}$ by applying the second Reidemeister move in one place (any of two places) is orientable.*

Exercise 22.3. *Prove the lemma.*

According to Lemma 22.2, the homology $\mathrm{Kh}(\widetilde{K})$ is well defined.

Therefore, by Lemma 22.3 the Khovanov homology of a "covered" link does not change under applying the Reidemeister move to the initial knot. This leads us to the following

Theorem 22.4. *The map $K \mapsto \mathrm{Kh}(\widetilde{K})$ gives a well-defined invariant of virtual links.*

Remark 22.5. *Note that only the second Reidemeister move Ω_2 can change the type of the corresponding atom (i.e. it can convert a non-orientable atom to an orientable one and vice versa). If, for example, we have an orientable atom $\mathrm{At}(K)$ and two components of the atom $\widetilde{At}(K)$, then the application of the second Reidemeister move (non-admissible version) to K can "connect" these components into one (this corresponds to the fact that after applying the second Reidemeister move, the atom may become non-orientable).*

Herewith the moves Ω_1, Ω_3 preserve the orientability of the atom.

Now let the atom corresponding to a diagram K be orientable. Then $\widetilde{K}$ consists of two copies of the atom corresponding to K. Since F is a field, we have $\mathrm{Kh}_{\mathrm{F}}(\widetilde{K}) = \mathrm{Kh}_{\mathrm{F}}(K)^{\otimes 2}$.

Therefore, the homology $\mathrm{Kh}(K)$ is obtained from the invariant homology $\mathrm{Kh}(\widetilde{K})$ by "extracting of the tensor square root". In the case when the ring of coefficients is a field, we have the Poincaré polynomial $\mathfrak{P}$ in two variables with all integer non-negative coefficients. From this polynomial we have to extract the "square root"; i.e. to find the Laurent polynomial $\mathfrak{Q}$ in the same two variables with integer non-negative coefficients (coefficients are non-negative since they are the ranks of Khovanov homology groups) such that the equality $\mathfrak{Q}^2 = \mathfrak{P}$ holds. It is obvious that if we can do this, then it can be done uniquely. Since this operation is unique, if it exists, we get the claim of Theorem 22.1.

Moreover, from these discussions we get the following

Theorem 22.5. *Let* F *be a field, and let for a virtual diagram* K *the graded homology* $\mathrm{Kh}_\mathrm{F}(\widetilde{K})$ *cannot be represented as the tensor square. Then* K *has no diagram with an orientable atom. In particular, the virtual link generated by* K *is not classical.*

It is natural that the Khovanov complex constructed in this section cannot detect non-triviality of the virtual knot depicted in Fig. 18.11, since this knot is obtained from the unknot by generalised Reidemeister moves and virtualisations.

The question about whether two non-isotopic classical links can be obtained from each other by a finite sequence of generalised Reidemeister moves and virtualisations is an important and interesting conjecture (*virtualisation conjecture*). Note that the virtualisation conjecture is true for the unknot (i.e. if a classical diagram of a knot is obtained from a diagram of the unknot by applying a finite sequence of the generalised Reidemeister moves and the virtualisation, then the classical diagram represents the unknot), since the Khovanov homology detects the unknot, see [KrMr4]. The Khovanov complex gives a partial answer to this question.

From Theorem 22.1 and the invariance of the Khovanov homology under virtualisation, we have the following theorem.

Theorem 22.6. *If a classical link is obtained from a classical link by applying generalised Reidemeister moves and virtualisations, then these links have the same Khovanov homology with coefficients from any preassigned field.*

Later in this chapter we shall show that this theorem is true for arbitrary coefficients (e.g. from the ring $\mathbb{Z}$), see Theorems 22.11 and 22.12.

22.6   Khovanov homology and parity

Assume there exists a map $\widetilde{f}$ sending the set of diagrams of virtual knots into itself and having the following properties:

1. for each virtual diagram K the diagram $\widetilde{f}(K)$ is a virtual diagram with an orientable atom;

2. if a diagram K has an orientable atom, then $\widetilde{f}(K) = K$;

3. if two diagrams K and K' are equivalent by means of Reidemeister moves, then $\widetilde{f}(K)$ and $\widetilde{f}(K')$ are equivalent by means of Reidemeister moves, where all intermediate diagrams connecting the diagrams $\widetilde{f}(K)$ and $\widetilde{f}(K')$ have orientable atoms.

Theorem 22.7. *The map* $K \mapsto \mathrm{Kh}(\widetilde{f}(K))$ *is an invariant of virtual links.*

22.7 Khovanov homology for virtual links

22.7.1 Atoms and twisted virtual knots

Bifurcations of types $2 \to 1$ and $1 \to 2$ in the Khovanov complex will (see Section 22.7.2) correspond to *partial differentials* ∂'; the differential ∂ consists of (see below); the bifurcation of type $2 \to 1$ corresponds to the multiplication m, and the bifurcation of type $1 \to 2$ corresponds to the comultiplication Δ.

The complete information about the number of circles in states of the diagram can be extracted from the corresponding atom. In other words, the state cube can be completely restored from the atom.

An actual problem is the problem of finding the the minimal genus of atoms corresponding to diagrams of the virtual link. Classical link diagrams of genus zero are the connected sums of alternating diagrams.

This genus is called the *virtual link genus* or the *Turaev genus* due to [Tur1], cf. Definiton 16.1. It turned out [Low] that this genus had an important significance in studying Heegaard–Floer homology of classical knots.

We shall construct the Khovanov complex starting with a given virtual link diagram. As we shall see, the homology of the complex constructed in this way really depends only on the corresponding atom. Thus the homology will be invariant under virtualisation. This supports the virtualisation conjecture mentioned above.

Twisted virtual knots [Bour, Vir2] are close relatives of virtual knots. They are represented by knots in oriented thickenings of not necessarily orientable surfaces modulo stabilisation/destabilisation.

A particular case of the theory of twisted virtual knots is the theory of knots in $\mathbb{R}P^3$ (cf. Chapter 23).

Definition 22.2. An *orientable thickening* of a two-dimensional surface M is an orientable three-dimensional I-bundle over M, where I is a segment.

Let us consider a non-orientable surface S and construct the canonical oriented I-bundled over it. It represents a three-dimensional manifold $S \widetilde{\times} I$ with boundary.

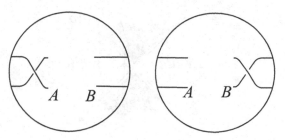

FIGURE 22.5: A branch AB forms overcrossing in the left picture and undercrossing in the right picture.

A nice example of such a thickened surface is $\mathbb{R}P^2 \widetilde{\times} I$, which is homeomorphic to $\mathbb{R}P^3 \backslash \{*\}$. Thus, by constructing the Khovanov homology for such knots, we shall get the Khovanov homology theory for knots in $\mathbb{R}P^3$.

Given a surface M and its thickening $M \widetilde{\times} I$, then links in $M \widetilde{\times} I$ can be considered by means of their diagrams: projections on M.

There are two types of stabilisation/destabilisation of such thickening surfaces: along orienting cycles and along non-orienting cycles. In the second case, we add/remove a thickened Möbius band.

In general position, a projection is a framed 4-graph. In order to restore the link, one should indicate for each crossing how the two branches behave in a neighbourhood of this crossing. In the orientable case, one just indicates which branch should be over, and which branch should be under. However, in the non-orientable case this indication is relative. While walking along a non-orienting circuit, the direction upwards changes to the direction downwards. So, for example, knots in $\mathbb{R}P^3 \backslash * = \mathbb{R}P^2 \widetilde{\times} I$ can be represented by diagrams in $\mathbb{R}P^2$ such that all crossings lie inside the disc $D^2 \subset \mathbb{R}P^2$; when passing the boundary of the disc the direction changes; see Fig. 22.5. To handle this, we choose an affine chart such that the complement to this chart in S is one-dimensional. For this chart we have a well-defined notion of an over/undercrossing.

Note that links in such surfaces are well described by atoms. Indeed, fix (once for all) an orientation on $M \widetilde{\times} I$. Now, for a link diagram in M, we already have the frame of the atom: a framed 4-graph.

Now, the way for attaching black cells is the following (see Fig. 22.6). For a vertex v, we take two emanating non-opposite half-edges a and b. The corresponding virtual link contains two points projected in the vertex v, one of which is incident to the edge corresponding to a, and the other one is incident to the edge corresponding to b. In a neighbourhood of v, denote by c the small vector going from a point on the edge a to a point on b. If the basis (a, b, c) is positively oriented in our three-dimensional manifold, then the angle between half-edges a and b is decreed to be white, as well as the opposite angle. Otherwise they are both black.

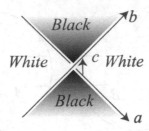

FIGURE 22.6: Constructing the atom from a diagram.

Note that this choice does not depend on the ordering of the pair (a, b), nor on their directions.

In the case of general virtual links which are a particular case of twisted virtual links, the way of pasting black cells described above is agreed with the way described in Chapter 16.

This leads to the following theorem.

Theorem 22.8. *There is a well-defined map from the set of twisted virtual knots to the set of virtual knots modulo virtualisation.*

Knots in such surfaces were considered by Asaeda, Przytycki and Sikora in [APS], and Viro [Vir2] (Bourgoin first considered stabilisations that led to twisted virtual knots). In [APS] a Khovanov homology theory for such surfaces was constructed by using an additional topological information coming from surfaces. See also [MN].

From Theorem 22.8 and the invariance of the Khovanov homology under virtualisation (Lemma 22.5, see below), it follows immediately that the Khovanov homology constructed below can be generalised for twisted virtual knots.

22.7.2   Khovanov complex for virtual knots

Our aim is to define a homology theory for virtual knots (with arbitrary atoms) over an arbitrary ring in such a way that:

1. the homology we are defining is invariant under the (generalised) Reidemeister moves;

2. for the case of virtual knots with orientable atoms (also known as *alternatible virtual knots*) this homology theory coincides with the one constructed in the previous sections;

3. the tensor product of the complex with $\mathbb{Z}_2$ coincides with the theory constructed in Sec. 22.3;

4. the graded Euler characteristic of the complex which will be constructed coincides with the Jones polynomial.

The invariant of Rushworth [Rush] violates the last condition.

Remark 22.6. *The coefficient ring might be an arbitrary abelian group with unit, for example,* $\mathbb{Z}$.

For the sake of simplicity we shall sometimes abuse the notation and call modules over rings "linear spaces", not depending on whether the ring is a field or not.

If no $1 \to 1$-bifurcations occur, we may construct the Khovanov cube by using the standard differentials, the multiplication m (for $2 \to 1$-bifurcations) and the comultiplication Δ (for $1 \to 2$-bifurcations).

The situation with the $1 \to 1$-bifurcation (the essential phenomenon of the theory of virtual knots appearing because of the existence of non-orientable atoms) makes the problem more complicated. Indeed, if we wish to construct a grading-preserving theory without introducing any new grading, this partial differential should be identically equal to zero because of the grading reasons (there should be a map from V to V that lowers the grading by one). In the space V, the basis consists of two elements with gradings $+1$ and -1. If we set this partial differential to be equal to zero with all other differentials (m and Δ) defined in the standard way, we get a straightforward generalisation for the $\mathbb{Z}_2$ case.

Below we involve two additional structures: The basis change in the space V corresponding to a circle and generated by $\{1, X\}$ (the homology group of the unknot) while passing from one crossing to another and the exterior product of "circles" instead of their usual tensor products.

Notational agreement. Given an unordered set of vector spaces, enumerate them arbitrarily: $V_1, \ldots, V_n$. We shall define a new space not depending on the ordering of the spaces, which will be denoted[1] by $V_1 \wedge V_2 \wedge \cdots \wedge V_n$ as follows. Consider all possible tensor products of these spaces and identify them according to the following rule. Let $x_i \in V_i$, $i = 1, \ldots, n$. We set $x_{\sigma_1} \otimes \cdots \otimes x_{\sigma_n} = \text{sign}(\sigma) \, x_1 \otimes \cdots \otimes x_n$.

We shall denote such tensor product $x_1 \otimes \cdots \otimes x_n$ of elements $x_i \in V_i$ by $x_1 \wedge x_2 \wedge \cdots \wedge x_n$. We call this space the *ordered tensor product*.

Remark 22.7. *To avoid confusion, note that, in writing* $X \wedge X$, *we always assume that the first* X *and the second* X *belong to different (but possibly isomorphic) spaces; thus* $X \wedge X$ *is not zero (unlike the wedge product of 1-forms).*

Let us consider a virtual diagram K.

To handle it and to make the whole cube anticommutative we have to add two ingredients, sensitive to orientability of the atom.

---

[1] In the case of coincidence of the linear spaces $V = V_1 = \cdots = V_n$ we shall use also the notation $V^{\wedge n}$.

1. With each circle C in each state we associate a vector space of graded dimension[2] equal to $q + q^{-1}$. Namely, given an orientation o of the circle C; we associate with this circle the graded vector space generated by elements 1 and $X_{C,o}$ of gradings 1 and -1, respectively. The orientation change of the circle (passing to $-o$) leads to $X_{C,-o} = -X_{C,o}$.

2. Given a state s of a virtual link diagram K having l circles $C_1, \ldots, C_l$, with this state, we associate an ordered tensor product $V^{\wedge l}$; as a basis of this product we take the product $(p^1)_{C_{a_1}} \wedge (p^2)_{C_{a_2}} \wedge \cdots \wedge (p^l)_{C_{a_l}}$, where $(p^i)_{C_{a_i}}$ represents an element from $V_{C_{a_i}}$.

Thus, we have defined the chain space of the complex corresponding to the virtual diagram K. We denote it by $[[K]]$. All the basis elements of this space correspond to some states of K with an additional choice of the elements ± 1 or $\pm X$. Let s be a state of K with the set of circles $C_1, \ldots, C_l$, whence for these circles we have chosen elements $\gamma_1, \ldots, \gamma_l$, each of them being ± 1 or $\pm X$. Then these elements form a chain of the complex $[[K]]$ having the height h, where h is the number of B-smoothings of s, and the grading which is equal to $h + \#1 - \#X$, where $\#1$ is the number of elements of type ± 1 among $\gamma_1, \ldots, \gamma_l$, and $\#X$ is the number of elements $\pm X$ among $\gamma_1, \ldots, \gamma_l$.

Our next goal is the description of the differential ∂ in this complex, which increases the height by one and does not change the grading.

Set $n_+ =$ the number of crossings , $n_- =$ the number of crossings .

Denote by $\mathcal{C}(K)$ the complex obtained from $[[K]]$ by the height shift and the grading shift: $\mathcal{C}(K) = [[K]]\{n_+ - 2n_-\}[-n_-]$; i.e. the height of each chain decreases by n_-, and the grading increases by $(n_+ - 2n_-)$; all differentials change respectively. Here we assume that $[[K]]$ is a complex, this fact will be proved below.

Whatever the differential ∂ is, from the construction of chains of the complex $\mathcal{C}(K)$ follows Theorem 22.9.

Theorem 22.9. *For any virtual diagram K we have $\chi(\mathcal{C}(K)) = \hat{J}(K)$.*

We shall think of all classical crossings as oriented upwards: and .

Consider a state s of a diagram of an oriented virtual link. Choose a classical crossing and consider all circles of the state s incident to this crossing. There are one or two such circles. Fix orientations on these circles according to the orientation of the edge emanating upwards to the right (and opposite to the orientation of the edge incoming to the crossing from the bottom left; see Fig. 22.7, upper part). As we shall see further, in the case of one circle, these two orientations defined locally can be uncoordinated, but this case can be treated easily.

---

[2]From now on, we have passes from the notation v_+ and v_- to the notation 1 and X (before v_+ play the role of unity). This leads to the same homology theory up to a grading shift and a normalization. In the sequel we should not pay attention to these normalizations and shifts, this agrees with [MI] in verbatim.

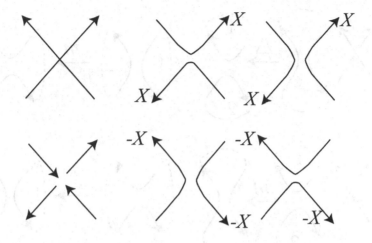

FIGURE 22.7: Definition of a basis at a crossing.

Thus, the orientations of these circles of the state s locally agree with the orientation of the edge emanating upwards to the right (as well as with the edge incoming from the bottom-right) and disagree with the orientation on the left side. We orient the half-edges as shown in the lower-left part of Fig. 22.7. Thus, we have fixed a choice of the generator X for any circle incident to a given crossing. Note that for another crossing for the same circle the choice of X may differ from this one by a sign.

Differentials will be defined according to the orientations of circles at classical crossings and local orderings of components with the following rule.

The orientations described above are well defined unless the case when the edge corresponding to the crossing of the diagram bifurcates one circle to one circle. *In such cases, we set the partial differential to be zero.*

Assume we have a $1 \to 2$ or $2 \to 1$-bifurcation at a crossing.

If we deal with two circles incident to the crossing from the opposite sides, we order them in such a way that the upper (respectively, left) circle is locally first; the lower (respectively, right) one is thus, the second. In the sequel, when defining partial differentials we assume that all circles are ordered in such a way that the circles we deal with are in the very first position in our tensor product; this can always be obtained by means of a permutation, which might lead to a sign change. The map on the other circles is identical.

Let there be given an edge of the bifurcation cube where the number of circles is changed by one. This bifurcation corresponds to a certain crossing; we have two options $2 \to 1$ or $1 \to 2$. In those states when we have two circles incident to the crossing, the circles are ordered. Moreover, all three circles are oriented, thus, we have chosen a basis for the space corresponding to each of these circles.

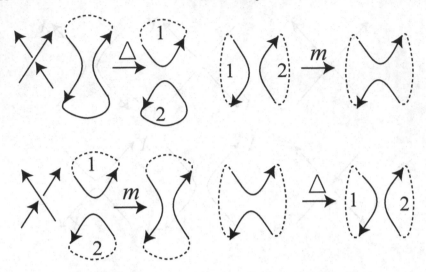

FIGURE 22.8: Defining operations m and Δ.

Now we define the maps $\Delta\colon V \to V \wedge V$ and $m\colon V \wedge V \to V$ locally according to the prescribed choice of generators at the crossing and local ordering (see Fig. 22.8):

$$\Delta(1) = 1_1 \wedge X_2 + X_1 \wedge 1_2; \quad \Delta(X) = X_1 \wedge X_2 \qquad (22.1)$$

and

$$m(1_1 \wedge 1_2) = 1; \quad m(X_1 \wedge 1_2) = m(1_1 \wedge X_2) = X; \quad m(X_1 \wedge X_2) = 0. \qquad (22.2)$$

Note that the map m is surjective and the map Δ is injective.

If we have some circles $C_1, \ldots, C_l$ not incident to the crossing in question, and elements $\gamma_1, \ldots, \gamma_l$ on them, the formulae for the partial differentials ∂' are written as:

$$\partial'(1 \wedge \gamma_1 \wedge \cdots \wedge \gamma_l) = \Delta(1) \wedge \gamma_1 \wedge \cdots \wedge \gamma_l$$
$$= 1_1 \wedge X_2 \wedge \gamma_1 \wedge \cdots \wedge \gamma_l + X_1 \wedge 1_2 \wedge \gamma_1 \wedge \cdots \wedge \gamma_l, \qquad (22.3)$$
$$\partial'(X \wedge \gamma_1 \wedge \cdots \wedge \gamma_l) = \Delta(X) \wedge \gamma_1 \wedge \cdots \wedge \gamma_l = X_1 \wedge X_2 \wedge \gamma_1 \wedge \cdots \wedge \gamma_l$$

(in the case of a $1 \to 2$-bifurcation) and

$$\partial'(1_1 \wedge 1_2 \wedge \gamma_1 \wedge \cdots \wedge \gamma_l) = m(1_1 \wedge 1_2) \wedge \gamma_1 \wedge \cdots \wedge \gamma_l$$
$$= 1 \wedge \gamma_1 \wedge \cdots \wedge \gamma_l,$$
$$\partial'(X_1 \wedge 1_2 \wedge \gamma_1 \wedge \cdots \wedge \gamma_l) = \partial'(1_1 \wedge X_2 \wedge \gamma_1 \wedge \cdots \wedge \gamma_l)$$
$$= m(X_1 \wedge 1_2) \wedge \gamma_1 \wedge \cdots \wedge \gamma_l \qquad (22.4)$$
$$= m(1_1 \wedge X_2) \wedge \gamma_1 \wedge \cdots \wedge \gamma_l$$
$$= X \wedge \gamma_1 \wedge \cdots \wedge \gamma_l,$$
$$\partial'(X_1 \wedge X_2 \wedge \gamma_1 \wedge \cdots \wedge \gamma_l) = m(X_1 \wedge X_2) \wedge \gamma_1 \wedge \cdots \wedge \gamma_l = 0$$

(in the case of a $2 \to 1$-bifurcation).

After that we define the differential ∂ on the chain space corresponding to the state s as the sum of partial differentials acting on the state s.

Example 22.3. *Thus, if we wish to comultiply the second factor X_2 in $X_1 \wedge X_2$, we get $X_1 \wedge X_2 = -X_2 \wedge X_1 \to -X_2 \wedge X_3 \wedge X_1 = -X_1 \wedge X_2 \wedge X_3$, where X_3 belongs to the newborn third component (under the condition that at the crossing of splitting the circle X_2 is locally first (i.e. upper and left), and the circle X_3 is locally second).*

Given an oriented diagram K of a virtual link, we have constructed a set of bigraded groups with the differential ∂. Denote the set of groups by $[[K]]$. The differential increases the height and does not change the grading.

Our goal is to prove the main theorem.

Theorem 22.10. *The set of groups $[[K]]$ together with the differential ∂ is a well-defined bigraded complex, i.e. $\partial^2 = 0$. Herewith the differential preserves the grading and increases the height by one.*

The complex $\mathcal{C}(K)$ is obtained from $[[K]]$ by the height shift and grading shift. From the constructions it will follow that the homology of the complex $\mathcal{C}(K)$ coincides with the homology constructed for the case of virtual knots with orientable atoms.

Further, from the proof of Theorem 22.10 the claim of Theorem 22.9 follows by construction.

The complex with coefficients in $\mathbb{Z}_2$ coincides with the complex over $\mathbb{Z}_2$ described in Sec. 22.3.

Theorem 22.11. *The homology of the bigraded complex $\mathcal{C}(K)$ is an invariant of the virtual link K under generalised Reidemeister moves.*

We first prove Theorem 22.10. After that, we shall prove Theorem 22.11; its proof will be more technical and it will follow the standard scheme described above some additional sign checks for partial differentials, appearing while ordering and orienting the circles, will be needed. We shall also show that the homology of $\mathcal{C}(K)$ coincides with the homology constructed for the case of virtual knots with orientable atoms.

We first prove two lemmas that establish some properties of our complex $\mathcal{C}(K)$ and simplify further arguments.

Let K be a virtual diagram. Consider a classical crossing v of it. Let the diagram K' be the diagram obtained from K by the virtualisation of v. Then there exists a one-to-one correspondence between the sets of classical crossings of the diagrams K and K'. It generates a one-to-one correspondence ϕ between the states (for the corresponding vertices we have either A-smoothings or B-smoothings). Note that such a bijection does not change the number of circles in the states; it follows from the fact that all states can be restored from the atom, and the atom does not change under virtualisations. Let us orient circles of corresponding states *identically* outside the crossing v. This identification defines the map $g\colon [[K]] \to [[K']]$ of the chain spaces according to the following rule. For any state s and the corresponding state $\phi(s)$, the diagrams K and K' look identical outside a neighbourhood of v. Thus, we can establish the bijection between oriented circles of s and oriented circles of $\phi(s)$, that leads to the definition of g. We shall use the same notation g for maps of vector spaces (modules) corresponding to the circles in states s and $\phi(s)$.

Let C_s be the subspace of the space $[[K]]$ associated with a state s of the diagram K. Denote the corresponding space for K' by $C_{s'}$.

Lemma 22.5. *Let K, K' be two diagrams obtained one from another by the virtualisation. Then there is a grading-preserving chain map $f\colon [[K]] \to [[K']]$ that maps C_s isomorphically to $C_{s'}$ and commutes with the local differentials.*

In particular, if $[[K]]$ is a well-defined complex, then so is $[[K']]$; herewith their homology groups are isomorphic.

Proof. Suppose the diagram K' is obtained from the diagram K by the virtualisation at a crossing v.

The map f is constructed according to the crossing type of v ($\times$ or $\times$). By construction, partial differentials of the complex $[[K']]$ coincide with the images of partial differentials of $[[K]]$ under g, except, maybe, those partial differentials corresponding to the crossing v. Furthermore, differentials corresponding to v split our cube to the "lower subcube" and the "upper subcube", as shown in Fig. 22.9.

Now, the remaining partial differentials differ possibly by signs on edges corresponding to the crossing v. Our goal is to show that they either all agree or all differ by -1 sign, as shown in Fig. 22.9.

Indeed, the bases at all crossings but v agree for K and K'. This leads to the identification of chains of the corresponding complexes. For this isomorphism for every circle C incident to v and the circle $g(C)$ corresponding to it in the corresponding state of the diagram K' we have $g(X_{C,o_K}) = -X_{g(C),o_{K'}}$, where o_K and $o_{K'}$ are the orientations of the circles C and C' at the crossing v of the diagrams K and K' chosen according to the rule depicted in Fig. 22.7. The latter identity holds because in any state s the circle C that tends from the upper-right to the crossing v of K, corresponds to the circle $\phi_*(C)$ in the state $\phi(s)$ that tends to v from the upper-left, this corresponds to the change

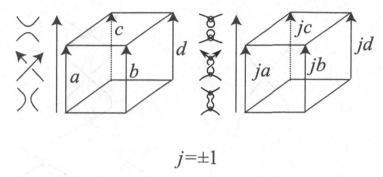

$$j=\pm1$$

FIGURE 22.9: The behaviour of the cube under the virtualisation.

X to $-X$ in the local basis of spaces V corresponding to circles of the state incident to the given crossing; see Fig. 22.7. If we dealt with the usual tensor product case regardless of the circle ordering, the transformation $X \to -X$ would leave m invariant and change Δ to $-\Delta$.

Assume now that the crossing v is positive (⟨image⟩). All maps of type m corresponding to v, represent bifurcations of two circles (a left one and a right one) into one circle. After the virtualisation, the circles interchange their roles; see Fig. 22.10.

Globally we get a sign change for all m-type partial differentials. For partial differentials of type Δ we have one circle that bifurcates into two ones, the upper one, and the lower one; the "up-down" position remains unchanged under virtualisation, that preserves all Δ-type partial differentials. The first component is shown locally by solid line, whence the second component is shown by a dashed line.

Summing up (and recalling the sign change of the partial differential Δ because of passing $X \to -X$), we see that the virtualisation of a positive crossing changes the signs of all partial differentials corresponding to this crossing.

Now divide the chain space $[[K]]$ and $[[K']]$ into two parts each, according to the smoothing of v; we call one part of the cube "upper", the remaining part being lower. Now set $f \colon [[K]] \to [[K']]$ as g for all elements from the lower subcube and as $-g$ for the upper subcube.

Evidently, this map commutes with partial differentials. Indeed, the commutativity of the map f with partial differentials inside one of the subcubes follows from the fact that the map g is anticommutative; therefore, the map f commutes.

Thus if the initial cube were anticommutative, then the constructed map would be an isomorphism in homology.

Similar arguments show that the virtualisation of a negative crossing does not change the cube at all. The minus sign that appears on edges correspond-

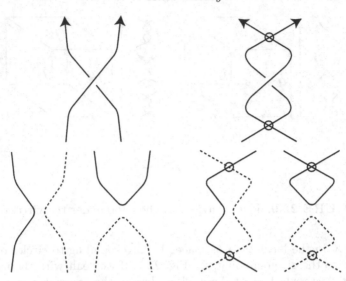

FIGURE 22.10: Virtualisation.

ing to Δ is canceled by the minus sign caused by the permutation of circles (the right one and the left one). This completes the proof of the lemma. $\square$

This lemma means that the homology of a virtual diagram with two classical crossings (if well defined) can be restored from an atom endowed with an orientation of the link components.

Thus, to prove that the cube $[[K]]$ anticommutes, we can make some preliminary virtualisations for classical crossings of K and consider the analogous question for the obtained diagram K'.

To check the anticommutativity of the cube $[[K]]$ we have to consider all 2-faces of it. Each 2-face is represented by fixing a way of smoothing some $(n-2)$ classical crossings of K; see Fig. 22.11. The remaining two crossings can be smoothed arbitrarily; the four possibilities correspond to the vertices of the 2-face.

In Fig. 22.11 the bifurcation cube is shown in the left part and the 2-face and the corresponding atom are shown in the right part. The atom can be restored from a knot diagram, as described above in Chapter 16.

Now, for these four states, there are some "common" circles which do not touch any of the two vertices in question (in the case depicted in Fig. 22.11 there are no such circles). After removing these circles, we get an atom with two vertices.

What we actually have to check is that any face corresponding to any possible atom with two vertices anticommutes.

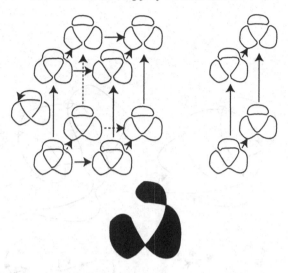

FIGURE 22.11: A 2-face generates an atom.

For the two vertices of such an atom, we have some local orientations of the link at each of these vertices; they are needed to fix the local ordering of components (see Fig. 22.7) when defining the differentials.

Note that globally these orientations might not agree on the circles; namely, an edge of the atom with two vertices consists of several edges of the diagram which might have opposite orientations; see Fig. 22.12.

It turns out, however, that these local orientations can be chosen arbitrarily without losing the anticommutativity property and without changing the homology.

Namely, fix an atom with two vertices. All possible occurrences of this atom in the cube correspond to local orientations of edges at these vertices. Fix an orientation for one crossing v_1 and choose two distinct orientations for the second crossing v_2 that differ from each other by the clockwise $\frac{\pi}{2}$-turn of the arrows; see Fig. 22.13. Thus, we get two pictures and two two-dimensional discrete cubes, Q_1 and Q_2. These cubes coincide as sets of linear spaces. Let V_s and $V_{s'}$ be linear spaces of Q_1 and Q_2 corresponding to some fixed state s and the state s' corresponding to it.

Lemma 22.6. *If Q_1 is anticommutative, then so is Q_2. Moreover, there exists a grading preserving chain map $f: Q_1 \to Q_2$ that takes V_s isomorphically to $V_{s'}$ and commutes with partial differentials.*

Proof. The proof of Lemma 22.6 is very much similar to that of Lemma 22.5.

A sketch of the proof goes as follows. After rotating all arrows at v_2 in the counterclockwise direction, we get the local sign change of X for all circles incident to this crossing. Analogously to Lemma 22.5, we consider two com-

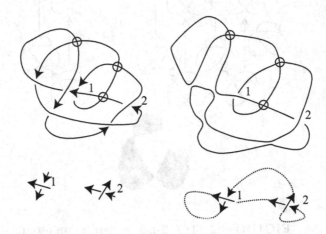

FIGURE 22.12: Orientation for atom crossings.

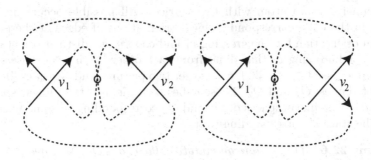

FIGURE 22.13: Q_1 and Q_2.

plexes and identify their chain spaces by means of the map g (analogous to the map g from Lemma 22.5) in such a way that the differentials corresponding to any other crossing coincide.

After that we correct g, as in Lemma 22.5, to get a map f that commutes with all partial differentials, which would yield the statement of the lemma. If we dealt with the usual unordered tensor product, this would lead to the sign change of all partial differentials of type Δ corresponding to v_2.

Furthermore, in the case of a positive crossing, all differentials of type m corresponding to this crossing, change their sign, too.

In the case of negative crossings, partial differentials of type $2 \to 1$ do not change, and $1 \to 2$-bifurcations change the sign again. Thus, we have the same situation as in Lemma 22.5, which completes the proof of Lemma 22.6. $\quad\square$

Let us continue the proof of Theorem 22.10.

Lemma 22.6 means that in order to check the anticommutativity of all possible faces, it is sufficient to enumerate all atoms with two vertices and check the anticommutativity for each of them. We first fix a representation of such an atom in $\mathbb{R}^2$ (i.e. an immersion of its frame preserving the A-structure); such immersions differ by a possible virtualisation which does not change the complex (up to isomorphism) by Lemma 22.5; then we choose a local orientation, which does not matter either by Lemma 22.6.

Note that among atoms with two vertices there are disconnected atoms; i.e. those for which each edge connects some vertex with itself. For such atoms in the case of ordinary tensor product we get by evident reasons commutative 2-faces. In the case of ordered tensor products the corresponding faces will obviously anticommute.

Some (connected) atoms with two vertices are inessential in the following sense. We have set the $1 \to 1$ differential to be zero. By parity reasons, in the 2-face of any atom there might be 0, 2 or 4 such edges. The case when we have no such edges is orientable. When we have four edges representing differentials of type $1 \to 1$, then the proof follows from the identity $0 = 0$. The same takes place in the case when in the diagram, the anticommutativity of which we prove, we have two compositions of maps and one of the maps at each composition is zero.

There are some inessential atoms, where two vertices are not connected to each other. For any of them, anticommutativity is obvious. There are six essential connected atoms with two vertices, as shown in Fig. 22.14. All these atoms except the first one are orientable.

For the first one, an accurate calculation corresponding to Fig. 22.15 shows that both compositions give zero.

Indeed, the lower composition is zero. Substituting X into the upper composition, we get $\pm X \wedge X$ at the first step and zero at the second step. If we start with 1, we get $1_{1,o_{v_1}} \wedge X_{2,o_{v_1}} + X_{1,o_{v_1}} \wedge 1_{2,o_{v_1}}$ at the first step; here the first index is the local number of the circle (the first circle is big and the second one is small), and the second index is the name of the vertex. When

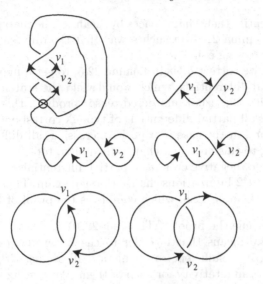

FIGURE 22.14: Essential atoms with two vertices.

passing to the second vertex v_2, the first and second circles change their roles: The circle number 1 becomes the lower one and number 2 becomes the upper one. Also, for the big circle, X changes to $-X$. Thus we get $-X \wedge 1 + 1 \wedge X$ which is mapped by m to zero.

Let us now check orientable atoms. For any of them, we fix an orientation as shown in Fig. 22.14. Such an orientation gives a coordinated orientation of circles at two crossings which are under consideration in the sense of Fig. 22.7. After that, we can fix the bases $\{1, X\}$ for all circles at vertices according to the rule shown in Fig. 22.7.

Now, the anticommutativity is checked as follows. If we dealt with the usual (unordered) tensor product case, everything would commute. Now, the enumeration of circles might cause minus signs on some edges. We have to check that for any of these five atoms the total sign would be minus.

For instance, in Fig. 22.16 we have an oriented atom with two vertices. The analogous check of the unordered tensor product case means the usual associativity $m \circ (m \otimes \mathrm{Id}) = m \circ (\mathrm{Id} \otimes m)$, where the circles are enumerated from the left to the right. In the left part of the figure, one pair of numbers of the circles 1 and 2 is drawn upside down to underline which circle is assumed to be locally the first (left); the other one is the second (right).

Here we have to take into account the global ordering of the components. Note that for three components, we always have to apply $m \wedge \mathrm{Id}$ first, taking those components to be multiplied with the first and second positions.

Thus, $m \circ (m \wedge \mathrm{Id})$ applied to $A_1 \wedge A_2 \wedge A_3$ gives us $m(m(A_1, A_2), A_3) = -(A_1 \cdot A_2 \cdot A_3)$; here $\cdot$ means the usual multiplication in Khovanov's sense:

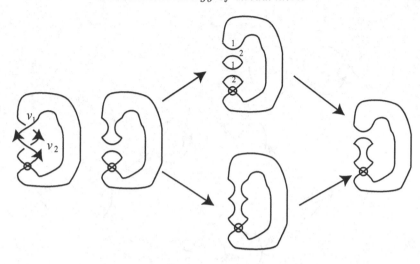

FIGURE 22.15: The non-orientable atom.

$X \cdot X = 0$; $X \cdot 1 = 1 \cdot X = X$; $1 \cdot 1 = 1$. Here the minus sign appears at the second crossing; we have two branches oriented downwards; thus, the rightmost circle occurs to be locally the left one.

On the other hand, if we consider the second crossing first, we get $A_1 \wedge A_2 \wedge A_3 = (A_2 \wedge A_3) \wedge A_1 = -(A_3 \wedge A_2) \wedge A_1 \rightarrow -(A_2 \cdot A_3) \wedge A_1 = A_1 \wedge (A_2 \cdot A_3)$. Applying m to that, we get $A_1 \cdot A_2 \cdot A_3$.

All other atoms are checked analogously. Note that our setup gives directly an anticommutative cube, unlike the Khovanov original setup, where we got an anticommutative cube from a commutative one by adding some minus signs on edges. Thus, Theorem 22.10 is proven. Therefore, Theorem 22.9 is also proven.

Let us prove Theorem 22.11.

Remark 22.8. *Throughout the rest of the proof of Theorem 22.11, we shall not care about height and degree shifts. The proof of their coincidence for diagrams differed by Reidemeister moves repeats verbatim that in the classical case.*

Proof of Theorem 22.11. First, note that the complex $C(K)$ itself does not change at all if we perform the detour move. Therefore, the homology does not change.

In the case of classical Reidemeister moves, the proof goes along the line of the proof for classical links.

Let us be more specific. The case of the first Reidemeister move is evident (see Theorem 7.8).

As in the case of the first Reidemeister move, the invariance under the sec-

I

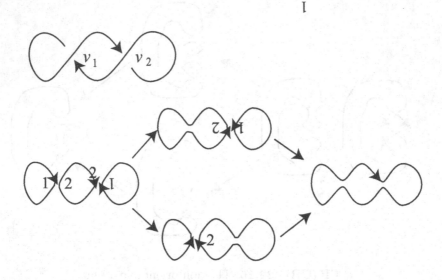

FIGURE 22.16: An orientable two-vertex atom.

ond Reidemeister move repeats the proof given in the classical case (see Theorem 7.8).

In addition to the classical case, we should pay attention to orientations of circles when we prove the invariance under the second Reidemeister move. But for the second Reidemeister move, we can choose orientations of all circles incident to a given crossing locally agreed (such that under passing along one circle from one crossing to the other one the variable X does not change the sign); see Fig. 22.17.

Let us now consider the third Reidemeister move shown in Fig. 7.5 (page 98).

It is well known (see, e.g. [Oht]) that any variant of the third Reidemeister move can be obtained as a composition of Ω_1, Ω_2 and one prefixed version of the third Reidemeister moves, in which a choice for over/undercrossing and orientations of edges is chosen. Consider only one case, shown in Fig. 22.19, with crossing smoothings as in Fig. 7.5.

At any crossing in Fig. 22.19 there is a local rule for orientations for all edges incident to it, according to the rule shown in Fig. 22.7. If two crossings are adjacent, the orientation might or might not be coordinated. We see that the orientation (defined according to Fig. 22.7) in the third crossing (left picture) does not agree with the orientations in the first and second crossings analogously; for the right picture, the second crossing disagrees with the first one and with the third one. Note that the rule in Fig. 22.7 does not depend on types of crossings, but does depend on the orientations of branches.

Apply virtualisations to crossings 1, 2 of the first diagram and to the sec-

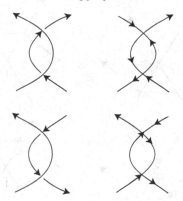

FIGURE 22.17: Orientations of upper-right agrees for Ω_2.

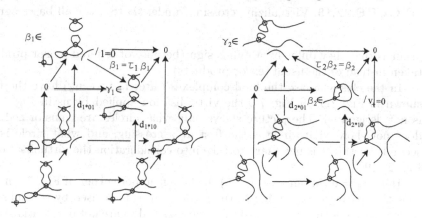

FIGURE 22.18: The diagrams after the virtualisation.

ond crossing of the second diagram; after that, all local orientations (in the sense of variable X) will be coordinated; see Fig. 22.18.

The positive smoothings at crossing 1 are the same (up to virtualisations) for both diagrams. The negative smoothing of them gives rise to two pictures obtained one from another by a sequence of (virtualisations and) two classical Reidemeister moves.

Thus, the complexes of the two diagrams in question can be rearranged to have coinciding bottom levels, and top levels having the same homology (in both cases we applied Ω_2).

The main thing to check is that the differentials going upwards agree for these complexes; i.e. the "upwards" maps in both cases either coincide or differ by a sign. These complexes are shown in Fig. 7.6.

In our situation, the only difference from the classical case which may

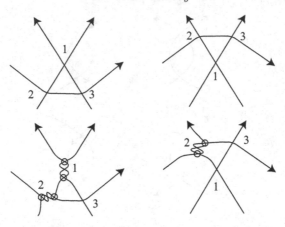

FIGURE 22.19: Virtualizing crossings under Ω_3 to make all bases agree.

occur is that they differ by a minus sign (because of ordered tensor products taken instead of the usual tensor products).

In the classical case the final complexes (after factorising) have the form shown in Fig. 7.6. In Fig. 7.6 the virtualisation applied by us in Fig. 22.19 is not designated. The picture shows only what circles are transformed, but does not show what circle is the first at a crossing, and what circle is the second (for this it is necessary to take into consideration the virtualisation in Fig. 22.19.

Here $v_+ = 0$ (in our case $1 = 0$) in the left upper corner of Fig. 7.6 means that the space corresponding to the given state is factorised by the subspace where the small circle is marked by 1. Here τ_1 and τ_2 are not differentials; they are chain maps taking an element to the element which is minus homologous to the initial one.

To establish the isomorphism in homology, it is sufficient to show that $\tau_1 \circ d_{1*01} = d_{2*01}$ and $d_{1*10} = \tau_2 \circ d_{2*10}$. In this case we shall show that all the maps "upwards" in both complexes differ by a sign (since in both cases τ_i is minus the identity in homology). After that the homotopy equivalence of the two complexes corresponding to the third Reidemeister move is proved as in Lemma 22.5: By means of a natural map that identifies lower subcubes and minus that map that corresponds to the complex which the upper subcube is reduced to.

The ordered tensor product case differs from the usual one, possibly, by signs on edges.

Let us check that the signs agree in our setup. We shall show that $\tau_1 \circ d_{1*01} = d_{2*01}$ (the remaining case $d_{1*10} = \tau_2 \circ d_{2*10}$ is completely analogous).

Let us view Fig. 7.6 and take into account the virtualisation of the right and left diagrams at crossings. The required identity will look like $p = q \circ \Delta^{-1} \circ \Delta$; see Fig. 22.20.

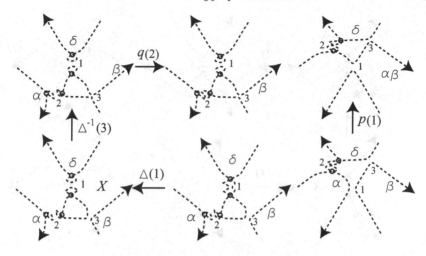

FIGURE 22.20: Checking the invariance under Ω_3.

Here d_{1*01} is a $1 \to 2$-bifurcation (we denoted it by Δ); $\tau_1 = \nu \circ \Delta^{-1}$, where ν is a partial differential and Δ^{-1} is assumed as an operation inverse to Δ (note that the space in the upper-left corner in which the element β_1 stays is factorised by $1 = 0$; i.e. the space associated with the small circle C, is one-dimensional with generator X). Then, the comultiplication map for which C is a resulting circle becomes an isomorphism.

View Fig. 22.20. For each of the maps in the brackets the number of a crossing is indicated which this map is applied to.

The maps p and q are just the usual local differentials, either both multiplications, or both comultiplications, or both zeros.

If $p = q = 0$, there is nothing to prove.

Consider the remaining cases. We have three fragments of circles α, β, δ. In the very initial state (which the map p in the right picture and Δ in the left picture are applied to) they may belong to one, two or three different circles. We shall first consider the case when all fragments containing α, β, δ belong to three different circles.

For simplicity we denote the elements of the algebra V (of type 1 or $\pm X$) related to these circles, by the same letters as fragments α, β, δ.

In our case, both operations p and q are multiplications.

Starting with $\alpha \wedge \beta \wedge \delta$, we get on the right picture the map d_{2*01}:

$$p: \alpha \wedge \beta \wedge \delta \to (\alpha \cdot \beta) \wedge \delta,$$

where $(\alpha \cdot \beta)$ means an ordinary product in the Frobenius algebra.

On the left picture we have:

$$\alpha \wedge \beta \wedge \delta = \delta \wedge \alpha \wedge \beta \overset{\Delta}{\to} \delta \wedge X \wedge \alpha \wedge \beta.$$

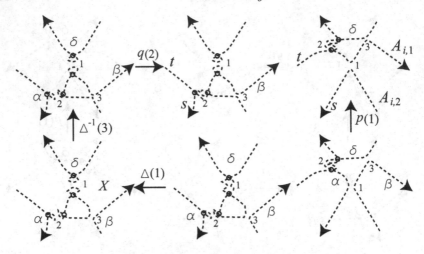

FIGURE 22.21: Checking the invariance under Ω_3.

Here we applied the comultiplication to δ to get two circles at the crossing number 1; the two resulting circles are denoted by δ (the upper one) and X (the lower one).

Now, $\delta \wedge X \wedge \alpha \wedge \beta = -\beta \wedge X \wedge \alpha \wedge \delta$. We then perform Δ^{-1} at crossing 3. This map joins the two circles marked by β and X.

At this crossing the generator X is related to the left circle, and β is related to the right circle. Thus, we have

$$-\beta \wedge X \wedge \alpha \wedge \delta = X \wedge \beta \wedge \alpha \wedge \delta \overset{\Delta^{-1}}{\to} \beta \wedge \alpha \wedge \delta.$$

Now, the operation q is the comultiplication at crossing 2, where the circle marked by β is the first (upper), and the one marked by α is the second one (lower). Thus, we get: $(\alpha \cdot \beta) \wedge \delta$.

Now assume that α and β form one circle (in the initial state), and δ forms a separate circle. Denote the mark (an element from V) corresponding to the first circle by A, and the mark corresponding to the second circle by δ.

The map p looks like:

$$A \wedge \delta \overset{\Delta}{\to} \sum_i A_{i,1} \wedge A_{i,2} \wedge \delta,$$

where $\sum_i A_{i,1} \otimes A_{i,2}$ is the result of application of the comultiplication to A in the ordinary sense (in the case of unordered tensor product); see Fig. 22.21.

In the further proof for simplicity of writing we shall not use the sum sign $\sum_i$.

In the left picture we have

$$A \wedge \delta = -\delta \wedge A \to -\delta \wedge X \wedge A$$

(at the first crossing the marking δ corresponds to the upper circle and X corresponds to the lower circle).

Then for the map Δ^{-1} at crossing 3 we have

$$-\delta \wedge X \wedge A = -X \wedge A \wedge \delta \to -A \wedge \delta$$

(here X was on the left side, and A was on the right side).

Finally, the map q at crossing 2 gives us

$$-A \wedge \delta \to -A_{i,1} \wedge A_{i,2} \wedge \delta.$$

Here $A_{i,1}$ corresponds to the locally upper component s at crossing 2, and $A_{i,2}$ is locally lower component t. But, in the right picture they have opposite ordering. More precisely, we have

$$-A_{i,1,s} \wedge A_{i,2,t} \wedge \delta.$$

In the first case (the map p) we had

$$A_{i,1,t} \wedge A_{i,2,s} \wedge \delta = -A_{i,2,s} \wedge A_{i,1,t} \wedge \delta.$$

These two results coincide because of cocommutativity of Δ in the ordinary case.

One can consider the remaining cases analogously.

Suppose that α and δ belong to one circle (the corresponding element being denoted by α), and β belongs to another circle. Then we have the following maps.

In the simplest case (the map p) we have

$$\alpha \wedge \beta \to (\alpha \cdot \beta).$$

On the left picture we have

$$\alpha \wedge \beta \to \alpha \wedge X \wedge \beta = X \wedge \beta \wedge \alpha \to \beta \wedge \alpha \to (\beta \cdot \alpha).$$

Consider the case of multiplication when β and δ form one circle (the corresponding element being denoted by β). We get:

$$\alpha \wedge \beta \to (\alpha \cdot \beta)$$

on the right picture (the map p) and

$$\alpha \wedge \beta = -\beta \wedge \alpha \to -\beta \wedge X \wedge \alpha = X \wedge \beta \wedge \alpha \to \beta \wedge \alpha \to (\beta \cdot \alpha)$$

on the left picture.

Finally, consider the case when at the beginning we have exactly one diagram, we get two comultiplications:

$$A \to A_{i,1,t} \wedge A_{i,2,s}$$

in the simplest case (the map p) and

$$A \to A \wedge X = -X \wedge A \to -A \to -A_{i,1,s} \wedge A_{i,2,t} = A_{i,2,t} \wedge A_{i,1,s}.$$

Thus, we have proved the equality $\tau_1 \circ d_{1*01} = d_{2*01}$. The proof of the equality $d_{1*10} = \tau_2 \circ d_{2*10}$ is completely analogous. $\qquad\square$

Theorem 22.12. *Let K be a virtual diagram for which the corresponding atom is orientable. Then the homology $\mathrm{Kh}(K)$ coincides with the Khovanov homology constructed in Lemma 22.2.*

During the proof of this theorem, we denote our complex and our homology by $\mathcal{C}(K)$ and $\mathrm{Kh}(K)$, respectively, and the ones constructed in Section 22.4 by $\mathcal{C}'(K)$ and $\mathrm{Kh}'(K)$ respectively.

Proof of Theorem 22.12. First we note that the shifts for $\mathcal{C}$ and $\mathcal{C}'$ are performed in the same manner. Thus, we can forget about additional normalizations of type $[-n_-]\{n_+ - 2n_-\}$.

First, we assume the diagram of K is chosen in such a way that all X's for all crossings and circles agree (that is, for a given state circle, while passing from one classical crossing P to another one Q, we get $X_{C,o_P} = X_{C,o_Q}$, not $X_{C,o_P} = -X_{C,o_Q}$). This is possible since the atom corresponding to K is orientable. Indeed, since the atom corresponding to K is orientable, we can globally define the orientation of all edges to be compatible with the orientation of the circles in each state. At each crossing of K this orientation may agree or disagree with the local orientation of edges determined by Fig. 22.7 (the orientation originates from the source–sink structure). Let us apply the virtualisation to all crossings of K where these orientations disagree. By Lemma 22.5, the homology of the complex $\mathcal{C}(K)$ remains the same, and the orientations of circles given locally at crossings according to the rule in Fig. 22.7 become compatible.

After that, we should just care about signs of local differential and enumeration of circles for any crossing.

We construct a homology-preserving map between two cubes. Fix an enumeration of the classical crossings of K. Let us associate the spanning tree for the cubes $\mathcal{C}(K)$ and $\mathcal{C}'(K)$ as follows. This tree consists of all edges of the form $(\alpha_1, \ldots, \alpha_l, *, 0, \ldots, 0), \alpha_j \in \{0, 1\}$; i.e. an edge in the direction x_{l+1} belongs to this tree if all the coordinates of $x_{l+2}, \ldots, x_n$ vanish; see Fig. 22.22.

With each state s of the complex $\mathcal{C}(K)$ we associate the ordered tensor power $V^{\wedge l}$, and with the corresponding state for the complex $\mathcal{C}'(K)$ we associate $V^{\otimes l}$, where l is the number of circles in the state s. Enumerate the circles in the A-state in some way. Then the ordering determines a map between the space corresponding to the A-state s in $\mathcal{C}(K)$ and the space corresponding to some state $g(s)$ of the complex $\mathcal{C}'(K)$. After that we can successively renumber the circles at all vertices of the tree in order that the identification of the chains in the corresponding states of the complexes $\mathcal{C}(K)$ and $\mathcal{C}'(K)$ commute with the partial differentials acting along the edges of the spanning tree. Thus

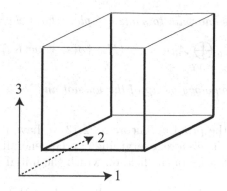

FIGURE 22.22: Choosing a spanning tree.

we have constructed a map between the whole chain space of $\mathcal{C}(K)$ and the chain space of $\mathcal{C}'(K)$.

This map g commutes with all the partial differentials for the following reasons. Let ∂', ∂'' be the partial differentials corresponding to the same edge of the complexes $\mathcal{C}$ and $\mathcal{C}'$. Then we have $g \circ \partial' = \pm \partial'' \circ g$.

If the compatibility holds for three of four edges of some two-dimensional face, then it also holds for the fourth edge, since both complexes are anticommutative and no one of the partial differentials is the identical zero.

To complete the proof, we note that all the edges of the cube can be exhausted if we start from the spanning tree and successively add the missing edges of the two-dimensional faces (add the fourth edge provided that we have three). $\qquad\square$

As it was done in Definition 7.6 we call by the *height* $h(\text{Kh}(K))$ of the Khovanov homology of a virtual link K the difference between the leading and lowest non-zero quantum gradings of non-zero Khovanov homology groups of the virtual link K. From Theorem 22.12 it follows that the definition given in Sec. 22.3 (using Khovanov homology for orientable atoms) is agreed with the definition for the ordinary case based on the construction of the present section.

22.8 Spanning tree for Khovanov complex

Spanning tree decomposition for Khovanov homology considered in Chapter 7 (Theorem 7.9) remains valid for virtual links.

Theorem 22.13. *The non-normalised Khovanov complex of a diagram K of* *a virtual link is isomorphic to some complex whose chain group looks like*

$$\bigoplus_{s\in\mathcal{V}_1} \mathcal{A}[\beta(s) + w(K_s)]\{\beta(s) + 2w(K_s)\}, \tag{22.5}$$

where $\mathcal{A}$ is the homology group of the unknot and $\mathcal{V}_1$ is the set of states with one circle.

Note that in the proof of Theorem 22.13 we have not used the fact that a link is classical. Therefore, everything can be generalised word by word for virtual links in the case of the field on which the initial Khovanov complex is well defined.

We assert that this proof fits for all models of the Khovanov complex of virtual knots in those cases when this complex is well defined.

22.9 The Khovanov polynomial and Frobenius extensions

The Khovanov theory of virtual knots described earlier in this chapter is not unique to what one can get by looking at the Kauffman model and the (anti)commutative state cube. The present section is devoted to a generalisation of the Khovanov theory which uses Frobenius extensions for classical and virtual links.

Below, we show that Khovanov's universal construction $(\mathcal{A}, \mathcal{R})$ (see Sect. 7.6.3) works in the case of orientable atoms straightforwardly, and write down the algebraic equations the partial differentials have to satisfy for the case of arbitrary virtual links.

22.9.1 Geometrical generalisations by means of atoms

With each virtual link diagram having an orientable atom, the universal $(\mathcal{R}, \mathcal{A})$-construction associates some bifurcation cube, the bigraded chain space with partial differentials, whose homology leads to an invariant of virtual links (after a normalization).

Here, with the state cube and the bifurcation cube we associate bigraded complexes with tensor powers of the ring $\mathcal{A}$ over the ring $\mathcal{R}$ staying in vertices of the cube; the tensor power corresponds to the number of circles in the given state; partial differentials in these cubes are defined by using m and Δ, and differentials are sums of partial differentials with signs.

From Khovanov's theory [Kho2] it follows that there exists a *local* proof of the invariance for the universal $(\mathcal{R}, \mathcal{A})$-construction; i.e. there is a number of algebraic steps (equivalences, analogous to the cancellation principle and short exact sequences) which leads to the following.

Let us fix a classical Reidemeister move Ω_i. Then for any classical diagrams K and K' which differ locally by a Reidemeister move Ω_i, there exists, see ahead, a consequence of algebraic transformations taking $\mathrm{Kh}_U(K)$ to $\mathrm{Kh}_U(K')$ and not depending explicitly on the behaviour of partial differentials of the Khovanov complexes for K and K' except for those whose explicit form (μ or Δ) follows from the structure of our Reidemeister move Ω_i.

This argument leads to the fact that the universal $(\mathcal{R}, \mathcal{A})$-construction can be generalised for virtual diagrams with orientable atoms. Namely, given a diagram K with an orientable atom, we can construct the corresponding bifurcation cube with differentials, corresponding to the multiplication and comultiplication operations (with signs) and calculate its homology. Furthermore, if two diagrams K, K' have orientable atoms and are obtained from each other by some classical Reidemeister move Ω_i, then according to the principle described above, there is an isomorphism between the graded homologies $\mathrm{Kh}_U(K) \cong \mathrm{Kh}_U(L')$. Since the universal $(\mathcal{R}, \mathcal{A})$-construction is tautologically invariant under the detour move (the bifurcation cube does not change), the following analogue of Lemma 22.3 holds.

Lemma 22.7. *Let K, K' be two diagrams with orientable atoms such that K' differs from K by an application of a detour move or one of the three classical Reidemeister moves. Then $\mathrm{Kh}_U(K) \cong \mathrm{Kh}_U(K')$.*

This argument together with Lemmas 22.2, 22.4 yields that the universal $(\mathcal{R}, \mathcal{A})$-construction works for

- the construction of the Khovanov homology theory Kh_U for framed virtual links by taking the $2l$ parallel copies;

- the construction of the Khovanov homology theory Kh_U for virtual knots by taking two-sheeted orienting coverings over the corresponding atoms.

- the construction of the Khovanov homology theory Kh_U for virtual knots obtained by taking parity projections, see [Man27].

More precisely, the following theorem holds.

Theorem 22.14. *1 Let l be a natural number. Then $\mathrm{Kh}_U(D_{2l}(K))$ is an invariant of the framed virtual link K.*
2 The map $K \mapsto \mathrm{Kh}_U(\widetilde{K})$ gives a well-defined invariant for virtual links.

22.9.2 Algebraic generalisations

As we have shown above, for virtual knots with orientable atoms the Khovanov homology with $\mathbb{Z}_2$-coefficients can be defined straightforwardly if we set all partial differentials of type $1 \to 1$ to be zero.

Let us now consider the universal $(\mathcal{R}, \mathcal{A})$-construction (see page 104), and let us generalise it for the case of virtual knots.

Note that if with each knot we associate a well-defined complex, then the homology of this complex will be automatically invariant under classical Reidemeister moves (according to the locality of the invariance proof) and the detour move (there is nothing to prove in this case).

Thus, we have reduced the problem of finding an extension for the ring $\mathcal{A}$ in order to construct the Khovanov homology theory for arbitrary virtual link diagrams, to the following problem. Find an operator (a homomorphism of $\mathcal{R}$-modules) $\mathfrak{I} \colon \mathcal{A} \to \mathcal{A}$ corresponding to maps of type $1 \to 1$ in such a way that for every virtual diagram the bifurcation cube with partial differentials obtained from m, Δ, $\mathfrak{I}$, is anticommutative.

Thus, we require the commutativity of the cube in order to turn it into an anticommutative cube (just as it was done in the usual case).

This problem is purely algebraic. In order to solve it, one has to consider all possible 2-faces of the bifurcation cube for a diagram K; there are finitely many such types (with each face, one associates some atom with two vertices). For each face, one has to check some algebraic conditions for the maps $\mathfrak{I}$, Δ and m.

For the space $\mathcal{A}$, let us take the basis $\{1, X\}$, and for the space $\mathcal{A} \otimes \mathcal{A}$ we take the basis $\{1 \otimes 1, 1 \otimes X, X \otimes 1, X \otimes X\}$.

Then in these bases the maps Δ and m are represented by the following matrices:

$$\Delta = \begin{pmatrix} -h & t \\ 1 & 0 \\ 1 & 0 \\ 0 & 1 \end{pmatrix}, \quad m = \begin{pmatrix} 1 & 0 & 0 & t \\ 0 & 1 & 1 & h \end{pmatrix}.$$

After that we shall use the sign of matrix multiplication instead of the composition of the operators. So, for example, we write $\mu \cdot \Delta$ instead of $\mu \circ \Delta$. One of the particular cases given here, is considered in detail in [TT].

We look for a matrix

$$\mathfrak{I} = \begin{pmatrix} p & q \\ r & s \end{pmatrix},$$

which corresponds to bifurcations of type $1 \to 1$ and gives, at the same time, the (anti)commutativity of the bifurcation cube.

Let a coefficient ring $\mathcal{R}$ containing elements h and t with gradings 2 and 4, respectively, be given. Denote the obtained bifurcation cube by $[[K]]_{\mathcal{R}}$. Let us define the differential as the sum of the partial differentials corresponding to edges (of type m, Δ, $\mathfrak{I}$) with signs arranged as it was done on page 403.

Lemma 22.8. *The bifurcation cube $[[K]]_{\mathcal{R}}$ is anticommutative if and only if the following properties hold:*

$$m \cdot \Delta = (\mathfrak{I})^2,$$

$$\Delta \cdot \mathfrak{I} = (\mathfrak{I} \otimes 1) \cdot \Delta = (1 \otimes \mathfrak{I}) \cdot \Delta, \tag{22.6}$$

$$\mathfrak{I} \cdot m = m \cdot (\mathfrak{I} \otimes 1) = m \cdot (1 \otimes \mathfrak{I}). \tag{22.7}$$

Proof. To check the (anti)commutativity of the state cube it is necessary for us to consider all possible sorts of faces of the cube. Later on, we disregard additional signs on edges and prove the commutativity.

In the "simple" case where we have the field $\mathbb{Z}_2$ and null-differentials corresponding to bifurcations of type $1 \to 1$, everything was reduced to the "classical" cases, and as well as to the case depicted in Fig. 22.1.

For the $(\mathcal{R}, \mathcal{A})$-theory we have to check more cases, since maps of type $1 \to 1$ are not assumed to be zero, and the maps m (multiplication) and Δ (comultiplication) are more complicated than in the case of the homology Kh.

Each two-dimensional face of the cube represents a collection consisting of four states, see Sec. 22.4. When pasing from one state to another, some circles are reconstructed and the others persist. Denote these four states by s_{00}, s_{01}, s_{10} and s_{11} depending on the values of two changing coordinates. Delete "common components" of the states s_{ij}; i.e. those components of the state s_{00} which do not connect to the crossings at which the substitution of the smoothing occurs. Then the given two-dimensional face of the cube will represent some virtual knot and, therefore, the atom corresponding to it. This atom will have exactly two vertices. If the atom is height, then the corresponding diagram is realized by a bifurcation of embedded circles into the plane, thus, the (anti)commutativity of the corresponding face belongs to the number of classical cases checked in [Kho4, TT].

For atoms with disconnected frames the check is obvious. Further, each orientable atom with the connected frame having two vertices is height. Thus, the required verification is reduced to sorting out unoriented atoms with two vertices (all of them by definition are not height). Sorting out these atoms, eventually we shall come to relations which are satisfied identically, e.g. $\Im \circ \mu = \Im \circ \mu$; see Fig. 22.23. Three atoms giving non-trivial relations pointed out in the claim of the lemma are given in Figs. 22.24, 22.25 and (the example considered above), Fig. 22.1. $\square$

We met the first equation already in the case of the general Khovanov homology $\mathcal{C}$ (there the composition $m \cdot \Delta$ looks simpler). In the case of the universal $(\mathcal{R}, \mathcal{A})$-theory we have:

$$m \cdot \Delta = \begin{pmatrix} -h & 2t \\ 2 & h \end{pmatrix}.$$

If we want to construct a $\mathbb{Z}$-*graded* theory, then it is necessary for us that the matrix $\Im$ increases the grading of elements of the ring $\mathcal{R}$ by one. This means that all elements $p, q, r, s \in \mathcal{R}$ should be homogeneous. In this case deg $p = 1$, deg $q = 2$, deg $r = 0$, deg $s = 1$; herewith it is possible that any of the elements p, q, r, s are equal to zero (in this case the grading is not defined). Then from the equality $(\Im)^2 = m \cdot \Delta$ it follows deg $(2t) =$ deg $t = 3$, which leads us to a contradiction, if $2 \neq 0$.

Thus (as well as in the case of the general Khovanov homology), under

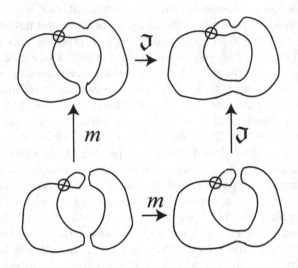

FIGURE 22.23: Bifurcation corresponding to tautological relation.

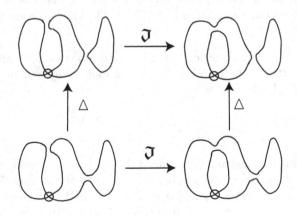

FIGURE 22.24: Relations $\Delta \cdot \mathfrak{I} = (\mathfrak{I} \otimes 1) \cdot \Delta = (1 \otimes \mathfrak{I}) \cdot \Delta$.

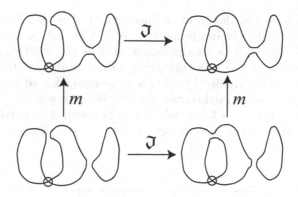

FIGURE 22.25: Relations $\mathfrak{J} \cdot m = m \cdot (\mathfrak{J} \otimes 1) = m \cdot (1 \otimes \mathfrak{J})$.

this approach the $\mathbb{Z} \oplus \mathbb{Z}$-bigraded homology theory is possible only in the case of a field of characteristic two.

Let us consider the case of a field of characteristic two. It turns out that in this case we have a simple non-trivial solution. Namely, in the case $2 = 0$ the matrix $m \cdot \Delta$ is turned into the diagonal matrix

$$m \cdot \Delta = \begin{pmatrix} h & 0 \\ 0 & h \end{pmatrix}.$$

Let us add to the ring $\mathcal{R}$ a new element $u = \sqrt{h}$, deg $u = 1$. Now set $\mathcal{R}' = \mathbb{Z}_2[u, t]$, herewith the algebra $\mathcal{A}$ takes the form $\mathcal{A}' = \mathcal{R}'[x]/(X^2 - u^2 X - t)$, where deg $X = 2$, deg $t = 4$, deg $u = 1$.

Set

$$\mathfrak{J} = \begin{pmatrix} u & 0 \\ 0 & u \end{pmatrix}. \tag{22.8}$$

In this case the matrix $\mathfrak{J}$ is scalar, and Eqs. (22.6) and (22.7) are satisfied automatically.

Thus, we conclude with the following theorem.

Theorem 22.15. *Over the field $\mathbb{Z}_2$ the pair of algebras $(\mathcal{R}', \mathcal{A}')$ together with multiplication m, comultiplication Δ defined by $\Delta(1) = 1 \otimes X + X \otimes 1 - u^2 \cdot 1 \otimes 1$, $\Delta(X) = X \otimes X + t1 \otimes 1$ and the scalar map $\mathfrak{J}$ looking like (22.8), gives an invariant homology theory for virtual links.*

In the general case; i.e. in the case of the Khovanov homology for virtual links, we have the following.

Theorem 22.16. *The restriction of Khovanov's universal theory for the case $h = 0$ (no restrictions on t) can be extended to virtual links by the method suggested in Sec. 22.7.*

The main idea of the proof of Theorem 22.16 is the following. Δ and m behave nicely under the involution $I\colon 1 \mapsto 1$, $X \mapsto -X$ that takes place while inverting the circle: The multiplication m does not change, and Δ changes the sign. Note that this takes place only for $h = 0$ (for arbitrary t). The case when $h \neq 0$ can be handled by using a more sophisticated twisting.

This generalises straightforwardly for the case when $h = 0$ (where all differentials of type $1 \mapsto 1$ are assumed to be zero). As a particular case, this leads to an analogue of Lee's theory, see [Lee1, Lee2, Rush].

22.10   Minimal diagrams of links

In the classification and tabulation of (virtual) knots the important step is to describe diagrams having a minimal number of (classical) crossings. One of the main achievements in the development of knot theory is Kauffman–Murasugi–Thistlethwaite theorem (Theorem 7.5) and the classification of alternating links by Menasco and Thistlethwaite [MT] following from this theorem.

In this section we shall mention theorems establishing the minimality of virtual and classical diagrams, see also [JS, Man17]. The proofs are analogous to the classical case (see Section 7.6.4).

Remember that the thickness $T(K)$ (see Definition 7.8) of a virtual diagram K measures the number of diagonals between the two extreme diagonals in the Khovanov homology of K.

Lemma 22.9. *For any diagram K (with a connected atom) of a virtual link we have: $T(K) \leqslant g(K)+2$, where $g(K)$ is the genus of the atom corresponding to K.*

Definition 22.3. Let us call a virtual diagram K *2-complete*, if $T(K) = g(K) + 2$.

Now we have the following

Theorem 22.17. *Let $T(K) = g + 2$, span $\langle K \rangle = s$. Then the number of classical crossings of a connected diagram of the virtual link generated by K cannot be smaller than $s/4 + g$.*

In particular, if a diagram with n crossings and the atom with genus g is 1-complete and 2-complete, then it is minimal.

Theorem 22.17 holds in any category in which the Khovanov complex is well defined and invariant. So, if we are interested in the invariance of a classical diagram in the category of classical diagrams, we can consider the thickness in the classical category.

Part V

Knots, 3-manifolds, and Legendrian knots

Chapter 23

3-Manifolds and knots in 3-manifolds

In the present chapter, we shall give an introduction to the theory of three–manifolds. The aim of this chapter is to show how different branches of low–dimensional topology can interact and give very strong results. In the first part, we shall describe a sympathetic theory of knots in $\mathbb{R}P^3$. The second part will be devoted to the deep construction of Witten–Reshetikhin–Turaev invariants of three–manifolds. We are also going to describe some fundamental constructions of three–dimensional topology, such as the Heegaard decomposition, and the Kirby moves. On one hand, they are necessary for constructing the Witten theory; on the other hand, they have their own remarkable interest. The theory of Witten–Reshetikhin–Turaev invariants, which is based on the Kauffman bracket on one hand and the Kirby theory on the other, is very deep. In fact, it leads to the theory of invariants of knots in three-manifolds. Many proofs will be sketched or omitted (e.g. the Kirby theorem). For the study of three-manifolds, we recommend Matveev's book [Mat5]; see also [Ch, DFN, Mat3, Mat4, Pra1, SeSm, Tur2, Vas4].

The Witten-Reshetikhin-Turaev invariants described in the present book are the first evidence of how *quantum invariants of knots can be generalised for 3-manifolds*. To pass from knots to 3-manifolds, one has to restrict the value of the variable (in our case, a for the Kauffman bracket) to some roots of unity.

For the general theory of quantum invariants of links and 3-manifolds, see monographs [Oht, Kas].

23.1 Knots in $\mathbb{R}P^3$

Here we are going to present a method of encoding knots and links in $\mathbb{R}P^3$ and a generalisation of the Jones polynomial for the case of $\mathbb{R}P^3$ proposed by Yu.V. Drobotukhina, see [Dro].

The projective space $\mathbb{R}P^3$ can be defined as the sphere S^3 with identified opposite points. The sphere S^3 consists of two hemispheres; each of them is homeomorphic to the ball D^3. Thus, the space $\mathbb{R}P^3$ can be obtained by identifying the opposite points of the boundary $S^2 = \partial D^3$.

Consequently, any link in $\mathbb{R}P^3$ can be defined as a set of closed curves and

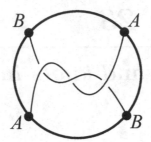

FIGURE 23.1: A diagram of a link in $\mathbb{R}P^3$

arcs in D^3 such that the set of endpoints of arcs lies in ∂D^3; and this set is centrosymmetrical in D^3. Without loss of generality, one can think that the intersection of the link with ∂D^3 lies on the "equator" of the ball. Thus, a link in $\mathbb{R}P^3$ can be represented by a diagram in D^2; the points of these diagrams lying at the boundary $S^1 = \partial D^2$, have to be centrosymmetrical.

An example of such a diagram is shown in Fig. 23.1.

Exercise 23.1. *Consider an arbitrary link diagram in $\mathbb{R}P^3$. Describe the pre-image of this diagram with respect to the natural covering $p : S^3 \to \mathbb{R}P^3$.*

We shall consider only diagrams for $\mathbb{R}P^3$ in general position. This means that the edges of the diagram are smooth, all crossings are double and transverse, and no crossings are available at the boundary of the circle, and no branch of the diagram is tangent to the circle.

We recall that the singularity conditions for the case of links in $\mathbb{R}^3$ (or S^3) lead to the moves Ω_2, Ω_3. Analogously, two new moves for diagrams of links in $\mathbb{R}P^3$ generate two new moves called Ω_4 and Ω_5; see Fig. 23.2.

Theorem 23.1. *Two link diagrams generate isotopic links in $\mathbb{R}P^3$ if and only if one can be transformed to the other by means of isotopies of D^2, classical Reidemeister moves $\Omega_1, \Omega_2, \Omega_3$, and moves Ω_4 Ω_5.*

Proof. The proof of this theorem is quite analogous to that of the classical Reidemeister theorem. It is left to the reader as an exercise. $\square$

Now we are ready to define the analogue of the Jones–Kauffman polynomial for the case of oriented links in $\mathbb{R}P^3$.

First, for a given diagram of an unoriented link $L = \langle\!\!\times\!\!\rangle$ in $\mathbb{R}P^3$, let us define an analogue of the Kauffman bracket satisfying the following axioms:

$$\langle L \rangle = a \langle \asymp \rangle + a^{-1} \langle)(\rangle; \tag{23.1}$$

$$\langle L \sqcup \bigcirc \rangle = (-a^2 - a^{-2})\langle L \rangle; \tag{23.2}$$

$$\langle \bigcirc \rangle = 1, \tag{23.3}$$

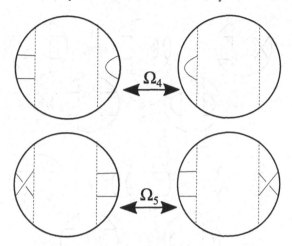

FIGURE 23.2: Moves $\Omega_{4,5}$

where $\bigcirc$ means a separated unknotted circle.

The proof of the existence and uniqueness of $\langle L \rangle$ is completely analogous to that for the Kauffman bracket defined on links in $\mathbb{R}^3$ (or S^3).

Also, we can analogously prove the invariance of $\langle L \rangle$ under Ω_2, Ω_3.

To prove the invariance of $\langle L \rangle$ under Ω_4, let us use the following formula

$$\langle D \rangle = \sum_s a^{\alpha(s)-\beta(s)}(-a^2 - a^{-2})^{\gamma(s)-1}, \tag{23.4}$$

where $\gamma(s)$ is defined as the number of circles of L in the state s.

Actually, for each state s of the diagram L, the numbers $\alpha(s), \beta(s)$ and $\gamma(s)$ are invariant under Ω_4.

The invariance of the bracket $\langle L \rangle$ under Ω_5 follows from the equalities shown in Fig. 23.3.

As in the case of links in $\mathbb{R}^3$, the bracket $\langle L \rangle$ is not invariant under Ω_1; this move multiplies the bracket by $(-a)^{\pm 3}$.

Now, for a diagram of an oriented link L, let us define $w(L)$ as in the usual case.

Definition 23.1. The *Drobotukhina polynomial* of the oriented link L in $\mathbb{R}P^3$ is defined as

$$X(L) = (-a)^{-3w(L)}\langle |L| \rangle,$$

where $|L|$ is obtained from L by "forgetting" the orientation.

Theorem 23.2. *The polynomial $X(L)$ is an invariant of oriented links in $\mathbb{R}P^3$.*

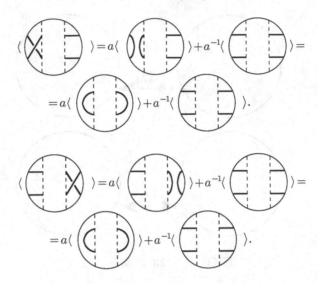

FIGURE 23.3: Invariance of bracket under Ω_5

Proof. The proof is quite analogous to the invariance proof of the Jones–Kauffman polynomial for links in S^3. It follows immediately from Theorem 23.1, the definition, invariance of the bracket $\langle \cdot \rangle$ under the moves $\Omega_2 - \Omega_5$ and its behaviour under Ω_1 which respects the behaviour of $w(\cdot)$.   □

By using this polynomial, Yu.V. Drobotukhina classified all links in $\mathbb{R}P^3$ having diagrams with no more than six crossings up to isotopy and proved an analogue of the Murasugi theorem for links in $\mathbb{R}P^3$.

23.2   An introduction to the Kirby theory

Kirby theory is a very interesting way for encoding three–manifolds. In the present book, we shall give an introduction to the theory of the Witten–Reshetikhin–Turaev invariants via Kirby theory. However, Kirby theory is interesting in itself. It can be used for constructing other invariants of three–manifolds (Witten, Viro, Reshetikhin, Turaev, Murakami, Ohtsuki, Yamada et al.)

23.2.1   The Heegaard theorem

Below, we shall use the result of Moise [Moi] that each three–manifold admits a triangulation.

Definition 23.2. The *Heegaard decomposition* of an orientable closed compact manifold M^3 without boundary is a decomposition of M^3 into the union of some two handlebodies — interiors of two copies of S_g attached to each other according to some homeomorphism of the boundary.

There are different ways to attach a handlebody to another handlebody (or to its boundary — a sphere with handles). However, it is obvious that the way of attaching each handlebody can be characterised by a system of meridians — contractible curves in the handlebody.

So, in order to construct a three–manifold, one can take two handlebodies M_1 and M_2 with the same number g of handles and the abstract manifold N that is homeomorphic to ∂M_1 (and hence, to ∂M_2). This manifold N should be endowed with some standard system of meridians $m_1, \ldots, m_g$ (say, it is embedded in $\mathbb{R}^3$ and meridians are taken to be in the most natural sense); see Fig. 23.4.

After this, we fix two systems of meridians $u_1, \ldots, u_g$ for M_1 and $v_1, \ldots, v_g$ for M_2 and two maps $f_1 : \partial M_1 \to N$ and $f_2 : \partial M_2 \to N$, respectively.

Definition 23.3. The system of curves $\{f_1(u_i)\}$ and $\{f_2(v_i)\}$ is said to be *the Heegaard diagram* for this attachment.

Each three–manifold obtained by such an attachment has some Heegaard diagram. While reconstructing the manifold by a Heegaard diagram, we have some ambiguity: the images of meridians do not define the map completely.

However, the following theorem takes place.

Theorem 23.3. *If two three–dimensional manifolds M and M' have the same Heegaard diagram then they are homeomorphic.*

Proof. Without loss of generality, suppose the manifold M consists of M_1 and M_2 and M' consists of M_1' and $M_2' = M_2$.

Let $f_i : \partial M_i \to N, i = 1, 2$, be the gluing homeomorphisms for the manifold M; for M' we shall use $f_i', i = 1, 2$.

Since M and M' have the same Heegaard diagram, then f and f' are homeomorphisms from ∂M and $\partial M'$ to N such that the images of meridians u_i and u_i' coincide for all $i = 1, \ldots, g$. Let us show that in this case the identical homeomorphism $h_2 : M_2 \to M_2'$ can be extended to a homeomorphism $h : M \to M'$. The homeomorphism $(f_1')^{-1} \circ f_1 : \partial M_1 \to \partial M_1'$ takes each meridian of the manifold M_1 to the corresponding meridian of the manifold M_1'. Since each homeomorphism of one circle to another one can be extended to a homeomorphism of whole discs (along radii), then the homeomorphism of meridians can be extended to the homeomorphism of discs bounded by these meridians in M_1 and M_1'. After cutting M_1 and M_1' along these discs, we obtain manifolds D_1 and D_1' both homeomorphic to the three–ball.

Because each homeomorphism between two 2–spheres can be extended to the homeomorphism between bounded balls (along the radii), then the homeomorphism between boundaries of D_1 and D_1' can be extended to the homeomorphism between the balls.

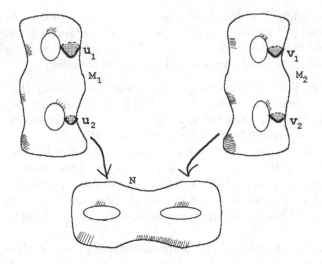

FIGURE 23.4: A Heegaard diagram

Thus, we have constructed a homeomorphism between M and M'. This completes the proof of the Theorem. □

Now we are ready to formulate the Heegaard theorem

Theorem 23.4. *Each orientable three–manifold M without boundary admits a Heegaard decomposition.*

Below, we give just a sketchy idea of the proof. It is well known (see, e.g. [Moi]) that each three–manifold can be triangulated. So, let us triangulate the manifold M. Now, we construct M_1 and M_2 from parts of tetrahedra representing the constructed triangulation. Each tetrahedron T will be divided into two parts T_1 and T_2 as shown in Fig. 23.5.

Now, the manifold M_1 will be constructed of T_1's, and M_2 will be constructed of T_2's. It is easy to see that each of these manifolds is a handlebody (the orientability follows immediately from that of M). The number of handles of M_1 and M_2 is the same because they have a common boundary.

Obviously, the only manifold that can be obtained by attaching two spheres is S^3. The spaces that can be obtained from two tori are S^3, $\mathbb{R}P^3$, and so-called lens spaces. They are closely connected with toric braids. For their description see [BZ], see also [Tur2].

23.2.2 Constructing manifolds by using framed links

In the present section, we shall give a very sketchy introduction to the basic concepts of Kirby theory — how to encode three–manifolds by means of knots (more precisely, by framed links).

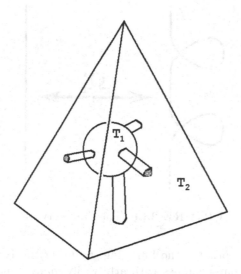

FIGURE 23.5: Dividing tetrahedra

Definition 23.4. A *framed* link is a link in $\mathbb{R}^3$ to each component of which an integer number is associated.

Each framed link can be represented as a band: for each link component we construct a band in its neighbourhood in such a way that the linking coefficient between the boundaries equals the framing of the component. One boundary component of the band should lie in the toric neighbourhood of the second one and intersect each meridional disc of the latter only once.

Definition 23.5. Two framed links are called *isotopic* if the corresponding bands are isotopic.

Having a framed link L, one can construct a three–manifold as follows. For each component K of L, we consider its framing $n(K)$. Now, we can construct a band: we take a knot K' collinear to K such that the linking coefficient between K and K' equals n. Note that this choice is well defined (up to isotopy): while changing the orientations of K and K' simultaneously, we do not change the linking coefficient.

Thus, we have chosen a curve for (each) link component.

The next step of the construction is the following. We cut all full tori, and then attach new ones such that their meridians are mapped to the selected curves which are called *longitudes*.

Theorem 23.5. *For each three–manifold M, there exists a manifold M' homeomorphic to M that can be obtained from a framed link as described above.*

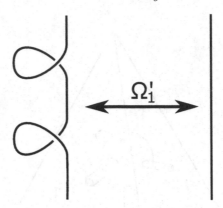

FIGURE 23.6: The double twist

The idea of the proof is the following: by using the Heegaard decomposition, we can use handlebodies with arbitrarily many handles; each homeomorphism of the S_g can be considered locally and be reduced to "primitive homeomorphisms"; the latter can be realised by using toric transformations as above. Here one should mention the following important theorem.

Theorem 23.6 (Dehn–Lickorish). *Each orientation–preserving homeomorphism of S_g can be represented as a composition of Dehn twistings and homeomorphisms homotopic to the identity.*

The first proof of this theorem (with some gaps) was proposed by Dehn in [Dehn1]. The first rigorous proof was found by Lickorish [Lic1].

23.2.3   How to draw bands

The framed link can be easily encoded by planar link diagrams: here the framing is taken to be the self-linking coefficient of the component (it can be set by means of moves Ω_1; each such move changes the framing by ± 1). If we want to consider links together with framing, we must admit the following moves on the set of planar diagrams: the moves Ω_2, Ω_3, and the double twist move Ω_1'; see Fig. 23.6.

23.2.4   The Kirby moves

Obviously, different links may encode the same three–manifolds (up to a diffeomorphism). It turns out that there exist two moves that do not change the three–manifold.

The first Kirby move is an addition (removal) of a solitary circle with framing ± 1, see Fig 23.7.

The second move is shown in Fig. 23.8. Let us be more detailed. While performing this move, we pull one component with framing k along the other

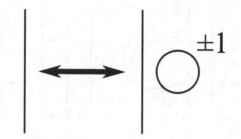

FIGURE 23.7: The first Kirby move

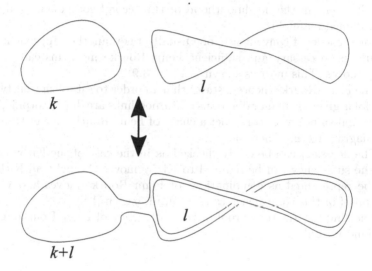

FIGURE 23.8: The second Kirby move

one (with framing l that stays the same). The transformed component would have framing $k + l$.

Remark 23.1. *Note that the component might be linked or not.*

The planar diagram formalism allows us to describe this move simply without framing numbers.

Theorem 23.7 (The Kirby theorem). *Two framed links generate one and the same 3-manifold if one can be transformed to the other by a sequence of Kirby moves and isotopies.*

The theorem says that these two moves are necessary and sufficient. The necessity of the first Kirby move is obvious: we cut one full torus off and attach a new torus almost in the same manner. The necessity of the second Kirby move can also be checked straightforwardly: one should just look at

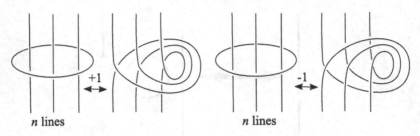

FIGURE 23.9: The Fenn–Rourke move

what happens in the neighbourhood of the second component and compare the obtained manifolds.

For the sake of convenience, one usually takes another approach: one uses the unique necessary and sufficient Fenn–Rourke move instead of the two Kirby moves. This move is shown in Fig. 23.9.

The Fenn–Rourke theorem states that in order to establish that two three–manifolds given by planar diagrams of framed links are diffeomorphic, it is necessary and sufficient to construct a chain of Fenn–Rourke moves transforming one diagram to the other one.

The necessity can be easily checked as in the case of the Kirby theorem.

The sufficiency can be reduced to Kirby moves. Namely, all Kirby moves can be represented as combinations of Fenn–Rourke moves and vice versa. The proof of this equivalence can be found, e.g. in [PS].

The simplest rigorous proof of the sufficiency of Fenn–Rourke moves can be found in [Lu].

23.3   The Witten–Reshetikhin–Turaev invariants

In the present section, we are going to give a sketchy introduction to the famous theory of Witten–Reshetikhin–Turaev invariants. The first article by Witten on the subject [Wit2] was devoted to the construction of invariants of links in three–manifolds. However, mathematicians were not completely satisfied by the strictness level of this work. The mathematical foundations of this theory are due to Viro, Turaev and Reshetikhin [RT, TV]. Here we follow the work of Lickorish [Lic2], where he simplified the work of these three authors for the case of three–manifolds (without links embedded in them).

The basic idea of the description is to use the Kauffman bracket (which is invariant under band isotopies) and apply it to some sophisticated combinations of links in such a way that the obtained result is invariant under the Kirby moves.

23.3.1 The Temperley–Lieb algebra

The Temperley–Lieb algebra is a classical object in operator algebra theory. It has much in common with the Hecke algebra. However, it is realised as a *skein algebra* (the name has come from skein relations). Here we are going to describe how the Kauffman bracket can be used for the construction of the three–manifold invariants via the Temperley–Lieb algebra.

Let M^3 be a manifold represented by means of a framed link in S^3. We shall use bands on the plane (or, simply, planar diagrams) in order to define the framing. The isotopies of these diagrams are considered up to Ω_2, Ω_3 and the double twist Ω_1', shown in Fig. 23.6.

The Temperley–Lieb algebra is a partial case of so–called *skein spaces*. Below, we shall give some examples that will be useful in the future.

With each diagram D of an oriented link, one associates the Kauffman bracket $\langle D \rangle$ in the variable a. For the concrete value $a_0 \in \mathbb{C}$ of a, the value of $\langle D \rangle$ can be calculated just as the unnormalised Kauffman bracket evaluated at a_0 (with the condition that $\langle \bigcirc \rangle$ is equal to one and not to $-a_0^2 - a_0^{-2}$. Let V be the linear space over $\mathbb{C}$ that consists of finite linear combinations of link diagrams. In the space V, consider the subspace V_0 generated by vectors of the type

$$\{ \diagdown\!\!\!\!\diagup - a_0 \asymp - a_0^{-1})(, \; D \sqcup \bigcirc + (a_0^{-2} + a_0^2)D \}.$$

Set $S = V/V_0$. The set S is called *the skein space* for $\mathbb{R}^2$. Under the natural projection $p : V \to V/V_0 = S$, the element D is mapped to λe_1, where λ is the value of the polynomial $\langle D \rangle$ evaluated in a_0, and $e_1 \in S$ is the image of the diagram consisting of one circle.

In fact, let E_i be the diagram consisting of i pairwise non–intersecting circles. It follows from the construction of $\langle D \rangle$ that $D = \sum \lambda_i E_i + w_1$ and $\sum \lambda_i E_i = \lambda E_1 + w_2$, where $w_1, w_2 \in V_0$, and λ is the value of $\langle D \rangle$ at a_0.

Now, let e be a non-zero element of S. Then, $p(D) = f(D)e$, where $f(D) \in \mathbb{C}$. For e, it is convenient to choose the element e_0, corresponding to the empty diagram. Such a choice of the basic element will correspond to multiplication of f with respect to the disconnected sum operation.

Exercise 23.2. *Prove this fact.*

Consequently,

$$f(D_1 \cup D_2) = \sum_{i,j} (c^i)^{\lambda_i} \times (c^j)^{\mu_j}$$

$$= \sum_i (c^i)^{\lambda_i} \times \sum_j (c^j)^{\mu_j} = f(D_1)f(D_2),$$

which completes the proof.

The construction of S admits the following generalisation. Let F be an oriented 2–surface with boundary ∂F (possibly, $\partial F = \emptyset$). Fix $2n$ points on ∂F (in the case of empty ∂F, $n = 0$).

By a *diagram* on F we mean a graph (possibly, disconnected or containing circles) Γ on F such that $\Gamma \cap \partial F$ consists precisely of $2n$ chosen points; these points are the only vertices of Γ of valency one; all the other vertices have valency four and are endowed with crossing structures (as in the case of knot diagrams). These diagrams are considered up to isotopy (not changing the combinatorial structure and fixing the endpoints).

Fix $a_0 \in \mathbb{C}$, and consider the vector space $V(F, 2n)$ over $\mathbb{C}$ consisting of finite linear combinations of diagram isotopy classes. Let $V_0(F, 2n)$ be the subset of $V(F, 2n)$, generated by vectors of the type

$$\{ \diagramcross - a_0 \diagramA - a_0^{-1} \diagramB, \ D \sqcup \bigcirc + (a_0^{-2} + a_0^2) D \}.$$

Put $S(F, 2n) = V(F, 2n)/V_0(F, 2n)$. For the sake of simplicity, denote $S(F, 0)$ by $S(F)$.

Remark 23.2. *It is essential to consider oriented surfaces in order to be able to distinguish between crossings* $\diagramcrossA$ *and* $\diagramcrossB$.

Theorem 23.8. *The image $P(D)$ of the diagram D under the natural projection to $S(F, 2n)$ is invariant under Ω_1' (see Fig. 23.6), Ω_2, Ω_3.*

The proof of this theorem is completely analogous to the invariance proof for the Kauffman bracket. It is left for the reader as a simple exercise.

Theorem 23.9. *The basis of $S(F, 2n)$ consists of images (under the natural isomorphism) of all isotopy classes of diagrams D_i containing neither crossings nor compressible curves.*

The proof is left to the reader.
The following proposition is obvious.

Proposition 23.1. $S(S^2) \cong S(\mathbb{R}^2) \cong \mathbb{C}$.

Proposition 23.2. *The space $S(I \times S^1)$ has the natural algebra structure over $\mathbb{C}$. This algebra is isomorphic to $\mathbb{C}[\alpha]$.*

Proof. By Theorem 23.9, as a basis one can take the family of diagrams where each of these diagrams consists of n circles parallel to the base of the cylinder $I \times S^1$. The multiplication is defined by attaching the lower base of one cylinder to the upper base of the other cylinder and rescaling the height of the cylinder. Let α be the image (under the natural projection) of the diagram consisting of one circle going around the parallel of the cylinder. The remaining part of the proposition is evident. $\square$

Now, let us consider the space $S(D^2, 2n)$. By Theorem 23.9, the basis of this space consists of pairwise non–intersecting arcs with endpoints on ∂D. The number of elements of the basis is equal to the n-th Catalan number c_n. These numbers have the following properties:

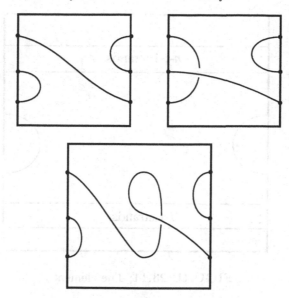

FIGURE 23.10: Product in the Temperley–Lieb algebra

1. $c_{n+1} = \sum_{i=0}^{n} c_i c_{n-i}$;

2. $c_n = \frac{1}{n+1} C_{2n}^n$.

Exercise 23.3. *Prove these two statements.*

The space $S(D^2, 2n)$ has an algebra structure, however, this structure is not canonical. To define this structure, one should consider the disc D^2 with fixed $2n$ points as a square having n fixed points on one side and the other points on the other side. After this, the multiplication is obtained just by attaching one square to the other and rescaling; see Fig. 23.10.

Definition 23.6. The algebra defined above is called the *Temperley–Lieb algebra*.
    Notation: TL_n.

Let e_i be the element of TL_n corresponding to the diagram shown in Fig. 23.11.
    Then, let 1_n be the element of TL_n corresponding to the diagram consisting of n parallel arcs.

Theorem 23.10. *The element 1_n is the unity of TL_n. The elements $e_1, \ldots, e_{n-1}$ represent a multiplicative basis of the algebra TL_n.*

The proof is evident.
    It is also easy to see that the following set of relations is a sufficient generating set for TL_n:

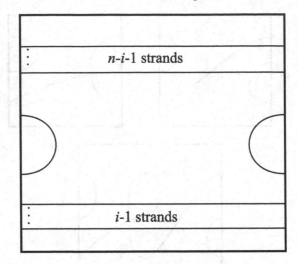

FIGURE 23.11: The element e_i

1. $e_i^2 = e^i(-a_0^2 - a_0^{-2})$;

2. $e_i e_{i+1} e_i = e_i$.

The proof is left for the reader.

23.3.2   The Jones–Wenzl idempotent

In order to go on, one should introduce an idempotent element $f^{(n)}$ in the Temperley–Lieb algebra TL_n that is called the *Jones–Wenzl idempotent*.

Let A_n be the subalgebra in TL_n generated by $e_1, \ldots, e_n$ (without 1_n). Then the element $f^{(n)}$ is defined by the following relations:

$$f^{(n)} A_n = A_n f^{(n)} = 0; \qquad (23.5)$$

$$1_n - f^{(n)} \in A_n. \qquad (23.6)$$

Such an element exists only if we make some restrictions for $a_0 \in \mathbb{C}$.

Theorem 23.11. *Let for each $k = 1, \ldots, n-1$, $a_0^{4k} \neq 1$. Then there exists a unique element $f^{(n)} \in TL_n$ satisfying the properties described above. This element is idempotent: $f^{(n)} \cdot f^{(n)} = f^{(n)}$.*

Exercise 23.4. *Calculate $f^{(2)}$.*

Note that the *existence* of $f^{(n)}$ yields uniqueness and idempotence. Namely, the uniqueness means the uniqueness of the unit element $1_n - f^{(n)}$ in A_n. Moreover, since $1_n - f^{(n)}$ is a unit element of A_n, we have $(1 - f^{(n)})(1 - f^{(n)}) = (1 - f^{(n)})$. From this equation we easily deduce the idempotence of $f^{(n)}$.

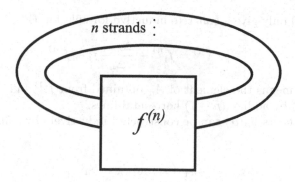

FIGURE 23.12: The element Δ_n

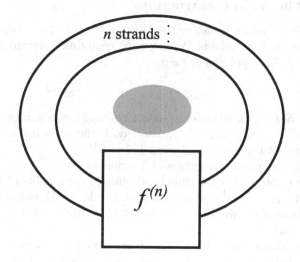

FIGURE 23.13: The element S_n

The construction of $f^{(n)}$ uses induction of n.

We shall need one more construction and some properties of it. Define the element $\Delta_n \in S(\mathbb{R}^2)$ as the *closure* of $f^{(n)}$; see Fig. 23.12. More precisely, $f^{(n)}$ is a linear combination of some diagrams; one should take their closures as shown in Fig. 23.12 and then take the corresponding linear combination.

Analogously, one defines the element $S_n(\alpha) \in S(S^1 \times I)$: this is a "closure" of $f^{(n)}$ in the ring; see Fig. 23.13. Here α denotes the generator of the ring of polynomials $S(S^1 \times I)$. Obviously, Δ_n is obtained from S_n by natural projecting $S(S^1 \times I) \to S(\mathbb{R}^2)$ (gluing the interior circle by a disc).

The elements Δ_n can be defined inductively as $\Delta_{n+1} = (-a_0^{-2} - a_0^2)\Delta_n - \Delta_{n-1}$.

Obviously, $\Delta_1 = -a_0^2 - a_0^{-2}$. Besides, if $a_0^{4(n+1)} \neq 1$ then $\Delta_n \neq 0$; more precisely, $\Delta_n = \frac{(-1)^n(a_0^{2(n+1)} - a_0^{-2(n+1)})}{a_0^2 - a_0^{-2}}$

We shall only give a concrete inductive formula for $f^{(n)}$:

$$f^{(n+1)} = f_1^{(n)} - \frac{\Delta_{n-1}}{\Delta_n} f_1^n e_n f_1^{(n)},$$

where $f_1^{(n)}$ means the element of A_n obtained from $f^{(1)}$ (all summands with coefficients) by adding $(n-1)$ horizontal lines.

The elements $S_n(\alpha)$ can be constructed inductively by using the following formula:

$$S_{n+1}(\alpha) = \alpha S_n - S_{n-1}.$$

23.3.3 The main construction

Now, we are ready to define the Witten–Reshetikhin–Turaev invariant of three–dimensional manifolds. Associate to each link diagram D with components $K_1, \ldots, K_n$ a polylinear map

$$\langle \cdot, \cdots, \cdot \rangle_D : S_1 \times \cdots \times S_n \to S(\mathbb{R}^2),$$

where $S_i \cong S(I \times S^1)$. In order to define this map, it is sufficient to define the elements $\langle \alpha_1^k, \ldots, \alpha_n^{k_n} \rangle_D \in S(\mathbb{R}^2)$, where α_i is the generator of S_i corresponding to the generator α of the algebra $S(I \times S^1)$.

We deal with planar diagrams of unoriented links.

In order to obtain the diagram (and, finally, the number) corresponding to the element $\langle \alpha_1^{k_1}, \ldots, \alpha_n^{k_n} \rangle_D$, we consider the knots B_i corresponding to K_i, and on each knot B_i, we draw k_i copies of the non-intersecting curves parallel to its boundary.

Suppose framed diagrams D and D' are equivalent by means of $\Omega_1', \Omega_2, \Omega_3$. Then the diagrams $\langle \alpha_1^{k_1}, \ldots, \alpha_n^{k_n} \rangle_D$ and $\langle \alpha_1^{k_1}, \ldots, \alpha_n^{k_n} \rangle_{D'}$ can also be obtained from each other by means of Ω_1', Ω_2 and Ω_3. Thus, the images of these two diagrams in $S(\mathbb{R}^2)$ coincide.

Hence, **the polylinear map constructed above is a framed link invariant**.

Now, we are going to construct the further invariant by using $\langle \cdot, \cdots, \cdot \rangle$. But since we are constructing an invariant of three–manifolds, we should also care about the Kirby moves.

To obtain the invariance under the second Kirby move, we shall use the element

$$\omega = \sum_0^{r-2} \Delta_n S_n(\alpha) \in S(I \times S^1),$$

where $r \geq 3$ is an integer.

Theorem 23.12. *Suppose a_0 is such that a_0^4 is the primitive root of unity of degree r. Suppose diagrams D and D' are obtained from each other by the second Kirby move. Then we have:*

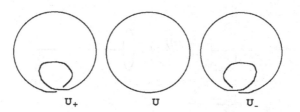

FIGURE 23.14:

$$\langle \omega, \omega, \ldots, \omega \rangle_D = \langle \omega, \omega, \ldots, \omega \rangle_{D'}.$$

The main idea is the following: while performing the second Kirby move, the difference between the obtained elements contains a linear combination of only those terms of $S(S^1 \times I)$ containing $f^{(r-1)}$ as a sub diagram. Since $a_0^{4r} = 1$, we have $\Delta_{r-1} = 0$ and all these terms vanish.

Now, let us consider the first Kirby move. We shall need the *linking coefficient matrix*.

For any n–component link diagram L, one can construct a symmetric $(n \times n)$–matrix of linking coefficients (previously, we orient all link components somehow). For the diagonal elements, we take the self-linking coefficient (which is equal to the framing). Denote the obtained matrix by B. Since B is symmetric, all eigenvalues of this matrix are real. Let b_+ be the number of positive eigenvalues of B, and b_- be the number of negative eigenvalues of B. For constructing the link invariants, we shall need not the matrix B itself, but only b_+, b_-.

It is easy to see that the change of the orientation for some components of B leads to a transformation $B \to B' = X^T B X$ for some orthogonal X, thus it does not change b_+ and b_-.

Let us see now that b_+ and b_- are invariant under the second Kirby move. To do this, it is sufficient to prove the following theorem.

Theorem 23.13. *Under the second Kirby move, the matrix B is changed as follows: $B' = X^T B X$, where X is some non-degenerate real matrix.*

This theorem is left to the reader as an exercise.

Consider the three standard framed diagrams U_+, U_-, U_0 shown in Fig. 23.14. They represent the unknotted curves with framings $1, 0$, and -1, respectively. In the case when $\langle \omega \rangle_{U_+} \langle \omega \rangle_{U_-} \neq 0$, for each diagram D, one can consider the following complex number:

$$I(D) = \langle \omega, \ldots, \omega \rangle_D \langle \omega \rangle_{U_+}^{-b_+} \langle \omega \rangle_{U_-}^{-b_-}.$$

Proposition 23.3. *The complex number $I(D)$ is a topological invariant of the three–manifold defined by D if $\langle \omega \rangle_{U_+} \langle \omega \rangle_{U_-} \neq 0$.*

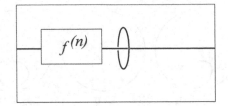

FIGURE 23.15:

To meet the above conditions, one should make the following restrictions on a_0. Namely, the conditions hold if a_0 is a primitive root of unity of degree $4r$ or a primitive root of unity of degree $2r$ for odd r. In both cases, a_0^4 is a primitive root of unity (so that it satisfies the condition of Theorem 23.12).

The proof of Proposition 23.3 is very complicated and consists of many steps. Here, we are going to give the only step that occurs in this proof.

Lemma 23.1. *Suppose a_0 is a complex number as above. Then we have:*

1. *for $1 \leq n \leq r - 3$ each diagram containing D_n (see Fig. 23.15) equals zero;*

2. *for $n = r - 2$ in the case when a_0 is a primitive root of degree $4r$, each diagram containing D_n, equals zero as well, and in the other case (degree $2r$ for odd r) the diagram D_n can be replaced with $\langle \omega \rangle_U f^{(n)}$.*

(Here by "diagrams" we mean their images in $S(\mathbb{R}^2)$.)

Collecting the results of Theorems 23.12 and 23.13 and Proposition 23.3, we obtain the main theorem.

Theorem 23.14. *If $a_0 \in \mathbb{C}$ is either the primitive root of unity of degree $4r$ or the primitive root of unity of degree $2r$ for odd r and $r \geq 3$ then*

$$W(M^3) = I(D) = \langle \omega, \dots, \omega \rangle_D \langle \omega \rangle_{U_+}^{-b_+} \langle \omega \rangle_{U_-}^{-b_-}$$

is a topological invariant of any compact three–manifold without boundary (here D is a framed link diagram representing M^3).

This invariant is called *the Witten–Reshetikhin–Turaev invariant of three–manifolds.*

Examples 23.1. *The sphere S^3 can be represented by the empty diagram; thus $I(S^3) = 1$.*

The manifold $S^1 \times S^2$ is represented by the diagram U. The linking co-efficient matrix for U is $B = (b_{11})$, where $b_{11} = 0$. Thus, $b_+ = b_- = 0$. Consequently, $I(S^1 \times S^2) = \langle \omega \rangle_U$.

23.4   Invariants of links in three–manifolds

The construction of invariants of three–manifolds proposed in the present chapter can also be used for more sophisticated objects, knots in three–manifolds. We shall describe these invariants following [PS].

Consider a compact orientable manifold M^3 without boundary. The manifold M^3 can be obtained by reconstructing S^3 along a framed link. This means that there exists a homeomorphism

$$f : S^3 \backslash L_M \to M^3 \backslash \tilde{L}_M,$$

where $\tilde{L}_M$ is a link in M^3. Without loss of generality, we can assume that the link $\tilde{L}_M$ does not intersect the given link L (this can be done by a small perturbation). Let $L_L = f^{-1}(L)$ be the pre-image of $L \subset M^3$ in the sphere S^3. Suppose if L is framed then L_L is framed as well. Thus, a framed link L in M^3 can be given by a pair of framed links (L_L, L_M) in the sphere S^3; the number of components of L_L coincides with that of L.

Let us discuss the following question: when do two couples (L_L, L_M) and (L'_L, L'_M) generate the same framed link L in M^3? The surgery of the sphere along framed links L_M and L'_M must generate the same manifold M^3; thus, L'_M can be obtained from L_M by Kirby's moves and isotopies. Moreover, during the isotopy, the link L_M should not intersect L_L. The point is that the isotopy that takes L_M to L'_M induces the homeomorphism

$$f : S^3 \backslash L_M \to M^3 \backslash \tilde{L}_M.$$

It is clear that in a general position the Kirby move does not touch a small neighbourhood of the link L_L. After we have performed the necessary Kirby moves and isotopies, we may assume $L'_M = L_M$. After this, we can apply Kirby moves and isotopies to $L_L \cup L_M$, whence the link L_L can only be isotoped. Thus, we have to clarify the connection between L_L and L'_L in the case when $L_M = L'_M$. In this case, after performing a surgery of S^3 along L_M, we obtain a manifold M^3, where the images of the links L_L and L'_L are isotopic. During this isotopy, the images of L_L and L'_L may intersect the image of L_M. Thus, the desired surgery is not reduced to the isotopy of L_L and L_M in the sphere S^3. More precisely, if in M^3, a component K_K of the image of K_L intersects a component K_M of the image of L_M, then the sphere S^3 undergoes the second Kirby move: namely, a band parallel to K_M is added to the band K_L.

Let us point out the following two important circumstances:

1. the numbers of components of L_L and L'_L are equal;

2. under second Kirby moves, we never add bands which are parallel to components of L.

Now, with the framed link $L_L \cup L_M$ we can associate a framed diagram $D_L \cup D_M$. For the link L_M, consider the matrix B of link coefficients. Let b_+ and b_- be the numbers of positive and negative eigenvalues of this matrix; let n be the number of components of L. Let us fix arbitrary elements

$$p_1(\alpha), \ldots, p_n(\alpha) \in S(I \times S^1) \cong \mathbb{C}[\alpha]$$

(here we use the previous notation).

Theorem 23.15. *Let $r \geq 3$ be an integer, a be a complex number and a_0 be such that: either a_0 is the primitive root of unity of degree $4r$ or r is odd and a_0 is the primitive root of unity of degree $2r$.*

Then the complex number

$$W(M^3, L) = \langle p_1(\alpha), \ldots, p_n(\alpha), \omega, \ldots \omega \rangle_{D_L \sqcup D_M} \langle \omega \rangle_{U_+}^{-b_-} \langle \omega \rangle_{U_-}^{-b_-} \qquad (23.7)$$

is an isotopy invariant of the framed link L in M^3.

Proof. First, note that n is invariant: it is the number of components of L. The proof of invariance of the number (23.7) under admissible transformations of the link $L_L \cup L_M$ is almost the same as the proof of invariance of Witten's invariant. The only additional argument is the following. While performing second Kirby moves, one should never add bands parallel to L_L. This allows us to use alternative marking: they can be marked as $p_1(\alpha), \ldots p_n(\alpha) \in S(I \times S^1)$, not only by ω. $\qquad \square$

Remark 23.3. *The formula (23.7) defines not a unique invariant, but an infinite series of invariants with the following parameters: $\alpha_0, p_1, p_2, \ldots, p_n$; the number α_0 should satisfy the conditions above.*

23.5 Virtual 3–manifolds and their invariants

Just recently, L. Kauffman and H. Dye [DK1, Kau8], (see also H. Dye's thesis [Dye]) generalised the Kirby theory for the case of virtual knots and constructed so–called "virtual three–manifolds" and generalised the invariants described here for the virtual case. Another definition of "virtual three–manifolds" was given by S.V. Matveev [Mat6].

The main idea is that all the constructions described above generalise straightforwardly.

In order to define a virtual three–manifold, one considers virtual link diagrams (just as in the classical case). After this, one allows the following equivalence between them: two diagrams are called *equivalent* if one can be obtained from the other by a sequence of generalised Reidemeister moves (all

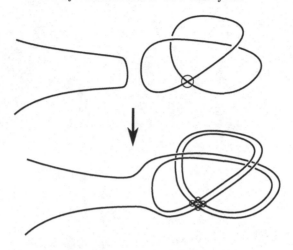

FIGURE 23.16: Virtual handle slide move

but the first classical one, Ω_1) and the two Kirby moves. As for the first Kirby move, the situation is quite clear: we are just adding a new circle with framing one.

As for the second Kirby move, the only thing we should do accurately is to define it for the component having virtual crossing. This move is called *the virtual handle slide move*; it is shown in Fig. 23.16.

These equivalence classes generate *virtual three–manifolds*. After such a definition, the following question arises immediately (stated by the author of this book).

Now, quite an interesting point arises.

First, the Poincaré conjecture evidently fails in this sense.

Furthermore, given two classical diagrams K_1 and K_2 representing the equivalent manifolds (in the sense of virtual Kirby theory); do they actually represent the same classical three–manifolds in the ordinary sense?

Now, what we have to do is to define the Witten–Reshetikhin–Turaev invariant for virtual knots. In fact, we already have all that we need for the definition of it: namely, we have to use the normalised Kauffman bracket and the Jones-Wenzl idempotent, see Theorem 23.14 and the formula therein. For more details, see original work [Dye].

Chapter 24

Heegaard–Floer homology

In the present chapter, we review the Heegaard-Floer homology, due to Peter Ozsváth and Zoltán Szabó [OS5]. The Heegaard-Floer homology is a homological invariant of 3-*manifolds* and *knots in 3-manifolds*.

To imagine the first step for construction of the Heegaard-Floer homology, let us take a Heegaard splitting of a closed 3-manifold M^3 into two handlebodies $N_\alpha \cup N_\beta$ glued along their common boundary Σ_g with sets of meridians $\{\alpha_1, \ldots, \alpha_g\}$ and $\{\beta_1, \ldots, \beta_g\}$.

The whole 3-manifold can be reconstructed by means of *handle gluing*: first we start with a standard 3-ball B, then we add g 1-handles corresponding to meridians $\alpha_1, \cdots, \alpha_g$; adding 2-handles along $\beta_1, \cdots, \beta_g$ returns us to a 4-ball which can be then capped with a 3-handle.

Thus we can construct a Morse function on M^3 with one minimum, one maximum, g critical points of index 1 and g critical points of index 2. If we just want to calculate the homology groups $H_1(M), H_2(M)$, it suffices to know single intersections of some α-curve with some β-curve.

However, we can pass to some more sophisticated configuration spaces that we can construct from M. This idea goes back to Floer [Flo], and it was realized by P.Ozsváth and Z.Szabó. In this setting, generators correspond to some *pairing* between α_j and $\beta_{\sigma(j)}$, where σ is some permutation of indices $\{1, \cdots, g\}$.

The differential will lead to a graded homology theory, and the invariance of this theory can be proved in a self-contained way just by checking moves on Heegaard-diagrams.

Now, if we have a *knot* in a manifold[1] M^3, it can be presented in a nice way with respect to a Heegaard splittings. Looking at the same generators of chain groups, we can treat this presence of a knot as a source for *new grading*. Then we get a homology theory, whose invariance can be also proved by purely combinatorial means.

The knot (and link) Floer homology was constructed by Ozsváth, Szabó and, independently, by J.Rasmussen [Ras1].

A crucial observation is that *the Euler characteristic of the Heegaard-Floer homology for knots in S^3 coincides with the Alexander polynomial*. This can be seen by considering some standard Heegaard decomposition of S^3 related to a particular diagram of a knot with a fixed system of meridians.

---

[1]In fact, not all M^3 work here; we shall concentrate ourselves just with S^3; see below.

Thus, the knot Heegaard-Floer homology is a *categorification* of the Alexander polynomial as the Khovanov homology is a *categorification* of the Jones polynomial.

However, as we saw before, the differential in the Khovanov complex is defined purely combinatorially though the differential in the Heegaard-Floer complex requires some sophisticated structures (complex structures, spin-c structures etc). Hence, since the Heegaard-Floer homology was created, the question about *purely combinatorial* representation of it arose. It was successfully constructed by C.Manolescu, P.Ozsváth, S.Sarkar and D.Thurston [MOS, MOST] by using *grid diagram presentation of knots and links*.

This lead to various questions about categorifications of the *HOMFLY-PT* polynomial; we send the interested author to the review [GS].

In the present chapter, we follow closely the survey of P.Ozsváth and Z.Szábo [OS5] and L.H. Kauffman [Kau9]. The chapter is organized as follows. The first section is devoted the definition of Heegaard–Floer homology of 3-manifolds. The second section is devoted the definition of Heegaard–Floer homology of knots. The third section is devoted the definition of Heegaard–Floer homology of links. The fourth section recalls the Alexander polynomial invariant and describes how it relates to Heegaard–Floer homology.

24.1 Heegaard-Floer homology of 3-manifolds

Floer homology was initially introduced by A. Floer to study questions in Hamiltonian dynamics [Flo]. He started with a symplectic manifold (M, ω) and a pair of Lagrangian submanifolds L_0 and L_1. His invariant, *Lagrangian Floer homology*, is the homology group of a chain complex generated freely by intersection points between L_0 and L_1, endowed with a differential which counts *pseudo-holomorphic discs*. This chain complex arises from a suitable interpretation of the Morse complex in a certain infinite-dimensional setting.

Soon after formulation of the Lagrangian Floer homology, Floer realised that his basic principles could also be used to construct a three-manifold invariant, *instanton Floer homology*, closely related to Donaldson's invariants for four-manifolds. In this version, the basic setup involves a closed oriented three-manifold[2] Y. One forms a chain complex, but this time the generators are $SU(2)$-representations of the fundamental group of Y.

Here we shall describe an adaptation of the Lagrangian Floer homology, the *Heegaard-Floer homology*, which gives rise to a closed three-manifold invariant. This invariant also fits into a four-dimensional framework. There is a related invariant of smooth four-manifolds, and indeed relative invariants for this four-

---

[2]Usually, Ozsváth and Szabó use the letter Y to denote 3-manifolds and the letter X to denote 4-manifolds because these letters Y and X have three and four ends, respectively

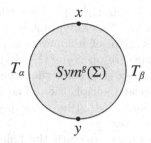

FIGURE 24.1: Whitney disk

manifold invariant take values in the Heegaard-Floer homology groups of its boundary.

As usual, a *Heegaard diagram* (cf. Section 23.2.1) is a triple consisting of an oriented two-manifold Σ of genus g and a pair of g-tuples of embedded disjoint homologically linearly independent curves $\alpha = \{\alpha_1, \ldots, \alpha_g\}$ and $\beta = \{\beta_1, \cdots, \beta_g\}$. A Heegaard diagram uniquely specifies a three-manifold obtained by gluing two genus 2 handlebodies U_α and U_β. In U_α, the curves α_i bound discs, while in U_β, the curves β_i bound discs. We associated to this data a suitable version of Lagrangian Floer homology.

Our ambient manifold in this case is the *g-fold symmetric product of* Σ, the set of unordered g-tuples of points in Σ. This space inherits a natural complex structure from a complex structure over Σ. Inside this manifold, there is a pair of g-dimensional real tori, $\mathbb{T}_\alpha = \alpha_1 \times \cdots \alpha_g$ and $\mathbb{T}_\beta = \beta_1 \times \cdots \beta_g$.

We fix also a reference point

$$w \in \Sigma \setminus (\alpha_1 \cup \cdots \cup \alpha_g \cup \beta_1 \cup \cdots \cup \beta_g).$$

This gives rise to a subvariety $V_w = \{w\} \times Sym^{g-1}(\Sigma) \subset Sym^g(\Sigma)$. We consider the chain complex generated by the intersection points $\mathbb{T}_\alpha \cap \mathbb{T}_\beta$. Concretely, an intersection point of $\mathbb{T}_\alpha$ and $\mathbb{T}_\beta$ corresponds to a permutation σ in the symmetric group on g letters, together with a g-tuple of points $\mathbf{x} = (x_1, \ldots, x_g)$ with $x_i \in \alpha_i \cap \beta_{\sigma(i)}$.

The differential again counts holomorphic discs; but some aspects of the homotopy class of the disc are recorded. Namely, for fixed $\mathbf{x}, \mathbf{y} \in \mathbb{T}_\alpha \cap \mathbb{T}_\beta$, let $\pi_2(\mathbf{x}, \mathbf{y})$ denote the space of homotopy classes of Whitney discs connecting $\mathbf{x}$ to $\mathbf{y}$; i.e., continuous maps of the unit disc $\mathbb{D} \subset \mathbb{C}$ into $Sym^g(\Sigma)$, mapping the part of the boundary of $\mathbb{D}$ with negative, resp., positive real part to $\mathbb{T}_\alpha$, resp., $\mathbb{T}_\beta$, and mapping i, resp., $-i$, to $\mathbf{x}$, resp., $\mathbf{y}$ (see Fig. 24.1). The algebraic intersection number of $\phi \in \pi_2(\mathbf{x}, \mathbf{y})$ with the subvariety V_w determines a well defined map

$$n_w : \pi_2(\mathbf{x}, \mathbf{y}) \to \mathbb{Z}.$$

It is also useful to think of the two-chain $\mathcal{D}(\phi)$, which is gotten as a formal sum of regions in $\Sigma \setminus (\alpha_1 \cup \cdots \cup \alpha_g \cup \beta_1 \cup \cdots \cup \beta_g)$, where a region is counted with multiplicity $n_p(\phi)$, where $p \in \Sigma$ is any point in this region. Given a Whitney disc, we can consider its space of holomorphic representatives $\mathcal{M}(\phi)$, using the induced complex structure on $Sym^g(\Sigma)$. If this space is non-empty for all choices of almost-complex structure, then the associated two-chain $\mathcal{D}(\phi)$ has only non-negative local multiplicities. The group $\mathbb{R}$ acts on $\mathcal{M}(\phi)$ by translation. The moduli space $\mathcal{M}(\phi)$ has an expected dimension $\mu(\phi)$, which is also called the *Maslov index*.

It is sometimes necessary to perturb the holomorphic condition to guarantee that moduli spaces are manifolds of the expected dimension. It is useful (though slightly imprecise) to think of a holomorphic disc in $\mathcal{M}(\phi)$ as a pair consisting of a holomorphic surface F with marked boundary, together with a degree g holomorphic projection map π from F to the standard disc, and also a map f from F into Σ.

Here f maps π^{-1} of the subarc of the boundary of $\mathbb{D}$ with negative resp. positive real part into the subset $\alpha_1 \cup \cdots \cup \alpha_g$, resp., $\beta_1 \cup \cdots \cup \beta_g$.

We now consider the complex $CF^-(Y)$ which is the free $\mathbb{Z}[U]$-module generated by $\mathbb{T}_\alpha \cap \mathbb{T}_\beta$, with differential given by

$$\partial \mathbf{x} = \sum_{\mathbf{y} \in \mathbb{T}_\alpha \cap \mathbb{T}_\beta} \sum_{\{\phi \in \pi_2(\mathbf{x},\mathbf{y}) | \mu(\phi)=1\}} \# \left(\frac{\mathcal{M}(\phi)}{\mathbb{R}} \right) U^{n_w(\phi)} \mathbf{y}. \qquad (24.1)$$

In the case where Y is an integral homology sphere, the above sum is finite. (In the case where the first Betti number is positive, some further constraints must be placed on the Heegaard diagram).

One can see that $\partial^2 = 0$. Roughly speaking, if one can get from a point $\mathbf{x}$ to a point $\mathbf{z}$ in two steps, then there are two ways to do it $\mathbf{x} \to \mathbf{y}_1 \to \mathbf{z}$ and $\mathbf{x} \to \mathbf{y} \to \mathbf{z}$ and these two ways lead to opposite signs; see Fig. 24.2.

The homology groups $HF^-(Y)$ of $CF^-(Y)$ are topological invariants of Y. Indeed, the chain homotopy type of $CF^-(Y)$ is a topological invariant, and, since $CF^-(Y)$ is a module over $\mathbb{Z}[U]$, there are a number of other associated constructions.

For example, we can form the chain complex $CF^\infty(Y)$ obtained by inverting U; i.e., a chain complex over $\mathbb{Z}[U, U^{-1}]$, with differential as in equation (24.1).

The quotient of $CF^\infty(Y)$ by $CF^-(Y)$ is a complex $CF^+(Y)$ which is often more convenient to work with. The corresponding homology groups are denoted by $HF^\infty(Y)$ and $HF^+(Y)$, respectively. Also, there is a chain complex $\widehat{CF}$ obtained by setting $U = 1$; explicitly, it is generated freely over $\mathbb{Z}$ by $\mathbb{T}_\alpha \cap \mathbb{T}_\beta$ and endowed with the differential

$$\hat{\partial} = \sum_{\mathbf{y} \in \mathbb{T}_\alpha \cap \mathbb{T}_\beta} \sum_{\{\phi \in \pi_2(\mathbf{x},\mathbf{y}) | \mu(\phi)=1, n_w(\phi)=0\}} \# \left(\frac{\mathcal{M}(\phi)}{\mathbb{R}} \right) \mathbf{y} \qquad (24.2)$$

and its homology also a topological invariant of Y is denoted by $\widehat{HF}(Y)$.

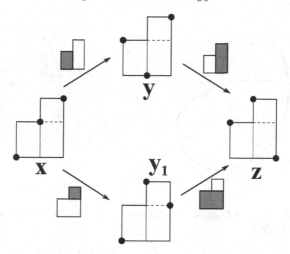

FIGURE 24.2: Differential property $\partial^2 = 0$.

The invariants $HF^-(Y), HF^\infty(Y)$, and $HF^+(Y)$, together with the exact sequence connecting them, are crucial ingredients in the construction of a Heegaard–Floer invariant Φ for closed, smooth four-manifolds.

24.2 Knot Heegaard-Floer homology

Heegaard-Floer homology for 3-manifolds has a refinement to an invariant for null-homologous knots in a three-manifold (as defined independently by Ozsváth, Szabó and Rasmussen [OS5, Ras1]).

A knot K in a 3-manifold Y is specified by a Heegaard diagram (Σ, α, β) for Y, together with a pair w and z of basepoints in Σ. The knot K is given as follows. Connect w and z by an arc ξ in $\Sigma \setminus (\alpha_1 \cup \cdots \cup \alpha_g)$ and an arc η in $\Sigma \setminus (\beta_1 \cup \cdots \cup \beta_g)$. The arcs ξ and η are then pushed into U_α and U_β, respectively, so that they both meet Σ only at w and z, giving new arcs ξ' and η'. Our knot K, then, is given by $\xi' - \eta'$. For simplicity, we restrict ourselves to the case where the ambient manifold Y is S^3.

Below we present a specific representative of the Heegaard splitting of the sphere which "agrees" with a planar knot diagram; all β-curves and all α-curves (except one) correspond to crossings. As we shall see later, this will correspond to the categorification of the Alexander polynomial. For the trefoil, view Fig. 24.3.

In Fig. 24.4, we show the local behaviour of α-curves in the neighbourhood of crossings.

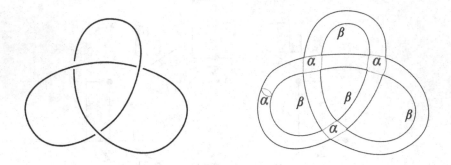

FIGURE 24.3: α-curves and β-curves for the handlebody in the neighbourhood of the trefoil

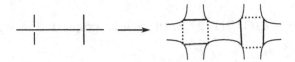

FIGURE 24.4: α-curves in neighbourhoods of crossings

The new basepoint z gives the Heegaard-Floer complex with a filtration. Specifically, we can construct a map

$$F : \mathbb{T}_\alpha \cap \mathbb{T}_\beta \to \mathbb{Z}$$

by setting

$$F(\mathbf{x}) - F(\mathbf{y}) = n_z(\phi) - n_w(\phi), \qquad (24.3)$$

where $\phi \in \pi_2(\mathbf{x}, \mathbf{y})$. It is easy to see that this quantity is independent on the choice of ϕ, depending only on $\mathbf{x}$ and $\mathbf{y}$. Moreover, if $\mathbf{y}$ appears with non-zero multiplicity in $\hat{\partial}(\mathbf{x})$, then $F(\mathbf{x}) \geq F(\mathbf{y})$. This follows from the fact that there is a pseudo-holomorphic disc $\phi \in \pi_2(\mathbf{x}, \mathbf{y})$ with $n_w(\phi) = 0$ and also $n_z(\phi) \geq 0$, since a pseudo-holomorphic disc meets the subvariety V_z with non-negative intersection number.

Equation (24.3) defines F uniquely up to an overall shift. This indeterminacy can be removed as follows. The filtered chain homotopy type of this filtered chain complex is an invariant of the knot K. For example, the homology of the associated graded object, the *knot Floer homology* is an invariant of $K \subset S^3$, defined by

$$\widehat{HFK}(S^3, K) = \oplus_{s \in \mathbb{Z}} \widehat{HFK}(S^3, K, s), \qquad (24.4)$$

where $\widehat{HFK}(S^3, K, s)$ is the homology group of the chain complex generated by intersection points $\mathbf{x} \in \mathbb{T}_\alpha \cap \mathbb{T}_\beta$ with $F(\mathbf{x}) = s$ endowed with the differential

$$\partial \mathbf{x} = \sum_{\mathbf{y} \in \mathbb{T}_\alpha \cap \mathbb{T}_\beta} \sum_{\{\phi \in \pi_2(\mathbf{x},\mathbf{y}) \mid \mu(\phi)=1, n_w(\phi)=n_z(\phi)=0\}} \# \left(\frac{\mathbf{M}(\phi)}{\mathbb{R}} \right) \mathbf{y}. \qquad (24.5)$$

Informally speaking, our differential counts only those holomorphic disks which behave nicely with respect to the knot.

The graded Euler characteristic of this theory is the Alexander polynomial of the knot K, in the sense that

$$\Delta_K(t) = \sum_{s \in \mathbb{Z}} \chi(\widehat{HFK}_*(K,s)) \cdot t^s.$$

We shall touch on this topic again in Section 24.4.

This formula can be used to get rid of the additive indeterminacy of F: we require that F be chosen so that the graded Euler characteristic is the symmetrized Alexander polynomial. In fact, this symmetry has a stronger formulation, as relatively graded isomorphism

$$\widehat{HFK}_*(K,s) \cong \widehat{HFK}_*(K,-s).$$

24.3 Link Heegaard-Floer homology

Heegaard-Floer homology groups of knots can be generalised to the case of links in S^3. For an l-component link, we consider a Heegaard diagram with genus g Heegaard surfaces, and two $(g+l-1)$-tuples attaching circles $\boldsymbol{\alpha} = \{\alpha_1, \cdots, \alpha_{g+l-1}\}$ and $\boldsymbol{\beta} = \{\beta_1, \cdots, \beta_{g+l-1}\}$. We require $\{\alpha_1, \cdots, \alpha_{g+l-1}\}$ to be disjoint and embedded, and to span a g-dimensional lattice in $H_1(\Sigma; \mathbb{Z})$. The same is required of the $\{\beta_1, \cdots, \beta_{g+l-1}\}$. Clearly, $\Sigma \setminus (\alpha_1 \cup \cdots \cup \alpha_{g+l-1})$ consists of l components $A_1, \cdots, A_l$. Similarly $\Sigma \setminus (\beta_1 \cup \cdots \cup \beta_{g+l-1})$ consists of l components $B_1, \cdots, B_l$. We assume that this Heegaard diagram has the special property that $A_i \cap B_i$ is non-empty. Indeed, for each $i = 1, \cdots, l$, we choose basepoints w_i and z_i to lie inside $A_i \cap B_i$. We call the collection of data $(\Sigma, \boldsymbol{\alpha}, \boldsymbol{\beta}, \{w_1, \cdots, w_l\}, \{z_1, \cdots, z_l\})$ a $2l$-pointed Heegaard diagram .

A link can now be constructed in the following manner. Connected w_i and z_i by an arc ξ_i in A_i and an arc η_i in B_i. Again, the arc ξ_i (resp., η_i) is pushed into U_α (resp., U_β) to give rise to a pair of arcs ξ_i' and η_i'. The link L is given by $\cup_{i=1}^{k} \xi_i' - \eta_i'$. For a $2l$-pointed Heegaard diagram for S^3 $(\Sigma, \boldsymbol{\alpha}, \boldsymbol{\beta}, \{w_1, \cdots, w_l\}, \{z_1, \cdots, z_l\})$, if L is the link obtained in this manner, we say that the Heegaard diagram is *compatible* with the link L.

We will need to make an additional assumption on the Heegaard diagram. A *periodic domain* is a two-chain in Σ of the form $\sum c_i(A_i - B_i)$ where $c_i \in \mathbb{Z}$. Our assumption is that all non zero periodic domains have both positive

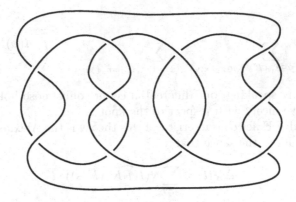

FIGURE 24.5: The Conway link

and negative local multiplicities c_i. This assumption on the pointed Heegaard diagram is called *admissibility*.

Let $L \subset S^3$ be an l-component link, suppose that L is embedded so that the restriction of the height function to L has b local maxima, then we can construct a compatible $2l$-pointed Heegaard diagram with Heegaard genus $g = b - l$.

For example, consider the two-component "Conway link" shown in Fig. 24.5.

For this link, $b = 4$, and hence we can draw it on a surface of genus 2 as shown in Fig. 24.6. Here we consider the surface of genus 2 as the result of attaching two handles to a sphere.

The *link Floer homology* $\widehat{HFL}(S^3, L)$ of the link L is defined as the homology group of the chain complex generated by intersection points $\mathbf{x} \in \mathbb{T}_\alpha \cap \mathbb{T}_\beta$ endowed with the differential

$$\partial \mathbf{x} = \sum_{\mathbf{y} \in \mathbb{T}_\alpha \cap \mathbb{T}_\beta} \quad \sum_{\{\phi \in \pi_2(\mathbf{x},\mathbf{y}) \mid \mu(\phi)=1, n_{w_i}(\phi)=n_{z_i}(\phi)=0, \ i=1,...,l\}} \# \left(\frac{\mathbf{M}(\phi)}{\mathbb{R}} \right) \mathbf{y}. \quad (24.6)$$

Theorem 24.1. *Link Floer homology* $\widehat{HFL}(S^3, L)$ *is a link invariant.*

The proof of the theorem can be found in [OS5].

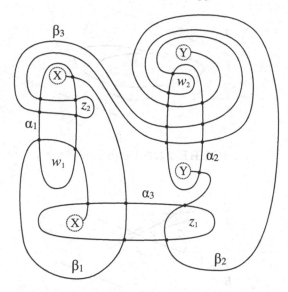

FIGURE 24.6: Pointed Heegaard diagram for the Conway link. The disks labeled by X and Y are the ends of the handles.

24.4 Heegaard splitting of S^3 and the Alexander polynomial

The standard description of the Alexander polynomial given in Chapter 5 is closely related to the standard Wirtinger presentation of the fundamental group. With a link diagram, we associate generators to arcs, and take the relation $bab^{-1} = c$ in the case of groups and $c = ta + (1 - t)b$ in the Alexander module. The same group and module can be presented from another point of view. Here we follow closely [Kau9].

Namely, let D be a link diagram on the plane with n crossings lying on the plane P. Then by Euler characteristic argument this diagram tiles the sphere (1-point compactification of the plane) into $n + 2$ cells, including the infinite one. Thus, we have $n + 1$ finite regions $r_1, \cdots, r_{n+1}$.

Let us fix two points A and B over the plane of projection and under the plane of projection. We denote the link itself by L and we may think that it lies in the neighbourhood of P.

We may fix an path p from B to A pickling the plane P in the infinite region.

Then the group $\pi_1(\mathbb{R}^3 \setminus L)$ can be presented in the following way. We take A to be the base point. Now, with all finite regions $r_1, \cdots, r_{n+1}$ we associate generators $g_1, \cdots, g_{n+1}$. Each generator g_j pickles the plane in the finite region

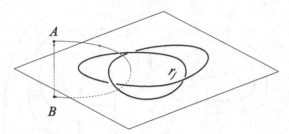

FIGURE 24.7: Dehn generator g_j

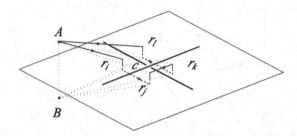

FIGURE 24.8: Crossing relation $g_i g_j^{-1} g_k g_l^{-1} = 1$

from A to B and then goes back through the infinite region as indicated in Fig. 24.7.

Then, one can easily see that for each crossing c where some four regions r_i, r_j, r_k, r_l meet, the relation $g_i g_j^{-1} g_k g_l^{-1} = 1$ holds; see Fig. 24.8.

One can easily check that this gives rise to a presentation of the fundamental group:

$$\pi_1(\mathbb{R}^3 \setminus L) = \langle g_1, \ldots, g_{n+1} \,|\, g_i g_j^{-1} g_k g_l^{-1} = 1 \text{ for all crossings } c\rangle. \qquad (24.7)$$

This presentation is called *the Dehn Presentation* [Dehn2].

Now, let us pass to the definition of the Alexander module and the Alexander polynomial via Dehn's presentation.

Four regions meet locally at a given crossing. Letting these be labeled generically $\{A, B, C, D\}$, as shown in Figure 24.9, we have an equation in the Alexander module

$$xA - xB + C - D = 0 \qquad (24.8)$$

to that crossing. Here A, B, C, D go cyclically around the crossing, starting at the top dot. In this way the two regions containing the dots give rise to the two occurrences of x in the equation. If some of the regions are the same at the crossing, then the equation is simplified by that equality. For example, if $A = D$ then the equation becomes $xA - xB + C - A = 0$. Each crossing in a diagram K gives an equation involving the regions of the diagram. Let us associate *the Dehn presentation Alexander matrix* to the diagram K —

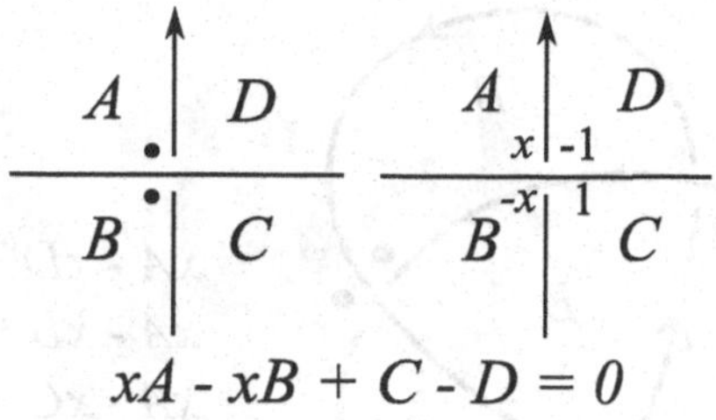

$$A \quad D \qquad\qquad A \quad D$$
$$\qquad\qquad\qquad x \mid -1$$
$$\overline{\qquad\qquad\qquad} \qquad \overline{\qquad\qquad\qquad}$$
$$\qquad\qquad\qquad -x \mid 1$$
$$B \quad C \qquad\qquad B \quad C$$

$$xA - xB + C - D = 0$$

FIGURE 24.9: Alexander labeling

a matrix M_K whose rows correspond to the crossings of the diagram, and whose columns correspond to the regions of the diagram. Each nodal equation gives rise to one row of the matrix where the entry for a given column is the coefficient of that column (understood as designating a region in the diagram) in the given equation. If R and R' are adjacent regions, let $M_K[R, R']$ denote the matrix obtained by deleting the corresponding columns of M_K. Finally, define the *Alexander polynomial* $\Delta_K(x)$ by the formula

$$\Delta_K(x) \doteq Det(M_K[R, R']). \tag{24.9}$$

The notation $a \doteq b$ means that $a = \pm x^n a$ for some integer n. It is proved in [Ale1], that this polynomial is well-defined, independent of the choice of adjacent regions and invariant under the Reidemeister moves up to $\doteq$.

Exercise 24.1. *Check that the Alexander polynomial is invariant under the Reidemeister moves up to multiplication by $\pm x^n$.*

Exercise 24.2. *Prove that the formula (24.9) coincides with the definition of Alexander polynomial of the Chapter 5.*
Hint. Use the skein relation of Alexander polynomial.

In Figure 24.10 we have shown the calculation of the Alexander polynomial of the trefoil knot using this method. Figures 24.10—24.14 are borrowed from Kauffman [Kau3].
In this figure we show the diagram of the knot, the labelings and the resulting full matrix and the square matrix resulting from deleting two columns corresponding to a choice of adjacent regions. Computing the determinant, we find that the the Alexander polynomial of the trefoil knot is given by the equation $\Delta \doteq x^2 - x + 1$.

Exercise 24.3. *Prove that $M[A, B]$ does not depend on the choice of B.*

24.4.1 Reformulating the Alexander polynomial as a state summation

In the present chapter, we follow Kauffman's formal knot theory [Kau3] by the following reasons.

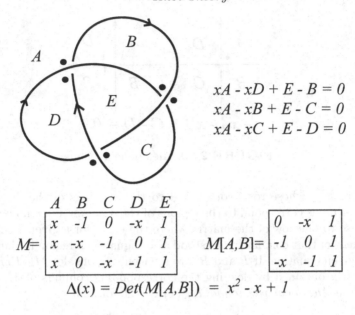

$$xA - xD + E - B = 0$$
$$xA - xB + E - C = 0$$
$$xA - xC + E - D = 0$$

$$M = \begin{array}{c} \\ \end{array} \begin{array}{ccccc} A & B & C & D & E \\ \hline x & -1 & 0 & -x & 1 \\ x & -x & -1 & 0 & 1 \\ x & 0 & -x & -1 & 1 \end{array}$$

$$M[A,B] = \begin{array}{ccc} 0 & -x & 1 \\ -1 & 0 & 1 \\ -x & -1 & 1 \end{array}$$

$$\Delta(x) = Det(M[A,B]) = x^2 - x + 1$$

FIGURE 24.10: The Alexander polynomial

Firstly, at the level of the Alexander polynomial, we shall see a combinatorial reformulation in terms of configurations related to knot and link diagrams.

On the other hand, by looking at these configurations, one can easily see that *the Alexander polynomial is the Euler characteristic of the Heegaard–Floer complex*: when looking at Fig. 24.15, we see that the terms of the Alexander polynomial correspond to the generators of the chain group in the Heegaard Floer complex.

It is an interesting and challenging problem tackled by Kauffman, Przytycki, Yongwu Rong and others: to construct a purely combinatorial presentation of the Heegaard Floer complex in terms of just planar diagrams of links.

Unfortunately, this theory (also called *clock homology theory*) is not invariant: though the generators of the Heegaard-Floer complex *can* be defined in terms of the knot or link diagram, the differentials still depend on some additional *complex* (or spin-*c*) structure.

Roughly speaking, when tiling a 2-surface into regions and looking at differentials corresponding to $2k$-gons for $k \in \mathbb{N}$, the differential does not depend on further geometry if and only if we deal with bigons or quadrilaterals.

The "clock homology theory" deals with differentials corresponding to polygons with a larger number of edges; hence, it is not invariant.

The solution of the problem "how to construct a purely combinatorial theory" came from the *grid diagrams* when one tiles $\mathbb{R}^3$ into a union of two full tori.

Another intriguing question which arises here is the following. As we shall see soon, all "clock" states are nothing but "single circle states" (cf. Chapter 6) which means that both the Alexander polynomial and the Jones polynomial can be expressed by a summation over the same set of states. What about the HOMFLY-PT polynomial? What about the HOMFLY-PT homology theory and superpolynomials?

Given a square $n \times n$ matrix M_{ij}, we consider the expansion formula for the determinant of M:

$$Det(M) = \Sigma_{\sigma \in S_n} sgn(\sigma) M_{1\sigma_1} \cdots M_{n\sigma_n}. \tag{24.10}$$

Here the sum runs over all permutations of the indices $\{1, 2, \ldots, n\}$ with $sgn(\sigma)$ being the sign of the permutation σ. In terms of the matrix, each product corresponds to a *choice* by each column of a single row such that each row is chosen exactly once. The order of rows chosen by the columns (taken in standard order) gives the permutation whose sign is calculated.

Consider our description of Alexander's determinant as given by the formula (24.10). Each crossing is labeled with Alexander's dots so that we know that the four local quadrants at a crossing are each labeled with x, $-x$, 1 or -1. The matrix has one row for each crossing and one column for each region. Two columns corresponding to adjacent regions A and B are deleted from the full matrix to form the matrix $M[A, B]$, and we have the Alexander polynomial $\Delta_K(X) \doteq Det(M[A, B])$. A generalisation of this approach to knots and links in thickened surfaces was considered by M. Zenkina [MZ, Zen1, Zen2].

In the Alexander determinant expansion the *choice* of a row by a column corresponds to *a region choosing a crossing* in the link diagram. The only crossings that a region can choose giving a non-zero term in the determinant are the crossings in the boundary of the given region. Thus the terms in the expansion of $Det(M[A, B])$ are in one-to-one correspondence with decorations of the flattened link diagram (i.e. we ignore the over and under crossing structure) where each region (other than the two deleted regions corresponding to the two deleted columns in the matrix) labels one of its crossings. We call these labeled flat diagrams the *states* of the original link diagram (cf. Chapter 7). See Figure 24.11 for a list of the states of the trefoil knot. In this figure we show the states and the corresponding matrix forms with columns choosing rows that correspond to each state.

At this point we have an almost complete combinatorial description of Alexander's determinant. The only thing missing is the permutation signs. One can pick up the permutations from the state labeling, but there is a better way. Call a state marker (label at a crossing as shown in Figure 24.11) *negative* if it labels a quadrant where both oriented segments point toward the crossing; see Fig. 24.12.

Let S be a state of the diagram K. Set $(-1)^{b(S)}$ where $b(S)$ is the number of negative markers in the state S. Then it turns out that with up to one global sign ϵ depending on the ordering of crossings and regions, we have

$$(-1)^{b(S)} = \epsilon \, sgn(\sigma(S)),$$

where $\sigma(S)$ is the permutation of crossings induced by the choice of ordering

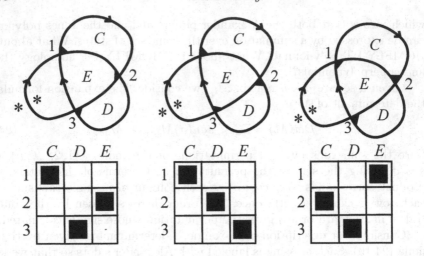

FIGURE 24.11: States with markers

FIGURE 24.12: A negative marker

of the regions of the state. This gives a purely diagrammatic access to the sign of a state and allows us to write

$$\Delta_K(x) \doteq \Sigma_S \langle K|S \rangle (-1)^{b(S)}, \tag{24.11}$$

where S runs over all states of the diagram for a given choice of deleted adjacent regions, and $\langle K|S \rangle$ denotes the product of the Alexander nodal labels at the quadrants indicated by the state labels in the state S. We call $\langle K|S \rangle$ the *product of the vertex weights*. Thus we have a precise reformulation of the Alexander polynomial as a state summation.

Let us mention that every term in the Dehn presentation Alexander matrix corresponds to a single-circle state, see Fig. 24.13.

In Figure 24.14 we illustrate the calculation of the Alexander polynomial of the trefoil knot using this state summation. Here we show the contributions of each state to a product of terms and in the polynomial we have followed the state summation by taking into account the number of negative markers in each state. Thus we get

$$(-1)^{b(S)} = \epsilon \; sgn(\sigma(S)).$$

Theorem 24.2. *The Euler characteristic of the knot Heegaard–Floer homology coincides with the Alexander polynomial of the knot.*

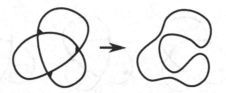

FIGURE 24.13: A single circle state

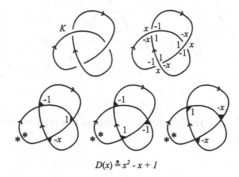

$$D(x) \stackrel{\cdot}{=} x^2 - x + 1$$

FIGURE 24.14: State sum calculation of Alexander polynomial

Sketch of the proof. Using the correspondence between the generators of the Heegaard–Floer complex (Fig. 24.1) and the diagram states (Fig. 24.11); see Fig. 24.15, one shows that Euler characteristic of the Heegaard–Floer complex is the Alexander polynomial.

24.5 Applications of the Heegaard–Floer homology

Heegaard–Floer homology is well suited to problems in knot theory and 3-manifold topology which can be formulated in terms of the existence of four-dimensional cobordisms; one of the most striking applications is that the Heegaard–Floer homology is an unknot detector. Some years after this fact was established, Peter Kronheimer and Tomasz Mrówka proved that Khovanov homology is also an unknot detector [KrMr4] by constructing a spectral sequence starting from Khovanov homology and converging to (a version of) the Heegaard-Floer homology. Namely, let $K \subset S^3$ be a knot. Recall that the *Seifert genus* of K, denoted by $g(K)$, is the minimal genus of any embedded surface $F \subset S^3$ with boundary K. Clearly, $g = 0$ if and only if K is the unknot. According to [OS2], knot Floer homology detects the Seifert genus of a

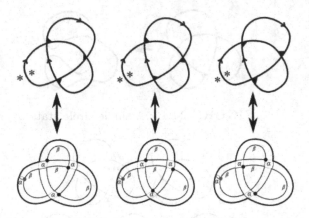

FIGURE 24.15: Correspondence between generators of the Heegaard–Floer complex and diagram states

knot by the property that

$$g(K) = max\{s | H\hat{F}K(K, s) \neq 0\}. \tag{24.12}$$

Chapter 25

Legendrian knots and their invariants

The theory of Legendrian knots first introduced by Dmitry Fuchs and Serge Tabachnikov [FT], lies at the juncture of knot theory, the theory of wave fronts and contact geometry. D.B. Fuchs calls the "Fuchs–Tabachnikov moves" "Swiatkowski moves" according to [Swi].

Legendrian knots in $\mathbb{R}^3$ represent the one–dimensional case of Legendrian manifolds (in the general case, a Legendrian manifold is a k–dimensional submanifold in a $(2k+1)$–dimensional manifold satisfying some conditions). The Legendrian knot theory is interesting because it allows us to introduce a new equivalence for knots: besides topological isotopy, one can consider a more subtle isotopy in the space of Legendrian knots.

25.1 Legendrian manifolds and Legendrian curves

One of the main questions of the theory of differential equations is to find an enveloping curve for the family of straight lines on the plane. It is well known that in the smooth case, this problem has a solution according to the existence and uniqueness theorem.

If we consider the field of, say, planes in $\mathbb{R}^3$, or, more generally, hyperplanes in odd–dimensional space $\mathbb{R}^{2n+1}$, the enveloping surface does not always exist.

In the general position called "the maximally non–integrable case", the maximal dimension of a surface tangent to these hyperplanes at each point equals n.

For the case $n = 1$ we obtain just curves in $\mathbb{R}^3$; i.e., knots (which can be tangent to the given family of planes).

Maximally non–integrable fields of hyperplanes are closely connected to contact structures.

25.1.1 Contact structures

Let us first introduce the notion of contact structure, see, e.g., [Arn2, Thu].

Definition 25.1. A *contact structure (form)* on an odd–dimensional manifold

M^{2n+1} is a smooth 1–form ω on M such that $\omega \wedge \underbrace{d\omega \wedge \cdots \wedge d\omega}_{n}$ is a non–degenerate form.

Having a contact form, one obtains a hyperplane in the tangent space at each point: the plane of vectors v such that $\omega(v) = 0$.

Let us consider the case of $\mathbb{R}^3$ and the form $\omega = -xdy + dz$. Obviously, at each point (x, y, z) the incident plane is generated by the two vectors $(1, 0, 0)$ and $(0, 1, x)$. Denote this field of planes by τ.

Definition 25.2. A *Legendrian link* is a set of non-intersecting oriented curves in $\mathbb{R}^3$ such that each link at each point is tangent to τ.

Since a Legendrian link is a link, one can consider its projection onto planes; i.e. planar diagrams. It turns out that projections on different planes have interesting properties.

Note that there are non-equivalent contact structures even in $\mathbb{R}^3$, the *tight* one and the *overtwisted one*. With respect to each contact structure, there is its own theory of Legendrian knots.

We shall not go into the detail and we shall restrict ourselves to the structure given above.

25.1.2   Planar projections of Legendrian links

First, let us consider a projection of a Legendrian link L to the Oyz plane. Let γ be the projection of one component of L. Consider $\dot{\gamma} = (0, \dot{y}, \dot{z})$. By definition $x\dot{y} = \dot{z}$, and we conclude that *the coordinate x of the Legendrian curve L equals the fraction* $\frac{\partial z}{\partial y}$, or, in other words, the abscissa equals the tangent of the tangent line angle.

The only inconvenience here is that $\dot{y}$ cannot be equal to zero.

This effect can be avoided by allowing x to be equal to ∞; i.e., by considering $\mathbb{R}^2 \times S^1$ instead of $\mathbb{R}^3$. Indeed, in this case there arises a beautiful theory of Legendrian links (as well as in any three–manifold that is a bundle over a two–surface M^2 with fiber S^1, for example, the unit bundle UT_*M of M). However, here we consider knots and links in $\mathbb{R}^3$; thus we must take the restriction $\dot{y} \neq 0$.

This means that on the plane Oyz our curve has no "vertical" tangent lines. So, the only possibility to change the sign of $\dot{y}$ is the existence of a cusp; see Fig. 25.1.

Generically, the cusp has the form of a semi cubic parabola: the curve $(\frac{3t}{2}, t^2, t^3)$ is a typical example of such a curve (the cusp takes place at $(0,0,0)$).

Having any piecewise–smooth oriented curve (smooth everywhere except cusps where $\dot{y}$ changes the sign), we can easily restore a Legendrian curve in $\mathbb{R}^3$ by putting

$$x = \frac{\dot{z}}{\dot{y}}.$$

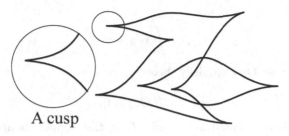

A cusp

FIGURE 25.1: A cusp of the front projection

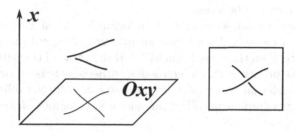

FIGURE 25.2: Restoring crossing types from the front projection

Obviously, taking a curve γ in general position (with only double transverse intersection points), one gets a link L having projection γ. So, in order to construct a shadow of a link isotopic to L, one should just smooth all cusps.

Besides this, the transverse intersection points define uniquely the crossing structure of the link. Namely, the x–coordinate is greater for the piece of curve where the tangent is greater.

One can slightly deform the projection γ (without changing the isotopy type of the corresponding Legendrian link) so that the two intersecting pieces of the curve $\tilde{\gamma}$ have directions northwest–southeast and northeast–southwest; see Fig. 25.2

In this case, the branch northeast–southwest will form an overcrossing.

Hence, we know how to construct planar link diagrams from diagrams of Legendrian links projections to Oyz.

The inverse procedure is described in the proof of the following

Theorem 25.1. *For each link isotopy class there exists a Legendrian link L representing this class.*

Now, let us look at what happens if we take the projection to Oxy.

It is more convenient in this case to consider each link component separately.

Let γ be a curve of projection. If we take an interval of this curve starting from a point A and finishing at a point B, we deduce from $\dot{z} = x\dot{y}$ that $z_A - z_B = \int_A x\,dy$. If we take an integral along all the curve γ (from A to A), we see that

$$\oint_\gamma x\,dy = 0,$$

or by the Gauss–Ostrogradsky theorem,

$$S(M_\gamma) = 0, \tag{25.1}$$

where S means oriented area, and M_γ is the domain bounded by γ (with signs).

Equation (25.1) is the unique necessary and sufficient condition for the Legendrian curve to be closed.

If we want to get some crossing information for a Legendrian knot, we should take a crossing P and take an integral along γ from P lying on one branch of γ to P on the other branch of γ. If the value of this integral is positive then the first branch is an undercrossing; otherwise it is an overcrossing.

If we consider an n–component link, one can easily establish the crossings for each of its components. Then one has $n - 1$ degrees of freedom in posing the components.

So, for the case of a link, the crossing structure cannot be restored uniquely.

25.2   Projections of Legendrian knots and their properties

Here we shall use the notation from [Che2, ChPu]. We are going to give definitions only for the case of knots; analogous constructions for links can be presented likewise.

Consider a smooth knot in the standard contact space $\mathbb{R}^3 = \{q, p, u\}$ with the contact form $\alpha = du - p\,dq$ (we introduce the new coordinates u, p, q instead of z, x, y respectively).

Definition 25.3. A smooth knot L is called *Legendrian* if the restriction of α to L vanishes.

Definition 25.4. Two Legendrian knots are called *Legendrian isotopic* if one can be sent to the other by a diffeomorphism g of $\mathbb{R}^3$ such that $g^*\alpha = \phi\alpha$, where $\phi > 0$.

There are two convenient ways of representing Legendrian knots by projecting them on different planes. The projection $\pi : \mathbb{R}^3 \to \mathbb{R}^2, (q, p, u) \to (q, p)$ is called *the Lagrangian projection*, and the projection $\sigma : (q, p, u) \to (q, u)$ is called *the front projection*.

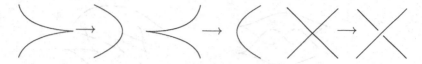

FIGURE 25.3: Restoring a diagram from a front

FIGURE 25.4: Constructing a front by a diagram

25.2.1 The front projection

Having a front projection, we can restore the Legendrian knot as follows. We just smooth all cusps, and set the crossings according to the following rule: *the upper crossing is the branch having greater tangency.* Since the front has no vertical tangent lines, the choice of crossing is well defined; see Fig. 25.3.

The front projection is called σ-*generic* if all self–intersections of it have different q coordinates.

The inverse procedure can be done as follows. After a small perturbation we can make a diagram having no vertical tangents at crossings. Now, let us replace all neighbourhoods of points with vertical tangents by cusps. All "good" crossings are just replaced by intersections. Besides, all double points with "bad" crossings are to be replaced just as shown in Fig. 25.4.

Thus, we have proved that each knot can be represented by a front; i.e., **each knot has a Legendrian representative**. Thus, we have proved Theorem 25.1.

In Fig. 25.5, we show diagrams and fronts for the two trefoils. The assymmetry of these two diagrams follows from the convention concerning the "good" crossings. It is quite analogous to the assymmetry of the simplest d–diagrams of the trefoils.

25.2.2 The Lagrangian projection

In the normal case, the Lagrangian projection is smooth (has no cusps or other singularities unlike the front projection). As in the case of front projection, the Lagrangian projection allows us to restore the crossing structure, and hence, the topological knot itself..

More precisely, a Lagrangian projection is called π-*generic* if all self–intersections of it are transverse double points.

Namely, having a planar diagram L of a knot, let us fix some point P of it different from a crossing and fix $u(P) = 0$. Now, we are able to restore the

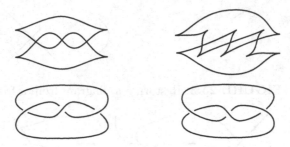

FIGURE 25.5: Left trefoil and right trefoil

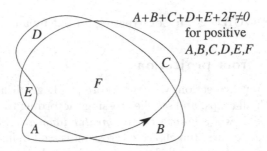

$A+B+C+D+E+2F \neq 0$
for positive
A,B,C,D,E,F

FIGURE 25.6: A braided diagram is not Lagrangian

coordinate u for all the points of L. Taking into account that $du = pdq$ along the curve, we see that the difference $u_A - u_B$ equals the oriented area of the domain restricted by the part of the curve from A to B.

Thus, if we have some projection L (combinatorial knot diagram) and we want it to be Lagrangian, we should check the following condition: if we go along the knot from some point A to itself, we obtain an equation on areas of domains cut by L. All crossing types are regulated by equations.

Exercise 25.1. *Write these equations explicitly*

This shows that some projections cannot be realised as Lagrangian ones.

For instance, having a braided diagram (see Fig. 25.6), we cannot realise it as Lagrangian because the equation to hold will consist only of positive numbers whose sum with positive coefficients should be equal to zero: we go around each area a positive number of times and each area is positive. Thus, it cannot be equal to zero.

In this sense, Lagrangian diagrams are opposite to braided diagrams.

Remark 25.1. *In fact, the third projection* $(q, p, u) \to (p, u)$ *is not interesting.*

A Legendrian knot $L \subset \mathbb{R}^3$ is said to be π–*generic* if all self intersections of the immersed curve $\pi(L)$ are transverse double points. In this case, this projection endowed with over– and undercrossing structure represents a knot diagram that is called the *Lagrangian diagram*.

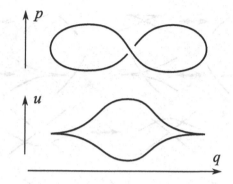

FIGURE 25.7: Lagrangian projection and front projection

Of course, not every abstract knot diagram in $\mathbb{R}^2$ is a diagram of a Legendrian link, or is oriented diffeomorphic to such a one.

For a given Legendrian knot $L \subset \mathbb{R}^3$, its σ–projection or *front* $\sigma(L) \subset \mathbb{R}^3$ is a singular curve with no vertical tangent vectors.

25.3 Fuchs–Tabachnikov (Swiatkowski) moves

Legendrian knots and links in their frontal projection admit a combinatorial interpretation like ordinary knots and links. Namely, there exists a set of elementary moves transforming one frontal projection of a Legendrian link to each other projection of the same link.

In fact, the following theorem holds.

Theorem 25.2 (Fuchs, Tabachnikov [FT]). *Two fronts represent Legendrian–equivalent links if and only if one can be transformed to the other by a sequence of moves 1–3 shown in Fig. 25.8.*

By admitting the 4th move, we obtain the ordinary (topological) equivalence of links.

Note that the tangency move (when two lines pass through each other) is not a Legendrian isotopy: this tangency in $\mathbb{R}^2$ means an intersection in $\mathbb{R}^3$ (the third coordinate is defined from the tangent line). Thus, this move changes the Legendrian knot isotopy type; it is not the second Reidemeister move.

Note that the *grid diagram presentation* for links allows one to get a nice way for encoding Legendirian links: the set of diagrams is the same, and in the set of moves we just forbid a certain sort of stabilisation.

Knot Theory

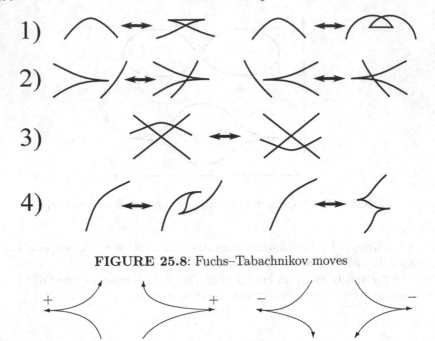

FIGURE 25.8: Fuchs–Tabachnikov moves

FIGURE 25.9: Positive cusps   Negative cusps

25.4 Maslov and Bennequin numbers

The invariants named in the section title can be defined as follows. The *Bennequin number* (also called the *Thurston–Bennequin number*) $\beta(L)$ of L is the linking number between L and $s(L)$, where s is a small shift along the u direction.

The *Maslov number* $m(L)$ is the rotation number of the projection of L to the (q, p) plane.

The change of orientation on L changes the sign of $m(L)$ and preserves $\beta(L)$.

Both these invariants can be defined combinatorially by using the front projection of the Legendrian knot. Namely, the Maslov number is half of the difference between the numbers of positive cusps and negative cusps; see Fig. 25.9.

Exercise 25.2. *Prove the equivalence of the two definitions of the Maslov number.*

Definition 25.5. A crossing is called *positive* if the orientation of two

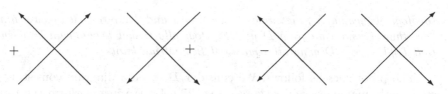

FIGURE 25.10: Positive crossings   Negative crossings

branches of the front have directions in different half–planes, and *negative* otherwise; see Fig. 25.10.

The *Bennequin number* is $\frac{1}{2}$(# cusps)+ (# positive crossings)–(# negative crossings).

Exercise 25.3. *Prove the equivalence of the two definitions of the Bennequin number.*

25.5   Finite–type invariants of Legendrian knots

By definition, each Legendrian knot is a topological knot; besides, two equivalent Legendrian knots are (topologically) equivalent knots. Thus, each knot invariant represents an invariant of Legendrian knots. So are finite-type invariants of knots. Moreover, one can easily define the finite type invariants of Legendrian knots and show that all finite-type invariants coming from "topological knots" have finite-type in the Legendrian sense.

Exercise 25.4. *Prove that the Maslov number and the Bennequin number are finite-type invariants: the first of them has order zero and the second one has order one.*

The most important achievement (classification) of the finite type invariants is described by the following.

Theorem 25.3 (Fuchs, Tabachnikov). *All Vassiliev invariants of Legendrian knots can be obtained from topological finite-type invariants and Maslov and Bennequin numbers.*

This means that the theory of finite–type invariants for Legendrian links is not so rich. As we are going to show, there are stronger invariants that cannot be represented in terms of finite order invariants.

This theorem comes from the following observation

Theorem 25.4. *For any two Legendrian knot diagrams L and L′ of the same*

*topological type having the same Maslov index and Thurston-Bennequin num-
bers there exists a diagram D of a topologically trivial Legendrian knot such
that $L\#D = L'\#D$ generate equivalent Legendrian knots.*

The proof goes as follows. We consider D to be a diagram consisting of
sufficiently many zigs and zags (see Fig. 25.8.4)) We just perform the topological isotopy between D and D' as classical knots. There are no problems
with third Reidemeister moves, but if we deal with the first Reidemeister move
(and the second Reidemeister move) we can need a cusp to perform it. This
cusp can be taken from D. We can take D large enough to be able to perform
all possible first and second Reidemeister moves.

After the whole isotopy is performed, we can take all "auxiliary" cusps
from D back to their position.

This theorem shows that from the point of view of Vassiliev's invariants
Legendrian knots do not differ so much from classical knots. In the next section
we shall see quite new and powerful invariants.

25.6   The differential graded algebra (DGA) of a Legendrian knot

In the present section, we shall speak about the differential graded algebra
(free associative algebra with the unit element) of Legendrian knots, proposed
by Chekanov. It turns out that all homologies of this algebra are invariants
of Legendrian knots. Now, we are going to work with Lagrangian diagrams of
Legendrian knots.

We associate with every π–generic Legendrian knot K a DGA (A, ∂) over
$\mathbb{Z}_2$ (see [Che]).

Remark 25.2. *A similar construction was given by Eliashberg, Givental and
Hofer, see [EGH, Eli]).*

Let L be a Lagrangian diagram of a Legendrian link. Denote crossings of
this diagram by $\{a_1, \ldots, a_n\}$.

We are going to denote a tensor algebra $T(a_1, \ldots, a_n)$ with generators
$a_1, \ldots, a_n$. This algebra is going to be a $\mathbb{Z}_{m(L)}$–graded algebra (free, associative and with unity).

First of all, let us define the grading for this algebra. Let a_j be a crossing
of L. Let z_+, z_- be the two pre-images of a_j in $\mathbb{R}^3$ under the Lagrangian
projection, whence z_+ has greater u–coordinate than z_-.

Without loss of generality, one might assume that the two branches of the
Lagrangian projection at a_j are orthogonal.

These points divide the diagram L into two pieces, γ_1 and γ_2, and we
orient each of these pieces from z_+ to z_-.

Now, for $\varepsilon \in \{1, 2\}$, the rotation number of the curve $\pi(\gamma_\varepsilon)$ has the form $\frac{N_\varepsilon}{2} + \frac{1}{4}$, where $N_\varepsilon \in \mathbb{Z}$. Clearly, $N_1 - N_2$ is equal to $\pm m(L)$. Thus N_1 and N_2 represent one and the same element of the group $\mathbb{Z}_{m(L)}$, which we define to be the degree of a_j.

Now, we are going to define the differential ∂. For every natural k, let us fix a curved convex k–gon $\Pi_k \subset \mathbb{R}^2$ whose vertices $x_0^k, \ldots, x_{k-1}^k$ are numbered counterclockwise.

The form $dq \wedge dp$ defines an orientation on the plane. Denote by $W_k(L)$ the collection of smooth orientation–preserving immersions $f : \Pi_k \to \mathbb{R}^2$ such that $f(\partial \Pi_k) \subset L$. Note that $f \in W_k(L)$ implies that $f(x_i^k) \in \{a_1, \ldots, a_n\}$.

Let us consider these immersions up to combinatorial equivalence (parametrisation) and denote the quotient set by $\tilde{W}_k(L)$. The diagram L divides a neighbourhood of each of its crossings into four sectors. Two of them are marked as positive (*opposite to the way used while we defined Kauffman's bracket*) and the other two are taken to be negative. For each vertex x_i^k of the polygon Π_k, a smooth immersion $f \in \tilde{W}_k(L)$ maps its neighbourhood in Π_k to either a positive or negative sector; in these cases, we shall call x_i^k *positive* or *negative*, respectively.

Define the set $W_k^+(L)$ to consist of immersions $f \in \tilde{W}_k(L)$ such that the vertex x_0^k is the only positive vertex for f; all other vertices are to be negative. Let $W_k^+(L, a_j) = \{f \in W_k^+(L) | f(x_0^k) = a_j\}$. Let $A_1 = \{a_1, \ldots, a_n\} \otimes \mathbb{Z}_2 \subset A$, $A_k = A_1^{\otimes k}$. Then $A = \oplus_{l=0}^\infty A_l$.

Let $\partial = \sum_{k \geq 0} \partial k$, where $\partial_k(A_i) \in A_{i+k-1}$ and

$$\partial_k(a_j) = \sum_{f \in W_{k+1}^+(L, a_j)} f(x_1^{k+1}) \cdots f(x_k^{k+1}).$$

Now, we can extend this differential for the algebra A by linearity and the Leibnitz rule. Now, the following theorem says that (A, ∂) is indeed a DGA.

Theorem 25.5. *The differential ∂ is well defined. We have $\deg\partial = -1$ and $\partial^2 = 0$.*

The main theorem on this invariant [Che], see also [Che2], is the following.

Theorem 25.6. *Let $(A, \partial), (A', \partial')$ be the DGA's of (π–generic) Legendrian knot diagrams L, L'. If L and L' are Legendrian isotopic then the homology rings $H(A, \partial)$ and $H(A', \partial')$ are isomorphic.*

The proof is a straightforward check.

25.7 Chekanov–Pushkar' invariants

The invariants described in this section are purely combinatorial. They are described in the terms of front projection. Though they are combinatorial

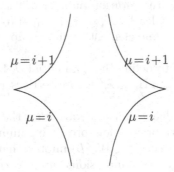

$\mu = i+1$ $\mu = i+1$

$\mu = i$ $\mu = i$

FIGURE 25.11: Jumps of the Maslov index near cusps

and the proof of their invariance can be easily obtained just by checking all Reidemeister moves, they have deep homological foundations.

Given a σ–generic oriented Legendrian knot L, let us denote by $C(L)$ the set of its points corresponding to cusps of σ_L. The *Maslov index* $\mu : L/C(L) \to \Gamma = \mathbb{Z}_m$ is a locally constant function uniquely defined up to an additive constant by the following rule: the value of μ jumps at points of $C(L)$ by ± 1 as shown in Fig. 25.11. A crossing is called a *Maslov crossing* if μ takes the same value on both its branches.

Assume that $\Sigma = \sigma(L)$ is a union of closed curves $X_1, \ldots, X_n$ that have finitely many self–intersections and meet each other at finitely many points. Then we call the unordered collection $\{X_1, \ldots, X_n\}$ a *decomposition* of Σ.

Now, a decomposition is called *admissible* if it satisfies some conditions.

1. Each curve X_i bounds a topologically embedded disk: $X_i = \partial B_i$.

2. For each i and $q \in \mathbb{R}$, the set $B_i(q) = \{u \in \mathbb{R} \,|\, (q, u) \in B_i\}$ is either a segment, or consists of a single point u such that (q, u) is a cusp of σ_L, or is empty.

Conditions 1 and 2 imply that each curve X_i has exactly two cusps (and hence the number of curves is half the number of cusps). Each X_i is divided by cusps into two pieces, on which the coordinate q is a monotone function. Near a crossing $x \in X_i \cap X_j$, the decomposition of Σ may look in one of the three ways represented in Figure 25.12. The first type of crossing shown in Fig. 25.12.a is automatically ruled out by conditions 1,2. The second type of decomposition (see Fig. 25.12.b) is called *switching*. The third type (Fig. 25.12.c) is called *non–switching*.

3. If $(q_0, u) \in X_i \cap X_j$ is switching for some $i \neq j$ then for each $q \neq q_0$ sufficiently close to q, the set $B_i(q) \cap B_j(q)$ either coincides with $B_i(q)$ or with $B_j(q)$, or is empty.

4. Every switching crossing of type shown in Fig. 25.12b is Maslov.

Definition 25.6. A decomposition is called *admissible* if it satisfies conditions 1–3 and *graded admissible* if it also satisfies condition 4.

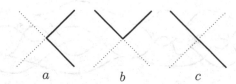

FIGURE 25.12: Three types of switchings

Note that there are three types of crossings with respect to the layout of X_i, shown in Fig. 25.12.

Denote by $Adm(\Sigma)$ (respectively, $Adm_+(\Sigma)$) the set of admissible (respectively, graded admissible) decompositions of Σ. Given $D \in Adm(\Sigma)$, denote by $Sw(D)$ the set of its switching points. Define $\Theta(D) = \#(D) - \#Sw(D)$.

Now, we are ready to formulate the main theorem on the Chekanov–Pushkar' invariants.

Theorem 25.7. *If the σ–generic Legendrian knots $L, L' \subset \mathbb{R}^3$ are Legendrian isotopic then there exists a one–to–one mapping*

$$g : Adm(\sigma(L)) \to Adm(\sigma(L'))$$

such that

$$g(Adm_+(\sigma(L))) = Adm_+(\sigma(L'))$$

and $\Theta(g(D)) = \Theta(D)$ for each $D \in Adm(\sigma(L))$.

In particular, the numbers $\#(Adm(\sigma(L)))$ and $\#(Adm_+(\sigma(L)))$ are invariants of Legendrian isotopy.

Proof. It is sufficient to establish a correspondence between the decompositions in the case when L and L' differ by a Fuchs–Tabachnikov move; see Fig. 25.8.

For the first move, the crossing of the move must be switching and the two cusps are paired.

For the second move, none of the crossings of the move can be switched so the structure of decompositions does not change.

For the third move, the natural bijection between the crossings of the diagrams L and L' induces a bijection between the decompositions. $\square$

25.8   Basic examples

Both Chekanov and Chekanov–Pushkar' invariants cannot be expressed in terms of finite type invariants.

To show this (in view of the Fuchs–Tabachnikov theorem) it suffices to present a couple of Legendrian knots which represent the same topological

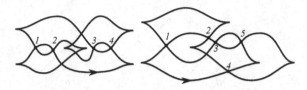

FIGURE 25.13: Chekanov's pair. The Maslov crossings are marked with dots and numbered.

knot (thus have all equal topological finite order invariants) and have equal Maslov and Bennequin numbers. This couple of knots is called a *Chekanov pair*. Their front projections are shown in Fig. 25.13. The left diagram has one graded admissible decomposition with switching crossings 1 and 4. The right diagram has two graded admissible decompositions with the sets of switching crossings $\{1, 2\}$ and $\{1, 2, 4, 5\}$.

Appendix A

Energy of a knot

There is an interesting approach to the study of knots, the knot energy. Such energies were first investigated by H.K. Moffat [Mof]. We shall describe the approach proposed by Jun O'Hara. In his work [O'Ha1] (see also [O'Ha2, O'Ha3]) he proposed studying the *energy of the knot*. First, it was thought to be an analogue of the Gauss electromagnetic function for links, but it has quite different properties.

An energy is a function on knots that has some interesting properties and some invariance (not invariant under all isotopies!) However, these properties are worth studying because some of them lead to some invariants of knots; besides, they allow us to understand better the structure of the space of knots. Here we shall formulate some theorems and state some heuristic conjectures.

In the present appendix, we shall give a sketchy introduction to the best-known energy, the *Möbius energy*.

Definition A.1. Let K be a knot parametrised by a natural parameter $r\colon S^1 \to \mathbb{R}^3$, where S^1 is the standard unit circle in $\mathbb{R}^2$. Then the Möbius energy of the knot K is given by:

$$E_f(K) = \int\int_{S^1 \times S^1} f(|r(t_1) - r(t_2)|, D(t_1, t_2))dt_1 dt_2,$$

where $f(\cdot, \cdot)$ is some function on the plane (one can take various functions, in these cases different properties of the energy may arise), $D(\cdot, \cdot)$ is the function on $S^1 \times S^1$ representing the distance between the two points along the circle (one takes the minimal length), and the integration is taken along the direct product of the knot by itself with two parameters t_1, t_2.

Physical matters (for example, a uniformly charged knot) lead to the formula with $f(x, y) = \frac{1}{x}$. However, such a function does not always converge. Thus, other cases are worth studying.

One often uses the function $f(x, y) = \frac{y^2}{x}$.

In this case, the Möbius energy has the following properties. The main property is that the integral converges for smooth knots.

Actually, the natural parametrisation is not always convenient to use. In fact, we shall study some transformations of knots where the natural parametrisation becomes unnatural.

So, it is convenient to define the Möbius energy for the arbitrary parametrisation $r(u) = r(t(u))$ as follows:

$$E(r) = \int \int \left(\frac{1}{|r(t(u)) - r(t(v))|^2} - \frac{1}{D^2(t(u), t(v))} \right) \cdot |\dot{t}(u)| \cdot |\dot{t}(v)| du dv.$$

1. The Möbius energy is invariant under homothetic transformations.

2. The Möbius energy tends to infinity while the knot is closing to a singular knot.

3. The Möbius energy is strictly positive.

4. The Möbius energy is invariant under Möbius transformations, namely, under inversions in spheres not centred at a point of the knot. If we invert in a sphere centred at a point of the knot K then we obtain a long knot K'; one can define the Möbius energy in the same way. In this case, $E(K) = E(K') + 4$.

5. The absolute minimum of the Möbius energy is realised on the standard circle.

6. The Möbius energy is smooth with respect to smooth deformations of the knot.

Let us consider these properties in more detail and prove some of them.

The first property easily follows from the form of the invariant: we have a double integration that is cancelled (while performing homothety) by the second power in the denominator. The invariance under shifts and orthogonal moves is evident.

The second property is obvious: one obtains a denominator that tends to infinity while the integration domain and coefficients remain separated from zero.

The third property follows from the fact that the distance between two points in $\mathbb{R}^3$ does not exceed the distance along the arc of the circle for the corresponding parametrising points (in the natural parametrisation case).

The fourth and the sixth property follow from a straightforward check.

Let us now discuss the fifth property in more detail and establish some more properties of the Möbius energy. First, let us prove the "long knot property": if K is a knot and X is a point on K then for any sphere centred at X, for the long knot K' obtained from K by inversion in this sphere, we have $E(K) = E(K') + 4$.

Rather than proving this property explicitly, we shall prove that the energy of the circle equals four. Then, from some reasonings this property will follow immediately.

In fact, for the circle in the natural parametrisation, the length along the circle is $(t_1 - t_2)$ if $0 \le (t_1 - t_2) \le \pi$. The distance is then $2 \sin \left(\frac{t_1 - t_2}{2} \right)$.

After a suitable variable change $u = t_1 - t_2$, the integral is reduced to the single integral multiplied by 2π. Besides, for symmetry reasons, is sufficient to consider u from zero to π, and then double the obtained result. Then we get:

$$4\pi \int_0^\pi \left(\frac{1}{4\sin^2(u/2)} - \frac{1}{u^2} \right) du.$$

The latter equals

$$\lim_{\varepsilon \to 0} 4\pi(-\frac{1}{2} \cot(u/2) + \frac{1}{u})|_\varepsilon^{\pi-\varepsilon},$$

which is equal to four.

Now, after a small perturbation, each knot can be considered as a knot containing a small piece of a straight line with arbitrarily small change of energy.

Taking into account that for any long knot different from the straight line, we see that the only knot having energy four is the circle; all other knots have greater energies. This proves the fifth property.

It is easy to see that for long knots the energy is non–negative as well. Thus, we immediately see that $E(K) > 4$ for ordinary knots.

Furthermore, for the straight line, the energy equals zero, whence for any other long knot it is strictly greater than zero. So, the energy of the circle equals four; this realises the minimum of the energy for all (classical) knots.

This property is very interesting. Thus one can consider the knot energy as the starting point of *Morse theory* for the space of knots. One considers the space of all (smooth knots) and studies the energy function and its properties. For instance, for each knot there exists its own knot theory with minima, maxima and other critical points.

The property described above shows that for the unknot there exists only one minimum, namely, four. This minimum is realised by the circle. The circle is considered up to moves of the space $\mathbb{R}^3$ (orthogonal and shifts) and homotheties. So, in some sense one can say that there exists *only one minimum* of the energy function on the space of knots.

So, if we consider knots up to orthogonal moves of $\mathbb{R}^3$, shifts, and homotheties, then the circle is the unique minimum for the unknot.

We see that the energy of each long knot is positive, so the energy of each closed knot is greater than or equal to four.

One can ask the question whether for each knot there exists a *normal form*; i.e., a representative with the minimal energy. The natural questions (conjectures) are:

1. Does such a normal form (realising the minimum) exist?

2. Is this unique?

Both these questions were stated by Freedman, and now none of them has any satisfactory solution except for the case of the unknot. Moreover, these conjectures are not strictly stated: one should find the class of deformation that defines "the same knot".

However, the set of minima and minimal values of the energy function can be considered as a knot invariant. Since we have many energies (for different functions f), these invariants seem to be quite strong, so that they seem to differ for all non-isotopic knots. This is another conjecture.

The only thing that can be said about all knot isotopy classes is that there exists only a *countable number of energy minima*.

The absolute minimum of any nontrivial knot has not yet been calculated. However, the following theorem holds.

Theorem A.1 (Freedman [FHW]). *If $E(K) \leq 6\pi + 4$, then K is the unknot.*

There is another interesting result on the subject, namely, the *existence theorem*:

Theorem A.2. *Let K be a prime knot isotopy class. Then there exists a minimal representative K_γ of K such that for each other representative K', we have $E(K') \geq E(K_g)$.*

Note that the analogous statement for knots which are not prime is not proved.

Appendix B

The A-polynomial

Now we are going to construct one more powerful knot invariant, the A-polynomial. It was defined in [CCGLS] by Cooper et al. This polynomial recognizes the unknot, which was proved independently by N.Dunfield and S.Garoufalidis in [DG] and, independently, by S.Boyer and X.Zhang, [BoZh].

In the present appendix we follow [DG]. The explanation here is however not self-contained for it relies on a deep result due to P.Kronheimer and T.Mrowka, whose proof is beyond the scope of our book.

Roughly speaking, the A-polynomial of a knot K in S^3 describes $SL(2, \mathbb{C})$-representations of the knot complement, as viewed from the boundary. The authors [CCGLS] initially defined the A-polynomial also for knots in homologically trivial spheres, but we shall restrict ourselves to knots in S^3. Let $M = S^3 \setminus N(K)$ be the complement to K. The boundary of M is the torus $\partial M = \partial \widehat{M}(K)$, whose fundamental group $\pi_1(\partial M) = \mathbb{Z}^2$ is generated by the meridian μ and the longitude λ. As we know from Chapter 4, if K is not trivial the natural map of fundamental groups $\pi_1(T^2) \to \pi_1(M(K))$ is an inclusion. Consider a representation $\rho : \pi_1(M) \to SL(2, \mathbb{C})$. The restriction of ρ to $\pi_1(\partial M)$ has a simple form, since a pair of commuting 2×2-matrices are typically simultaneously diagonalisable; i.e., up to a conjugation we have

$$\rho(\mu) = \begin{pmatrix} M & 0 \\ 0 & M^{-1} \end{pmatrix} \quad \text{and} \quad \rho(\lambda) = \begin{pmatrix} L & 0 \\ 0 & L^{-1} \end{pmatrix}. \tag{B.1}$$

The possible eigenvalues (M, L) of $(\rho(\mu), \rho(\lambda))$ as ρ varies form a complex algebraic subvariety of $\mathbb{C}^2$. The A-*polynomial* is *the defining equation for the* 1-*dimensional part of this subvariety;* that is, it describes a plane curve whose points correspond to the restrictions to $\pi_1(\partial M)$.

As the group of isometries of hyperbolic 3-space in $PSL(2, \mathbb{C})$, the A-polynomial is connected to the study of deformation of (incomplete) hyperbolic structures on M. For example, the variation of the volume of hyperbolic structures on M depends only on their restriction on the boundary torus, and it is controlled entirely by the A-polynomial. Topologically, the sides of the Newton polygon of the A-polynomial give rise to incompressible surfaces in M.

The A-polynomial of the unknot is simply $L - 1$. The A-polynomial always contains a factor $L - 1$ coming from reducible representations; we say that the A-polynomial is *non-trivial* if it has an additional factor.

Theorem B.1. *A non-trivial knot in S^3 has a non-trivial A-polynomial. Moreover, the A-polynomial of a non-trivial knot is not a power of $(L-1)$.*

We deduce Theorem B.1 as a direct corollary from the following theorem due to P.Kronheimer and T.Mrowka [KrMr2]:

Theorem B.2. *Let K be a non-trivial knot in S^3. For $r \in \mathbb{Q}$, let M_r be the 3-manifold which is the r-Dehn surgery on K. If $|r| \leq 2$, then there exists a homomorphism $\pi_1(M_r) \to SU(2)$ with non-cyclic image.*

Their proof uses Gauge theory; we shall not present it here. It was previously known for all non-satellite knots for simple geometric reasons, as we now describe. When M is hyperbolic, we have the *holonomy representation* $\pi_1(M) \to SL(2,\mathbb{C})$ of the complete hyperbolic structure; Thurston showed in his Hyperbolic Dehn Surgery Theorem that this representation has a complex curve of deformations which change the holonomy along the boundary [Thi].

Thus, in this case, the A-polynomial is non-trivial. Non-hyperbolic knots are torus knots or satellites. For torus knots, the non-triviality of the A-polynomial was shown in [CCGLS] by direct calculation. Satellite knots are those which have closed incompressible tori in their complements. One can look at the resulting geometric decomposition, and try to understand how the representation of each piece could glue together to give a representation of the whole fundamental group $\pi_1(M)$; however, this seems to be quite difficult in general.

Since the proof of Theorem B.1 is based on the existence of $SU(2)$-representations, we really show that if one looks only at representations $\rho : \pi_1(M) \to SU_2$, then the eigenvalues (M,L) of $(\rho(\mu), \rho(\lambda))$ sweep out a real 1-dimensional subset of the unit torus in $\mathbb{C}^* \times \mathbb{C}^*$. This is interesting even in the case of hyperbolic knots.

The key ingredient in the reduction of Theorem B.1 to Theorem B.2 is the Proposition B.1.

B.1   Connection to the Jones polynomial

While the A-polynomial arose from the study of hyperbolic geometry, it turns out to have connections to seemingly disparate parts of low-dimensional topology, including the Jones polynomial. As we will now explain, the non-triviality of the A-polynomial of a knot has implications to the strength of the coloured Jones function. The latter is essentially the sequence of Jones polynomials of a knot and its connected parallels. In [GL], it was proved that the coloured Jones function of a knot is a sequence of Laurent polynomials which satisfy a q-difference equation. It was observed by Garoufalidis that one can choose the q-difference in a canonical manner. The corresponding operator to this q-difference equation is an element of the non-commutative ring

$$\mathbb{Z}[q^{\pm 1}]\langle Q^{\pm 1}, E^{\pm 1}\rangle(EQ - qQE)$$

of Laurent polynomials in E and Q which satisfy the commutation relation $EQ = qQE$.

This operator defines the so-called non-commutative A-polynomial of a knot. In [Garo], Garoufalidis conjectured that specializing the non-commutative polynomial at $q = 1$ coincides with the A-polynomial of a knot after a change of variables $(E, Q) = (L, M^2)$ (there may also be changes in the multiplicities of factors and polynomials in Q). This is called the AJ-conjecture, and if it holds then Theorem B.1 would imply

Corollary B.1. *If the AJ Conjecture holds then the coloured Jones polynomial distinguishes the unknot.*

B.2   Connection to contact homology

Another surprise is that the A-polynomial is connected with contact geometry.

L.Ng proved that the A-polynomial can be derived from the simplest piece of the framed knot contact homology. He constructed a homology theory for knots in S^3, the *framed knot contact homology.*

Theorem B.3. *[NgL] The framed knot contact homology distinguishes the unknot from any other contact knot in S^3.*

Now we pass to the proof of Theorem B.1.

We begin by reviewing the definition of the A-polynomial for a compact 3-manifold M whose boundary is a torus. Let $R(M)$ denote the set of representations $\pi_1(M) \to SL(2, \mathbb{C})$, which is an affine algebraic variety over $\mathbb{C}$. It is natural to study representations over tiny automorphisms of $SL(2, \mathbb{C})$, so let us consider the *character variety*, $X(M)$, which is the quotient of $R(M)$ by the action of $SL(2, \mathbb{C})$ conjugation. Technically, one has to take the algebro-geometric quotient to deal with orbits of reducible representations which are not closed; in this way $X(M)$ is also an affine complex variety.

To define the A-polynomial, we first need to understand the character variety $X(\partial M)$ of the torus ∂M. We know that $\pi_1(\partial M) = \mathbb{Z} \times \mathbb{Z}$; let us fix its generators μ, λ. Since this group is commutative, any representation $\rho : \pi_1(\partial M) \to SL(2, \mathbb{C})$ is reducible, that is, has a global fixed point for the Möbius action on $\mathbb{C}P^1$. Moreover, if no element of $\rho(\pi_1(\partial M))$ is parabolic, then ρ is conjugate to a diagonal representation as in (B.1). As such, $X(\partial M)$ is approximately the whole of $\partial C, \partial C$ with coordinates being the eigenvalues of (M, L). This is not quite right, as switching (M, L) with (M^{-1}, L^{-1}) gives a conjugate representation. In fact, $X(\partial M)$ is exactly the quotient of $\mathbb{C}^* \times \mathbb{C}^*$ under the involution $(M, L) \mapsto (M^{-1}, L^{-1})$.

Now the inclusion $i : \partial M \to M$ induces a regular map $i^* : X(M) \to X(\partial M)$ via restriction of representations from $\pi_1(M)$ to $\pi_1(\partial M)$. Let V be the (complex) 1-dimensional part of $i^*(X(M))$. More precisely, take V to be the union of the 1-dimensional $i^*(X)$, where X is an irreducible component of $X(M)$. The curve V is used to define the A-polynomial. To simplify things, we look at the plane curve $\hat{V}(M)$ which is the inverse image of V under the quotient map $\mathbb{C}^* \times \mathbb{C}^* \to X(\partial M)$. The A-polynomial is the *defining equation for* $\hat{V}(M)$: it is a polynomial in M, L. Since all the maps involved are defined over $\mathbb{Q}$, the A-polynomial can be normalised to have integral coefficients.

In the definition of the A-polynomial, we looked only at those irreducible components where $i^*(X)$ is 1-dimensional. In the proof of Theorem B.1, we shall need the following standard lemma.

Lemma B.1. *Let X be an irreducible component of $X(M)$. Then $i^*(M)$ has dimension* 0 *or* 1.

Proof. There are various proofs of this theorem in the literature; we sketch two proofs of this.

We need to rule out the possibility that $i^*(X)$ is 2-dimensional and thus, Zariski-open subset of $X(\partial M)$. The approach from [CCGLS] is to introduce the notion of the volume of a representation $\rho : \pi_1(M) \to SL(2, \mathbb{C})$. This gives a natural function $Vol : X(M) \to R$. Then Schläfli's formula for the change of volume of a family of polyhedra in $\mathbb{H}^3$ shows that the derivative of Vol depends only on the restriction of representations to $\partial_1(\partial M)$. This leads to an 1-form on $X(\partial M)$ which must be exact on $i^*(X(M))$. This form is not exact on any Zariski-open subset of $X(\partial M)$, and hence $i^*(X)$ is at most 1-dimensional.

The other argument is to observe that if i^* were 2-dimensional, it would then let us construct ideal points of $X(M)$ where the associated surface has whatever boundary slope we want. This would contradict Hatcher's finiteness theorem for boundary slopes. In more detail, start with a slope $\alpha \in \pi_1(\partial M)$ and let β be a complementary slope. Choose a complex number c so that the curve Y in $X(\partial M)$ given by $tr_a = c$ has $i^*(X) \cap Y$ dense in Y. Choose a curve $\tilde{Y} \subset X$ whose image under i^* is dense in Y. Since tr_α is constant on Y, an incompressible surface associated to the ideal point p must have boundary slope α. But A. Hatcher showed that there are only finitely many α which are boundary slopes of incompressible surfaces [Hat], a contradiction. $\square$

Let us proceed now with the proof of Theorem B.1. When M is the exterior of a knot in S^3, then, up to orientation convention, there is a canonical meridian-longitude basis (μ, λ) for $\pi_1(\partial M)$, and one uses this basis for writing the A-polynomial. Since we are interested in the non-triviality of the A-polynomial, we need to discuss the conventions for dealing with the reducible representations. When M is the exterior of a knot in S^3, one has $H_1(M, \mathbb{Z}) = \mathbb{Z}$, and so there are many reducible representations which factor through $\pi_1(M) \to Z \to SL_1(\mathbb{C})$. Irreducible components of $X(M)$ either consist of solely reducible representations, or have a Zariski-open subset of irreducible representations. In the case of the exterior of a knot S^3, there is a single irreducible component of $X(M)$ consisting entirely of reducible representations.

This component contributes a factor $(L - 1)$ to the A-polynomial. Some authors exclude it from the A-polynomial, and define the curve V above to be the image under i^* of the irreducible components of $X(M)$ which contain an irreducible representation. To say that the A-polynomial is non-trivial, we mean that it does not just consist of the $L - 1$ coming from the irreducible representations. We shall now show that the A-polynomial of a non-trivial knot in S^3 is non-trivial, and, moreover, that it is not just a power of $L - 1$.

Let M be the exterior of K. Let $X'(M)$ denote $X(M)$ minus the component consisting of reducible representations, and let V' be the union of the 1-dimensional $i^*(X)$ where X is an irreducible component of $X'(M)$. The main part of the theorem is to show that V' is non-empty. To this end we shall prove the following

Proposition B.1. *There exists an infinite collection of irreducible representations $\rho_n : \pi_1(M) \to SL(2, \mathbb{C})$ whose restrictions to $pi_1(\partial M)$ are all distinct in $X(\partial M)$.*

Before proving this claim, let us deduce $V' \neq \emptyset$ from it. Assuming the claim, as a 0-dimensional algebraic variety consists of finitely may points, there must be some irreducible X in $X'(M)$ so that the dimension of $i^*(X)$ is at least 1. Since we eliminated 2-dimensional components, the dimension of i^* must be exactly one, and so, $V' \neq 0$.

To prove the claim, we use the $SU(2)$ representation given in B.2. Let $M_{\frac{1}{n}}$ be the $\frac{1}{n}$ filling of M. By Theorem B.2, for each non-zero $n \in Z$ we have a representation $\rho_n : \pi_1(M_{\frac{1}{n}}) \to SU(2)$ with non-cyclic image. First, we claim that the ρ_n are irreducible as representations into a larger group of $SL_2(\mathbb{C})$. Suppose rho_n were reducible. Since $H_1(M_{\frac{1}{n}}, \mathbb{Z})) = 0$, the group $G = \pi_1(M_{\frac{1}{n}})$ coincides with $[G, G]$. As ρ_n is reducible, and commutators of elements of $SL_2(C)$ with common fixed point are parabolic with trace 2, it follows that $tr(\rho_n(\gamma)) = 2$ for all $\gamma \in G$. But the only element of $SU(2)$ with trace 2 is the identity, and so ρ_n would be trivial, a contradiction. So, ρ_n is irreducible.

As $\pi_1(M_{\frac{1}{n}})$ is a quotient of $\pi_1(M)$, we shall regard ρ_n as a representation of $\pi_1(M)$ inot $SU(2) \leq SL(2, \mathbb{C})$. To prove Proposition B.1, we need to show that the restrictions of the ρ_n to $\pi_1(\partial M)$ gives an infinite collection of points on $X(\partial M)$. Two representations of $\pi_1(\partial M)$ into $SU(2)$ which correspond to the same point in $X(\partial M)$ are actually conjugate, because they both must be conjugate to the same diagonal representation. Because of this, the proof of the Proposition B.1 reduces to the fact that the kernels K_n of the ρ_n give rise to an infinite collection of different subgroups of $\pi_1(\partial M) = \mathbb{Z}^2$.

For α is a slope in ∂M, note that ρ_n extends to $\pi_1(M_n)$ if and only if $\alpha \in K_n$. As ρ_n comes from $M_{\frac{1}{n}}$, we have $(1, n) \in K_n$ for each $n \neq 0$. As $(1, 0)$-filling gives S^3, which is simply connected, we have $(1, 0) \neq K_n$.

Because of this, the desired statement follows directly from the following Lemma with γ being the line $x = 1$:

Lemma B.2. *Suppose γ is a line in $\mathbb{R}^2$ which contains infinitely many lattice points of $\mathbb{Z}^2$ and does not contain 0. Consider a collection K_n of subgroups of $\mathbb{Z}^2$ whose union, K, contains all but finitely many of the lattice points on γ. Suppose, in addition, that there is a lattice point on γ which is not in K. Then there are infinitely many distinct K_n.*

Proof. Assume that there are finitely many K_n. If K_n has rank less than 2, then K_n is contained in a line through the origin, and so K_n intersects γ in at most one point. So, we can remove all such K_n of rank less than 2.

So, we can assume the quotient group $\mathbb{Z}^2/K_n$ to be finite for each n. Let L be the intersection of K_n; as there are finitely many groups K_n, the subgroup L is also a finite-index subgroup of $\mathbb{Z}^2$. Now, let γ' be the line parallel to γ which passes through the origin. As $\mathbb{Z}^2/L$ is finite, the subgroup $H = \gamma' \cap L$ is infinite. Let v_0 be the given point in $\gamma \backslash K$. Then if $h \in H$, we have that $v_0 + h$ is also in $\gamma \backslash K$ since if $v_0 + h$ is in some K_n then so is $v_0 = (v_0 + h) - h$. But H is infinite, and thus so is $\{v_0 + h\}$ which contradicts that $\gamma \backslash K$ is finite. Thus we must have an infinite collection of distinct K_n.

$\square$

To complete the proof of Theorem B.1, we need to show that the A-polynomial M is not a power of $L - 1$. Assume the contrary. Consider the point $p_n = (m_n, l_n) \in \mathbb{C}^* \times \mathbb{C}^*$ corresponding to the restriction of ρ_n to $\pi_1(\partial M)$. As ρ_n comes from the $(1, n)$-filling of M, we have $m_n l_n^n = 1$. By the above argument, all but finitely many of the pairs satisfy the A-polynomial equation, and hence $l_n = 1$. Then for such n, the relation $m_n l_n^n = 1$ implies that $m_n = 1$. As ρ_n has image in $SU(2)$, this implies that ρ_n is trivial when restricted to $\pi_1(\partial M)$. But then ρ_n factors over to the S^3-surgery, and we get a contradiction. Thus the A-polynomial is not a power of $L - 1$.

Appendix C

Garside's normal form

Below we present a normal form due to Garside [Gars] and give a solution to the conjugacy problem for classical braids. Very briefly, up to a high power of a kernel element Δ^p, all elements become positive, and the equivalence of positive elements with no negative generators involved can be checked "by hand". Note that since the first solution due to Garside, there have been many other solutions to the conjugacy problem, see e.g., [Deh2, KT]. We shall restrict ourselves with the historically first case, which treats the usual Artin braid group only; the reader interested in the solution to the conjugacy problem in a more general situation is referred to [KT].

We deal with braid words in letters $\alpha_1^{\pm 1}, \ldots, \alpha_n^{\pm 1}$, hence, with letters generating B_{n+1}. We say that a braid word is *positive* if it contains no negative exponents α_j^{-1}.

Words. If A, B are words in the generators and their inverses, then $A = B$ will mean that A can be transformed into B by using the defining relations, whence $A \sim B$ will mean the two words are identical letter by letter, and $A \cong B$ means that A is *conjugate* to B. A word consisting of an ordered sequence of the generators only, in which no inverse of any generator occurs will be called a *positive word*. We shall denote by $L(W)$ the *word-length* of a word W.

For positive words we introduce the notation $A \doteq B$ if the word A can be transformed into the word B by elementary braid-moves within the set of positive words. In the sequel, whenever we use the symbol $\doteq$, we deal with positive braids only.

If $A \doteq B$, then $L(A) = L(B)$.

Theorem C.1 (Theorem H). *If $\alpha_i X \doteq \alpha_k Y$, either of the following holds:*

1. $k = i \implies X \doteq Y$;

2. $|k - i| \geq 2 \implies X \doteq \alpha_k Z, Y \doteq \alpha_i Z$

3. $|k - i| = 1 \implies X \doteq \alpha_k \alpha_i Z, Y \doteq \alpha_i \alpha_k Z$.

for some Z.

The proof of Theorem H is given below.

The next theorem is just the "mirror image" of theorem H, so it's proof is almost identical.

Theorem C.2 (Theorem K). *In $B(n+1)$ for $i, k = 1, 2, \ldots, n$ if $X\alpha_i \doteq Y\alpha_k$, then:*

1. *If $k = i$ then $X = Y$;*

2. *If $|k - i| \geq 2$ then $X \doteq Z a_k, Y \doteq Z a_i$ for some Z*

3. *If $|k - i| = 1$ then $X \doteq Z a_i a_k, Y \doteq Z a_k a_i$.*

Let us denote by L the braid length (sum of exponents).

Corollary C.1. *In the group $B(n + 1)$ if $A \doteq P, B \doteq Q, AXB \doteq PYQ$, $(L(A) \geq 0, L(B) \geq 0)$, then $X = Y$.*

This corollary immediately follows from Theorems H and K. The latter means that the monoid S is *cancellative*.

Now let us introduce the *fundamental word* Δ.

We denote the word $\alpha_r \alpha_{r+1} \cdots \alpha_s$ (resp., $\alpha_r \alpha_{r-1} \cdots \alpha_s$) with increasing (res., decreasing) order of generators, by $(\alpha_r \cdots \alpha_s)$. By Π_s we shall denote the word $(\alpha_1 \cdots \alpha_s)$.

In the group B_{n+1} if $\mathfrak{R}$ is a map of the set $(\alpha_1, \ldots, \alpha_n)$ onto itself given by $\mathfrak{R}\alpha_i = \alpha_{n+1-i}$, then it is easy to see that the map $\mathfrak{R}$ can be extended to an automorphism of B_{n+1}. Note also that if $P \doteq Q$, then $\mathfrak{R}P \doteq \mathfrak{R}Q$.

Set

$$\Delta_r \sim \Pi_r \Pi_{r-1} \cdots \Pi_1.$$

We shall call Δ_r the *fundamental word of order $r + 1$.*

Later on we shall often omit the index r and write simply Δ.

The key properties of the element Δ are:

1. The square of this element belongs to the centre of the braid group (hence, for the solution of the conjugacy problem we can multiply by even powers of Δ);

2. each braid becomes positive after multiplying by Δ^N for some sufficiently large N.

Hence, after some effort, we can reduce the braid conjugacy problem for arbitrary braids to the conjugacy problem for positive braids; if we conjugate positive braids by positive braids, the problem becomes finite.

To realise the above plan, we start with some lemmas

Thus, we get the following:

Lemma C.1 ("inverting lemma"). *In B_{n+1} for $1 < s \leq t \leq n$ we have $a_s \Pi_t = \Pi_t a_{s-1}$.*

The proof follows from a direct check.

Analogously one can check the following

Lemma C.2. *In B_{n+1}, the following equalities hold:*

1. $a_1 \Delta_t \doteq \Delta_t a_t$ $(t = 1, 2, \ldots, n)$;

2. $a_s \Delta \doteq \Delta \Re a_s$;

3. $a_s^{-1} \Delta \doteq \Delta (\Re a_s)^{-1}$;

4. $a_s \Delta^{-1} \doteq \Delta^{-1} \Re a_s$;

5. $a_s^{-1} \Delta^{-1} \doteq \Delta^{-1} (\Re a_s)^{-1}$.

From these lemmas the reader immediately deduces the following

Theorem C.3. *1. $P \Delta^{2m} \doteq \Delta^{2m} P$, $P \Delta^{2m+1} \doteq \Delta^{2m+1} \Re P$ for all positive words P $(m \geq 0)$*

2. *$Q \Delta^{2m} = \Delta^{2m} Q$, $Q \Delta^{2m+1} = \Delta^{2m+1} \Re Q$ for all words Q, m positive or negative.*

From easy calculations using the above techniques, one gets the following

Lemma C.3. *In the group B_{n+1} we have*

1. $\Re \Delta \doteq \Delta$

2. $rev \Delta \doteq \Delta$.

Here rev is the reverse operator: $rev(x_1 x_2 \cdots x_t) = x_t \cdots x_2 x_1$ where $x_1, x_2, \ldots, x_t$ are generators or their inverses.

The proof of 1) is left to the reader as an exercise.

The proof of 2) is by induction. Assume that for any particular r we have $rev \Delta_r \doteq \Delta_r$. Then

$$rev \Delta_{r+1} \sim rev\{(a_1 \cdots a_{r+1}) \Delta_r\}$$

$$\sim rev \Delta_r rev(a_1 \cdots a_{r+1})$$

$$\doteq \Delta_r (a_{r+1} \cdots a_1),$$

using the induction hypothesis; i.e.,

$$rev \Delta_{r+1} \doteq \Pi_r \Pi_{r-1} \cdots \Pi_1 (a_{r+1} \cdots a_1).$$

Now a_{r+1} commutes with $\Pi_1, \Pi_2, \ldots, \Pi_{r-1}$; a_r commutes with $\Pi_1, \Pi_2, \ldots, \Pi_{r-2}, \ldots$, etc.

Hence

$$rev \Delta_{r+1} \doteq \Pi_r a_{r+1} \Pi_{r-1} a_r \cdots \Pi_2 a_3 \Pi_1 a_2 a_1 \sim \Delta_{r+1}.$$

The induction is now established, since the hypothesis is clearly true for $r = 1$ and the result follows.

Lemma C.4. *In B_{n+1}, there exist positive words X_r, Y_r such that*

$$a_r X_r \doteq \Delta \doteq Y_r a_r \quad (r = 1, 2, \ldots, n).$$

By definition $\Delta \sim \Pi_n \Pi_{n-1} \cdots \Pi_2 \Pi_1$, i.e. $\Delta \doteq Y_1 a_1$, where $Y_1 \sim \Pi_n \Pi_{n-1} \cdots \Pi_2$.

We now observe that if $f(a_2, a_3, \ldots, a_t)$ is any positive word involving the generators $a_2, a_3, \ldots, a_t$ only, then by Lemma C.1, we have

$$\Pi_1 f(a_1, a_2, \ldots, a_{t-1}) \doteq f(a_2, a_3, \ldots, a_t) \Pi_t.$$

Let a_t by any particular one of the generators $a_2, \cdots, a_n$. Then, denoting $\Pi_{t-1} \Pi_{t-2} \cdots \Pi_1$ by $f(a_1 a_2 \cdots a_{t-1})$, we have

$$\Delta \sim \Pi_n \Pi_{n-1} \cdots \Pi_t f(a_1, a_2, \cdots, a_{t-1})$$

$$\doteq \Pi_n \Pi_{n-1} \cdots \Pi_{t+1} f(a_2, a_3, \cdots, a_t) \Pi_t$$

$$\doteq \Pi_n \Pi_{n-1} \cdots \Pi_{t+1} f(a_2, a_3, \cdots, a_t)(a_1 \cdots a_{t-1}) a_t$$

$$\sim Y_t a_t,$$

say.

Now putting $X_r = rev Y_r$, we have, for $r = 1, 2, \ldots, n$

$$a_r X_r \sim a_r rev Y_r \sim rev(Y_r a_r) \doteq rev \Delta \doteq \Delta.$$

Hence words X_r also exist, and the proof is complete.

Corollary C.2. *In B_{n+1}, if A is any positive word, then for $r = 1, 2, \ldots, n$, there exist words A_r such that $\Delta A \sim A_r a_r$.*

For

$$\Delta A \doteq (\mathfrak{R}A) \Delta \doteq (\mathfrak{R}A) Y_r a_r \sim A_r a_r,$$

say.

Lemma C.5. *Let a_i be any one of the n generators of the group B_{n+1}, and let $x_1, x_2, \cdots, x_t$ be generators, not necessarily distinct, such that each x_r commutes with a_i. Then if $a_i P \doteq x_1 x_2 \cdots x_l Q$, then there exists some R such that $Q \doteq a_i R$.*

We have $a_i P \doteq x_1 x_2 \cdots x_t Q$. Hence, by making successfull applications of theorem H (ii), we have $x_2 x_3 \cdots x_l Q \doteq a_i R_2$ for some R_2; $x_3 x_4 \cdots x_t Q \doteq a_i R_3$ for some R_3; $\ldots$ $x_t Q \doteq a_i R_t$ for some R_t; and finally $Q \doteq a_i R$ for some R, as required.

Lemma C.6. *In B_{n+1}, if $a_{i+1} P \doteq \Pi_i Q$, then $Q \doteq a_{i+1} a_i R$, for some R ($i = 1, 2, \ldots, n-1$).*

By hypothesis $a_{i+1}P \doteq a_1 a_2 \cdots a_i Q$, and hence

$$a_i Q \doteq a_{i+1} T$$

for some T. Hence, by Theorem H 3), it follows that $Q \doteq a_{i+1} a_i R$ for some R, as required.

Theorem C.4. *If W is any positive word in B_{n+1} such that either*

$$(i) \quad W \doteq a_1 X_1 \doteq a_2 X_2 \doteq \cdots \doteq a_n X_n,$$

or

$$(ii) \quad W \doteq Y_1 a_1 \doteq Y_2 a_2 \doteq \cdots \doteq Y_n a_n$$

then $W \doteq \Delta Z$ for some Z.

Proof. The proof is by induction. Let $r \le n - 1$ be a positive integer. Then as our induction hypothesis we assume that, in B_{n+1}, if

$$W \doteq a_1 X_1 \doteq a_2 X_2 \doteq \cdots \doteq a_r X_r,$$

then $W \doteq \Delta_r P_r$ for some P_r.

Now suppose that $W \doteq a_1 X_1 \doteq a_2 X_2 \cdots \doteq a_r X_r \doteq a_{r+1} X_{r+1}$.

Then from the above formula and from the induction hypothesis it follows that

$$a_{r+1} X_{r+1} \doteq W \doteq \Delta_r P_r \sim (a_1 \cdots a_r) \Delta_{r-1} P_r.$$

Hence, by Lemma C.6,

$$\Delta_{r-1} P_r \doteq a_{r+1} a_r Q_r$$

for some Q_r, so that $W \doteq (a_1 \cdots a_r) a_{r+1} a_r Q_r$, or putting

$$T \sim a_r Q_r$$

$$W \doteq (a_1 \cdots a_{r+1}) T \sim \Pi_{r+1} T.$$

Now we have

$$a_{i+1} X_{i+1} \doteq (a_1 \cdots a_i)(a_{i+1} \cdots a_{r+1}) T, \ i = 1, 2, \ldots, r - 1,$$

so that, by Lemma C.6,

$$(a_{i+1} \cdots a_{r+1}) T \doteq a_{i+1} a_i S_i$$

for some S_i.

Therefore, by Corollary C.1, $(a_{i+2} \cdots a_{r+1}) T \doteq a_i S_i$.

Applying Lemma C.5 it follows that for some Q_i

$$T \doteq a_i Q_i \ (i = 1, 2, \ldots, r-1).$$

This leads us to

$$T \doteq \Delta_r P_{r+1}$$

for some P_{r+1} and hence to

$$W \doteq \Pi_{r+1} \Delta_r P_{r+1} \sim \Delta_{r+1} P_{r+1}.$$

Remarking that the induction hypothesis is clearly true for $r = 1$, the induction is now established and the result follows.

The case $ii)$ is easily obtained from $i)$ by applying rev.

$\square$

Lemma C.7. *If X, Y are any two positive words in B_{n+1}, then there exist words U, V, such that $UX \doteq VY$.*

Proof. Indeed, let $X \sim r_1 r_2 \cdots r_l$, $Y \sim s_1 s_2 \cdots s_m$ be any two positive words where r_i and s_i are generators, not necessarily distinct. Then, by repeated application of Corollary C.2, we have

$$\Delta^m X \doteq \Delta^{m-1} A_1 s_m \doteq \Delta^{m-2} A_2 s_{m-1} s_m \doteq \cdots \doteq A_m Y.$$

The result follows on putting $U \sim \Delta^m, V \sim A_m$.

$\square$

The following fact is crucial

Theorem C.5. *In B_{n+1} if two positive words are equal then they are positively equal.*

Proof. Let S be the semi-group generated by $a_1, a_2, \cdots, a_n$ subject to the following relations

$$a_i a_{i+1} a_i = a_{i+1} a_i a_{i+1} \ (1 \le i \le n-1)$$

$$a_i a_k = a_k a_i, (|i - k| \ge 2).$$

By Lemma C.7, S is right-reversible. By Corollary C.1, it is cancellative. It is a standard fact that the map from a right-reversible and cancellative monoid to the group with the same generators and relations is an inclusion [Öre].

$\square$

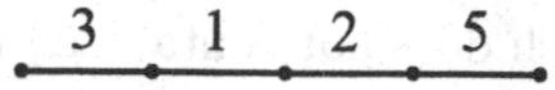

$$\underset{3}{\bullet} \quad \underset{1}{\bullet} \quad \underset{2}{\bullet} \quad \underset{5}{\bullet}$$

FIGURE C.1: Cayley graph of the word $a_3 a_1 a_2 a_5$

C.1 Cayley diagrams

The braid group with its standard presentation with generators and defining relations can be represented by its *Cayley graphs*. In the sequel, we shall considerably use the Cayley graph.

Links. We shall call two successive generators of a positive word *links*. Thus the initial link of the word $a_3 a_1 a_2 a_5$ is a_3; the third link is a_2; etc. No arrow will be put in the Cayley graph, as it will be always understood that the positive direction is from the left to the right. The drawn figure will show the initial link on the left, the successive other links extending in order to the right; see Fig. C.1.

Diagram. Let W be any positive word, and let $W, W_1, W_2, \cdots, W_m$ be the complete set of distinct words which are *positively* equal to W. Then we shall refer to this set as *the diagram of* W, and write $D(W)$. Clearly, $D(W) \sim D(W_1) \sim \cdots \sim D(W_m)$. The words $W, W_1, \cdots, W_m$ will be called *routes* of $D(W)$. The process of enumerating routes of $D(W)$ will be called *drawing the diagram* $D(W)$. In the drawn figure the diagram of W is the Cayley diagram of all words positively equal to W. The name Cayley will be omitted from now on.

Nodes of $D(W)$. Let W be any positive word, and $D(W)$ its diagram. If A, X are any two positive words such that $W = AX$ $(0 \leq L(A), L(X))$, then we shall call $D(A)$ *a node of* $D(W)$. When we are considering nodes we shall frequently write the node $D(A)$ as the node A or simply A. If $L(A) = t$, we shall say that the node A is of order t.

Sub-routes of $D(W)$. If $W \doteq AXB (L(A) \geq 0, L(B) \geq 0)$, we shall say that X is a *sub-route* of $D(W)$. If $L(A) = 0$, we shall say that X is an *initial sub-route of* $D(W)$. If $W \doteq PXQ(L(P) \geq 0, L(Q) \geq 0)$, we shall say that the sub-route X *starts* at P. If $W \doteq RQ \doteq PXQ$, we shall say that the sub-route X *ends* at R.

Incidence. If the link a_r either i) starts at P or ii) ends at P, we shall say that the link a_r is *incident* at P. If the links a_r, a_s are both incident at P, we shall say that they *meet* at P. We shall also say that P is the *meet* of the links a_r, a_s. If a link a_r ends at P and a link a_r starts at P, we shall say that the link a_r is *repeated* at P.

If any sub-route of $D(W)$ is Δ; i.e., if $W = A\Delta B$, $(L(A) \geq 0, L(B) \geq 0)$, we shall say that Δ is a *factor* of W or simply that W *contains* Δ. From Theorem C.3 it follows that if W contains Δ, then $W \doteq \Delta X$ for some X. If W

is any positive word which does not contain Δ, we shall say that W is *prime to* Δ.

Base of $D(W)$. In B_{n+1} suppose W is of word-length L, and suppose $D(W)$ consists of the t words $W_1 \sim a_i a_j a_k \cdots , W_2 = a_p a_q a_r \cdots , \ldots W_t \sim a_x a_y a_z \cdots$. Then there is a one-to-one correspondence between the words $W_1, W_2, \cdots , W_t$ and the set of numbers $P_1 = ijk \cdots , P_2 = pqr \cdots , P_t = xyz \cdots$, where each number P is expressed in the scale of $n + 1$, and consists of L digits. The numbers P are all distinct. Suppose the smallest is P_r. Then the corresponding word W_r, which is uniquely defined, will be called the *base* of $D(W)$. If A is a positive word prime to Δ, we shall sometimes denote the base of A by $\bar{A}$. The use of this notation will imply that A is positive and prime to Δ.

Lemma C.8. *The diagram of any positive word W in B_{n+1} can be systematically drawn, and is finite.*

Let the set of all distinct words positively equal to W through a transformation of chain-length 1 be $W_1, \ldots , W_t$. It is clear that this set can be enumerated, and is finite. Now consider the set of words positively equal to W_1 by a transformation of chain-length 1. Denote those which are distinct from $W, W_1, \ldots , W_t$ and from each other, by $W_{t+1}, W_{t+2}, \ldots$. Continue to repeat the process for $W_2, W_3, \ldots$. Clearly the number of positive words of word-length equal to $L(W)$ is finite, and hence the set of words positively equal to W is finite. Hence the sequence $W, W_1, \ldots$, ultimately terminates. It is clear that any word which is positively equal to W must ultimately be reached through the process outlined above, and the lemma is proved.

C.2   Solution to the word problem

The following theorem is crucial for the solution of the word problem and conjugacy problem for braids.

Theorem C.6. *In B_{n+1}, every word W can be expressed* uniquely *in the form* $\Delta^m \bar{A}$.

Proof. First suppose P is any positive word. From the set $D(P)$ select any route starting with as many consecutive sub-routes Δ as possible equal to t, say, $t \geq 0$. Suppose, $P = \Delta^t A$. Then A is prime to Δ as otherwise there would be a route of $D(P)$ starting with more than t consecutive sub-routes Δ. Denoting the base of A by $\bar{A}$, we have $P = \Delta^t \bar{A}$.

Now let W be any word in B_{n+1}. Then clearly we may put

$$W = W_1(x_1)^{-1} W_2(x_2)^{-1} \cdots (x_s) W_{s+1},$$

where each W_r is a positive word of word-length ≥ 0, and the x_r are generators.

Note for each x_r there exists a positive word X_r such that $x_r X_r \doteq \Delta$, so that $(x_r)^{-1} = X_r \Delta^{-1}$, and hence

$$W = W_1 X_1 \Delta^{-1} W_2 X_2 \Delta^{-1} \cdots W_s X_s \Delta^{-1} W_{s+1}.$$

Hence, moving the factors Δ^{-1} to the left, we get $W = \Delta^{-s} P$, where P is positive. Now, we have $P = \Delta^t \hat{A}$ for maximum possible t, hence $W = \Delta^{-s} \Delta^t \bar{A}$, or putting $t - s = m$, we get

$$W = \Delta^m \bar{A}.$$

Now it remains to show that this form is unique.

Suppose $\Delta^m \bar{A} = \Delta^p \bar{B}$.

First suppose $p < m$, and let $m - p = t$, where $t > 0$. Then $\Delta^t \bar{A} = \bar{B}$, and hence $\Delta^t \bar{A} \doteq \bar{B}$. Hence $\bar{B}$ contains Δ, which is impossible. Therefore (by symmetry) $p = m$. This means $\bar{A} = \bar{B}$, hence $\bar{A} \doteq \bar{B}$. But any positive word has one and only one base. Hence $\bar{A} \sim \bar{B}$, and the uniqueness is established.

$\square$

Definition C.1. Any word W of B_{n+1} expressed in the unique form $\Delta^m \bar{A}$ will be said to be in the *standard form*. The index m will be called the *power* of W.

Theorem C.7. *The necessary and sufficient condition that two words in B_{n+1} are equal is that their standard forms are identical.*

C.2.1 The center of B_{n+1}

Theorem C.8. 1. *When $n = 1$ the centre B_{n+1} is generated by Δ*

2. *When $n > 1$, the centre of B_{n+1} is generated by Δ^2.*

Proof. The first case is trivial.

Let $n > 1$ and W be any word in the centre. Then, if X is any word in B_{n+1}, $X^{-1} W X = W$, so that

$$WX = XW$$

. There are three possible forms for W: (a) $W = \Delta^p \bar{A}$, where $L(\bar{A}) > 0$; (b) $W = \Delta^{2m+1}$; (c) $W = \Delta^{2m}$. We proceed to consider each in turn.

a) $W = \Delta^p \bar{A} (L(\bar{A}) > 0)$.

Let $\bar{A} \doteq a_i A_i$, $L(A_i) \geq 0$. Let $|s - i| = 1$.

Considering first the case when p is even, put $X \sim a_s a_i$. Then we have

$$\Delta^p a_i A_i a_s a_i = a_s a_i \Delta^p a_i A_i = \Delta^p a_s a_i a_i A_i.$$

Hence

$$a_i A_i a_s a_i = a_s a_i a_i A_i,$$

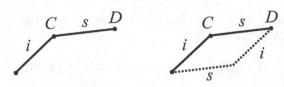

FIGURE C.2: Case (i)

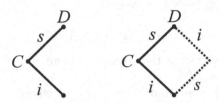

FIGURE C.3: Case (ii)

therefore

$$a_i A_i a_s a_i \doteq a_s a_i a_i A_i,$$

hence by Theorem H, $a_i a_i A_i \doteq a_i a_s A_s$ for some A_s,

$\square$

C.2.2   The structure of $D(\Delta)$

Theorem C.9. *In B_{n+1}, if $W \doteq \Delta V$ is any positive word containing Δ, then each of the n links a_r $(r = 1, 2, \ldots, n)$ is incident at each node of $D(W)$.*

Proof. We know that $W \doteq a_1 W_1 \doteq a_2 W_2 \doteq \ldots \doteq a_n W_n$, so the theorem is certainly true for the initial node.

The proof of the theorem will be by induction. As our induction hypothesis we assume the theorem is true for all nodes of $D(W)$ of order $\leq m$. Let C be any node of order m, and let a_s be any link of the diagram starting at C and ending at D.

(a) We first consider the links a_i, $|i - s| \geq 2$. By the induction hypothesis $D(W)$ includes either (i) a link a_i ending at C or (ii) a link a_i starting at C, or (iii), both (i) and (ii) are true.

(i) a_i ends at C ($|i - s| \geq 2$). The diagram $D(W)$ includes Fig. C.2(left). By the defining relations this implies Fig. C.2(right); i.e., $D(W)$ includes a link a_i ending at D.

(ii) a_i starts at C ($|i - s| \geq 2$). The diagram $D(W)$ includes Fig. C.3(left). By Theorem H this implies Fig. C.3(right); i.e., $D(W)$ includes a link a_i starting at D.

(iii) If (i) and (ii) are both true then $D(W)$ must include both a link a_i ending at D and a link a_i starting at D.

Hence in all cases, for $|i - s| \geq 2$, at least one link a_i is incident to D.

(b) It remains to consider the links a_t where $|t - s| = 1$. The proof will be omitted. If follows the same pattern as (a) above. In all cases, if $|t - s| = 1$, at least one link a_i is incident to D.

Now, by hypothesis there is a link a_s ending at D. Hence, by (a) and (b) together we see that the n links a_r $(r = 1, 2, \ldots, n)$ are incident at D. The induction is established, and the result follows.

$\square$

Theorem C.10. *In B_{n+1} every node of $D(\Delta)$ is the meet of the n links $a_1, a_2, \ldots, a_n$. Furthermore only n links are incident at each node.*

By the previous theorem it follows at once that each node of $D(\Delta)$ is the meet of the n links $a_1, \cdots, a_n$. It therefore remains only to prove that we cannot have a repeated link at any node. For suppose the contrary is true, so that for some A, r, B we have $\Delta \dot{=} A a_r a_r B$. Then

$$A a_r a_r B \Re A \dot{=} \Delta \Re A = A\Delta \text{ and } a_r a_r X \dot{=} \Delta,$$

where $X \sim B\Re A$.

This leads us to

$$a_r a_r X \dot{=} a_1 A_1 \dot{=} \cdots \dot{=} a_{r-1} A_{r-1} \dot{=} a_{r+1} A_{r+1} \dot{=} \cdots \dot{=} a_n A_n,$$

and Theorem H now gives

$$a_r X \dot{=} a_1 B_1 \dot{=} \cdots \dot{=} a_{r-1} B_{r-1} \dot{=} a_{r+1} B_{r+1} \dot{=} \cdots \dot{=} a_n B_n.$$

Hence, $a_r X$ contains Δ, which is impossible since $L(a_r X) < L(\Delta)$. The theorem therefore follows.

C.3   Solution to the conjugacy problem in B_{n+1}

C.3.1   Index length

The algebraic sum of the indices of any given word will be called the *index length*. For instance, $(\alpha_2)^2(\alpha_3)^{-1}\alpha_1^5$ has index length 4.

The following lemma is crucial for solving the conjugacy problem.

Lemma C.9 (Crucial Lemma). *In B_{n+1} the number of words in standard form of index length t and power $\geq p$ is finite.*

Let $\Delta^m \bar{A}$ be any word satisfying the conditions. Then if $L(\Delta) = d$, we have

$$m \geq p \tag{C.1}$$

$$t = md + L(\bar{A}) \qquad\qquad (C.2)$$

Since $L(\bar{A}) \geq 0$ and d is positive, the last equation gives

$$m \leq \frac{t}{d}. \qquad\qquad (C.3)$$

(C.1) and (C.3) together show that the number of values of m is finite. (C.2) shows that for any fixed m, $L(\bar{A})$ is constant, and so the number of possible values of $\bar{A}$ is finite. This leads us to Lemma C.9.

Definition C.2. In the diagram $D(\Delta)$ in B_{n+1} let α be any initial sub-route so that $\Delta \doteq \alpha X$ $(0 \leq L(X) \leq L(\Delta))$. We shall call such a sub-route an α-route. If W is any word in B_{n+1}, then the word $\alpha^{-1}W\alpha$ reduced to the standard form, will be called an *alpha-transformation* of W. If $\tilde{\alpha}$ is the base of any α-route α, then we shall call $\tilde{\alpha}$ an $\tilde{\alpha}$-route and the transformation $\tilde{\alpha}^{-1}W\tilde{\alpha}$ an $\tilde{\alpha}$-transformation of W. It is clear that any α-transformation is equal to the corresponding $\tilde{\alpha}$-*transformation*.

C.3.2 Summit form. Summit set. Summit. Summit power

Let W be any word in B_{n+1} with standard form $\Delta^m\bar{A} = W_1$, say. Consider now the following chain of α-transformations of W.

Take all the α-transformations of W_1, and let those which are of power $\geq m$ and which are distinct from W_1 and from each other, be $W_2, W_3, \ldots, W_l$. Now, repeat the process for each of the words $W_2, W_3, \ldots, W_l$, in turn, denoting successively by $W_{l+1}, W_{l+2}, \cdots$, any new words occurring, the condition being always that each new word must be of power $\geq m$. Continue to repeat the process for every new distinct word arising, as the sequence $W_1, W_2, \ldots, W_{l+2}, \ldots$, expands. Now, each word of the sequence is of the same index length as W. Hence, by Crucial Lemma C.9, the sequence is finite, and ultimately a stage must be reached when further application of the process will yield no new words.

Suppose that in the set $W_1, W_2, \cdots$, the highest power reached in s, and that the words of power s from the subset $V_1, V_2, \cdots$. Then any V_r will be said to be a *summit form* of W . The set $V_1, V_2, \cdots$, will be called the *summit set* or simply the *summit* of W. The power s of any summit form will be called the *summit power* of W. It is clear from the definitions given above that no single α-transformation of a summit from can be a power greater than the summit power.

Lemma C.10. *In B_{n+1}, if $W = \Delta^m V$, where V is positive, and P is a positive word such that $P^{-1}WP$ is of power $m + r$ $(r > 0)$, then VP contains Δ.*

By hypothesis, $P^{-1}\Delta^m VP = \Delta^{m+r}\hat{Q}$, so that

$$VP = \Delta^{-m}P\Delta^{m+r}\hat{Q} \tag{C.4}$$

Put $\tilde{P} \sim P$ for $(m+r)$ is even, $\tilde{P} \sim \mathfrak{R}P$ $(m+r$ is odd). Then, $VP = \Delta^r\tilde{P}\hat{Q}$, which means that $VP \dot{=} \Delta^r\tilde{P}\hat{Q}$. Hence VP contains Δ.

Lemma C.11. *In B_{n+1}, if $W \cong V$, then there exists a positive word X such that $X^{-1}WX = V$.*

By hypothesis there exists a word A such that $A^{-1}WA = V$. Let $A = \Delta^m\bar{P}$. Then

$$\bar{P}^{-1}\Delta^{-m}W\Delta^m\bar{P} = V.$$

If m is even, we get $\bar{P}^{-1}W\bar{P} = V$ ($\bar{P}$ positive). If m is odd, we may apply the same trick with $\Delta\bar{P}$ instead of just $\bar{P}$.

Lemma C.12. *In B_{n+1}, suppose (i) that $W \sim \Delta^p\bar{P}$ is a summit form of any given word A, (ii) that X is any positive word such that $X^{-1}WX = \Delta^q\hat{Q}$, where $q \geq p$, and (iii) that $X = uY$, where u is an α-route of maximum length. Then $u^{-1}Wu$, reduced to standard form is a summit form of A.*

Theorem C.11. *In B_{n+1}, $A \cong B$ if and only if their summit sets are identical.*

Proof. (i) If the condition is satisfied, let C be any member of the common summit set. Then $A \cong C, B \cong C$. Hence $A \cong B$, so that the condition is certainly sufficient.

(ii) We now proceed to show that the condition is necessary. Suppose

$$A \cong B \tag{C.5}$$

Let

$$\Delta^p\hat{P} \cong A \text{ be any summit form of } A \tag{C.6}$$

$$\Delta^q\hat{Q} \cong B \text{ be any summit form of } B \tag{C.7}$$

First suppose $q \geq p$. Clearly $\Delta^p\hat{P} \cong \Delta^q\hat{Q}$, and hence, there exists a positive word X such that

$$X^{-1}\Delta^p\hat{P}X = \Delta^q\hat{Q} \quad (q \geq p).$$

Let $X \dot{=} u_1X_1, X_1 \dot{=} u_2X_2, \ldots$, etc., and finally $X_s \dot{=} u_{s+1}$, where $u_1, u_2, \ldots$, are defined successively as α-routes of maximum length, and $X_1, X_2, \ldots$, are words of steadily reducing length, so that the final word X_{s+1} is the empty word.

Then

$$X \doteq u_1 u_2 \cdots u_{s+1}.$$

Using the last equation, the formula $X^{-1} \Delta^p \hat{P} X = \Delta^q \hat{Q}$ may be regarded as the product of the $s+1$ successive transformations $(u_1)^{-1} \Delta^p \hat{P} u_1 = \hat{W}_1$, say, in standard form: $u_2^{-1} W_1 u_2 = W_2$, say, in standard form; $\cdots$, $(u_{s+1})^{-1} W_s u_{s+1} = \Delta^q \hat{Q}$ are each summit forms of A.

Hence we cannot have $q > p$ and similarly we cannot have $p > q$. Hence $p = q$ and by the argument above $\Delta^q \hat{Q} \sim \Delta^p \hat{Q}$ is a summit form of A. We have thus proved that any summit form of B is a summit form of A. Similarly any summit form of A is a summit form of B; i.e., the summit sets of A and B are identical.

$$\square$$

C.4   The proof of Theorem H

The theorem for words X, Y of word-length s will be refereed to as H_s. For $s = 0, 1$, the theorem is obvious and is left to the reader.

The proof of the general theorem now follows by induction. For our induction hypothesis we assume

(α) H_s is true for $0 \leq s \leq r$ for transformations of all chain-lengths and
(β) H_{r+1} is true for all chain-lengths $\leq t$.

Let X, Y be of word-length $r + 1$, and let $a_i X \doteq a_k Y$ through a transformation of chain-length $t + 1$. Let the successive words of the transformation be $W_1 \sim a_i X, W_2 \sim \cdots, W_{t+2} \sim a_k Y$.

Choose arbitrarily any intermediate word $W_c \sim a_m W$, say, from the middle of the chain somewhere. The transformations $a_i X \to a_m W, a_m W \to a_k Y$ are each of chain-length $\leq t$, and we can therefore apply (β) to them. We have then

$$a_i X \doteq a_m W \doteq a_k Y.$$

For the complete proof of Theorem H we have to consider all possible variations in the values of i, m, k. The general pattern of the proof is, however, exactly the same for each variation, and it will be sufficient here to deal with two cases only, as typical examples of the common method of proof.

Case 1. $k = i, |m - i| \geq 2$.
We have:

$$a_i X \doteq a_m W; a_m W \doteq a_i Y \ (|m - i| \geq 2).$$

By (β) we have
$$X \doteq a_m P, W \doteq a_i P \text{ for some } P;$$

and

$$W \doteq a_i Q; Y \doteq a_m Q \text{ for some } Q.$$

Hence $X \doteq a_m P \doteq a_m Q \doteq Y$ as required.

Case 2. $|k - i| \geq 2, |m - i| \geq 2, |k - m| = 1$.

We have

$$a_i X \doteq a_m W; a_m W \doteq a_k Y.$$

By (β) we get

$$X \doteq a_m P; W \doteq a_i P \text{ for some } P;$$

and

$$W \doteq a_k a_m Q, Y \doteq a_m a_k Q \text{ for some } Q.$$

By (α), the two expressions for W give us

$$P \doteq a_k R, a_m Q \doteq a_i R \text{ for some } R.$$

The last equation now gives

$$Q \doteq a_i S, R \doteq a_m S \text{ for some } S.$$

Therefore

$$X \doteq a_m a_k a_m S, Y \doteq a_m a_k a_i S.$$

Hence, using the defining relations, we have

$$X \doteq a_k a_m a_k S, Y \doteq a_m a_i a_k S \doteq a_i a_m a_k S,$$

i.e., $X \doteq a_k Z, Y \doteq a_i Z$ as required, where $Z \sim a_m a_k S$.

The proofs of other variations in the values of i, m, k are similar.

Since H_{r+1} is true for chain length 1, an induction proves it for all chain lengths, and a further induction (on r) completes the proof of Theorem H.

Appendix D

Unsolved problems in knot theory

I. Below, we give a list of unsolved problems. While compiling the list, we mostly referred to the Robion Kirby homepage www.math.berkeley.edu/~kirby (his problem book on low–dimensional topology). We could not place here the whole list of problems listed there. We have chosen the problems with possibly easier formulations but having a great importance in modern knot theory. Besides, we used some "old" problems formulated in [CF] and still not solved, and problems from [CD3, Jon4, Mor2]. As for virtual knot theory, we mostly used the lists of problems [FKM, FIKM] by the author, Louis H. Kauffman, Roger A. Fenn and Denis P. Ilyutko.

If the problem was formulated by some author, we usually indicate the author's name. If the problem belongs to the author of the present book, we write (V.M).

1. How do we define whether a knot is invertible?

2. Let $C(n)$ be the number of simple unoriented knots in S^3 with minimal number of crossings equal to n. Describe the behaviour of $C(n)$ as n tends to ∞. Some bounds for $C(n)$ can be found in [Wel].

    The first values of $C(n)$ (starting from $n = 3$) are: $1, 1, 2, 3, 7, 21, 49, 165, 552, 2167, 9998$.

3. Similarly, let $U(n)$ be the number of prime knots K for which the unknotting number equals n. What is the asymptotic behaviour of $U(n)$?; of $C(n)/U(n)$?

4. Is it true that the minimal number of crossings is additive with respect to the connected sum operation: $c(K_1 \# K_2) = c(K_1) + c(K_2)$?

5. (de Souza) Does the connected sum of n knots (not unknots) have unknotting number at least n?

    The positive answer to this question would follow from the positive answer to the following question:

6. Is the unknotting number additive with respect to the connected sum: $u(K_1 \# K_2) = u(K_1) + (K_2)$?

7. The counterexample to the third Tait conjecture (see Chapter 7) was found in [HTW1]. It has 15 crossings.

Are there amphicheiral knots with arbitrary minimal crossing number ≥ 15?

8. Are there (prime, alternating) amphicheiral knots with every possible even minimal crossing number?

9. (Jones) Is there a non-trivial knot for which the Jones polynomial equals 1? Bigelow proved that this problem is equivalent to the existence of a kernel for the Burau representation of $Br(4)$.

   (Kauffman, "virtual reformulation") There are different methods for constructing virtual knots with the trivial Jones polynomial. Are there classical knots among them?

10. Are there infinitely many different knots with the same Kauffman two–variable polynomial?

    In fact, it is shown in [DH] that if such an example exists, the number of its crossings should be at least 18.

11. Find an upper bound on the number of Reidemeister moves transforming a diagram of a knot with n crossings to another diagram of the same knot with m crossing (a function of m, n).

12. Is the Burau representation of $Br(4)$ faithful?

13. (Vassiliev) Are the Vassiliev knot invariants complete ?

14. (Partial case of the previous problem) Is it true that the Vassiliev knot invariants distinguish inverse knots? In other words, is it true that there exists inverse knots K and K' and a Vassiliev invariant v such that $v(K) \neq v(K')$?

15. Is there a faithful representation of the virtual braid group $VB(n)$ for arbitrary n?

    O. Chterental [Cht1] constructed a faithful (nonlinear) representation of the virtual braid group into the automorphism group of "virtual curve diagrams" and used it to solve the word problem for virtual braids.

16. (Fox) A knot is called a *slice* if it can be represented as an intersection of some $S^2 \subset \mathbb{R}^4$ and some three–dimensional hyperplane.

    A knot is called a *ribbon* if it is a boundary of a ribbon disk. Obviously, each ribbon knot is a slice.

    If K is a slice knot, is K a ribbon knot?

17. Two knots given by maps $f_1, f_2 : S^1 \to S^3$ are called *concordant* if the maps $f_{1,2}$ can be extended to an embedding $F : S^1 \times I \to S^3$.

    Problem (Akbulut and Kirby) If 0–frame surgeries on two knots give the same 3–manifold then the knots are concordant.

Remark D.1. *The theory of knots, cobordisms and concordance is represented very well in the works [COT] by Cochran, Orr and Teichner and [Tei] by Teichner.*

18. (Jones and Przytycki) A *Lissajous knot* is a knot in $\mathbb{R}^3$ given by the following parametric equations:

$$x = cos(\eta_x t + \phi_x)$$

$$y = cos(\eta_y t + \phi_y)$$

$$x = cos(\eta_z t)$$

for integer numbers η_x, η_y, η_z. Which knots are Lissajous?

19. (V.M) Find an analogue of combinatorial formulae for the Vassiliev invariants for knots and links in terms of d–diagrams.

20. Does every non-trivial knot K have property P, that is, does Dehn surgery on K always give a non-simply connected manifold?

   To date, this problem has been solved positively in many partial cases.

21. (Kauffman) Find a purely combinatorial proof that any two virtually equivalent classical knots are classically equivalent.

22. (V.M) Can the forbidden move be used for constructing a system of axioms in order to obtain virtual knot invariants (like skein relations and Conway algebras).

23. Is there an algorithm for recognising virtual knots?

   This problem was essentially solved by the author in [MI] and the question is to find an explicit and constructive description for the algorithm.

24. Is there an algorithm that distinguishes whether a virtual knot is isotopic to a classical one?

25. Crossing number problems (R. Fenn, L. Kauffman, V. M).

   For each virtual link L, there are three crossing numbers: the minimal number C of classical crossings, the minimal number V of virtual crossings, and the minimal total number T of crossings for representatives of L.

   What is the relationship between the least number of virtual crossings and the least genus in a surface representation of the virtual knot?

   Is it true that $T = V + C$?

   Are there some (non-trivial) upper and lower bounds for T, V, C coming from virtual knot polynomials?

26. Let U_1, U_2 be some two virtual knots. Is it true that the triviality of $U_1 \# U_2$ implies the triviality of one of them, where by $U_1 \# U_2$ we mean

    any connected sum;

    arbitrary connected sum?

The same question about long virtual knots.

27. (X.–S. Lin) Suppose that oriented knots K_+ and K_- differ at exactly one crossing at which K_+ is positive and K_- is negative. If $K_+ = K_-$, does it follow that K_+ equals

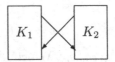

where either K_1 or K_2 could be the unknot?

28. There is a visible similarity between the behaviour of parity for classical knots and the behaviour of the Maslov index.

    Construct a parity theory for Legendrian knots in front projection. Use it to enhance it in a way similar to Chekanov-Eliashberg DGA.

II. In 2009, the author introduced the notion of parity into knot theory, see [Man22]. The main feature of parity is the possibility of constructing picture-valued invariants. The simplest key invariant is the parity bracket for free knots which is defined as the sum over all smoothings at even crossings. The smoothings themselves are framed 4-graphs since odd crossings remain untouched. The simplest parity bracket is considered as a $\mathbb{Z}_2$-linear combination of framed 4-graphs modulo second Reidemeister moves.

If a framed 4-graph K has no odd crossings then by definition

$$[K] = K,$$

since there are no crossings to smooth and K is the unique crossing of itself. If K admits no decreasing Reidemeister moves, then this means that K in the right hand side is the minimal diagram of K modulo Reidemeister moves, which, in turn, means that for any diagram K equivalent to K' we have

$$[K'] = K$$

which by construction means that K appears as a smoothing of K'.

This principle can be formulated as: *If K is complicated enough (in our case, odd and irreducible) then it realizes itself.*

The same principle works not only for odd diagrams, see [Man26, KM2, KM3].

Recently, Sam Nelson with coauthors [NOR] introduced various enhancements of quantum invariants by means of quandles and biquandles.

The pattern is as follows. We take a colouring of a knot diagram by elements of a finite (bi)quandle and then modify some known invariant pattern (say, skein-relation) by taking into account the colours. Then we take the sum over all admissible colourings.

This can be also done with the elements of the universal quandle.

Nelson's construction leads to a (huge) system of equations depending on colours which can be written down for each particular quandle. This system of equation a priori leads to a solution where coefficients do not depend on colours and allows one to restore the initial quantum invariant times the number of colourings.

Nelson's enhancements turn out to be a priori at least as powerful as many known classical invariants, and they indeed turn out to be more powerful since they detect some effects the initial classical invariants do not detect.

The main difference between the parity bracket and invariants constructed before is that the parity bracket is valued in pictures unlike the Kauffman bracket etc. We do not count the number of components after the smoothings; we count the result of smoothings as a graph (maybe, modulo some moves, which, however, turn out to be very easy).

The crucial observation of myself of 2016 realized in our joint paper with D. Ilyutko in January 2017 [IM2] is that

the Gaussian parity itself can be axiomatized by using a very simple biquandle.

Namely, having a framed 4-graph, we can colour its edges by 0 and 1 (after each crossing we switch from 0 to 1 and back). The crossing is *even* if two incoming edges have different labels (1 and 0) and *odd* otherwise.

Thus, the parity bracket can be written down as follows: we colour edges by elements of this simple biquandle, and associate with each even crossing the sum of two splittings with coefficient 1, and associate with each odd crossing the rigid crossing with coefficient 1. This sum is considered modulo 1, and graphs with rigid vertices are considered modulo the second Reidemeister moves.

Thus, we see, that a modification of Nelson's principle allows one to get the parity bracket. However, our graphs should be evaluated as graphs, not as numbers (polynomials, etc).

This allows one to axiomatize further, by using arbitrary biquandles or universal biquandles, etc. The axioms are written down in my joint paper with Ilyutko [IM2].

This system of axioms a priori has two "independent" solutions: Nelson's solution (when we drop graphs and consider numbers instead) which is extremely powerful for classical knots and my bracket which leads to pictures for virtual knots and allows one to realize the parity principle.

Problem. Does this system have any solution for classical knots (links) valued in pictures?

Or, globally, how can we get invariants of classical links valued in graphs related to classical link diagrams?

For example, can we get, say, Borromean rings with a non-zero coefficient (note that Borromean rings here are considered as a "doodle" not as a link)?

The main problem is that the computer evaluation of this system of equations even for 3-element biquandles requires a lot of memory time.

The invariant defined by Ilyutko and myself admits several variations.

1. One can consider graphs up to first and second Reidemeister moves (doodles)

2. One can forget about first and second Reidemeister moves if we know a priori that the invariant gets multiplied by some constant

3. One can use not just Kauffman bracket formalism, but other graph formalisms. In particular, if we take Kuperberg's brackets (three of them) [Kup2], then we can still realize the principle *"If a diagram is complicated enough then it realizes itself"* not exactly by 4-graphs but by using 3-graphs related to them.

   Here the word "complicated enough" leads to a larger class than just "odd irreducible diagrams".

It is known [Man26, KM2] that such diagrams almost completely classify for free knots, that is, one can prove that if diagram has no loops, bigons, triangles and quadrilaterals, then this diagram is minimal and this minimal representative is unique.

In particular, the problem of a complete classification of free knots remains open for a small class of diagrams.

Which is the geometrical interpretation of free knots?

Bibliography

[Ada] Adams C. (1994), *The knot book* (New York: Freeman & Co.).

[AH] Appel, K. I. and Haken, W. (1989), Every planar map is four colourable. *Contemp. Math.*, **98**, AMS, Providence.

[Akb] Akbulut, S. (2016), *4-manifolds, Oxford Graduate Texts in Mathematics*, **25**, Oxford University Press, Oxford, 262 pp.

[Ale1] Alexander, J. W. (1923), Topological invariants of knots and links. *Trans. AMS.*, **20**, pp. 257–306.

[Ale2] Alexander, J. W. (1923), A lemma on systems of knotted curves, *Proc. Nat. Acad. Sci. USA*, **19**, pp. 93–95.

[Ale3] Alexander, J.W. (1933), A matrix knot invariant. *Proc. Nat. Acad. Sci. USA*, **19**, pp. 222–275.

[Arn1] Arnol'd V.I. (1969), Cohomology ring of the pure braid groups. *Mathematical Notes*, **5** (2), pp. 227–231.

[Arn2] Arnol'd V.I. (1996), *Osobennosti kaustik i volnovykh frontov (Singularities of caustics and wave fronts)*, Moscow, Fasis (in Russian).

[Art1] Artin, E. (1925), Theorie der Zöpfe. *Abh. Math. Sem. Univ. Hamburg*, **4**, pp. 27–72.

[Art2] Artin, E. (1947), Theory of braids, *Annals of Mathematics* (2) **48**, pp. 101–126.

[APS] Asaeda, M., Przytycki, J. and Sikora, A. (2004). Categorification of the Kauffman bracket skein module of I-bundles over surfaces, *Algebr. Geom. Topol.* **4**, 52, pp. 1177–1210.

[Ash] Ashley, C. W. (1947), *The Ashley book of knots* (London: Coles Publ. Co. Ltd).

[Ati1] Atiyah M. (1989–90), The Jones–Witten invariants of knots, Seminaire Bourbaki, **715**.

[Ati2] Atiyah, M. (1990), *The geometry and physics of knots* (Cambridge: Cambridge University Press).

[Ber] Berger, M. (1994), Minimum crossing for 3–braids. *Journal of Physics. A: Math. Com.*, **27**, pp. 6205–6213.

[BF] Bartholomew, A. and Fenn, R. (2008), Quaternionic invariants of virtual knots and links, *J. Knot Theory Ramifications*, **17**(2), pp. 231–251.

[BGRT] Bar–Natan, D., Garoufalidis, S., Rozansky, L. and Thurston, D. (2003), Wheels, wheeling, and the Kontsevich integral of the unknot. *Israel Journal of Mathematics*, **119**, pp. 217–237.

[Big1] Bigelow, S. (2001), Braid groups are linear *J. Amer. Math. Soc.*, **14**, pp. 471–486.

[Big2] Bigelow, S. (2002), Does the Jones polynomial detect the unknot, *Journal of Knot Theory and Its Ramifications*, **11**, pp. 493–505.

[Bir1] Birman, J.S. (1974), *Braids, links and mapping class groups*, Princeton, NJ: Princeton Univ. Press, 1974 (Ann. Math. Stud., 1982).

[Bir2] Birman, J.S. (1993), New points of view in knot and link theory, *Bull. Amer. Math. Soc.*, **28**, pp. 253–287.

[Bir3] Birman, J.S. (1994). Studying links via closed braids, In: *Lecture Notes on the Ninth Kaist Mathematical Workshop*, **1**, pp. 1–67.

[Bjö] Björner, A. (1993), Subspace arrangements. In: *Proceedings of the First European Congress of Mathematicians (Paris 1992)*, Birkhäuser, Boston, pp. 321–370.

[BL] Birman, J.S. and Lin X.-S. (1993), Knot polynomials and Vassiliev's invariants, *Inventiones Mathematicae*, **111**, pp. 225–270.

[Bla] Blanchet, C. (2010). An oriented model for Khovanov homology, *J. Knot Theory Ramifications* **19**, 2, pp. 291–312.

[Ble] Bleiler, S. (1984), A note on the unknotting number, *Math. Proc. Camb. Phil. Soc.*, **96**, pp. 469–471.

[BLT] Bar–Natan, D., Le, T. and Thurston, D. (2003), Two applications of elementary knot theory to Lie algebras and Vassiliev invariants, *Geom. Topol.*, **7**, pp. 1–31.

[BM1] Birman, J.S. and Menasco, W. (1991), Studying links via closed braids II: On a theorem of Bennequin, *Topology and Its Applications*, **40**, pp. 71–82.

[BM2] Birman, J.S. and Menasco, W. (1993), Studying links via closed braids III: Classifying links which are closed 3–braids, *Pacific Journal of Mathematics*, **161**, pp. 25–113.

[BM3] Birman, J.S. and Menasco, W. (1990), Studying links via closed braids IV: Composite links and split links, *Inventiones Mathematicae*, **102**, pp. 115–139.

[BM4] Birman, J.S. and Menasco, W. (1992), Studying links via closed braids V: The unlink, *Transactions of the AMS*, **329**, pp. 585–606.

[BM5] Birman, J.S. and Menasco, W. (1992), Studying links via closed braids VI: A non–finiteness theorem, *Pacific Journal of Mathematics*. **156**, pp. 265–285.

[BM6] Birman, J.S. and Menasco, W. (2002), On Markov's theorem, *Journal of Knot Theory and its Ramifications*, 11(3), pp. 295–310.

[BN1] Bar–Natan, D. (1995), On the Vassiliev knot invariants, *Topology*, **34**, pp. 423–475.

[BN3] Bar–Natan, D. (1995), Vassiliev homotopy string link invariants, *Journal of Knot Theory and its Ramifications* 4 (1), pp. 13-32.

[BN4] Bar–Natan, D. (1997), Lie algebras and the four colour theorem, *Combinatorica*, **17** (1), pp. 43–52.

[BN5] Bar-Natan, D. (2002). On Khovanov's categorification of the Jones polynomial, *Algebr. Geom. Topol.* **2**(16), pp. 337–370.

[BN6] Bar-Natan, D. (2005). Khovanov's homology for tangles and cobordisms, *Geom. Topol.*, **9**, pp. 1443–1499.

[BNh] D. Bar–Natan's homepage, http://www.ma.huji.ac.il/~ drorbn

[BNk] D. Bar-Natan. The Knot Atlas. www.math.toronto.edu/ drorbn/ KAtlas/index.html, 2003.

[Bou] Bourbaki, N. (1982), *Groupes et Algèbres de Lie, IV,V,VI*, Masson, Paris.

[Bour] Bourgoin, M. O. (2008), Twisted link theory, *Algebr. Geom. Topol.*, **8**(3), pp. 1249–1279.

[BoZh] Boyer, S. and Zhang, X. (2005), Every nontrivial knot in S^3 has nontrivial A-polynomial, *Proc. Amer. Math. Soc.*, **133**(9), pp. 2813–2815.

[BPW] Bollobás B., Pebody, L. and Weinreich, D. (2002), A state space definition of the HOMFLY invariant, In: *Contemporary Combinatorics*, Budapest, pp. 139–184.

[BR] Bollobás, B. and Riordan, O. (2000), Linearized chord diagrmas and an upper bound for Vassiliev invariants, *Journal of Knot Theory and its Ramifications*, **9** (7), pp. 847–852.

[Bri1] Brieskorn, E.V. (1971), Die Fundamentalgruppe des Raumes der regulären Orbits einer endlichen komplexen Spiegelsgruppe, *Inventiones Mathematicae*, **12** (1), pp. 57–61.

[Bri2] Brieskorn, E.V. (1973), Sur les groupes des tresses (d'après V.I.Arnol'd). In: *Seminaire Bourbaki 1971-72*, **401**, *Lect. Notes Math.*, **317**, Springer, Berlin, pp. 21–44.

[Bru] Brunn, H. (1898), Über verknotene kurven, In: *Mathematiker–Kongress Zürich 1897*, Leipzig, ss. 256–259.

[Bura] Burau, W. (1936), Über Zopfgruppen und gleichzeitig verdrillte Verkettungen, *Abh. Math. Sem. Univ. Hamburg*, **11**, pp. 179–186.

[Burm] Burman, Yu. M. (1999), Dlinnyje krivye, Gaussovy diagrammy i invarianty (Long curves, Gauss diagrams, and invariants), *Mat. Prosveshcheniye*, **3**, pp. 94–115 (in Russian)

[BW] Björner, A. and Welker, V. (1995), The homology of "k–equal" manifolds and related partition lattices, *Adv. Math.*, **110** (2), pp. 277–313.

[BZ] Burde, G. and Zieschang, H. (2003), *Knots* (Berlin: Walter de Gruyter).

[Car12] Carter, J.S. (2012), A survey of quandle ideas, *Introductory lectures on knot theory*, Ser. Knots Everything, **46**, World Sci. Publ., pp. 22–53.

[Car93] Cartier, P. (1993), Construction combinatoire des invariants de Vassiliev–Kontsevich des nœuds, *C.R.Acad. Sci. Paris, Sér. I, Math.*, **316**, pp. 1205–1210.

[Cas] Cassel, Ch. (1995), *Quantum groups* (New York: Springer–Verlag).

[CCGLS] Cooper, D., Culler, M., Gillet, H., Long, D. and Shalen, P. (1994), Plane curves associated to character varieties of 3-manifolds, *Invent. Math.*, **118**, pp. 47–84.

[CD1] Chmutov, S. and Duzhin, S. (1994), An upper bound for the number of Vassiliev knot invariants, *Journal of Knot Theory and its Ramifications*, **2** (2), pp. 141–151.

[CD2] Chmutov, S. and Duzhin, S. (1999), A lower bound for the number of Vassiliev knot invariants, *Topology and its Applications*, **92**, pp. 201-223.

[CD3] Chmutov, S.V. and Duzhin, S.V. (1999), Uzly i ikh invarianty (Knots and their invariants), Mat. Prosveshcheniye, **3**, pp. 59–93 (in Russian)

[CDL] Chmutov, S.V., Duzhin, S.V. and Lando, S.K. (1994), Vassiliev knot invariants $I - III$, *Advances in Soviet Mathematics*, **21**, pp. 117–147.

[CDM] Chmutov, S., Duzhin, S. and Mostovoy, J. (2012), *Introduction to Vassiliev Knot Invariants*, Cambridge University Press, Cambridge.

[CES] Carter J. S., Elhamdadi, M. and Saito M. (2004), Homology theory for the set-theoretic Yang-Baxter equation and knot invariants from generalizations of quandles on sun, *Fund. Math.*, **184**, pp. 31–54.

[CF] Crowell, R.H. and Fox, R.H. (1963), *Introduction to knot theory*, (New York: Ginn & Co).

[CFM] Cheng, Z., Fedoseev, D.A. and Manturov, V.O. (2015), On marked braid groups, *J. Knot Theory Ramifications*, **24**(13), 1541005, 12 pp.

[Ch] Chernavsky, A. V. (1963), Geometricheskaja Topologija Mnogoobrazij (Geometric Topology of Manifolds), In: *Algebr. Top. 1962 (Itogi nauki i tekhniki. VINITI AN SSSR)*, pp. 161–187 (in Russian).

[Che] Chekanov, Yu. (2002), Differential algebras of Legendrian links, *Inventiones Mathematicae*, **150**(3), pp. 441–483.

[Che2] Chekanov, Yu. V. (2002), Invariants of Legendrian knots, In: *Proceedings of the International Congress of Mathematicians, Beijing*, **2**, pp. 385–394., (Beijing: Higher Education Press)

[ChFa] Farber, M. Sh. and Chernavsky A.V. (1985), Uzlov Teoriya (Knot Theory), In: *Mathematical Encyclopaedia*, **5**, (Moscow: Nauka, in Russian)

[ChKo] Champanerkar, A. and Kofman, I. (2006). Spanning trees and Khovanov homology, preprint, arXiv:math.GT/0607510.

[ChPu] Chekanov, Yu.V., Pushkar, P.E. (2005), Differential algebras of Legendrian links, *Russian Math. Surveys*, **60**(1), pp. 95–149.

[Chr] Chrisman, M.W. (2013), Prime decomposition and non-commutativity in the monoid of long virtual knots, arXiv:1311.5748.

[Cht1] Chterental, O. (2015), Virtual braids and virtual curve diagrams, *J. Knot Theory Ramifications*, **24**, 1541001.

[Cht2] Chterental, O. (2017), Distinguishing virtual braids in polynomial time, arXiv:1706.01273

[CM] Chrisman, M.W. and Manturov, V.O. (2012), Parity and exotic combinatorial formulae for finite-type invariants of virtual knots, *J. Knot Theory Ramifications*, **21**(13), 1240001, 27 pp.

[CMW] Clark, D., Morrison, S. and Walker, K. (2009), Fixing the functoriality of Khovanov homology, *Geom. Topol.*, **13**(3), pp. 1499–1582.

[Con] Conway, J.H, (1970), An enumeration of knots and links and some of their algebraic properties, In: *Computational Problems in Abstract Algebra* (New York, Pergamon Press), pp. 329–358.

[COT] Cochran, T.D., Orr, K.E. and Teichner, P. (1999), Knot concordance, Whitney towers and L^2–signatures, arXiv:math. GT/9908117, 22 Aug. 1999.

[Cro] Crowell, R. H. (1964), The derived group of a permutation representation, *Advances in Mathematics*, **53**, pp. 88–124.

[CS] Carter, J. S. and Saito, M. (1993) Reidemeister moves for surface isotopies and their interpretation as moves to movies, *J. Knot Theory Ramifications*, **2**, pp. 251–284.

[CV] Chmutov, S.V. and Varchenko, A.N. (1997), Remarks on the Vassiliev knot invariants coming from sl_2, *Topology*, **36**, pp. 153–178.

[Das] Dasbach, O. (2000), On the combinatorial structure of primitive Vassiliev invariants III — A lower bound, *Communications in Contemporary Mathematics*, **2**,(4), pp. 579–590.

[Deh1] Dehornoy, P. (1995), From large cardinals to braids via distributive algebra, *Journal of Knot Theory and its Ramifications*, **4**, pp. 33–79.

[Deh2] Dehornoy, P. (2002), Groupes de Garside, *Ann. Sci. École Norm. Sup. (4)*, **35**, pp. 267–306.

[Deh3] Dehornoy, P. (2002), Braids and self–distributivity, *Progress in Mathematics*, **192**, Birkhäuser.

[Dehn1] Dehn, M. (1910), Über die Topologie des dreidimensionalen Raumes, *Mathematische Annalen*, **69**, ss. 137–168.

[Dehn2] Dehn, M. (1914), Die beiden Kleeblattschlingen, *Mathematische Annalen*, **102**, ss. 402–413.

[DF] Dollard, J. D. and Friedman, Ch. N. (1979), *Product integration with applications to differential equations* (London: Addison–Wesley Publishing Company).

[DFN] Dubrovin, B.A., Fomenko A.T and Novikov, S.P. (1984,1985,1990), *Modern Geometry–Methods and Applications*, Parts I–III (New York: Springer–Verlag).

[DG] Dunfield, N.M. and Garoufalidis, S. (2004), Non-triviality of the A-polynomial for knots in S^3, *Algebr. Geom. Topol.*, **4**, pp. 1145–1153.

[DH] Dasbach and O.T., Hougardy, S. (1997), Does the Jones polynomial detect unknottedness?, *Journal of Experimental Mathematics*, **6** (1), pp. 51–56.

[DK1] Dye, H.A. and Kauffman, L.H. (2005), Virtual knot diagrams and the Witten-Reshetikhin-Turaev invariant, *J. Knot Theory Ramifications*, **14**(8), pp. 1045–1075.

[DK2] Dye, H.A. and Kauffman, L.H. (2005), Minimal Surface Representation of Virtual Knots and Links, *Algebr. Geom. Topol.*, **5**, pp. 509–535.

[DK3] Dye, H.A. and Kauffman, L.H. (2009), Virtual crossing number and the arrow polynomial, *J. Knot Theory Ramifications*, **18**(10), pp. 1335–1357.

[DKK] Dye, H.A., Kaestner, A. and Kauffman, L.H. (2017), Khovanov homology, Lee homology and a Rasmussen invariant for virtual knots, *Journal of Knot Theory and Its Ramifications*, **26**(3), 1741001.

[DKM] Dye, H. A., Kauffman, L. H. and Manturov, V. O. (2010), On two categorifications of the arrow polynomial for virtual knots, in: *The Mathematics of Knots*, Contributions in the Mathematical and Computational Sciences, **1**, Springer, Berlin, pp. 95–124.

[Dri1] Drinfel'd, V.G. (1985), Hopf algebras and quantum Yang–Baxter equations, *Doklady Mathematics*, **32**(1), pp. 254–258.

[Dri2] Drinfeld, V.G. (1986), Quantum Group, *Proc. Intl. Congress Math.*, Berkeley, USA, pp. 789–820.

[Dro] Drobotukhina, Yu. V. (1991), An analogue of the Jones polynomial for links in $\mathbb{R}P^3$ and a generalisation of Kauffman–Murasugi theorems, *Algebra and Analysis*,**2**(3), pp. 613–630.

[Duz] Duzhin, S. (1996), A quadratic lower bound for the number of primitive Vassiliev invariants. Extended abstract, *KNOT'96 Conference/Workshop report, Waseda University, Tokyo*, pp. 52–54.

[Dye] Dye, H.A. (2003), Detection and Characterization of Virtual Knot Diagrams, Ph.D. Thesis, University of Illinois at Chicago.

[ECH...] Epstein D., Cannon J. W, Holt D. F, Levy, S.V.S, Paterson M. S. and Thurston W.P.(1992), *Word Processing in Groups*, Jones and Barlett Publ.

[EGH] Eliashberg, Ya., Givental, A. and Hofer, H. (2002), An introduction to symplectic field theory, *Geom Funct. Anal.*, Special Volume, Part II, pp. 560–673.

[Eli] Eliashberg, Ya. (1993), Legendrian and transversal knot invariants in tight contact 3-manifolds *Topological methods in modern mathematics*, Publish or Perish, Houston, pp. 171–193.

[EKT] Eliahou, Sh., Kauffman, L.H. and Thistletwaite, M. (2003), Infinite families of links with trivial Jones polynomial, *Topology*, **42**(1), pp. 155–169.

[FeMa1] Fedoseev, D.A. and Manturov, V.O. (2016), Parities on 2-knots and 2-links, *J. Knot Theory Ramifications*, **25**(14), 1650079, 24 pp.

[FeMa2] Fedoseev, D.A. and Manturov, V.O. (2017), A sliceness criterion for odd free knots, arXiv:1707.04923.

[FIKM] Fenn, R., Ilyutko, D.P., Kauffman, L.H. and Manturov, V.O. (2014), Unsolved problems in virtual knot theory and combinatorial knot theory. *Knots in Poland III. Part III*, Banach Center Publ., **103**, Polish Acad. Sci. Inst. Math., Warsaw, pp. 9–61.

[FKM] Fenn, R., Kauffman, L.H. and Manturov, V.O. (2005), Virtual knot theory — unsolved problems, *Fund. Math.*, **188**, pp. 293–323.

[Flo] Floer, A. (1988), Morse theory for Lagrangian intersections, *J. Differential Geom.*, **28**, pp. 513–547.

[FFG] Fomenko, A.T, Fuchs, D.B and Gutenmacher, V. (1986), *A course of homotopy topology*, Homotopic topology, (Budapest: Akademiai Kiado, Publishing House of the Hungarian Academy of Sciences).

[FHW] Freedman, M.H., He, Z.-X. and Wang, Z. (1994), Möbius energy for knots and unknots, *Annals of Mathematics*, **139**(2), pp. 1–50.

[FJK] Fenn, R., Jordan-Santana, M. and Kauffman, L.H. (2004), Biracks and virtual links, *Topology Appl.*, **145**(1–3), pp. 157–175.

[FM] Fomenko, A.T and Matveev, S.V. (1997), *Algorithmic and computer methods for three-manifolds, Mathematics and its Applications*, **425**, Kluwer Academic Publishers, Dordrecht.

[FMT] *Tensor and vector analysis. Geometry, mechanics and physics* (1998), edited by A. T. Fomenko, O. V. Manturov, and V.V. Trofimov, (London: Gordon and Breach Science Publishers).

[Fom] Fomenko A. T. (1991), The theory of multidimensional integrable hamiltonian systems (with arbitrary many degrees of freedom). Molecular table of all integrable systems with two degrees of freedom, *Adv. Sov. Math*, *6*, pp. 1-35.

[FR] Fenn, R., Rourke, C. (1992), Racks and links in codimension two, *J. Knot Theory Ramifications*, 4, pp. 343–406.

[Fra] Francis, G. K. (1987–1988), *A topological picturebook*, (New York: Springer–Verlag).

[Fre] Freedman, M. (1982), A surgery sequence in dimension four; the relations with knot concordance, *Invent. Math.*, **68**(2), pp. 195–226.

[FRR] Fenn, R, Rimanyi, P. and Rourke, C.P.(1997), The braid–permutation group, *Topology*, **36**(1), pp. 123–135.

[FT] Fuchs, D. and Tabachnikov, S. (1997), Invariants of Legendrian and transverse knots in the standard contact space, *Topology*, **36**, pp. 1025–1053.

[Gari] Garity, D.J. (2001), Unknotting numbers are not realized in minimal projections for a class of rational knots, *Rend. Inst. Mat. Univ. Trieste*, Suppl. 2., Vol. XXXII, pp. 59–72.

[Garo] Garoufalidis, S. (2004), A conjecture on Khovanov's invariants, *Fund. Math.*, **184**, pp. 99–101.

[Gars] Garside, F.A. (1969), The braid group and other groups, *Quart. J. Math. Oxford*, **20**, pp. 235–254.

[Gau] Gauss, C.F. (1877), Zur Mathematischen Theorie der electrodynamischen Wirkungen, Werke Köningl. Gesell. Wiss. Göttingen **5**, s. 605

[GL] Gordan., C. McA, and Luecke, J. (1989), Knots are determined by their complements, *J. Amer. Math. Soc.*, **2**, pp. 371–415.

[GM] Gaifullin, A.A. and Manturov V.O. (2002), On the recognition of Braids, *Journal of Knot Theory and Its Ramifications*, **11** (8), pp. 1193–1209.

[GMM] Goda, H., Matsuda, H. and Morifuji, T. (2005), Knot Floer homology of (1, 1)-knots, *Geom. Dedicata*, **112**, pp. 197–214.

[GoSt] Gompf, R.E. and Stipsicz, A.I. (1999), *4-manifolds and Kirby calculus*, *Graduate Studies in Mathematics*, **20**, American Mathematical Society, Providence, 558 pp.

[GPV] Goussarov M., Polyak M., and Viro O.(2000), Finite type invariants of classical and virtual knots, *Topology* **39**, pp. 1045–1068.

[GS] Gukov, S. and Saberi, I. (2014), Lectures on knot homology and quantum curves, in *Topology and Field Theories* (S. Stolz, ed.), AMS Contemporary Mathematics, **613**.

[Gus] Goussarov M.N (1991), Novaja forma polinoma Jonesa–Conwaya dlja orientirovannykh zatseplenij (A new form of the Jones–Conway polynomial for oriented links), *LOMI scientific seminars*, **193**, Geometry and Topology, 1, pp. 4–9, 161, (in Russian).

[Hak] Haken, W. (1961), Theorie der Normalflächen, *Acta Mathematicae*, **105**, pp. 245–375.

[Hat] Hatcher, A. (1982), On the boundary curves of incompressible surfaces, *Pacific J. Math.*, **99**, pp. 373–377.

[Haz] Hazewinkel, M. (1995), Multiparameter quantum groups and multiparameter R–matrices, *Acta Applicandae Mathematicae*, **47**, pp. 51–98.

[Hem] Hemion, G. (1992), *The classification of knots and 3–dimensional spaces*, (Oxford: Oxford Univ. Press).

[Hil] Hillman, J. A. (2002), *Four-manifolds, geometries and knots. Geometry & Topology Monographs*, **5**, Geometry & Topology Publications, Coventry, 379 pp.

[HK] Hrencecin, D. and Kauffman, L.H. (2003), On Filamentations and Virtual Knots, *Topology Appl.*, **134**(1), pp. 23–52.

[HL] Habegger, N., Lin, X.-S. (1990), The classification of links up to link-homotopy, *J. Amer. Math. Soc.*, **3**(2), pp. 389–419.

[HOMFLY] Freyd, P., Yetter, D., Hoste, J., Lickorish, W.B.R, Millett, K.C. and Ocneanu A. (1985), A new polynomial invariant of knots and links, *Bull. Amer. Math. Soc.* **12**, pp. 239–246.

[Hre] Hrencecin, D. (2001), *On Filamentations and Virtual Knot Invariants*, Thesis, www.math.uic.edu/ ∼dhren/FINALCOPY.ps

[HTW1] Hoste, J., Thistletwaite, M., and Weeks, J., Table of knots and the KnotScape program, http://www. math. utk. edu/∼ morwen (Morwen Thistletwaite's homepage).

[HTW2] Hoste, J., Thistletwaite, M. and Weeks, J. (1998), The first 1,701,936 knots, *Mathematical Intelligencer*, **20**, pp. 33–48.

[IKK] Ishii, A., Kamada, N. and Kamada, S. (2008), The virtual magnetic Kauffman bracket skein module and skein relations for the f-polynomial. *J. Knot Theory Ramifications*, **17**(6), pp. 675–688.

[IM1] Ilyutko, D.P and Manturov, V.O. (2015), A parity map of framed chord diagrams, *J. Knot Theory Ramifications*, **24**(13), 1541006, 15 pp.

[IM2] Ilyutko, D.P and Manturov, V.O. (2015), Picture-valued biquandle bracket, arXiv:1701.06011.

[IMN1] Ilyutko, D.P., Manturov, V.O. and Nikonov, I.M. (2013), Parity in knot theory and graph links, *J. Math. Sci.*, **193**(6), pp. 809–965.

[IMN2] Ilyutko, D.P., Manturov, V.O. and Nikonov, I.M. (2014), Virtual knot invariants arising from parities, *Knots in Poland. III. Part 1*, Banach Center Publ., **100**, Polish Acad. Sci. Inst. Math., Warsaw, pp. 99–130.

[IMN3] Ilyutko, D.P., Manturov, V.O. and Nikonov, I.M. (2015), *Parity and patterns in low dimensional topology*, Reviews in mathematics and mathematical physics, Cambridge Scientific Publ., 166 pp.

[Jac] Jacobsson, M. (2004), An invariant of link cobordisms from Khovanov's homology theory, *Algebr. Geom. Topol.*, **4**, pp. 1211–1251.

[JKS] Jaeger, F., Kauffman, L.H., and Saleur, H. (1994), The Conway Polynomial in S^3 and Thickened Surfaces: A New Determinant Formulation, *J. Combin. Theory. Ser. B.*, **61**, pp. 237-259.

[Jon1] Jones, V. F. R. (1985), A polynomial invariant for links via Neumann algebras, *Bull. Amer. Math. Soc.*, **129**, pp. 103–112.

[Jon2] Jones, V. F. R. (1987), Hecke algebra representations of braid groups and link polynomials, *Annals of Mathematics*, **126**, pp. 335–388.

[Jon3] Jones, V. F. R. (1989), On knot invariants related to some statistical mechanical models, *Pacific Journal of Mathematics*, **137** (2), pp. 311-334.

[Jon4] Jones, V. F. R. (2000), Ten Problems, In: *Mathematics: frontiers and perspectives*, AMS, Providence, RI, pp. 79–91.

[Joy] Joyce, D. (1982), A classifying invariant of knots, the knot quandle, *Journal of Pure and Applied Algebra*, **23** (1), pp. 37–65.

[JP] Jeong, M.J. and Park, C.Y. (2002), The critical properties for fractional chromatic Vassiliev invariants and knot polynomials. *Topology and Its Applications*, **124** pp. 505–521.

[JS] Jablan, S. and Sazdanovic, R. (2007). *LINKNOT. Knot Theory by Computer, Series on Knots and Everything*, **21**, World Scientific, 500 pp.

[Kad] Kadison, L. (1999). *New Examples of Frobenius Extensions, University Lecture Series*, **14**, American Mathematical Society, Providence, RI, 84 pp.

[Kal] Kalfagianni, E. (1998), Finite type invariants of knots in 3–manifolds, *Topology*, **37**(3), pp. 673–707.

[KamS1] Kamada, S. (2000), Braid presentation of virtual knots and welded knots, arXiv:math. GT/0008092 v1, 2000.

[KamS2] Kamada, S. (2002), *Braid and knot theory in dimension four*, Surveys and monographs, **50**, AMS.

[KamN1] Kamada, N. (2002). On the Jones polynomial of checkerboard colourable virtual knots, *Osaka J. Math.* **39**, 2, pp. 325–333.

[KamN2] Kamada, N. (2005). A relation of Kauffman's f-polynomials of virtual links, *Topol. Appl.* **146–147**, pp. 123–132.

[Kas] Kassel, C. (1995), *Quantum groups*, Graduate Texts in Mathematics, **155**, Springer New York.

[Kau1] Kauffman, L.H. (1983), Combinatorics and knot theory, *Contemporary Mathematics*, **20**, pp. 181–200.

[Kau2] Kauffman, L.H. (1987), *On Knots*, (Annals of Math Studies, Princeton University Press).

[Kau3] Kauffman, L.H. (1987), State Models and the Jones Polynomial, *Topology*, **26**, pp. 395–407.

[Kau4] Kauffman, L.H. (1991), *Knots and Physics*, (Singapore: World Scientific).

[Kau5] Kauffman, L. H. (1997), Virtual Knots, talks at MSRI Meeting in January 1997 and AMS meeting at University of Maryland, College Park in March 1997.

[Kau6] Kauffman, L. H. (1999), Virtual knot theory, *European Journal of Combinatorics* **20**(7), pp. 662-690.

[Kau7] Kauffman, L.H. (2001), Detecting virtual knots, *Atti. Sem. Math. Fis., Univ. Modena*, **49**, pp. 241–282.

[Kau8] Kauffman, L.H. (2003), e-mail to the author, May, 2003.

[Kau9] Kauffman, L.H. (2006), *Formal Knot Theory*, Dover Publications, 272 pp.

[Kau10] Kauffman, L.H., *Diagrammatic Knot Theory* (in Preparation).

[Kaw] Kawauchi, A. (1996), *A survey of Knot theory*, (Basel: Birkhäuser).

[Kel] Thomson, W. (Lord Kelvin) (1867), Hydrodynamics, *Proc. Royal. Soc. Edinburgh*, **41**, pp. 94–105.

[Kho1] Khovanov, M. (1997), A categorification of the Jones polynomial, *Duke Math. J.*, **101**, pp.359–426.

[Kho2] Khovanov, M. (2002), A functor–valued invariant of tangles, *Algebr. Geom. Topol.*, **2**, pp. 665–741.

[Kho3] Khovanov, M. (2003), Patterns in knot cohomology I, *Experiment. Math.*, **12**(3), pp. 365–374.

[Kho4] Khovanov, M. (2004). Link homology and Frobenius extensions, preprint, arXiv:math.GT/0411447.

[Kho5] Khovanov, M. (2005). Categorifications of the coloured Jones polynomial, *J. Knot Theory Ramifications* **14**, 1, pp. 111–130.

[KhR1] Khovanov, M. and Rozansky, L. (2004). Matrix factorisations and link homology, preprint, arXiv:math.GT/0401268.

[KhR2] Khovanov, M. and Rozansky, L. (2005). Matrix factorisations and link homology II, preprint, arXiv:math.GT/0505056.

[KhR3] Khovanov, M. and Rozansky, L. (2007). Virtual crossings, convolutions and a categorification of the SO(2N) Kauffman polynomial, preprint, arXiv:math.GT/0701333.

[Kim] Kim, S.G. (2000), Virtual knot groups and their peripheral structure, *J. Knot Theory Ramifications*, **9**(6), pp. 797–812.

[Kis] Kishino, T. and Satoh, S. (2004), *Journal of Knot Theory and Its Ramifications*, **13** (7), pp. 845–856.

[KK] Kamada, N. and Kamada, S. (2000), Abstract link diagrams and virtual knots, *Journal of Knot Theory and Its Ramifications*, **9** (1), pp. 93–109.

[KM1] Kauffman, L.H. and Manturov, V.O. (2005), Virtual biquandles, *Fund. Math.*, **188**, pp. 103–146.

[KM2] Kauffman, L.H. and Manturov, V.O. (2014), A graphical construction of the $sl(3)$ invariant for virtual knots, *Quantum Topol.*, **5**(4), pp. 523–539.

[KM3] Kauffman, L.H. and Manturov, V.O. (2015), Graphical constructions for the $sl(3)$, C_2 and G_2 invariants for virtual knots, virtual braids and free knots, *J. Knot Theory Ramifications*, **24**(6), 1550031, 47 pp.

[Kne] Kneissler, J. (1997), The number of primitive Vassiliev invariants up to degree twelve. arXiv:q-alg/9706022.

[KnotPlot] KnotPlot http://knotplot.com.

[KnotScape] KnotScape   http://www.math.utk.edu/   morwen/knotscape.html.

[KNS] Kamada, N., Nakabo, S. and Satoh, S. (2002). A virtualized skein relation for Jones polynomial, *Illinois J. Math.* **46**, 2, pp. 467–475.

[Kon1] Kontsevich, M. (1993), Vassiliev's knot invariants, Adv. in Soviet Math. **16**(2) (1993), pp. 137–150.

[Kon2] Kontsevich, M. (1994), Feynman diagrams and low–dimensional topology, *First European Congress of Mathematics, Vol. II (Paris, 1992)*, Progr. Math., **120**, Birkhauser, Basel, pp. 97–121.

[KR] Kauffman, L.H. and Radford, D. (2002), Bi-Oriented Quantum Algebras and a Generalized Alexander Polynomial for Virtual Links, *AMS Contemp. Math*, **318**, pp. 113–140.

[Kra1] Krammer, D. (2000), The braid group B_4 is linear, *Invent. Math.*, **142**(3), pp. 451–486.

[Kra2] Krammer, D. (2002), Braid groups are linear, *Ann. of Math.*, **2** (155), pp. 131–156.

[KrMr1] Kronheimer, P.B. and Mrówka, T.S. (1993), Gauge theory for embedded surfaces. I, *Topology*, **32**, pp. 773–826.

[KrMr2] Kronheimer, P. B. and Mrówka, T. S. (2004), Dehn surgery, the fundamental group and $SU(2)$, *Math. Res. Lett.*, **11**(5–6), pp. 741–754.

[KrMr3] Kronheimer, P.B. and Mrówka, T.S. (2007), *Monopoles and three-manifolds, New Mathematical Monographs*, **10**, Cambridge University Press, Cambridge, 796 pp.

[KrMr4] Kronheimer, P. B. and Mrówka, T. S. (2011), Khovanov homology is an unknot–detector, *Publ. Math. Inst. Hautes Études Sci.*, **113**, pp. 97–208.

[KT] Kassel, C. and Turaev, V. (2008), Braid groups, *Graduate Texts in Mathematics*, **247**, Springer, New York, 340 pp.

[KZ] Knizhnik, V.G. and Zamolodchikov (1984), A.B., Current algebra and Wess–Zumino model in two dimensions, *Nucl. Phys.*, **B247**, pp. 83–103.

[Kup1] Kuperberg, G. (2002), What is a Virtual Link?, *Alg. Geom. Topol.*, **3**, pp. 587–591.

[Kup2] Kuperberg, G. (1994), The quantum G_2 link invariant, *Internat. J. Math.*, **5**(1), pp. 61–85.

[Lan] Lando, S.K. (2004), Vassiliev knot invariants, Main constructions, In: Lando, S., Zvonkin, A., *Graphs on surfaces and their applications,Encyclopaedia of Mathematical Sciences*, **141**, Springer-Verlag, Berlin, 455 pp.

[Law] Lawrence, R. (1990), Homological representations of the Hecke algebra, *Comm. Math. Phys.*, **135**, pp. 141–191.

[Lee1] Lee, E. S. (2002), The support of the Khovanov's invariants for alternating knots, arXiv: math.GT/ 0201105

[Lee2] Lee, E.S. (2005), An endomorphism of the Khovanov invariant, *Adv. Math.*, **197**(2), pp. 554–586.

[Lev] Levine, J. (1967), A method for generating link polynomials, *Amer. J. Math.*, **89**, pp. 69–84.

[Lic1] Lickorish, W. B. R. (1962), A representation of orientable combinatorial 3–manifolds, *Annals of Mathematics*, **76**, pp. 531–540.

[Lic2] Lickorish, W. B. R. (1993), The skein method for three–manifold invariants,*Journal of Knot Theory and Its Ramifications*, **2**(2), pp. 171–194.

[Lin1] Lin, X.–S. (1992), Milnor link invariants are all of finite type, Columbia Univ., Preprint.

[Lin2] Lin, X.-S. (1994), Finite type link invariants of 3–manifolds, *Topology*, **33**, pp. 45–71.

[Liv] Livingston, C. (2004), Computations of the Ozsváth–Szabó knot concordance invariant, *Geom. Topol.*, **8**, pp. 735–742.

[LM] Le, T.Q.T and Murakami, J. (1995), Representations of the category of tangles by Kontsevich's iterated integral, *Comm. Math. Phys.*, **168**, pp. 535–562.

[Low] Lowrance, A. (2007). Heegaard–Floer homology and Turaev genus, preprint, arXiv:math.GT/0709.0720.

[Lu] Lu, N. (1992), A simple proof of the fundamental theorem of Kirby calculus on links, *Transactions of the American Mathematical Society*, **331**, pp. 143–156.

[Mak] Makanin, G.S. (1989), On an analogue of the Alexander–Markov theorem, Izvestiya Acad. Nauk SSSR, **53** (1), pp. 200–210.

[Maka] Makanina, G.A. (1992), Defining relation in the pure braid group, *Moscow Univ. Math. Bull.*, **3**, pp. 14–19.

[MaMa] Manturov, O. V. and Martynyuk, A.N, Ob odnom algoritme v teorii mul'tiplikativnogo integrala (On one algorithm in the multiplicative integral theory), Izvestiya VUZov. Matematika, N.5, pp. 26–33 (in Russian).

[Man1] Manturov, V.O. (2000), d–Diagrams, Chord Diagrams, and Knots, *POMI scientific seminars*, **267**, Geometry and Topology, 5, pp. 170–194.

[Man2] Manturov, V.O. (2000), Bifurcations, atoms, and knots, *Moscow Univ. Math. Bull.*,**1**, pp. 3–8.

[Man3] Manturov, V.O. (2000), The bracket semigroup of knots, *Mathematical Notes*, **67**(4), pp. 571–581.

[Man4] Manturov, V.O. (2002), On Invariants of Virtual Links, *Acta Applicandae Mathematicae*, **72**(3), pp. 295–309.

[Man5] Manturov, V.O.(2002), Knots and the Bracket Calculus, *Acta Applicandae Mathematicae*, **74**(3), pp. 293–336.

[Man6] Manturov, V.O. (2002) A combinatorial representation of links by quasitoric braids, *European Journal of Combinatorics*, **23**, pp. 203–212,

[Man7] Manturov, V.O. (2002), Invariants of Virtual Links, *Dokl. Math.*, **65**(3), pp. 329–331.

[Man8] Manturov, V.O. (2002), Invariant two–variable invariant polynomials for virtual links, *Russian Math. Surveys*, **57**(5), pp. 997–998.

[Man9] Manturov, V.O. (2003), Multivariable polynomial invariants for virtual knots and links, *Journal of Knot Theory and Its Ramifications*, **12**(8), pp. 1131–1144.

[Man10] Manturov V.O. (2003), Kauffman–like polynomial and curves in 2–surfaces, *Journal of Knot Theory and Its Ramifications*, **22**, 1350044, 20 p.

[Man11] Manturov, V.O. (2003), Curves on Surfaces, Virtual Knots, and the Jones–Kauffman Polynomial, *Dokl. Math.*, **67**(3), pp. 326-328.

[Man12] Manturov, V.O. (2003), Atoms and minimal diagrams of virtual links, *Dokl. Math.*, **391** (2), pp. 166–168.

[Man13] Manturov V.O. (2004), *Knot Theory*, Chapman & Hall/CRC, 400 p.

[Man14] Manturov, V. O. (2004). The Khovanov polynomial for virtual knots, *Dokl. Math.*, **70**(2), pp. 679–681.

[Man15] Manturov, V. O. (2005). On long virtual knots, *Dokl. Math.*, **71**(2), pp. 253–255.

[Man16] Manturov, V. O. (2005). The Khovanov complex for virtual links, *J. Math. Sci. (New York)*, **144**(5), pp. 4451–4467.

[Man17] Manturov, V. O. (2006). The Khovanov complex and minimal knot diagrams, *Dokl. Math.*, **73**(1), pp. 46–48.

[Man18] Manturov, V. O. (2007). Khovanov homology of virtual knots with arbitrary coefficients, *Izv. Math.*, **71**(5), pp. 967–999.

[Man19] Manturov, V. O. (2007). Khovanov homology for virtual links with arbitrary coefficients, *J. Knot Theory Ramifications*, **16**(3), pp. 345–377.

[Man20] Manturov, V. O. (2008). Additional gradings in Khovanov's complex for thickened surfaces, *Dokl. Math.*, **77**(3), pp. 368–370.

[Man21] Manturov, V. O. (2008). Additional gradings in Khovanov homology, in *Topology and Physics. Dedicated to the Memory of X.-S. Lin*, Nankai Tracts in Mathematics, Singapore: World Scientific, Singapore, pp. 266–300.

[Man22] Manturov, V.O. (2010), Parity in knot theory, *Sb. Math.*, **201**(5–6), pp. 693–733.

[Man23] Manturov V.O. (2011), Parity, free knots, groups, and invariants of finite type. *Trans. Moscow Math. Soc.*, pp. 157–169.

[Man24] Manturov V.O. (2012), Free knots and parity, *Introductory lectures on knot theory*, Ser. Knots Everything, **46**, World Sci. Publ., pp. 321–345.

[Man25] Manturov V.O. (2012), Parity and cobordisms of free knots, *Sb. Math.*, **203**(1–2), pp. 196–223.

[Man26] Manturov, V.O. (2013), An almost classification of free knots, *Dokl. Math.*, **88**(2), pp. 556–558.

[Man27] Manturov, V.O. (2013), Parity and projection from virtual knots to classical knots, *Journal of Knot Theory and Its Ramifications*, **12**(8), pp. 1145–1153.

[Man28] Manturov, V.O. (2015), Reidemeister moves and groups, *J. Knot Theory Ramifications*, **24**(10), 1540006, 14 pp.

[Man29] Manturov, V.O. (2015), One-term parity bracket for braids, *Journal of Knot Theory and Its Ramifications*, **24**, 1540007.

[Man30] Manturov, V.O. (2015), Monodromy, groups and invariants of links with values in the images, *Dokl. Math.*, **91**(3), pp. 376–378.

[Man31] Manturov, V.O. (2016), New parities and coverings over free knots, *J. Knot Theory Ramifications*, **25**(11), 1650077, 18 pp.

[Man'1] Manturov, V.O. (1998), Atoms, height atoms, chord diagrams, and knots. Enumeration of atoms of low complexity using Mathematica 3.0, In: *Topological Methods in Hamiltonian Systems Theory*, (Moscow: Factorial), pp. 203-212 (in Russian).

[Man'2] Manturov, V.O. (2001), *Lekcii po teorii uzlov i ikh invariantov (Lectures on the theory of knots and their invariants)*, (Moscow: URSS, in Russian).

[Man'3] Manturov, V.O (2003), O raspoznavanii virtual'nykh kos (On the Recognition of Virtual Braids), *POMI Scientific Seminars*, **299**. Geometry and Topology, **8** (in Russian).

[Man'4] Manturov, V. O. (2005). *Teoriya uzlov (Knot theory)*, Moscow–Izhevsk: RCD, 512 pp. (in Russian).

[Man'5] Manturov, V.O. (2016), An elementary proof of the embeddability of classical braids into virtual braids. *Dokl. Akad. Nauk*, **469**(5), pp. 535–538 (in Russian).

[ManO1] Manturov, O. V. (1985), Algebras with irreducible group of automorphisms, *Doklady Mathematics*, **281**(5), pp. 1048–1051.

[ManO2] Manturov, O. V. (1986), Mul'tiplikativnyi integral (Multiplicative Integral), *Itogi nauki*, **22**, pp. 167–215 (in Russian).

[ManO3] Manturov, O. V. (1986), Odnorodnye prostranstva i invariantnye tenzory (Homogeneous spaces and invariant tensors), *Itogi Nauki. Problemy geometrii*, **18**, pp. 105–142. (im Russian)

[ManO4] Manturov, O. V. (1991), *Kurs vysshey matematiki (A course of Higher Mathematics)*, (Moscow: Vysshaya shkola, in Russian).

[ManO5] Manturov, O. V. (1992), *Elementy tenzornogo ischisleniya (Elements of Tensor Calculus)*, (Moscow: Prosveshcheniye, in Russian)

[ManO6] Manturov, O. V. (2000), Polynomial invariants of knots and links, *POMI Scientific seminars*, **267**, *Geometry and Topology* (5), pp. 195–206.

[Mar] Markoff, A. A. (1936), Über die freie Äquivalenz der geschlossenen Zöpfe, *Mat. Sbornik*, **1**, pp. 73–78.

[Mar'] Markoff, A.A. (1945), Osnovy algebraicheskoy teorii kos (Fundaments of the algebraic braid theory), *Trudy MIAN*, **16**, pp. 1–54 (in Russian).

[Mat1] Matveev, S.V. (1981), Consturction of a complete algebraic knot invariant, *Chelyabinsk Univ.*, pp. 14.

[Mat2] Matveev, S.V. (1984), Distributive groupoids in Knot Theory, *Sb. Math.*, **47** , pp. 73–83.

[Mat3] Matveev, S.V. (1988), Transformations of special spines and Zeeman's conjecture, *Izvestiya Acad Sci. USSR*, **31**(2) , pp. 423–434.

[Mat4] Matveev, S.V. (1997), Classification of 3-manifolds which are sufficiently large, *Russian Mathematical Surveys*, **52** (5), pp. 147–174.

[Mat5] Matveev, S.V. (2003), *Algorithmic Topology and Classification of 3-Manifolds*, (Springer Verlag Berlin Heidelberg).

[Mat6] Matveev, S.V. (2009), Virtual 3-manifolds, *Siberian electronic mathematical reports*, **6**, pp. 518–521.

[McC] McCleary, J. (2001), *A User's Guide to Spectral Sequences (2nd ed.)*, Cambridge Studies in Advanced Mathematics, **58**, Cambridge University Press.

[McD] McDougall, A. (2009). A Diagramless link homology, preprint, arXiv:math.GT/0911.2518.

[MeMo] Melvin, P. M. and Morton, H.R. (1995), The coloured Jones function, *Comm. Math. Physics*, **169** , pp. 501–520.

[MI] Manturov, V.O. and Ilyutko, D.P. (2012), *Virtual knots: the state of the art*, World Scientific, 552 p.

[Mil1] Milnor, J. (1954), Link groups, Annals of Mathematics, **59**, pp.177-195.

[Mil2] Milnor, J. (1954), Isotopy of links. Thesis (Ph.D.)–Princeton University.

[MiMo] Milnor, J. and Moore, J. (1965), On the structure of Hopf algebras, *Annals of Mathematics*, **81**, pp. 211-264.

[Miy] Miyazawa, Y. (2006), Magnetic graphs and an invariant for virtual links, *J. Knot Theory Ramifications*, **15**, pp. 1319–1334.

[MKS] Magnus, W., Karras, A., and Solitar, D. (1966), *Combinatorial group theory: presentations of groups in terms of generators and relations*, (New York: Interscience Publ. Wiley & Sons).

[MM1] Manturov, O.V. and Manturov, V.O. (2010), Free knots and groups, *J. Knot Theory Ramifications*, **19**(2), pp. 181-186.

[MM2] Manturov, O.V. and Manturov, V.O. (2010), Free knots and groups, *Dokl. Math.*, **82**(2), pp. 697-700.

[MN] Manturov, V.O. and Nikonov, I.M. (2015), Homotopical Khovanov homology, *J. Knot Theory Ramifications*, **24**(13), 1541003, 17 pp.

[Mof] Moffat, H.K. (1985), Magnetostatic Equilibria and Analogous Euler Flows of Arbitrary Complex Topology. Part 1. Fundamentals, *Journal of Fluid Mechanics*, **35**, pp. 117–129.

[Moi] Moise, E.E. (1952), Affine structures in 3–manifolds. V. The triangulation theorem and Hauptvermutung, *Annals of Mathematics*, **57**, pp. 547–560.

[Moo91] Moody, J.A.(1991), The Burau representation of the braid group B_n is unfaithful for large n, *Bull. Amer. Math. Soc.*, **25**, pp. 379–284.

[Moo01] Moore, J.D. (2001), *Lectures on Seiberg-Witten invariants. Second edition. Lecture Notes in Mathematics*, **1629**, Springer-Verlag, Berlin, 121 pp.

[Mor1] Morton, H.R. (1986), Threading knot diagrams, *Math. Proc. Cambridge Phil. Soc.*, **99**, pp. 247-260,

[Mor2] Morton, H.R. (1988), Problems, In: *Braids* (Birman, J.S. and A.Libgober, eds), *Contemporary Math.*, **78**, pp. 557–574.

[MOS] Manolescu, C., Ozsváth, P., Sarkar, S. (2009), A combinatorial description of knot Floer homology, *Ann. of Math.*, **169**, pp. 633–645.

[MOST] Manolescu, C., Ozsváth, P., Szabó, Z. and Thurston, D. (2007), On combinatorial link Floer homology, *Geom. Topol.*, **11**, pp. 2339-2412.

[MT] Menasco, W. and Thistlethwaite, M. (1993), A classification of alternating links, *Annals of Mathematics*, **138**, pp. 113–171.

[Mur1] Murasugi, K. (1987), The Jones polynomial and classical conjectures in knot theory, *Topology*, **26**, pp. 187–194.

[Mur2] Murasugi, K. (1987), Jones polynomial and classical conjectures in knot theory *II*, *Math Proc. Cambridge Phil. Soc.*, **102**, pp. 317–318.

[MZ] Manturov, V.O. and Zenkina, M.V. (2011), An invariant of links in a thickened torus, *J. Math. Sci.*, **175**(5), pp. 501–508.

[Nak] Nakanishi, Y. (1983), Unknotting numbers and knot diagrams with the minimum crossings, *Math. Sem. Notes, Kobe Univ.*, **11**, pp. 257–258.

[Nel] Nelson, S. (2001), Unknotting virtual knots with Gauss diagram forbidden moves, *Journal of Knot Theory and Its Ramifications*, **10** (6), pp. 931–935.

[NgK] Ng, K. Y. (1998), Groups of Ribbon Knots, *Topology*, **37**, pp. 441–458.

[NgL] Ng, L. (2008), Framed knot contact homology, *Duke Math. J.*, **141**(2), pp. 365–406.

[NOR] Nelson, S., Orrison, M.E. and Rivera, V. (2017), Quantum enhancements and biquandle brackets, *J. Knot Theory Ramifications*, **26**(5), 1750034, 24 pp.

[Oni] Onishchik, A.L. (1985), Hopfa Algebra, Mathematical Encyclopaedia,5, (Moscow, in Russian).

[O'Ha1] O'Hara, J. (1991), Energy of a Knot, *Topology*, **30**, (2), pp. 241–247.

[O'Ha2] O'Hara, J. (1992), Family of energy functionals on knots, *Topology and Its Applications*, **48**(2), pp. 147–161.

[O'Ha3] O'Hara, J. (1994), Energy functionals on knots, *Topology and Its Applications*, **56**(1), pp. 45–61.

[Oht] Ohtsuki, T. (2002), *Quantum Invariants. A Study of Knots, 3-Manifolds, and Their Sets*,(Singapore: World Scientific).

[Öre] Öre, O. (1931) Linear equations in non-commutative fields, *Ann. Math.*, **32**, pp. 463–477.

[ORS] Ozsváth, P., Rasmussen, J. and Szabó, Z. (2007). Odd Khovanov homology, preprint, arXiv:math.QA/0710.4300.

[OS1] Ozsváth, P. and Szabó, Z. (2002), Heegaard Floer homology and alternating knots, *Geom. Topol.*, **7**, pp. 225–254.

[OS2] Ozsváth, P. and Szabó, Z. (2002), Holomorphic disks and knot invariants, *Adv. Math.*, **186**(1), pp. 58–116.

[OS3] Ozsváth, P. and Szabó, Z. (2003), Knot Floer homology and the four-ball genus, *Geom. Topol.*, **7**, pp. 615–639.

[OS4] Ozsváth, P. and Szabó, Z. (2004), Knot Floer homology, genus bounds, and mutation, *Topology Appl.*, **141**(1–3), pp. 59–85.

[OS5] Ozsváth, P. and Szabó, Z. (2006), Heegaard diagrams and Floer homology, *Proceedings of the 2006 ICM, Madrid*, **2**, pp. 1083–1099.

[Pap] Papakyriakopoulos, C.D. (1957), On Dehn's lemma and asphericity of knots., *Annals of Mathematics*, **66**, pp. 1–26.

[Per] Perko, K.A. (1973), On the classification of knots, *Notices Amer. Math. Soc.*, **20**, pp. 453–454.

[Piu] Piunikhin, S. (1995) Combinatorial expression for universal Vassiliev link invariant, *Comm. Mat. Phys.*, **168**, pp. 1–22.

[PP] Paoluzzi, L and Paris, L. (2002), A note on the Lawrence–Krammer–Bigelow representation, *Algebraic and Geometric Topology*, **2**, pp. 499–518.

[Pra1] Prassolov, V.V. (1995), *Nagljadnaja Topologija (Intuitive Topolgy)*, (Moscow: MCCME, in Russian).

[Pra2] Prassolov, V.V. (1999), Poverhnost' Seiferta (Seifert Surface), *Mat. Prosveshchenie*, **3**, pp. 116–126.

[Prz] Przytycki, J. H. (2017), *KNOTS:From combinatorics of knot diagrams to combinatorial topology based on knots*, Cambridge University Press, *Kobe J. Math.*, **4**, pp. 117–139.

[PS] Prasolov, V.V. and Sossinsky, A.B. (1997), *Knots, Links, Braids, and 3–Manifolds*, Translation Math. Monographs, (AMS, Providence, R.I.).

[PT] Przytycki, J. H. and Traczyk, P.(1987), Invariants of Conway type, *Kobe J. Math.*, **4**, pp. 117–139.

[PV] Polyak, M. and Viro, O. (1994), Gauss diagram formulae fot Vassiliev invariants, *Int. Math Research Notices*, **11**, pp. 445–453.

[Ras1] Rasmussen, J. A. (2003), Floer Homology and Knot Complements, PhD thesis, Harvard University, math.GT/0306378.

[Ras2] Rasmussen J. A. (2010), Khovanov homology and the slice genus, *Invent. Math.*, **182**(2), pp.419–447.

[Reid] K. Reidemeister (1948), *Knot Theory*, (New York: Chelsea Publ. & Co.).

[Rein] Reinhart, B.L. (1962), Algorithms for Jordan Curves on Compact Surfaces, *Annals of Mathematics*, **75**(2), pp, 209–222.

[Rol] Rolfsen, D. (1976), *Knots and Links*, Publish or Perish.

[Rou] Rourke, C. (2006), What is a welded link?, *Intelligence of low dimensional topology 2006. Ser. Knots Everything*, **40**, pp. 263–270.

[RT] Reshetikhin, N. and Turaev, V. (1990), Ribbon graphs and their invariants derived from quantum groups, *Comm. Math. Phys.*, **127**, pp. 1–26.

[Rush] Rushworth, W. (2016), On the virtual Rasmussen invariant, arXiv:1603.02893.

[Sat] Satoh, S. (2000), Virtual knot presentation of ribon torus–knots, *Journal of Knot Theory and Its Ramifications*, **9** (4), pp. 531–542.

[Sav] Saveliev, N. (2012), *Lectures on the topology of 3-manifolds. An introduction to the Casson invariant. Second revised edition*, De Gruyter Textbook, Walter de Gruyter & Co., Berlin, 207 pp.

[Saw] Sawollek, J. (2001), On Alexander–Conway Polynomial for Virtual Knots and Links, arXiC: math. GT/9912173, 6 Jan. 2001.

[Sch] Schubert, H. (1949), Die Eindeutige Zerlegbarkeit eines Knotens in Primknoten, *Sitzungsberichte der Heidelberger Akademie der Wissenschaften Mathematisch–Naturwissenschaftlische Klasse*, **3**, ss. 57–104.

[Sco] Scorpan, A. (2005), *The wild world of 4-manifolds*, American Mathematical Society, Providence, 609 pp.

[Sei1] Seifert, H. (1934), Über das Geschlecht von Knoten, *Mathematische Annalen*, **110**, 571–592.

[Sei2] Seifert, H. (1935), Die Verschlingunginvarianten der zyklischen knotenüberlagerungen, *Abh. Math. Sem. Hansischen Univ.*, **11**, ss. 84–101.

[Sem] Semyonov–Tian–Shansky, M.A. (1983), What is a classical R–matrix, *Functional Analysis and its Applications*, **17**(4), pp. 17–33.

[SeSm] Seidel, P. and Smith, I. (2004). A link invariant from the symplectic geometry of nilpotent slices, preprint, arXiv:math.SG/0405089.

[Shu1] Shumakovitch,   A.   (2003),   KhoHo   pari   package, www.geometrie.ch/KhoHo/.

[Shu2] Shumakovitch, A. (2004), Torsion of the Khovanov homology, preprint, arXiv:math.GT/0405474.

[Shu3] Shumakovitch, A. (2012), Khovanov homology theories and their applications, in: *Perspectives in Analysis, Geometry, and Topology, Progress in Mathematics*, **296**, pp. 403–430.

[Sto] Stoimenow, A. (1998), Enumeration of chord diagrams and an upper bound for Vassiliev invariants, *Journal of Knot Theory and Its Ramifications*, **7** (1), pp. 94–114.

[SW] Silver, D. and Williams, S. (2001), Alexander groups and virtual links, *J. Knot Theory Ramifications*, **10**, pp. 151–160.

[Swi] Swiatkowski, J. (1992), On the isotopy of Legendrian knots, *Ann. Glob. Anal. Geom.*, **10**, pp. 195–207.

[Tai] Tait, P.G. (1898), On Knots, In: *Scientific paper I*, pp. 273–274, (London: Cambridge University Press).

[Tei] Teichner, P. (2002), Knots, von Neumann Signatures, and Grope Cobordisms, In: *Proceedings of the International Congress of Mathematicians, Beijing*, **2**, pp. 437–446., (Beijing: Higher Education Press)

[Thi] Thistletwaite, M. (1987), A spanning tree expansion for the Jones polynomial, *Topology*, **26**, pp. 297–309.

[Thu] Thurston, W.P. (1997), Three–Dimensional Geometry and Topology, (New–Jersey: Princeton University Press).

[Tor] Torres, G. (1953), On the Alexander polynomial, *Annals of Mathematics*, **57** (1), pp. 57–89.

[Tra] Traczyk, P. (1998), A new proof of Markov's braid theorem, In: *Knot theory* (Warsaw,1995), Warsaw: Banach Center Publ., **42**, Polish Acad. Sci., pp. 409–419.

[Tro] Trotter, H.F. (1964), Non–invertible knots exist, *Topology*, **2**, pp. 275–280.

[TT] Turaev, V. G. and Turner, P. (2006). Unoriented topological quantum field theory and link homology, *Algebr. Geom. Topol.* **6**, pp. 1069–1093.

[Tur1] Turaev V.G. (1992), The Yang–Baxter equation and invariants of links, *Inventiones Mathematicae*, **3**, pp. 527–553.

[Tur2] Turaev V.G. (2001), *Introduction to combinatorial torsions*, Lectures in Mathematics, Birkäuser, ETH Zürich.

[Tur3] Turaev, V.G. (2002), Faithful linear representations of the braid group, *Astérisque*, **276**, pp. 389–409.

[TV] Turaev, V.G. and Viro, O.Ya. (1992), State sum invariants of 3-manifolds and quantum 6j-symbols, *Topology*, **31**(4), pp. 865–902.

[Tyu1] Tyurina, S.D. (1999), On formulas of Lannes and Viro–Polyak type for finite type invariants, *Mathematical notes*, **66** (3–4), pp. 525–530.

[Tyu2] Tyurina, S.D. (1999), Diagrammatic formulae of Viro–Polyak type for finite type invariants, *Russian Math. Surveys*, **54**(3), pp. 187–188.

[Vas1] Vassiliev, V. A. (1990), Cohomology of knot spaces, in Theory of Singularities and its applications, *Advances in Soviet Mathematics*, **1**, pp. 23–70.

[Vas2] Vassiliev, V. A. (1994), *Complements of Discriminants of Smooth Maps: Topology and Applications, Revised Edition*, Amer. Math. Soc., Providence, R. I.

[Vas3] Vassiliev, V. A. (1997), *Topologiya dopolneniy k diskriminantam (Topology of complement to discriminants)*, (Moscow: Phasis, in Russian).

[Vas4] Vassiliev, V. A. (1997), *Vvedenie v topologiyu (Introduction to topology)*, (Moscow: Fasis, in Russian).

[Ver] Vershinin, V. (2001), On Homology of Virtual Braids and Burau Representation, *Journal of Knot Theory and Its Ramifications*, **18**(5), pp. 795–812.

[Vir1] Viro, O. (2002), Remarks on definition of Khovanov Homology, arXiv: math. GT/0202199.

[Vir2] Viro, O. (2005). Virtual links and orientations of chord diagrams, in *Proceedings of the Gökova Conference*, 2005, International Press, pp. 187–212.

[Vog1] Vogel, P. (1990), Representations of links by braids: A new algorithm, *Comm. Math. Helvetici*, **65**, pp. 104-113.

[Vog2] Vogel, P. (1995), *Algebraic structures on modules of diagrams*, Institut de Mathématiques de Juissieu, Prépublication, 32.

[Vog3] Vogel, P. (1999), The Universal Lie Algebra, Preprint.

[Wal] Waldhausen, F. (1967), On irreducible 3-manifolds which are sufficiently large, *Annals of Mathematics*, **87**(1), pp. 56–88.

[Weh] Wehrli, S. (2008), A spanning tree model for the Khovanov homology, *J. Knot Theory Ramifications* **17**(12), pp. 1561–1574.

[Wei] Weinberg, N.M. (1939), Sur l'équivalence libre des tresses fermées, *C. R. (Doklady) Academii Nauk SSSR*, **23**, pp. 215–216

[Wel] Welsh, D. J. A. (1992), On the number of knots and links, *Colloq. Math. Soc. János Bolyai*, **60**, pp. 713–718.

[Wit1] Witten, E. (1986), Physics and geometry, In: *Proc. Intl. Congress Math., Berkeley, USA*, pp. 267–303.

[Wit2] Witten E. (1989), Quantum field theory and the Jones Polynomial, *Comm. Math. Phys.*, **121** (1989), pp. 351–399.

[Wit3] Witten E. (1990), Gauge theory, vertex models, and quantum groups, *Nucl. Phys. B.*, **330**, pp. 225–246.

[Zag] Zagier, D. (2001), Vassiliev invariants and a strange identity related to the Dedekind eta–function, *Topology*, **40**(5), pp. 945–960.

[Zen1] Zenkina, M.V. (2011), An invariant of links in a thickened torus, *Math. Notes*, **90**(1–2), pp. 227–237.

[Zen2] Zenkina, M.V. (2016), A knot invariant in thickened surfaces, *J. Math. Sci.*, **214**(5), pp. 728–740.

[Zhi] Zhiqing, Y. (2017), Knot invariant with multiple skein relations, arXiv:1703.05911 [math.GT].

Index